Journal of Chromatography Library — Volume 2

EXTRACTION CHROMATOGRAPHY

JOURNAL OF CHROMATOGRAPHY LIBRARY

Journal of Chromatography Library — Volume 2

EXTRACTION CHROMATOGRAPHY

Edited by

T. BRAUN

Institute of Inorganic and Analytical Chemistry
L. Eötvös University
Budapest, Hungary

and

G. GHERSINI

CISE
Centro Informazioni Studi Esperienze
Segrate–Milano, Italy

ELSEVIER SCIENTIFIC PUBLISHING COMPANY
AMSTERDAM — LONDON — NEW YORK
1975

Technical editor

E. K. Kállay

The distribution of this book is being handled by the following publishers:

for the U.S.A. and Canada

American Elsevier Publishing Company, Inc.
52 Vanderbilt Avenue
New York, New York 10017

*for the East European countries, China, North Korea,
Cuba, North Vietnam and Mongolia*

Akadémiai Kiadó, The Publishing House of the
Hungarian Academy of Sciences, Budapest

for all remaining areas
Elsevier Scientific Publishing Company

335 Jan van Galenstraat
P. O. Box 330, Amsterdam, The Netherlands

Library of Congress Card Number 74—84054

ISBN 0−444−99878−0

CONTRIBUTORS

AKAZA, I., Kanazawa Women's College,
Kanazawa, Japan

ALIMARIN, I. P., Institute of Analytical Chemistry, Moscow University,
Moscow, USSR

BOLSHOVA, A. T., Institute of Analytical Chemistry, Moscow University,
Moscow, USSR

BONNEVIE-SVENDSEN, M., Institute for Atomic Energy,
Kjeller, Norway

BRAUN, T., Institute of Inorganic and Analytical Chemistry, L. Eötvös University,
Budapest, Hungary

CERRAI, E., CISE, Centro Informazioni Studi Esperienze,
Segrate–Milano, Italy

DRENT, W., EUROCHEMIC, European Company for the Chemical Processing
of Irradiated Fuels, Mol, Belgium

ESCHRICH, H., EUROCHEMIC, European Company for the Chemical Processing
of Irradiated Fuels, Mol, Belgium

FARAG, A. B., Institute of Inorganic and Analytical Chemistry, L. Eötvös University,
Budapest, Hungary

FIDELIS, I., Department of Radiochemistry, Institute of Nuclear Research,
Warsaw, Poland

GHERSINI, G., CISE, Centro Informazioni Studi Esperienze,
Segrate–Milano, Italy

JOON, K., OECD Halden Reactor Project,
Halden, Norway

KATYKHIN, G. S., Research Institute for Physics, Leningrad University,
Leningrad, USSR

MARKL, P., Institute of Analytical Chemistry, University of Vienna,
Vienna, Austria

MÜLLER, W., European Institute for Transuranium Elements, Karlsruhe,
Federal Republic of Germany

ŠEBESTA, F., Department of Nuclear Chemistry, Technical University of Prague,
Prague, Czechoslovakia

SCHMID, E. R., Institute of Analytical Chemistry, University of Vienna,
Vienna, Austria

SIEKIERSKI, S., Department of Radiochemistry, Institute of Nuclear Research,
Warsaw, Poland

STRONSKI, I., Institute of Nuclear Physics,
Krakow, Poland

TESTA, C., Casaccia Nuclear Centre,
Casaccia–Roma, Italy

PREFACE

The birth of extraction chromatography dates back to about 16 years ago, when Siekierski and Winchester independently reported their results obtained with chromatographic columns in separating inorganic substances, taking advantage of the selectivity features of organic compounds normally used as liquid-liquid extractants in the reprocessing of spent nuclear fuel. A few isolated papers had appeared earlier, dealing with chelating compounds supported on different solid materials and used in columns for the separation of metals from aqueous solutions; these were, however, of limited interest because of a number of drawbacks incident to the chosen systems and problems unsolved at that time (see Introduction to Chapter 11).

Siekierski in particular has the merit of having recognized the possibilities of the method and having started a systematic demonstration of its versatility; therefore, extraction chromatography is rightfully considered to have originated from his works. After Siekierski, many other research groups all over the world were attracted by the promising features of the new separation method that rapidly developed, so that more than 600 original papers on the subject are available at present in the open literature (see Chapter 15). A number of reviews have also appeared, covering more or less critically some of the many aspects of the technique.

Growing interest in extraction chromatography called for a work collecting and organizing the whole available information, and discussing in detail the different aspects related to its theory and applications. This has been particularly felt in these years, when the method, originally developed in highly specialized nuclear laboratories, is gradually finding successful applications in more common analytical problems.

This book is aimed at filling such a gap, at least in *column extraction chromatography*. In effect, the very large amount of work already done in both column and laminar extraction chromatography suggested keeping the two systems definitely separated, focussing attention on columns because of their undoubtedly greater importance in quantitative analytical applications. In spite of this limitation, the task was still too difficult to be easily accomplished by few authors, and this book is the result of the collective work of many specialists, each one being responsible for one chapter where a definite aspect of column extraction chromatography is thoroughly presented and discussed.

The first two chapters deal with the basic aspects of extraction chromatography, whereas Chapter 3 is a guide to all technical aspects of the method.

The next two chapters are devoted to the two most peculiar components of the technique, i.e. the organic stationary phase and its supporting material; connected with them are Chapters 11 and 13 also, specifically dealing with chelating compounds as the stationary phase and with foamed materials as the support.

A survey of the available methods for the separation of elements, with particular reference to radiochemical problems and the separation of closely similar elements arranged in groups of the Periodic Table, is the subject of Chapter 6.

The following chapters deal with special aspects in those fields of analytical chemistry where extraction chromatography has proved particularly useful: the separation of actinides (Chapter 7); fundamental studies on the chemical behaviour of lanthanides, and their separation (Chapter 8); the separation of fission products from one another, and from the bulk fissionable material (Chapter 9); radiotoxicological determinations (Chapter 10); and the preconcentration of trace elements in various materials (Chapter 12).

In Chapter 14, the usefulness of paper and thin-layer chromatographic data for planning column separations is discussed, and a general outline is given on the available laminar results.

Finally, Chapter 15 is a bibliographic study of the extraction chromatographic works that can be found in the literature, covering both column and laminar systems, with particular attention to the former ones.

A last word must be devoted to the choice of the term "extraction chromatography", used throughout the book. Several different names have been proposed in the literature to define the separation of inorganic substances by means of chromatographic systems involving an organic compound as the supported stationary phase and an aqueous solution as the eluent (mobile phase), and no universally accepted term is yet available at present. We believe that nowadays the technique has reached sufficient interest and popularity to gain the privilege of being rightfully defined by a concise term without the need of specifying the details of its principles; we have preferred *"extraction chromatography"*, first proposed by Hulet* because it effectively points out its strict linkage to liquid-liquid extraction.

T. Braun

G. Ghersini

---

* "The term reversed-phase partition chromatography carries little meaning except within its historical backround. In the same mould as vapour-phase chromatography, ion-exchange chromatography, etc., the term extraction chromatography is suggested as a briefer and more descriptive expression." (J. Inorg. Nucl. Chem., 26 (1964) 1721).

CONTENTS

THEORETICAL ASPECTS OF EXTRACTION CHROMATOGRAPHY

S. SIEKIERSKI

1. GENERAL COMMENTS

Extraction chromatography is a particular form of liquid-liquid column chromatography. The difference between extraction chromatography and normal partition chromatography lies in the fact that in the process of partition the solute molecules undergo little, if any, chemical change apart from association or proton exchange, while extraction involves the transfer of the initially ionic solute from water into an organic phase, most often accompanied by complex chemical changes involving many interactions and equilibria. The term extraction chromatography is generally used when the stationary phase is an organic liquid or organic solution, and the mobile phase an aqueous solution.

Extraction chromatography couples the favourable selectivity features of the organic compounds used in liquid-liquid extraction, with the multistage character of a chromatographic process. The method is now competing favourably with ion-exchange chromatography in many separation problems, and is particularly advantageous when micro amounts are concerned, as is the case with radiochemical separations. The characteristic features of extraction chromatography can be best discussed, therefore by comparing it with ion-exchange chromatography with regard to both selectivity and column performance.

The selectivity of the ion-exchange process depends on the properties of the ion-exchanger used, and the composition of the aqueous phase. In the case of two ions having the same charge and very similar radii, the selectivity due to the properties of the ion-exchanger (such as acidity or basicity, and the degree of cross-linking) is not sufficient for ensuring effective separation. In such a case, an appropriate complexing agent has to be added to the aqueous phase: the selectivity attained is then due either to the differences in the stability constants or to the different charges or structures of the complexes formed.

Contrary to ion-exchange chromatography with non-specific ion-exchangers, the stationary phase in extraction chromatography most often has complex forming properties.

In spite of this, the difference between ion exchange and extraction, as far as the origin of the selectivity is concerned, is not so great as may appear at first sight. It will be shown in one of the following sections that in many cases extraction can be considered as consisting of two steps: complex formation with the dissolved extractant in the aqueous phase, and the transfer of the complex from the aqueous into the organic phase. According to this approach, when very similar ions are concerned selectivity in extraction mainly originates from complex formation in the aqueous phase. In the event of dissimilar ions, the characteristic features of each method, that is sorption of ions in one case, and partition of complexes in the other, play a more important role.

According to the two-step mechanism of extraction, the comparison of extraction chromatography with ion-exchange chromatography, in regard to the separation of very similar ions, resolves itself into the comparison of the selectivities of hydrophobic and hydrophilic ligands. A priori, it seems that there is no reason to assume that one of these two classes of ligands should be preferred. In Table 1 the mean separation factors of lanthanides with the three complexing agents most often used in the elution separation of lanthanides by ion-exchange chromatography are compared to the mean separation factors obtained with the three most popular extractants. The complexing agents are ethylenediaminetetraacetic acid (EDTA), α-hydroxy isobutyric acid (HIB) and lactic acid (Lac), and the extractants are tributyl phosphate (TBP), di(2-ethylhexyl) phosphoric acid (HDEHP) and 2-ethylhexyl phenyl-phosphonic acid (HEHΦP).

TABLE 1

Mean separation factors ($\bar{\beta}$) of lanthanides with
different complexing agents

	EDTA	HIB	Lac	TBP	HDEHP	HEHΦP
$\bar{\beta}\ \dfrac{Lu}{Ln}$	1.98	1.65	1.68	1.57	2.50	3.02

It follows from these data that the mean separation factors are higher for the best extractants than for the most commonly used water-soluble complexing agents. In such a comparison, however, it should be remembered that the mean separation factor in the case of lanthanides is only a rough measure of the selectivity because of the double-double effect (see Chapter 8). If the separation factors for individual pairs were compared rather than the mean values, then the selectivity order in the case of few pairs would be a little different. However, a detailed comparison of the separation factors for individual lanthanide pairs also shows that extraction chromatography is at the moment somewhat more selective than ion-exchange chromatography, one of the few exceptions is, the Pr—Nd pair.

When dissimilar ions are concerned, extraction chromatography offers many possibilities of achieving effective separations. Firstly, an extraction mechanism can be chosen which best makes use of the existing differences. Secondly, a water-soluble complexing agent may be applied: with dissimilar ions the probability that the complexing properties of the water-soluble ligand are quite different from those of the extractant is much

2

greater than in the case of similar ions, and in this case an additional selectivity can be obtained. Thirdly, a selectivity can be gained in the partition step due to the different charge or volume of the extracted species. This type of selectivity can be increased by applying a suitable diluent.

However, since with ion-exchange chromatography there are many possibilities for achieving a good separation, a comparison of the merits of the two methods is difficult when dissimilar ions are concerned.

Besides thermodynamic factors, such as the separation factor, dynamic factors too influence the resolution considerably in both methods. When plate height (HETP) values obtained in practice are compared, then the two methods appear to be approximately equivalent. It should be remembered, however, that the HETP depends on the chemical properties of the system, so that a general comparison cannot be made.

Since extraction chromatography consists in the application of the column chromatographic technique to the extraction process, the theoretical aspects of extraction chromatography resolve themselves into three main topics:

1. The influence of the support of the stationary phase on the thermodynamic activities of the extractant and extracted complex.

2. Extraction mechanism, with particular emphasis on the differences in the extractability of ions.

3. Dynamic factors.

These problems will be discussed in the following sections.

2. THE INFLUENCE OF THE SUPPORT ON THE THERMODYNAMIC ACTIVITIES OF THE EXTRACTANT AND EXTRACTED COMPLEX

The basic concept of extraction chromatography consists in the combination of liquid-liquid extraction with a chromatographic technique. In the course of the extraction chromatographic process many extractions and reextractions occur, but the question arises as to whether they are strictly equivalent to an extraction process with a free (not adsorbed) stationary organic phase. In other words, the question is whether the extraction coefficient calculated from the position of the chromatographic peak maximum is strictly equal to that determined from static experiments made with the same system. Since the position of the peak maximum depends on the thermodynamic activity of the organic phase (extractant) and on the activity coefficient of the extracted complex, the problem of the equivalence between dynamic and static experiments is decided by the question of whether these two activity terms change when the organic phase is adsorbed on a chromatographic support.

In order to provide an answer to this, consideration will be given to the changes in the properties of TBP adsorbed on two types of support, namely kieselguhr and silicagel, following the model outlined by Giddings[1]. These two supports have been chosen since they are frequently used, their structures are known, and they differ considerably in their specific surface area and pore size.

The organic phase retained on the support exists in two forms: as capillary liquid and as adsorbed liquid. Provided that the organic liquid does not flood the support (and this is the usual case in chromatographic support loading), an equilibrium exists between the capillary and adsorbed liquid. The chemical potential change $\Delta\mu$ for forming capillary liquid from free liquid is given by the following equation:

$$\Delta\mu = -\frac{2\gamma\bar{v}}{r}$$

where γ is the surface (interfacial) tension, $\bar{v}$ the molar volume, and r the radius of curvature of the concave liquid surface. Provided that the surface is completely wetted by the liquid, r equals the effective radius of the capillary or pore.

Because of the existing equilibrium the above expression also gives the chemical potential change for the formation of the adsorbed liquid from free liquid, this change being difficult to calculate directly. Substituting 7.4 dyne/cm for γ (the value of the interfacial tension between water and TBP), 274 cm^3 for $\bar{v}$, and 10^{-4} cm (in the case of kieselguhr) or $7.5 \cdot 10^{-7}$ cm (in the case of silicagel) for r, one gets $\Delta\mu = -1$ cal/mole and -130 cal/mole respectively, for the two supports. While the change of TBP activity is negligible in the case of kieselguhr, in the case of silicagel the thermodynamic activity (or activity coefficient) of TBP is 0.80.

When the activity of the extractant is the only activity term influencing extraction in the organic phase, then the extraction coefficient determined from column experiments using silicagel as support, would differ considerably from that determined from static extraction, particularly when several extractant molecules take part in complex formation. However, when only one extractable species is present, its concentration in the aqueous phase is negligible and the composition of the aqueous phase is constant, then it can easily be shown[2] that the following expression holds for the distribution coefficient (D):

$$D = A \frac{a_E^n}{f_c} = A \frac{x_E^n \cdot f_E^n}{f_c} \tag{1}$$

where A is a constant taking into account the composition of the aqueous phase, a_E is the activity of the extractant, n is the number of extractant molecules in the extracted complex, and f_E and f_c are the activity coefficients in the organic phase of the extractant and of the extracted complex, respectively. It has been shown[2] experimentally that the two activity coefficients do not change independently with the change of the diluent, and that the following approximate relation exists:

$$f_c \approx k \cdot f_E^n \tag{2}$$

where k is a proportionality constant. This relation also results from the theory of regular solutions, provided it can be assumed that $\bar{v}_c = n\bar{v}_E$, and $\delta_c = \delta_E$, where $\bar{v}$ is the molar volume and δ the solubility parameter. According to relation (2) the extraction coefficient changes only slightly with the diluent (at constant extractant concentration), and less than would be expected if only the changes of a_E were taken into account.

4

The above considerations and experimental results concern the changes in the activity terms due to the change of the diluent, but there are reasons for assuming that the same considerations are also valid in cases where other factors are responsible for the changes in the activity coefficients. In fact, equation (1) is a general one, and holds irrespective of the cause of the changes in the activity coefficients. One such cause may be the sorption on a solid surface: both relation (1) and relation (2) may be expected to hold in this case, since there should be no fundamental differences between the modes of interaction of the molecules of the extractant and the complex with the molecules of the diluent, and with the atoms or molecules of the solid surface. There should be a formal, and not only formal, analogy between the activity coefficient changes due to dilution and those due to adsorption or formation of capillary liquid, so that one would expect the ratio f_E^n/f_c, and consequently the extraction coefficient, not to change in the process of formation of the stationary phase.

To summarize, the extraction coefficient should not change with the formation of the stationary phase, either because the activity coefficients remain constant (kieselguhr) or because their ratio does not change (silicagel). This conclusion should be valid for each type of support, whereas the particular reason for the constancy of the extraction coefficient should depend on the pore size. Nevertheless the constancy of the ratio f_E^n/f_c, as found in extraction, is expected to exist only if there are no specific interactions either with the molecules of the extractant or with the extracted complex; in this respect, supports may differ according to their chemical properties. Specific interactions with some extracted species may exist in the case of kieselguhr or silicagel: the treatment of these supports with hydrophobyzing agents probably does not remove all the active sites, so that complexes of ions such as zirconium may additionally be retained due to adsorption on the surface of the support.

No direct experiments have been done to measure the thermodynamic activities of the extractants retained on the supports and only very few column experiments permit comparison of the extraction coefficients determined by liquid-liquid extraction with those obtained from column experiments. The main reason for this is that liquid-liquid extractions are generally performed with organic phases containing a diluent, whereas undiluted extractants are most often used in extraction chromatography. When such a comparison is possible, e.g. in the case of lanthanide extraction by means of TBP or other extractants, the agreement is quite good.

From the practical viewpoint, the influence of the support on the separation factors β is much more important than its influence on the extraction coefficient. A priori, one can expect that in the case of similar extracted ions, the changes in β should be even smaller than the changes in the extraction coefficient. Provided the number of extractant molecules attached to each of two ions is the same, it follows from equation (1) that $\beta_2^1 = f_{c_2}/f_{c_1}$, where f_{c_1}, and f_{c_2} are activity coefficients in the organic phase of complexes formed by the first and second ion, respectively. When two extracted ions are very similar in their chemical properties (as is the case for neighbouring lanthanides) the ratio of the activity coefficients, f_{c_2}/f_{c_1}, should nearly be independent of the properties of the organic phase, since the interactions of the two extracted complexes with either the molecules of the diluent or the support should be almost the same.

5

In fact, there is generally good agreement between separation factors determined from static and chromatographic experiments, although comparison is sometimes difficult owing to different experimental conditions. In particular it is worth pointing out that peculiar effects in the separation factors, such as the double-double effect in the case of lanthanides and actinides, are observed in both types of experiments. An exception to the general agreement between static and chromatographic experiments is reported by Ali and Eberle[3], who found that the separation factor for the Am—Ce pair in the system HDEHP — lactic acid + diethylenetriaminepentaacetic acid is much higher in the case of extraction chromatography. Further, the changes in β_{Ce}^{Am} with pH are completely different in the two methods. The cause of this unusual behaviour requires further investigation.

3. MAIN FACTORS INFLUENCING THE DISTRIBUTION COEFFICIENT AND SELECTIVITY

One of the differences between static extraction and extraction chromatographic processes consists in the change in the activities of the extractant and of the extracted complex due to the influence of the support. Since the change in the extractant activity can also be caused by several other factors, e.g. the presence of a diluent, it does not present a problem unique to extraction chromatography. Another difference lies in the non-attainment of thermodynamic equilibrium when the extraction process is carried out on a column; however, this does not influence the chemical steps of the transfer of the solute from phase to phase. All problems of the extraction mechanism are therefore the same, irrespective of whether the extraction is carried out under static or dynamic conditions.

In spite of this, it seems desirable to discuss in this section some problems of special importance for extraction chromatography. Since the aim of extraction chromatography is separation, primarily those factors will be discussed, that influence the magnitude of the distribution coefficient and the differences in the extractability of elements. As most of the methods of separation of ions are based on complex formation, attention will also be paid to the mutual relations between extraction and complex formation.

Although extraction of ions is a complex process involving many interactions and equilibria, its essential steps can be formulated in quite a simple way. Let us discuss the extraction of a metal ion, M^{+z}, from an aqueous phase into a non-miscible organic solvent. Because of the principle of the electroneutrality of the phases, an equivalent amount of anions A^- is also extracted (for the sake of simplicity only univalent inorganic or organic anions are considered). The overall extraction equation is then:

$$M_{aq}^{+z} + zA_{aq}^- \rightleftharpoons MA_{z,org} \tag{3}$$

The species formed, MA_z, can be solvated in the organic phase and can also contain some water molecules. In those cases when the anion (in its acidic form HA) is initially present in the organic phase, its distribution between the two phases also has to be taken into account:

$$HA_{org} \rightleftharpoons HA_{aq}$$

For a further discussion of the solvent extraction of ions, it is convenient to introduce two different models or approaches to the extraction process, in both of which the extraction is split into two steps but in different ways. In the first it is assumed that the extractable neutral species is formed in the aqueous phase and then transferred to the organic phase; according to this model, the overall equation (3) of extraction can be split into two equations:

$$M_{aq}^{+z} + zA_{aq}^{-} \rightleftharpoons MA_{z,aq} \tag{4}$$

$$MA_{z,aq} \rightleftharpoons MA_{z,org} \tag{5}$$

In order to obtain an extractable species, (and in particular when A^- is an inorganic anion) in many cases the complex MA_z has to be solvated by organic molecules B having electron donor properties. Equations (4) and (5) then read as follows:

$$M_{aq}^{+z} + zA_{aq}^{-} + yB_{aq} \rightleftharpoons MA_zB_{y,aq}$$

$$MA_zB_{y,aq} \rightleftharpoons MA_zB_{y,org}$$

Since B is initially present only in the organic phase, the following distribution must also be taken into account

$$B_{org} \rightleftharpoons B_{aq}$$

The second approach is based on the assumption that equivalent amounts of cations and anions are first transferred from the aqueous to the organic phase and then associated, at least partially, to form a neutral molecule. The equations describing these processes are as follows:

$$M_{aq}^{+z} + zA_{aq}^{-} \rightleftharpoons M_{org}^{+z} + zA_{org}^{-} \tag{6}$$

$$M_{org}^{+z} + zA_{org}^{-} \rightleftharpoons MA_{z,org} \tag{7}$$

It should be emphasized that neither model comes near to describing the real course of the extraction process. In the majority of cases the extractable species is formed at the phase boundary, so that complex formation or association proceeds simultaneously with the transfer process. Furthermore, the equilibrium concentrations of the extractable complex and the extractant in the aqueous phase are generally so low as to be devoid of any chemical meaning, as when the extractable complex or the extractant are large hydrophobic molecules.

Nevertheless, this shortcoming of the two models does not impair their usefulness. Both models are correct from the thermodynamic viewpoint, since thermodynamic functions for the overall process do not depend on the method of dividing the whole process into steps. They both provide too a simple description and understanding of the basic interactions determining the magnitude of extraction and the differences in the

7

extractability of different cations. The choice between the two approaches mainly depends on practical reasons. The former can be most profitably applied to the extraction of cations of high charge, whereas the latter is particularly suitable for the description of the extraction of ions of low charge and large radius. The latter model, in its simple form represented by equations (6) and (7), seems to have few practical applications, since most of the cations of practical interest for extraction have high charge. It has to be considered, that such cations are readily transformed through complexation with simple inorganic anions X^- into large low-charge anions:

$$M_{aq}^{+z} + nX_{aq}^- \rightleftharpoons MX_{n,aq}^{z-n} \tag{8}$$

The complex anion MX_n^{z-n} is then transferred into the organic phase together with an equivalent amount of large organic cation R^+:

$$MX_{n,aq}^{z-n} + (n-z) R_{aq}^+ \rightleftharpoons MX_{n,org}^{z-n} + (n-z) R_{org}^+$$

and in the organic phase the association of the ion pair takes place. Since the cation R^+ is originally present in the organic phase, its distribution (in the form of a salt) has to be taken into account:

$$RX_{org} \rightleftharpoons R_{aq}^+ + X_{aq}^-$$

It follows from the outlined presentation that complex formation is one of the two essential factors determining the magnitude of the distribution coefficient. The more the equilibrium is shifted towards the formation of the extractable complex, the greater is the extraction. Complex formation is also the main cause of the differences in the extractabilities of ions. At the same concentration of extractant in the aqueous phase, extraction should be greater for that ion which more readily forms an extractable complex. As far as selectivity is concerned therefore, the comparison of extraction chromatography with other methods based on complex formation, reduces simply in the majority of cases to the comparison of the selectivities of the complexing agents used in the two methods.

The second factor determining both the magnitude of the distribution coefficient and the separation factor is the partition of either the uncharged complex or of the ion pair. Since the uncharged complex has a hydrophobic surface in the majority of cases, its partition should closely resemble that of organic molecules. Partition of organic substances between water and a solvent is closely related to their solubility in water, because solubility can be considered as partition of a substance between water and its own liquid phase. It is well known that the solubility in water of hydrocarbons which form homologous series decreases with the number of carbon atoms, McAuliffe[4] has shown that for each group of hydrocarbons (paraffins, cycloparaffins, olefines, acetylenes, aromatic hydrocarbons) a linear relationship exists between the standard free energy of solubility and the molar volume of the dissolved substance. Moreover, from the existing data[5] it appears that the main contribution to the positive standard free energy change of solubility in water originates from the large negative entropy change. The low solubility of hydrocarbons in water and the entropic character of this process is a result of

8

the structure-forming behaviour of hydrocarbons. According to Frank and Evans[6], organic molecules dissolved in water increase the number of tetracoordinated water molecules. The formation of a cluster of tetracoordinated water molecules around the dissolved hydrocarbon or hydrocarbon-like molecule is equivalent to the increase of the "ice-like" structure of the water. This increase is accompanied by a substantial decrease in entropy, because the water molecules that form the cluster lose their translational degrees of freedom.

Since the transfer of an organic substance from water to a solvent is the reverse process to its solubility, then provided that the differences between the solvent and the liquid phase of the partitioned substance can be neglected, this process should be accompanied by a decrease in the free energy, due almost entirely to the increase in the entropy term. This should also be true for the transfer of complexes of metal ions with hydrophobic ligands from water to a solvent. In such complexes the metal ion is almost completely "buried" in the bulky ligands, and the interaction of the complex with both water and the organic solvent is mainly due to the hydrophobic parts of the ligands. The main conclusion, therefore, is that partition of a complex between the organic phase and water should increase with the molar volume of the complex. The greater the number of the ligands attached to the ion the greater the molar volume of the ligand, and the greater should be the overall extraction coefficient, provided the complex forming properties of the ligand do not change with increase in its size.

It follows from the above considerations that in addition to complex formation, the partition step can also be the source of the differences in the extractabilities of ions. For instance, it would be expected that, of two ions, that one would be the more readily extracted which forms the complex containing the greater number of given ligand molecules, because of its greater volume. Especially high partition would be expected for a complex with bidentate ligands formed by a cation which has a coordination number of twice the charge. In that case, the bidentate (monoacidic) ligands would neutralize the charge and at the same time remove all the water molecules from the coordination shell of the cation.

Like neutral molecules, large low-charge ions, and in particular large organic cations, also increase the number of tetracoordinated water molecules[7]. The decrease in the "ice-like" structure of the water and the ensuing increase in entropy are the main factors causing the transfer of the ion pair from water into an organic solvent. As is the case with neutral molecules, the transfer of an ion pair into the organic phase should increase with its molar volume, which is mainly determined by the molar volume of the organic cation. The transfer should also increase with the decrease of the charge on the complex anion, because of the unfavourable electrostatic free energy term. However, the effect of the charge can be neutralized, at least partially, since with a complex anion of charge $z = -2$, two (instead of one in the case of $z = -1$) large structure-forming cations of the type R_4N^+ are co-transferred into the organic phase.

The selectivity in the extraction of ion pairs depends primarily on the complex formation step described by equation (8). This is particularly true when two cations $^IM^{+z}$ and $^{II}M^{+z}$ are of the same charge and similar radius. In this case, the complex anions $^IMX_n^{z-n}$ and $^{II}MX_n^{z-n}$ are also of the same charge and have nearly the same radius. The free energies of transfer of ion pairs formed by the two complex anions with the same cation R^+ are then almost the same, so that no selectivity can result from the partition step. However, when the two cations differ in their charges and/or have significantly different radii, then the partition step too can be the source of selectivity. In this case, the complex anions of the type MX_n^{z-n} can differ in their charges because of different values of either z, or n, or both. As a result, different amounts of cations R^+ are co-transferred, and the free energies of transfer can differ in both the electrostatic and water structure forming terms.

4. DYNAMIC FACTORS IN EXTRACTION CHROMATOGRAPHY

The degree of separation of two ions by extraction chromatography can be measured by the resolution function which is defined by the following equation:

$$R_s = \frac{V_{R_1} - V_{R_2}}{2(\sigma_2 + \sigma_1)}$$

where V_R is the eluate volume to peak maximum, σ is the standard deviation of the Gaussian elution curve and the subscripts 1 and 2 refer to the first and second ion, respectively. It follows from the above equation that R_s depends on both thermodynamic and dynamic factors. The thermodynamic properties of the system are reflected in V_R because of the relation $V_R = V_m + DV_s$, where V_m and V_s are the volumes of the mobile and stationary phases, respectively, and D is the extraction coefficient. The main factors affecting the magnitude of the extraction coefficient, which is a specific property of the extracting system, have been discussed in the preceding section.

The dynamic parameters influencing the resolution are contained in the values of σ, and some of them are independent of the properties of the extracted ions and can be discussed in a more or less general way. The discussion which follows is based mainly on the general approach outlined by Giddings[8] and its recent application to extraction chromatography by Horwitz and Bloomquist[9]. Because of the relation

$$H = \frac{\sigma^2}{L},$$

where H is the plate height and L is the distance migrated by the centre of the zone, the discussion will be devoted to the various factors affecting the plate height.

According to Giddings[8], the general expression for the plate height is:

$$H = \frac{B}{v} + Cv + \Sigma \frac{1}{1/2\lambda_i \cdot d_p + D_m/\omega_i v \cdot d_p^2} \tag{9}$$

where v is the average mobile phase velocity, d_p is the diameter of the support particle, and D_m is the diffusion coefficient of the ion in the mobile phase, assumed to

10

be equal to 10^{-5} cm^2/sec. The meaning of the remaining constants will be explained in the discussion which follows. In extraction chromatography practice, supports have been used which differ considerably as regards particle sizes. Since the best results have been obtained with particles of diameters between 15 and 50 μm, the discussion will be limited to these particle sizes.

The first term in equation (9) describes the contribution to the total plate height of the longitudinal diffusion in both mobile and stationary phases. The separate terms are:

$$H_{1,m} = \frac{2\gamma_m \cdot D_m}{v}$$

and

$$H_{1,s} = \frac{2\gamma_s \cdot D_s}{v} \cdot \frac{1-R}{R}$$

where D_s is the diffusion coefficient in the stationary phase, R is the relative migration rate, and γ_m ($= 0.6$) and γ_s are the obstructive factors in the mobile and stationary phases, respectively. At $v = 10^{-2}$ cm/sec, which corresponds to a flow rate of 0.42 ml·cm^{-2}·min^{-1} (void fraction = 0.7), the contribution of the longitudinal diffusion in the mobile phase is about $1.2 \cdot 10^{-3}$ cm (at 25 °C). This is a small, but not negligible effect, particularly at higher temperatures when this contribution may amount to $2-3 \cdot 10^{-3}$ cm. There are indications[10] that the influence of longitudinal diffusion can be observed in extraction chromatography, although it does not seem possible that the observed effect can be attributed solely to it. According to Horwitz and Bloomquist[9], the diffusion coefficient of the Eu-HDEHP complex in the HDEHP phase is about 10^{-7} cm^2·sec^{-1} (at 25 °C), and thus the longitudinal diffusion contribution in HDEHP at $R = 0.2$ should be about 25 times smaller than that in the mobile phase (assuming $\gamma_s = \gamma_m$). However, in the case of extractants of smaller viscosity, like TBP, the contribution of the longitudinal diffusion in the organic phase may be of the same order of magnitude as that of the analogous factor in the mobile phase.

The second term in equation (9) describes the contribution to the plate height of the diffusion-controlled kinetics in the stationary phase and/or a slow chemical step. The term due to lateral diffusion in the stationary phase is:

$$H_d = \frac{qR\,(1-R)d_s^2 v}{D_s} \tag{10}$$

where d_s is the depth of the stationary phase, and q is a configuration factor which depends on the shape of the stationary phase itself. Assuming after Horwitz and Bloomquist[9], that $q = \frac{1}{4}$, $d_s = 4 \cdot 10^{-3}$ cm and $D_s = 10^{-7}$ cm^2/sec (the diffusion coefficient of the Eu complex in the HDEHP phase), and further that $R = 0.2$ and $v = 10^{-2}$ cm/sec, one gets $H_d = 6.5 \cdot 10^{-4}$ cm. The plate height due to the diffusion-controlled kinetics in the stationary phase, although very low at this particular mobile phase velocity, is a very important factor for two reasons: Firstly, since it is proportional

11

to v, its contribution to the total plate height increases with the mobile phase velocity and becomes one of the controlling factors at the highest flow rates. Secondly, the presence of this term is responsible for the changes in the total, experimentally observed plate height as a function of certain parameters. It follows from equation (10) that H_d is proportional to d_s^2, the square of the diffusion distance in the organic phase. Although d_s is probably not strictly proportional to the amount of the organic phase on the support, an increase in this amount should result in an increase in d_s, and consequently in the plate height. Many authors have observed that the height of the plate increases considerably with the amount of the stationary phase, but only when a certain critical amount has been reached[9,11–15]. This critical amount depends mainly on the support, the extractant, and the bed preparation technique. This behaviour can easily be explained in view of the fact that with small amounts of the stationary phase, the plate height contribution of the diffusion kinetics in the stationary phase is very small and the changes in it do not influence the overall plate height. Only when this contribution becomes comparable to other factors determining the height of the plate can a further increase in H_d be observed as a change in the total plate height. Since the diffusion distance d_s should also depend on the pore structure, supports should be preferred which have the same surface area per gram but smaller pores and hence a shorter diffusion distance.

It also follows from equation (10) that H_d decreases with increasing diffusion coefficient D_s, which in turn depends on the viscosity of the stationary phase and on the properties of the extracted complex. If one assumes that the diffusion coefficient of the Eu–TBP complex in the TBP phase is $\frac{50}{4} \cdot 10^{-7}$ cm$^2 \cdot$sec^{-1} ($\frac{50}{4}$ is the ratio of the viscosity of HDEHP to that of TBP), then under the same conditions as above the plate height due to this effect would be only $5.2 \cdot 10^{-5}$ cm, a completely negligible value. Since the contribution of diffusion kinetics is small even in the case of HDEHP, which has a very high viscosity, provided the amount of the stationary phase is small and the flow rate is low, no influence of physical properties of the extractant on the total plate height should be observed. It is difficult to verify this conclusion on the basis of existing experimental data, since experiments with different extractants have not been carried out under comparable conditions. It appears however that with well-prepared columns, plate heights as low as 0.1–0.2 mm can be obtained, irrespective of the properties of the extractant, provided the amount of extractant and the flow rate are low. Of course, at high flow rates, when the diffusion-controlled kinetics and chemical kinetics are important factors, substantial differences should be observed between the various extractants.

Assuming that the diffusion distance d_s is proportional to the amount of the stationary organic phase Q and that D_s is inversely proportional to the stationary phase viscosity η, the plate height contribution due to the lateral diffusion should be approximately proportional to the product $Q\eta$. As long as this product remains constant, the change of the extractant should not influence the height of the plate. This seems to be the case when TBP and HDEHP are compared: with both extractants plate heights as low as 0.1–0.15 mm can be obtained[14,16], but the amount of TBP can be about three times higher than that of HDEHP.

It follows from the above discussion that the influence of a diluent should be observed only at the highest flow rates and with highly viscous stationary phases, like HDEHP and other acidic phosphorus-based extractants. Grosse-Ruyken and Bosholm[12] found that at a mobile phase velocity of about $2 \cdot 10^{-2}$ cm/sec the height of the plate is lower in the case of HDEHP diluted with toluene than in the case of undiluted HDEHP. Sochacka and Siekierski[11], however, did not find the plate height to be influenced by dilution of HDEHP. On the other hand, according to Smułek and Siekierski[17] the height of the plate decreases when 2,6,8-trimethylnonyl-4 phosphoric acid is diluted with chlorobenzene. Obviously, the influence of the diluent deserves to be further studied under strictly comparable conditions.

The second term in equation (9) also contains the contribution of the extraction kinetics. This contribution cannot be discussed in general terms since it is highly specific and depends exclusively on the properties of the extraction system studied (ion, extractant, eluting acid). It should be born in mind however, that according to experimental results and theoretical calculations[9], extraction kinetics play a very important role in determining the height of the plate, particularly at high flow rates and low temperatures. In some cases[10,18], the contribution to the plate height of the extraction kinetics exceeds the contribution of all the other factors by one or even two orders of magnitude, and is then practically solely responsible for the experimentally observed plate height. This problem is discussed in more detail in the chapter devoted to the separation of lanthanides.

The last term in equation (9) is due to the coupling of the lateral diffusion of the solute molecules in the mobile phase with the flow pattern in the chromatographic column. According to Giddings[8], the flow pattern is independent of the mobile phase properties, and depends only on the structure of the support material and the incorporated flow space. Because of the coupling of the two factors (lateral diffusion and flow variations), the plate height is reduced below either of the limits resulting from each effect separately. It follows from equation (9) that the plate height decreases with d_p. The plate height also depends on the parameters λ_i and ω_i which are related to the bed structure and velocity inequalities between the various stream-paths. The magnitude of each of these parameters has to be calculated for each of the five effects: transchannel,

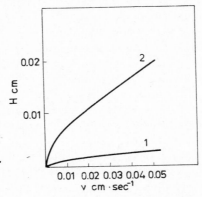

Fig. 1. The plate height due to the coupling of lateral diffusion with flow pattern. (1) $d_p = 15\ \mu m$; (2) $d_p = 50\ \mu m$.

transparticle, short-range interchannel, long-range interchannel and transcolumn. In order to illustrate the effect of the particle size, the plate height has been calculated as a function of the flow velocity for two particle sizes: $d_p = 15$ μm and $d_p = 50$ μm. The values of the parameters λ_i and ω_i calculated by Giddings[8] have been used. The transcolumn effect has not been taken into account, since it has been shown by Horwitz and Bloomquist[9] that column diameter has little influence on the plate height. The results of the calculations are shown in Fig. 1: it is evident that the plate height increases markedly with the flow rate and depends appreciably on the particle size. At $v = 0.01$ cm·sec^{-1} and for $d_p = 15$ μm, the contribution to the total plate height is about $1.2 \cdot 10^{-3}$ cm.

The results of theoretical predictions concerning the total plate height can now be compared with experimental data. In the case of $d_p = 15$ μm, $v = 10^{-2}$ cm/sec, $R = 0.2$, $D_m = 10^{-5}$ cm^2/sec and $D_s = 10^{-7}$ cm^2/sec, the total plate height (without the contribution of the extraction kinetics) should be $1.2 \cdot 10^{-3} + 6.5 \cdot 10^{-4} + 1.2 \cdot 10^{-3} \approx$ $\approx 3 \cdot 10^{-3}$ cm. The calculations of Horwitz and Bloomquist[9] suggest that the plate height contribution from extraction kinetics of Cm ions in the HDEHP $-$ HNO$_3$ system should be about $5 \cdot 10^{-4}$ cm at this flow rate and R value. Assuming the same value for a lanthanide ion, the total plate height in the case of lanthanide extraction chromatography with HDEHP or other similar extractants should be about $3.5 \cdot 10^{-3}$ cm. The lowest experimentally observed values of plate height for lanthanides are about 10^{-2} cm, that is 3 times greater than theoretically predicted. In the case of a larger particle size (35 μm), a comparison of theoretical predictions with experimental results has been made in a broad range of flow velocities and for different temperatures by Horwitz and Bloomquist[9]: here too the experimental value of the plate height was found to be about 2 times greater than theoretically predicted.

The data of several authors[10,18–20] indicate that the plate height in extraction chromatography increases almost linearly with the mobile phase velocity in the range 0.01–0.1 cm·sec^{-1} (flow rates about 0.4–4 ml·cm^{-2}·sec^{-1}). This is to be expected since in this region the main contributions to the plate height are those of lateral diffusion in the organic phase, from extraction kinetics, and the transparticle effect. The three contributions increase linearly with the flow rate (the contribution of the transparticle effect almost linearly) whereas the non-linear changes due to other effects are much smaller. As a result, the overall change in the plate height is almost linear with v.

All three terms in equation (9) are temperature dependent. The first term increases with temperature because of the increase of D_m and D_s whereas the remaining two terms decrease with increasing temperature. As a result, except at low flow rates, the total plate height decreases with increasing temperature, as found by many authors[9,10,12,20,21]. This decrease is particularly marked when slow extraction kinetics is the main factor determining the plate height: in this case the plate height at 75 °C might well be several times lower than that at 25 °C.

It now seems worthwhile to discuss the problem whether column performance in extraction chromatography can be improved. At high flow rates the main contributions to the total plate height are from transparticle effect, extraction kinetics, and from lateral diffusion in the organic phase. The two latter contributions decrease with increasing temperature, but this possibility of improving column performance has already been ex-

14

ploited, since most of the separations have been carried out at higher temperatures. The contribution of lateral diffusion in the stationary phase can be decreased by reducing the depth of the stationary phase on the support. In this respect much can be gained by using the recently introduced packing materials for high-performance liquid chromatography[22]. The transparticle effect can be reduced by decreasing the particle size. However, with the currently used irregular particles as the packing material, little − if anything − can be gained by decreasing the particle size below 15 μm since such a decrease would certainly result in a less uniform bed packing. This might not be the case with spherical particles, even of very low diameters.

At low flow rates an important contribution to the total plate height originates from the flow pattern and lateral diffusion in the mobile phase. Detailed calculations show that the main factors are short-range and long-range interchannel effects. Both effects are caused by non-uniformity of the bed packing, that is by the existence of channels in which the mobile phase velocity differs substantially from the mean value. The actually existing bed packing irregularities and the resulting velocity differences, particularly when small-particle packing material is used, are probably greater than those assumed by Giddings[8] and accounted for by his values of the λ_3, λ_4 and ω_3, ω_4 parameters. The improvement of the bed packing should therefore result in a substantial decrease in the plate height. It seems, however, that with packing materials of irregular particle shape, such as those that have been used in extraction chromatography up to now, more uniform packing cannot be obtained. Again, a substantial improvement can be expected at low flow rates from the application of the spherical low-diameter packing materials used in high-performance liquid chromatography. Even at low flow rates, however, there is naturally a limit to the improvement in column performance obtainable by decreasing the particle size and by increasing the uniformity of the bed packing. This limit is set at a value of about $3 \cdot 10^{-3}$ cm by the combined action of all the other effects.

REFERENCES TO CHAPTER 1

1. J. C. GIDDINGS, Anal. Chem., 34 (1962) 458.
2. S. SIEKIERSKI, J. Inorg. Nucl. Chem., 24 (1962) 205.
3. S. A.ALI, S. H. EBERLE, J. Inorg. Nucl. Chem. Letters, 7 (1971) 153.
4. C. McAULIFFE, J. Phys. Chem., 70 (1966) 1267.
5. A. D. NELSON, C. L. DE LIGNY, Rec. Trav. Chim. Pays Bas, 87 (1968) 528.
6. H. S. FRANK, M. W. EVANS, J. Chem. Phys., 13 (1957) 507.
7. H. S. FRANK, W. Y. WEN, Disc. Farad. Soc., 24 (1957) 133.
8. J. C. GIDDINGS, Dynamics of Chromatography, Part I, Marcel Dekker Inc., New York, 1965.
9. E. P. HORWITZ, C. A. A. BLOOMQUIST, J. Inorg. Nucl. Chem., 34 (1972) 3851.
10. S. SIEKIERSKI, R. SOCHACKA, J. Chromatog., 16 (1964) 385.
11. R. SOCHACKA, S. SIEKIERSKI, J. Chromatog., 16 (1964) 376.
12. H. GROSSE−RUYKEN, J. BOSHOLM, J. Prakt. Chem., 25 (1964) 79.
13. I. FIDELIS, S. SIEKIERSKI, J. Chromatog., 17 (1965) 542.
14. E. HERRMANN, J. Chromatog., 38 (1968) 498.
15. E. P. HORWITZ, C. A. A. BLOOMQUIST, D. J. HENDERSON, J. Inorg. Nucl. Chem., 31 (1969) 1149.
16. I. FIDELIS, S. SIEKIERSKI, J. Chromatog., 5 (1961) 161.
17. W. SMUŁEK, S. SIEKIERSKI, J. Chromatog., 19 (1965) 580.

18. H. F. ALY, M. RAIEH, Anal. Chim. Acta, 54 (1971) 171.
19. T. B. PIERCE, R. S. HOBBS, Anal. Chim. Acta, 12 (1963) 74.
20. E. P. HORWITZ, C. A. A. BLOOMQUIST, D. J. HENDERSON, D. E. NELSON, J. Inorg. Nucl. Chem., 31 (1969) 3255.
21. E. CERRAI, C. TESTA, C. TRIULZI, Energia Nucleare, 9 (1962) 377.
22. J. J. KIRKLAND, J. Chromatog. Sci., 7 (1969) 7.

CHAPTER 2

CORRELATION BETWEEN EXTRACTION CHROMATOGRAPHY AND LIQUID-LIQUID EXTRACTION

I. AKAZA

1. THEORETICAL CORRELATION BETWEEN LIQUID-LIQUID EXTRACTION AND EXTRACTION CHROMATOGRAPHY

Solvent extraction is nowadays commonly used as an analytical technique in various fields. Originally a simple batch method, it was also developed to give continuous or multiple extraction methods such as extraction with a continuous extractor, true countercurrent extraction with equipments in which two liquid phases move in opposite directions, or multiple discontinuous countercurrent extraction with Craig's apparatus. Such continuous methods are based on the theory of gas-liquid countercurrent distillation.

When the solute is distributed between two liquid phases in the batch method, a concentration equilibrium can be reached. According to the Nernst partition isotherm, the distribution of species B between two (organic and aqueous) liquid phases is expressed by:

$$[B]_{org}[B]_{aq} = \text{constant} = K$$

Here, brackets express concentrations and K is the distribution coefficient, and its relationship with the thermodynamic partition coefficient p is given by

$$p = \{B\}_{org}/\{B\}_{aq} = f_{org}[B]_{org}/f_{aq}[B]_{aq} = K(f_{org}/f_{aq})$$

where the braces express the activities of the species B, and f its activity coefficients.

For many practical applications, the ratio of the analytical concentrations (i. e. total concentrations) of the metal in both phases in equilibrium is expressed as the distribution ratio D (often called extraction coefficient E_a°). When various species B_1, B_2, etc. are present, all containing the metal M, D is expressed by:

$$D = [M]_{org}/[M]_{aq} = [B_1]_{org} + [B_2]_{org} + .../[B_1]_{aq} + [B_2]_{aq} + ...$$

When two metals (M_1 and M_2) have to be separated, and their separation factor, $\beta = D_1/D_2$ is close to 1, the above-mentioned continuous or multiple extraction must be applied for successful separation.

Partition chromatography, devoted to this field with such effect by Martin and Synge, is principally considered a kind of countercurrent extraction, where the components are distributed between two liquid phases, one of which is held on a solid support (this is the less polar solvent in reversed-phase methods), while the other one moves in a given direction. There have been[1—4a] various theoretical studies on column partition chromatography, based on the plate concept developed for distillation processes. The chromatographic column is divided into a number of sections compared to hypothetical plates, and it is assumed that on each of the plates, equivalent to one extraction vessel, the solute is distributed between the two phases, and that the mobile phase containing the solute shifts from one plate to the next one. The chromatographic theory based on this concept thus seems to be closely related to Craig's theory of countercurrent distribution. However, the true equilibrium that can be attained in the discontinuous process with Craig's apparatus is impossible on each plate in column of partition chromatography. To overcome this, Martin defined the plate in chromatography as the layer where the ratio between the average concentrations of the solute in the stationary phase and in the solution flowing out of the region corresponds to the ratio attained in true equilibrium. The height of the plate is denoted as HETP (height equivalent to one theoretical plate).

To describe the plate concept on the movement of the solute in the column, let v_s and v_m be the volumes of the stationary and mobile phases of each plate, respectively, and C_s and C_m the concentrations of the solute in the two phases. When a volume dv of the mobile phase flows from the $(n-1)$th to the nth plate, an amount $C_{m,(n-1)} \cdot dv$ of solute enters the nth plate, while at the same time the amount $C_{m,n} \cdot dv$ of solute is transferred from the nth to the $(n+1)$th plate. If the solute concentrations of the two phases in the nth plate are changed by $dC_{m,n}$ and $dC_{s,n}$, respectively, by the transfer of solute, the following equation holds:

$$(C_{m,(n-1)} - C_{m,n}) \cdot dv = v_m \cdot dC_{m,n} + v_s \cdot dC_{s,n}$$

and since $C_{s,n} = K*C_{m,n}$ the following equation can be derived:

$$dC_{m \cdot n}/dv = (C_{m,(n-1)} - C_{m \cdot n})/(v_m + K* \cdot v_s)$$

if $C_{m,o}$ is used for the initial concentration of the solute in the mobile phase and V for the total volume of the mobile phase used for the development of the chromatogram, and when the volume expressed in a unit of the effective plate volume, $(v_m + K* \cdot v_s)$, is

18

expressed as v_t, $V/(v_m + K^* \cdot v_s)$, and if it is assumed that at the beginning of the chromatographic process all the solute is held only in the first plate, the following equation is obtained:

$$C_n = e^{-v}t \cdot v_t^n /n!, \quad (C_n = C_{m \cdot n}/C_{m \cdot o}) \tag{1}$$

This equation describes the distribution of the solute concentration in the column, and corresponds to the expression for Poisson distribution. If the concentration of the solute in the effluent from a column with a total number of plates N is plotted against v_t, the peak maximum of the eluted substance will appear when the volume of mobile phase passed through the column corresponds to N times the effective plate volume, $v_t = n$, as explained by differentiating eq. (1). This volume is termed the retention volume V_R, and can be expressed as follows:

$$V_R = N(v_m + K^* v_s) = V_m + K^* V_s \tag{2}$$

where V_s and V_m are the total volumes of the stationary and mobile phases, respectively, if is the height of a plate is H, and the cross-sectional areas of the mobile and stationary phases are A_m and A_s respectively, equation (2) can be expressed as:

$$V_R = N(HA_m + K HA_s)$$

Assuming $D = K^*$, from equation (2) the distribution coefficient is:

$$D = (V_R - V_m)/V_s \tag{3}$$

The same treatment of column chromatography can also be applied to both paper and thin layer chromatography, where the speed of transfer of the solute is usually given in R_f values[5]:

$$R_f = \frac{\text{movement of band}}{\text{movement of advancing front of liquid}} = \frac{V_m}{V_m + K^* V_s} \tag{4}$$

When D instead of K^*, and if A_s and A_m are used, it follows that:

$$D = (1/R_f - 1) V_m/V_s = (1/R_f - 1) A_m/A_s = (1/R_f - 1)k \tag{5}$$

since A_m/A_s is assumed to be constant in the practical process of chromatography.

Equations (2) (3), (4) and (5) indicate the direct relationships between the distribution in solvent extraction and the quantities obtained from extraction chromatography, the former and latter pair of equations being related to the elution volume, and the R_f values, respectively. Through these equations, V_R or R_f can be predicted by using D values obtained from liquid-liquid extraction, and conversely D (or a separation factor) can be calculated from chromatographic data.

The efficiency of a column is given by N, the number of plates, and can therefore be also expressed by the above-mentioned HETP, which is calculated by dividing the height of the column by N. The N of the column is calculated by the ratio of the band width w to the elution volume, shown by the elution curve, and the following equation is normally used for the calculation:

$$N = 16\,(V_R/w)^2$$

The band width w expresses the distance between the intersections with the elution volume axis of the tangents to its inflection points. The equation is based on chromatograms whose peak shapes correspond to Poisson distribution, as seen in equation (1); the following equation derived from Gaussian distribution is also used:

$$N = 8\,(V_R/w)^2$$

where w is the peak width at $1/e$ of the peak height.

Various assumptions are involved in the plate theory for its simplification; namely, that the two phases should be in equilibrium in each plate; that the distribution coefficients should be constant throughout the column and independent of concentration; that diffusion between plates be negligible; that the solute be originally loaded on the first plate, and so on. Accordingly, it is obvious that some departures will certainly appear between the plate theory and chromatographic practice, but this concept of idealized plates enables one to understand the correlation between column chromatography and solvent extraction. An outline of the dynamic approach to plate heights is given in Chapter 1.

2. GENERAL CONSIDERATIONS ON THE COMPARISON
OF LIQUID-LIQUID EXTRACTION AND EXTRACTION CHROMATOGRAPHIC
RESULTS

The process will be considered here in which hydrated ions interact with reagents to form complexes, and in which the complexes are extracted into an organic solvent. The composition of a formed complex varies depending on the coordination numbers, the charge of the central atom, and the denticity and basicity of the reagent. When a formed complex is coordinatively unsaturated by the reagent, it will meet with difficulty in being extracted because of the otherwise vacant coordination sites being occupied by water molecules. The water molecules are displaced by ligands such as excess reagent molecules or coordinatively active extractants containing polar O, N and S, etc, and in some cases, are hydrogen-bonded to these ligands, and this makes extraction easier.

If one wishes to express the formation of any type of complex by a general formula which can be adapted to every special case by various combinations of the related factors, then the species present in the organic phase when both phases are in equilibrium, may be written as: $M_m A_n (HA)_a L_l X_x (OH)_h (H_2O)_w S_s$. In this formula, HA represents acidic extractants, X complexing agents, L adduct-forming agents, and S the organic solvent. Subscripts can assume zero or certain integral values, while the overall species must be electrically neutral.   Next the aqueous-phase species may be written as $M_{m'}A_{n}(HA)_{a'}L_{l'}X_{x'}(OH)_{h'}(H_2O)_{w'}S_{s'}$, where subscripts can again assume zero or integral values. It is presumed in this formula that a great number of concentration variables are involved in the equilibria relating to the formation and extraction of these complexes; the equilibria containing the hydrogen ion can be expressed in terms of the hydroxyl ion and the ionic product of water.

The distribution ratio, which must cover all species of each phase, can be written (by substituting concentrations for activities) as:

$$D = \frac{\sum\limits_{mnalxhws}[M_mA_n(HA)_aL_lX_x(OH)_h(H_2O)_wS_s]_{org}}{\sum\limits_{mnalxhws}[M_{m'}A_n(HA)_aL_lX_x(OH)_h(H_2O)_wS_s]_{aq}} \qquad (6)$$

The summations can be taken independently over all the variables. Besides the above-mentioned different concentration variables, other factors which affect solvent extraction of metal complexes seem mainly to be temperature and the time of contact of the two phases.

The comparison of the results of extraction chromatography and liquid-liquid extraction has usually been made both for the general behaviour of the different metals in the two techniques, and for the mechanism responsible for the extraction of the metals.

The change in the behaviour of a certain metal, that is to say the change of the distribution ratio in the chromatographic system due to the variation of the factors which affect the extraction can easily be foreseen qualitatively from liquid-liquid extraction data. Among the various possible relations of the magnitude of the distribution to the concentration variables, the relation of the distribution ratio or extraction percentage to the composition of the aqueous phase, especially the hydrogen ion concentration or acid concentration, is generally termed the „extraction curve" in liquid-liquid extraction; such curves have been systematically obtained for a great number of elements in most extraction systems; results for HDEHP, for example, have been published by Kimura[6,7], and for TBP [8],TOPO[9] and long chain amines[10] by Ishimori's group. In several books such as Refs,[11] and [11a], distribution charts are reported for various extraction systems, that easily allow one to obtain analogous information.

The comparison of distribution behaviours obtained from extraction chromatography with the liquid-liquid extraction curves is made on the major tendencies, on the values of each distribution ratio and on the values of the separation factors. Chromatographic distribution ratios as functions of the aqueous phase composition are available for a great number of elements, especially from thin layer and paper chromatographic experiments, where they are reported as "R_f spectra": they can easily be compared on major tendencies with the extraction curves. In the case of the rare earth elements, whose similar property make separation difficult, the distribution behaviours in the two techniques are compared on the basis of the separation factors of pairs of adjacent elements.

The extraction mechanisms in the two methods can be compared too. As shown in equation (6), the composition of the extracted species may be varied by changing the different variables, so that some complicated cases may be encountered in which several kinds of species coexist. In order to anticipate the chemical mechanism of the reaction, it is important to define the composition of the extracted species. Though there are several techniques for such a definition, the quantitative treatment of distribution equilibria based on the concept of an extraction constant is frequently employed in both liquid-liquid and chromatographic extraction. In fact, when the overall stability constant and partition coefficient of the extracted complex in equation (6) are expressed by β_m and P_m, respectively, and the overall stability constant of the species in the aqueous phase is expressed by $\beta_{m'}$, the following equation results:

$$D = \frac{\Sigma\beta_m P_m K_i^{a-a'} K_w^{h-h'} [M]_{aq}^{m-m'} [HA]_{org}^{n+a-n'-a} L_{org}^{l-l'} [S]_{aq}^{s-s'}}{\Sigma\beta_{m'} P_{HA}^{n+a-n'-a'} P_L^{l-l'} [H]_{aq}^{n+h-n'-h'} [X]_{aq}^{-x+x'}}$$

where K_i and P_L are the dissociation constant and the partition coefficient respectively, of the complexing agent, K_w is the ionic product of water and $[H_2O]$ is assumed to be constant. In a simple case in which only one principal species exists in each of the two phases, the above equation may be given as follows:

$$D = K_{ex} \frac{[M]_{aq}^{m-m'} [HA]_{org}^{n+a-n'-a'} [L]_{org}^{l-l'} [S]_{org}^{s-s'}}{[H]_{aq}^{n+h-n'-h'} [X]_{aq}^{-x+x'}} \tag{7}$$

where partition of the solvent is considered, and all constants are expressed collectively in the term K_{ex} (extraction constant).

In all the above equations, the activity of each variable should be employed. However, concentration can be usually employed in place of the activity for convenience, provided the experiments are performed at constant ionic strength.

When both sides of equation (7) are expressed in logarithmic form, a more advantageous equation is obtained for the analysis of the extraction systems. If the concentration of only one variable, is changed all other variables being kept constant, a plot of log D against the logarithm of the concentration of the changed variable will give a straight line with slope equal to the exponent of the term of that variable in equation (7). Information on the composition of the extracted species can thus be obtained, and accordingly the mechanism of extraction can be indirectly learned. By adopting the same treatment in extraction chromatography, a comparative study of the mechanisms in the two extraction techniques has been made. In some cases, furthermore, formation constants and apparent equilibrium constants were determined, and compared with the analogous constants obtained by liquid-liquid extraction. Batch extraction using the extractant held on a solid support was also utilized in correlating liquid-liquid extraction and chromatographic data.

Thus, the comparative study of the behaviours ranges from finding approximate analogies to the strict comparison of the distribution ratios and of the values of the equilibrium constants. However, it must be considered that there are more factors affecting the extraction involved in a dynamic chromatographic process than in liquid-liquid batch extraction. These factors include the particle size of the support in the column, the degree of packing, the flow rate, diffusion effects, the amounts of eluted elements, the volume of the feed solution, the capacity of the column, the amounts of extractants, temperature, the rate of formation of the extractable complex, and the rate at which the two phases reach equilibrium. These factors, which are in most cases correlated, affect the peak width and tailing of the elution curve, and accordingly influence the distribution ratio and HETP, that is the overall efficiency of the column. Furthermore, it is difficult to keep the chromatographic parameters A_m and A_s, and the other experimental conditions constant throughout the processes of dynamic chromatography. When this characteristic feature of chromatography is taken into consideration, a certain flexibility should be given in its comparison with liquid-liquid batch extraction.

3. SIMILARITIES IN THE PRACTICE OF LIQUID-LIQUID EXTRACTION AND EXTRACTION CHROMATOGRAPHY IN EACH TYPE OF EXTRACTION SYSTEM

While in Section 1 the theoretical correlation between liquid-liquid extraction and extraction chromatography has been described, in this section some examples are presented which reveal the similarities between the experimental results obtained by extraction chromatography and liquid-liquid extraction.

Though several classifications of extraction systems have been tried by many authoritative workers [12–15], the main extraction systems adopted in chromatography are broadly classified here according to the nature of the mechanism in the extraction process, since the comparison of such mechanisms is actually the objective in this section. All extraction systems are therefore classified into three groups: chelate and cation exchange systems, ion association systems, and solvation systems. For each a few appropriate data are chosen for comparison. In addition, reference is made to the "tetrad" or "double-double" effect relating to the behaviour of lanthanides and actinides in many extraction systems, and to synergistic extraction as a mixed extraction system.

3.1. "TETRAD EFFECT" OR "DOUBLE-DOUBLE EFFECT"

In an attempt to describe the similarities between the data obtained from liquid-liquid extraction and extraction chromatography, a comparison of the behaviours of series of closely similar elements such as the lanthanides and actinides, is more advantageous than that of one single element. At the same time, the correlation between the behaviours in several different extraction systems is more useful than that referred to only one kind of extractant. The "tetrad effect" which involves lanthanides and actinides and several different extractants, is therefore the most appropriate example (see also Chapter 8).

The chemical and physical properties of the 15 lanthanides and the actinides are similar. The properties usually change slightly in the order of the atomic number, and it was in these variations that Peppard's and Siekierski's groups recently found certain regularities termed "tetrad effect" or "double-double effect" according to whether the

23

results are obtained mainly from liquid-liquid extraction or from extraction chromatography. Namely, in these series of elements discontinuities appear in some properties for the atoms in which the 4f or 5f electron shells are a half, one-quarter, and three-quarters full. It follows that if the numerical measures of properties of the elements such as the extraction behaviours or certain physical characteristics are plotted against the atomic number, the overall pattern appears to be divided into four smooth curves, corresponding to four definite tetrads of elements[16]. At present, theoretical explanations are available for this effect from the viewpoints of thermodynamics and interelectronic repulsion energy [17—23].

After Peppard had pointed out that the plot of log D vs. Z appeared to be a straight line in the solvent extraction of rare earths by TBP[24], the „two octads" assumption[25,26] and „odd-even effect"[26—28] were proposed as representative of this be-

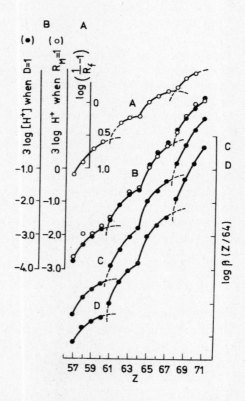

Fig. 1. Dependence of the separation factors on Z for adjacent lanthanides, in extraction chromatography and liquid-liquid extraction.
A — Paper chromatography using HTTA — TOPO — HCl (from Ref.[33], by courtesy of the author);
B — Column chromatography (o) and liquid-liquid extraction (•) using HDEHP — HClO₄ (from Ref.[32] by courtesy of the authors);
C — Liquid-liquid extraction using HDEHP — n-heptane — HCl (from Ref.[16],by courtesy of the authors)
D — Column chromatography using HEHΦP—HCl [16,30] (from Ref. [16], by courtesy of the authors)

haviour in the field of solvent extraction, but when more precise methods of determining distribution ratios were found, these assumptions proved to be incorrect, and in due course the "four tetrads" came to be considered more suitable[29]. Before Peppard's presentation of the "tetrad effect", based on liquid-liquid extraction data, a similar behaviour had been found by Siekierski's group in extraction chromatography results[30,31] and some regularities had also been observed in liquid-liquid extraction and chromatographic extraction by Pierce et al.[32,32a] The analogy of phenomena in liquid-liquid extraction and extraction chromatography played a very important role in confirming that this tetrad or double-double hypothesis (Siekierski) was of general application.

References [16] and [29] report curves obtained by the two methods, using several kinds of extractants. Peppard's presentation of the data of Siekierski's group [30] (D) and the results of Pierce et al. (B) [32a] are illustrated in Fig. 1. together with that of liquid-liquid extraction results (C, B). Curve (A) is a trial curve of the "tetrad effect" drawn from the extraction chromatography data for a synergistic extraction system (33), which will be discussed later. It is clear from the Figure that the variations of separation factor vs. Z are very similar to the liquid-liquid extraction and extraction chromatography data for the various systems; it is interesting to note that a similar behaviour is found in synergistic systems as well.

3.2. CHELATE AND CATION-EXCHANGE EXTRACTION SYSTEMS

3.2.1. CHELATE EXTRACTION SYSTEMS (SEE ALSO CHAPTER 11)

The theory of chelate extraction has been more thoroughly studied than those of other extraction systems, and it is well known that the mass action law can be applied most successfully to it, following the general treatment given in the previous section. Features that are known to be effective for liquid-liquid extraction separations with chelate systems, such as pH optimization and the use of masking agents, are also expected to allow good separations in extraction chromatography.

Chelating agents that have been used as stationary phases in extraction chromatography include dithizone [34,35], TTA[36], α-hydroxyoxime[37] and other oximes[38], dithiocarbamates[39,40], and N-benzoyl–N–phenylhydroxylamine[41]. They are solid in the main at ordinary temperatures, and the stationary phases generally consist of their solution in an organic solvent adsorbed on the support. There are some reports on chromatography with stationary phases the organic solvent of which had been evaporated, but it is evident that retention in this case is not based on distribution between two liquid phases, so that such systems cannot be regarded as involving extraction chromatography but may be similar to solid ion-exchange. In this section, the extraction chromatographic results are compared with the theory of solvent extraction with regard to several factors influencing the extraction of chelates.

Let us first consider the most general factors, the concentrations of hydrogen ion and of the reagent. If the extraction process is described by the equation

$$M^{n+}_{aq} + nHA_{org} \rightleftharpoons MA_{n,org} + n H^+_{aq}$$

the distribution ratio derived from equation (7), simply is

25

$$D = K_{ex} [HA]_{org}^n / [H^+]_{aq}^n$$

$$\log D = n\log [HA]_{org} - n\log[H^+]_{aq} + \log K_{ex} \tag{8}$$

If either $[HA]_{org}$ or $[H^+]_{aq}$ is kept constant the distribution ratio increases with increasing pH or reagent concentration, respectively.

According to the previous section and equation (8), the plots of log D vs. pH and log D vs. $\log[HA]_{org}$ should give straight lines with slope equal to the charge n of the metal. If the pH corresponding to 50% extraction (log D = 0) is expressed by $pH_{\frac{1}{2}}$, equation (8) becomes:

$$pH_{\frac{1}{2}} = -\frac{1}{n}\log K_{ex} - \log[HA]_{org}$$

if the reagent concentration is changed from [HA] to [HA]', the corresponding $pH_{\frac{1}{2}}$ is:

$$pH_{\frac{1}{2}'} = pH_{\frac{1}{2}} - \log[HA]_{org}' + \log[HA]_{org}$$

Hence when the reagent concentration of the organic phase increases by one order, $pH_{\frac{1}{2}}$ decreases by one unit, i. e. the plot of log D vs. pH shifts to lower pH.

Reference [34] is taken here as a practical example of reversed-phase extraction chromatography. In this reference, the distribution behaviour of zinc was studied on a column of dithizone $-$ CCl_4, using $HClO_4$ $-NaClO_4$ as aqueous solution; curves of log D vs. pH were given (see Fig. 1. in Chapter 11) at various dithizone concentrations. Straight lines of slope 2 were obtained at each concentration, corresponding to the charge of zinc ions. When the concentration of dithizone was changed from 10^{-3} M to $2 \cdot 10^{-4}$ M, $pH_{\frac{1}{2}}$ shifted from 2 to 2.7, that is by 0.7 unit. This value coincided with that theoretically predicted from the above equation. What is more, the mean value obtained for the logarithm of the extraction constant (log K_{ex} in equation (8)) was 2.07±0.04, in reasonable agreement with the value 2.2, obtained by the same authors for liquid-liquid extraction, and 2.0–2.3 in other publications. These results prove that the same distribution equilibrium is valid in liquid-liquid and chromatographic extraction.

Let us now consider the case of a complexing agent X present in the system, that forms a complex MX_x in the aqueous phase stepwise, but does not participate in the extraction. If MA_n is the only extracted species, the distribution coefficient is expressed as follows:

$$D = [MA_n]_{org}/[M^{n+}]_{aq}\left\{ 1 + \sum_1^x \beta_x[X]_{aq}^x \right\}$$

If the distribution coefficient obtained when no complexant is present in the system is expressed as D_o, it follows

$$D_o/D = 1 + \sum_1^x \beta_x[X]_{aq}^x \tag{9}$$

26

The reaction of the metal with the complexant results in the decrease of the extraction. From equations (8) and (9) we have

$$\log D = \log K_{ex} + n\log[HA]_{org} + npH - \log(1 + \sum_{1}^{x}\beta_x[X^-]_{aq}^x)$$

Comparison between this equation and equation (8) and of the corresponding equations for $pH_{\frac{1}{2}}$, reveals that the presence of a complexing agent decreases the slope of the extraction curve and at the same time shifts the extraction curve to a higher pH value. (see also Fig. 1. in Chapter 11).

This complexing agent can be effectively used as an eluent in extraction chromatography. In the previously mentioned work with zinc and dithizone, a complexing agent ($H_2C_2O_4 - Na_2C_2O_4$) was also used to elute two metals. As the above theories predict, the straight line on log D vx. pH plot was shifted to higher pH, that is, the same distribution ratio as obtained in the absence of $H_2C_2O_4$ was obtained at a higher pH value, while at the same time the slope decreased (see Fig. 1. in Chapter 11). When $(D_o/D - 1)/[C_2O_4^{2-}]$ is plotted vs. $[C_2O_4^{2-}]$ (see equation (9)), the stability constants of the oxalate complexes, β_1 and β_2, can be determined graphically from the intercept and the slope, respectively of the straight line by the limiting value method. In this way, the stability constants of $Zn(C_2O_4)$ and $Zn(C_2O_4)_2^{2-}$ were found as log $\beta_1 = 4.48$ and log $\beta_2 = 7.00$, consistent with the published values of log $\beta_1 = 4.68, 5.00$ and log $\beta_2 = 7.04, 7.36$, obtained by liquid-liquid extraction. These similarities suggest that it is possible, from the extraction constants obtained by liquid-liquid extraction and the known stability constants of the complexes to anticipate the possibility of extraction chromatographic separation of various metals by using complexing agents, and this has actually been applied for various practical column separations[35].

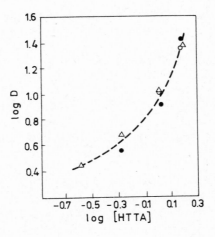

Fig. 2. Dependence of the distribution ratio of Ca on the concentration of HTTA in the organic phase, in the HTTA – MIBK–acetate buffer pH 5.5 system (from Ref. [36].)

O – Liquid-liquid extraction; Δ – Batch extraction with Kel–F; ● – Column extraction chromatography

The extraction is sometimes increased when an adduct $MA_n(HA)_a$ is formed by addition of the undissociated chelate agent to coordinatively unsaturated chelates at comparatively high reagent concentration; $(HA)_a$ in equation (6) represents the above undissociated reagent. A simple case may be represented by the following reaction:

$$M^{n^+}_{aq} + (n+a)HA_{org} \rightleftharpoons MA_n \ (HA)_{a\,org} + n \ H^+_{aq}$$

When the number of undissociated reagent molecules varies from zero to a, complexes MA_n, $MA_n(HA)_1$, $MA_n(HA)_2$ ..... are assumed to be formed, and the coefficient of HA on the left hand side of the above equation changes continuously via n, n+1, n+2, ... , the slope of the log D vs. log$[HA]_{org}$ plot therefore, gradually becomes steeper. Such adduct formation with the reagent was found in the study of the separation of alkaline earths in the extraction system HTTA—MIBK-acetate buffer. Figure 2 shows the log D vs. log$[HTTA]_{org}$ plots in liquid-liquid extraction [42] and column extraction chromatography [36], and the data of batch extraction using HTTA—MIBK loaded on Kel—F. The results of these methods are satisfactorily in agreement. The curves have slopes greater than 2 at the highest HTTA concentrations, suggesting gradual addition of HTTA molecule to the chelate. The coincidence of all results, including those obtained with different methods, suggest that almost all the HTTA in the extraction system, except that portion coordinating directly with the chela, participates indirectly in large aggregates in the increase of the distribution ratio.

The above chelate extraction systems can be used to advantage to illustrate a discussion as to the influence of the extraction equilibrium rates on the relation between liquid-liquid and column extraction. In fact, the distribution equilibria in such systems involve a number of processes such as the partition and dissociation of the reagent, the formation and partition of the chelates, and also processes such as the keto-enol tautomerism of the reagent, and so on. These processes are influenced by several factors, and the time needed to attain equilibrium depends strongly on the type of complex formed. In general, the rate determining step in extraction equilibria depends particularly on the process of formation of the compound to be extracted, where the nature of the metal, and reagent and the solvent play important roles. Some examples of the slow attainment of equilibrium are the extractions of dithizonates and of certain β-diketone complexes, and those of a number of elements (such as Cr(III), Co(III), and platinum metals) which are known to form inert complexes in which the ligand exchange is very slow. There are also examples, such as Co(III), in which the times required for the attainment of equilibrium in extraction and back extraction are somewhat different.

In chromatography, the aqueous phase begins to flow before the attainment of strict equilibrium between the two phases, and thus the factors which influence the rate of the process can be seen to be particularly important. The process of retention by the stationary phase is equivalent to normal extraction and washing in the liquid-liquid system, and the elution process is equivalent to back extraction: accordingly, the rates of both direct and back extractions are very important when liquid-liquid and chromatographic results are compared, therefore it is desirable to pay great attention to the back extraction curve.

On the assumption that the same extraction mechanism is operative in both chromatography and liquid-liquid extraction, it is possible to investigate the complex forming process in an extraction by taking advantage of the incomplete equilibrium features of chromatography: the HTTA—MIBK column, already referred to, is taken as example[36]. Figure 3 illustrates break-through curves resulting from the continuous feeding of calcium

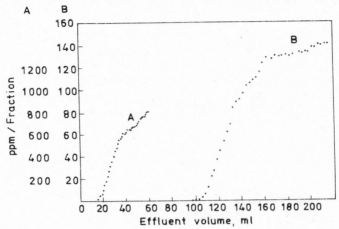

Fig. 3. Break-through curves of Ca on a HTTA — MIBK column. Feed solutions of pH = 5.5 contained 0.033 mmole/ml (curve A) and 0.0041 mmole/ml Ca (curve B)

solution into the column. These curves are neither sharp nor smooth, thereby differing from normal break-through curves. A gently continuous curve is usually believed to result from the slowness of the transfer between the stationary and mobile phases, or between the surface and the inner part of a stationary phase; in the Figure, however the curves clearly fit the broken lines. This fact may be expained as follows: in this column, as seen from the log D vs. log [HTTA] plot of Fig. 2, higher HTTA chelates are formed stepwise, and their formation is not considered to be very fast. In a chromatographic process where equilibrium cannot be attained, therefore this stepwise formation of the chelate results in a change of the apparent distribution ratio on the break-through curve. Such a phenomenon appears to be fairly useful, allowing the investigation of the process of extraction and some parts of its mechanism via the incomplete equilibrium conditions of chromatography.

Metal ions can be extracted and separated by means of exchange reactions with already formed chelates.

The reaction is expressed as follows:

$$nM_1 A_{m,org} + mM_2^{n+}{}_{aq} \rightleftharpoons mM_2 A_{n,org} + nM_1^{m+}{}_{aq}$$

and the exchange constant may be written as

$$K_{displ.} = \frac{[M_2 A_n]_{org}^m [M_2^{m+}]_{aq}^n}{[M_1 A_m]_{org}^n [M_2^{n+}]_{aq}^m} = \frac{K_{ex_2}{}^m}{K_{ex_1}{}^n} = \frac{(\beta_{m_2} P_{m_2})^m}{(\beta_{m_1} P_{m_1})^n}$$

29

It is clear that the actual exchange depends on the ratio of the extraction constants of the complexes of the two metals, that is on the stability constants and partition coefficients of these complexes. Accordingly, the choice of the type of metal chelates which are to be dissolved in the organic phase, and of the type of metals which are added to the aqueous phase for displacement, requires the consideration of these factors. In the case of displacement extraction chromatography [39] in which the exchange reaction was applied, a stationary phase of Zn-diethyldithiocarbamate-CHCl$_3$ was used, and the elements Mn, Zn, Cd, Ag, Hg and Co were separated substoichiometrically. The elements Cd, Cu, Ag, and Hg replaced Zn in the column and were therefore retained; successive replacements of Zn by Cd, Cd by Cu, Cu by Ag, and Ag by Hg were next performed. These exchange behaviours were based on the exchange series and on the values of K_{ex} acquired by liquid-liquid extraction, and the correlation of the two methods was acknowledged as well.

3.2.2. CATION-EXCHANGE EXTRACTION SYSTEMS

In cation-exchange extraction systems acidic organophosphorus compounds are generally used as extractants. Needless to say, these reagents are widely employed in liquid-liquid extraction.

For reversed-phase extraction chromatography, monobasic dialkyl phospates such as HDBP and HDEHP are commonly used. On the one hand, they have one hydrogen atom which can be displaced by metal ions, and that is why they are often called liquid cation-exchangers; and on the other hand, they have a basic P = O group which can coordinate with the metal, so that they do not only satisfy the charge of the metal, but also meet its coordination requirements. Such extraction agents are therefore formally similar to chelating agents, but whereas extraction with normal chelating agents exhibits ideal behaviour which is described by the simple mass-action law, the behaviour of acidic organophosphorus compounds often deviates from the ideal, since the states of the agents themselves and of the extraction species can vary remarkably, depending on the natures of the ligand and of the diluent, and on the concentration of the ligand. A great number of papers have appeared on this subject [43–49].

The extractants in general are dimers $(HX)_2$ at low concentrations in non-polar solvent; the proton of a singly-ionized dimeric dialkyl ester is displaced by the metal atom in the extraction process, but as the number of the extractant molecules in the extracted complex is in some cases considerably different from the coordination number to be expected from the charge of the metal ion, the extraction reaction can be better expressed by:

$$M^{z+}_{aq} + n(HX)_{2org} \rightleftharpoons MX_z \cdot (2n - z)HX_{org} + zH^+_{aq}$$

Reversed-phase paper chromatography with HDEHP of HDBP may profitably be compared here with liquid-liquid extraction; data obtained with the two techniques for the distribution equilibria of alkaline earths and lanthanides are listed in Table 1. The z and n in the Table represent the coefficients of H^+ and $(HX)_2$ respectively in the

TABLE 1
Acid (z) and extractant (n) dependencies in the extraction
of alkaline earth and rare earth metals in HDBP and HDEHP systems

Element	Extrac-tant	Aqueous medium	Acidity M	z	n	Method	Ref-er-ence
Ca	HDBP	HCl	log[H$^+$] -0.4- -1.6	1.5		Paper chrom.	50
Ca		HCl	0.05 log[H$^+$]		1.5	Paper chrom.	50
Be		HCl	0.5- -0.8	1.8		Paper chrom.	50
Be		HCl	0.5		1.5	Paper chrom.	50
Be		HNO$_3$	0.25	2.0	2.2	Extraction	51
Be		HNO$_3$	2	2.0	1.5−2.0	Extraction	51
Mg, Ca	HDEHP	HCl	0.1−0.001	1.5	3.0	Paper chrom.	52
Ca		HCl	0.001−0.1	1.5	3.0	Paper chrom.	53
Sr, Ba		HCl	0.001−0.01	2.0−1.5	3.0	Paper chrom.	53
			0.01−0.1	1.5	3.0		
Mg, Ca, Sr, Ba		HCl	0.01−1	2.0		Extraction	6
Mg, Sr, Ba / Ca		HCl	0.05		2.0 / <2.0	Extraction	7
Ca		Chloride	0.01−0.25	2.0		Extraction	54
Ca		HCl	0.01		3.0	Extraction	54
Be		HNO$_3$	0.25−3	2.0		Extraction	51
La	HDBP	HCl	0.35		1.5	Paper chrom.	50
Ce		HCl	0.35		1.5	Paper chrom.	50
Sc		HCl	6		1.5	Extraction	55
Sc		HCl	6		3.0(HDBP 0.1M)	Extraction	55
La~Lu		HClO$_4$	0.1	3.0	3.0	Extraction	56 48
Eu, Am		HNO$_3$	0.1	3.0	3.0	Extraction	57
Es	HDEHP	HCl HNO$_3$	0.1−1.0	3.0		Column chrom.	58
La~Sm		HCl	0.01−1	3−1.5		Paper chrom.	59
Lanthanides		HCl	0.1−10	3.0−<3.0		Paper chrom.	52
Lanthanides		HCl	0.1−10		3−1,<1	Paper chrom.	52
Lanthanides		HNO$_3$	0.2−4.0	3.0		Thin Layer c.	60
Lanthanides Actinides		HCl	0.74−1.15	3.0		Column chrom.	61
Lanthanides		HCl	0.1−10	3.0		Column chrom.	32
Lanthanides		ClO$_4^-$		3.0	3.0	Extraction	48
Lanthanides		HCl	log[H$^+$] -1.5−1.0	3.0		Extraction	32
Lanthanides Actinides		HCl	log[H$^+$] 0.18--0.32	3.0		Extraction	61

TABLE 2
Apparent equilibrium constants (logK)

	Reversed-phase paper extraction chromatography	Liquid-liquid extraction
Be^{2+}	2.58 (HDBP−CCl$_4$, HCl)	1.55 (HDBP-toluene, HNO$_3$)[51]
La^{3+}	1.59 (HDBP−CCl$_4$, HCl)	1.23 (HDBP-butylether,HClO$_4$)[48]
Ce^{3+}	1.83 (HDBP−CCl$_4$, HCl)	2.00 (HDBP-butylether,HClO$_4$)[48]

4

above extraction equilibrium; in paper chromatography they can be obtained from the slope of log-log plots, for which log $(1/R_f-1)$ is used instead of log D.

In liquid-liquid extraction the distribution ratio exhibits negative second power and negative third power dependences on $[H^+]$ for alkaline earths and lanthanides, respectively, with any extractant. The lower values of z found in some cases are interpreted via partial complex formation of the metal ion with the anion in the aqueous phase. The coefficient n in liquid-liquid extraction sometimes depends on the extractant concentration, tending to increase with the increase of the concentration, and it is also affected by the nature of the diluent. On the other hand, the coefficient n in paper and thin-layer chromatography is not valid in the strict sense, because the support treated by the extractant-diluent solution is normally used after its diluent is evaporated.

Taking the above mentioned comments into consideration comparison of the figures for each group in the Table reveals that a similarity really does exist between liquid-liquid extraction and the various kinds of chromatography. This similarity was also observed in the "apparent equilibrium constants". An extraction equilibrium was discussed in Ref. [50], cited in the Table, where a study is presented on paper treated with HDBP: the following expression was presented for beryllium:

$$\log (1/R_f-1) = \log [H_2A_2]_{org}^{3/2} / [H^+]_{aq}^2 + \log K_h \qquad (10)$$

where $\log K_h = \log K_e/k \cdot f(\gamma)$, while K_e is an equilibrium constant, k is A_m/A_s in the equation (5), and $f(\gamma)$ is the activity coefficients ratio. An apparent constant K_h was obtained by extrapolating to the position of $\log [H_2A_2]_{org}^{3/2} / [H^+]_{aq}^2 = O$ on the $\log (1/R_f-1)$ vs. $\log [H_2A_2]_{org}^{3/2}/[H^+]_{aq}^2$ plot of the chromatographic data; and the "apparent equilibrium constant" was determined via this value of K_h and a chromatographic constant k. The same treatment was also applied for tivalent metals: Table 2 shows the results together with those obtained in liquid-liquid extraction, considerable similarity being apparent.

All the above-mentioned examples indicate that the same mechanism holds in reversed-phase extraction chromatography and solvent extraction.

3.3. ION-ASSOCIATION EXTRACTION SYSTEMS

In this system, the extractable species is formed by interaction between an ion in the aqueous phase and an oppositely charged ion present in either the aqueous or the organic phase. However, the system cannot simply be described by the two ions forming an ionpair, and other considerations too must be taken into account.

Ion-pair formation may be treated as an equilibrium reaction, and according to Bjerrum's theory the association equilibrium constant of an ion association reaction is expressed as[62]

$$K_{ass} = \frac{4\pi N}{1000} \left(\frac{Z^2e^2}{\epsilon kT} \right)^2 Q(b); \quad b = e^2/akT\epsilon$$

where ϵ is the dielectric constant, Ze the charge, N the Avogadro's number, k Boltzmann's constant, T the absolute temperature, and a the distance of closest approach of

32

the two ions. The free energy of ion-pairing is therefore expressed as $\Delta F_{ip} = -RT \ln K_{ass}$. On the other hand, the free energy for the transfer of the ionic compound from the aqueous solution into the organic phase is expressed by Born's treatment[63] for the change of free energy in the hydration process, the following equation holding:

$$\Delta F_t = \frac{Ze^2}{2r} \left(\frac{1}{\epsilon_o} - \frac{1}{\epsilon_w} \right)$$

where r is the ionic radius and ϵ_w and ϵ_o are the dielectric constants of water and the organic solvent, respectively. It is generally agreed that both the above-mentioned changes of free energy should be considered for the processess of formation and extraction of the extractable species [14]. The results obtained from the Bjerrum and Born equations are over simplified, since only electrostatic free energy is taken into consideration; in these extraction systems more attention should be paid to the variables relating to the organic phase, that is to the nature, structure and size of the organic base, and to the concentration and type of the organic solvent used as diluent. In practice, however, the electrostatic interaction in ionpairing, and the size and degree of hydration of the ions, allow an approximate estimate of the extraction, so that it can be readily said that a singly-charged ion and one of opposite charge, that are large and not hydrated in the aqueous phase, can be extracted most easily as a pair into the organic solvent.

The extraction of metals through the formation of ion pairs belongs to one or other of the following cases:

(a) a simple cation is extracted into a polar solvent as a pair with a comparatively bulky anion;

(b) a complex anion or a negatively charged chelate complex produced from a metal is extracted together with a bulky pairing cation;

(c) a metal is extracted as a cationic chelate compound with a bulky pairing anion.

High-molecular weight amines and substituted ammonium salts, usually termed liquid anion-exchangers, are the most widely used cationic moiety for ion-association in these systems, but other onium bases, such as Ph_4As^+ or Ph_4P^+, and bulky cations of synthetic dyes are utilized. Systems such as Ph_4As^+ however, are hardly applicable to a reversed-phase technique, since they consist of water-soluble salts.

It is rather difficult to treat concisely the equilibria of formation and partition of extracted complexes in ion-association systems, though the formula of mass-action may also be applied as in chelate extraction systems. The reasons for this difficulty are that a large number of equilibria, such as higher aggregate formation, are involved in the ion-association system, and that most of the activity coefficients are uncertain because of the high concentrations of electrolytes employed.

The equilibrium of extraction of a metallic species M^{n+} by a long-chain aliphatic amine or ammonium compound from an aqueous phase containing a singly-charged anion X^- (when M^{n+} exists as an anionic complex and the amine RN is previously equilibrated with acid HX) may be expressed as follows:

$$x\,(RNHX)_{org} + MX^{x-}_{n+x,aq} \rightleftharpoons (RNH)_x MX_{n+x,org} + x\,X^-_{aq}$$

In extraction chromatography using amines or ammonium compounds, a great number of R_f spectra are available for different compositions of the aqueous solution used as eluent, and for almost all elements. Comparison of these R_f spectra and the extraction curves from liquid-liquid extraction has often been made.

In a paper concerning thin layer chromatography with Amberlite LA-1 and HCl, $R_M = \log(1/R_f-1)$ and log D obtained from liquid-liquid extraction were plotted against HCl concentration [64]. The two curves, shown in Fig. 4, exhibited an identical pattern. In another study on the separation of rare earth elements on paper treated with TNOA[65], R_M was plotted, vs. log [LiNO$_3$], vs. Z, and again the patterns were found to be similar to those obtained by liquid-liquid extraction. In the opinion of the author, the separation factors of adjacent lanthanides, given as a function of the atomic number in this last paper, follow a "tetrad effect" similar to that found in the other solvent extraction systems.

The separation of Cf and Es, carried out with Aliquat-336 nitrate[66], is one example of column work. The dependence of the distribution ratio of Cf on the nitrate concentration of the aqueous solution at constant acidity was studied both by the column method and by the corresponding liquid-liquid extraction system using xylene as the diluent. Figure 5 shows the data obtained; they demonstrate the very close parallelism between the two methods, in spite of the slight difference of the distribution ratios. Thus, if this difference is determined in advance as the paper describes, the elution volume from the column could easily be predicted from the liquid-liquid extraction data. In another work involving a column of Aliquat-336[67], Am and the rare earths all gave straight lines with positive slopes of approximately 2 in the logarithmic relationship between D and the thiocyanate concentration in the eluent, the distribution of the rare earths increasing with the atomic number. The logarithmic relationship between the extractant concentration on the column and the distribution coefficient showed that the ligand to metal ratio was equal to unity. All these results were identical to those obtained from liquid-liquid extraction. The acid dependence also showed a qualitative agreement.

In general, however, in the systems where high molecular weight amines are used as extractants, it seems to be difficult to compare the results of liquid-liquid extraction quantitatively with those of chromatography under the same experimental conditions. This

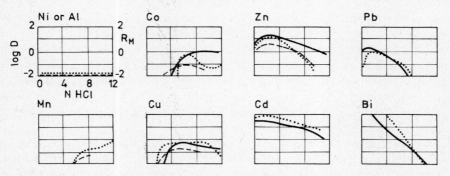

Fig. 4. Comparison of log D and R_M (= log (1/R_f-1)) for Amberlite LA-1 (from Ref. [64], by courtesy of the authors).

Dashed line – extraction using Amberlite in xylene; Dotted line – extraction using Amberlite in diethylbenzene; Solid line – R_M values from thin-layer chromatography

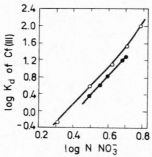

Fig. 5. Nitrate ion dependence of the extraction of Cf in the Aliquat-336 nitrate – HNO_3 system
(from Ref. [66], by courtesy of the authors).

O – Liquid-liquid extraction (in xylene); ● – Column extraction chromatography

stems from the fact that liquid-liquid extraction is mostly carried out in the presence of
a diluent, while in chromatography a support impregnated with the extractant solution
in an organic diluent is normally used after the diluents are evaporated. In fact, extraction of these ion-association systems in known to be strongly affected by the properties
of the diluent used. Attention should therefore be paid to the presence or absence of
diluent, in a comparison of the methods. Another type of comparison is found in a
paper which reports the separation of metals on a Corvic column impregnated with
TIOA [68]. In order to predict the elution behaviours of the metals on the column,
batch extractions were carried out using an undiluted TIOA-Corvic solid phase, so as to
have an organic phase closer to that of the column. The extraction isotherm obtained
for Zn showed good linear dependence over a wide zinc concentration range between
the zinc concentration in hydrochloric acid solution and that on the treated polymer,
this indicates that the distribution equilibrium also can be attained when no diluent is
present. Moreover, the extraction curves for a number of other elements, obtained by
this batch system, were helpful in devising conditions suitable for metal separations on
the column. The general patterns for metals in batch extraction with the supported
undiluted amine were also compared with the data obtained by using suitably diluted
TIOA, and a similar shape was found. In a thin layer chromatographic study on a liquid-
anion-exchanger [69], specific liquid-liquid extractions were carried out with the undiluted exchanger to get information on metal chloro anions in the organic phase through
spectral measurements of the organic extracts.

Brinkman has recently published general reviews on reversed-phase chromatography
with amines and ammonium salts [70,71].

Extraction chromatographic works involving systems where a simple cation is extracted into a polar solvent, forming an ion pair with a larger anion, includes a paper on the
separation of alkali metals in a polyiodide-nitrobenzene column [72]. The order of retention of the metals was found to be Cs > Rb > Na > Li; in other words, as predicted
by the general extraction theory previously outlined, the efficiency of extraction increased in accordance with the size of the cation and with the decrease of its hydration,
in agreement with the results from liquid-liquid extraction [73]. According as the amount
of the metals loaded in the column increased, the elution volumes, the distribution coef-

35

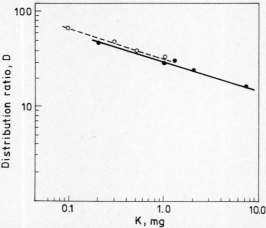

Fig. 6. Relation between the amount of potassium loaded and the distribution ratio in the ammonium polyiodide – nitrobenzene system (from Ref. [72]).
● – Column extraction chromatography; ○ – Liquid-liquid extraction.

ficients appreared to decrease. This was also found in solvent extraction: Fig. 6 gives a comparison of the behaviours of K in the two methods, and reveals the fairly close agreement between them. It was also found that the number of theoretical plates decreased as the distribution ratio decreased. A theoretical treatment on the solvent extraction of this system can be found in Ref. [74].

Another paper reports the use of an ammonium dipicrylaminate-nitrobenzene column for the separation of alkali metals [75], this can also be considered an ion-pair extraction of the metal cation and dipicrylaminate anion. Traces of cesium were enriched and separated from a large excess of alkali and alkaline earth metals in frontal chromatography. The most important equilibrium reaction in this case was taken for an extraction exchange, and the equilibrium constants K_M^{Cs} of the exchange reaction between each metal and cesium, obtained in liquid-liquid extraction, were found to be closely correlated to the column capacity obtained from the break-through curves for cesium in frontal chromatography, by using solutions containing each of other metals. These values for cesium relating to each of the metals were obtained in the order of $Ca > Li > Na$ in both methods. The same mechanism was therefore considered to be operative in liquid-liquid extraction and in extraction chromatography.

3.4. SOLVATION EXTRACTION SYSTEMS

In these systems the species extracted is solvated by the molecules of the organic solvent. The formation of such species can also be represented by equation (6), where the solvent is indicated by S. Such solvents include weakly basic carbonyl compounds, e. g. ethers, ketones and other oxygenated compounds, and also neutral organophosphorus compounds e. g. basic phosphates, phosphonates, phosphinates and phosphine oxides. All these neutral molecules, which contain $>O$, $>C=O$, or $\geqslant P=O$ groups, coordinate to a

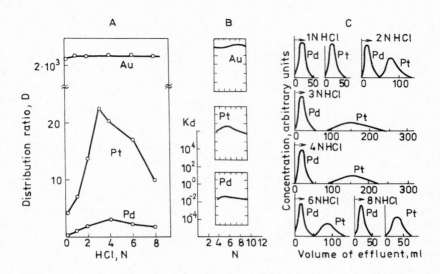

Fig. 7. Analogy between extraction chromatography and liquid-liquid extraction of Au, Pt and Pd in TBP – HCl system (from Ref. [76]).
A – Distribution curves for batch extraction with TBP – Daiflon (Pt, Pd) and by liquid-liquid extraction (Au); B – Distribution curves by liquid-liquid extraction (from Ref. [8], by courtesy of the authors); C – Chromatographic behaviours of individual ions

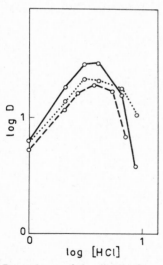

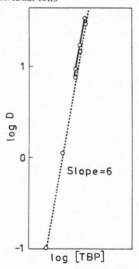

Fig. 8. Dependence of the distribution ratio of Pt on the concentration of HCl with different techniques in the TBP – HCl system (from Ref. [76]). Solid line – Column chromatography; dotted line – Batch equilibrium with TBP – Daiflon; dashed line – Liquid-liquid extraction

Fig. 9. Dependence of the distribution ratio of Pt on the concentration of TBP in the TBP – HCl system (from Ref. [76]). Solid line – Column chromatography; dotted line – Liquid-liquid extraction

37

central atom, their oxygen atoms generally displacing a water molecule from the coordination sphere but in some cases they coordinate through a water molecule bridge.

Acid extraction by these compounds takes place through association of the anion with the complex cation formed by protonation of the extractant: for example, the extraction of $HClO_4$ by TBP (diluted in CCl_4) is assumed to involve an ion-associated species $(mTBP \cdot H^+ \cdot xH_2O - ClO_4^-)_{org}$, and $H^+ \cdot xH_2O$ is considered to be present as the trihydrated hydronium ion at the highest TBP concentrations. That is why these extractants are sometimes classified as ion-association systems, however, a clear distinction is advisable between them and the more basic ones containing nitrogen, which produce ion pairs much more easily. The same treatment of extraction equilibrium as in the chelate extraction system may be applied to these systems, that leads to the solvation number.

Extraction chromatography with these compounds is of common use, and especially TBP has been widely applied from the early stages of the technique. One example of the relationship between extraction chromatography and solvent extraction, is a paper on the separation of Au, Pt, and Pd on a column of TBP on Daiflon, with HCl as the eluent[76], where the R_f spectra from thin layer chromatography[77], the results of countercurrent extraction[78], and the behaviours in liquid-liquid extraction[8] gave a useful information for the separation. Figure 7 illustrates the results obtained with the different techniques, and the good agreement is obvious. The use of 3–4 N HCl for good separation, suggested by the preliminary experiments, was successfully utilized for the practical column separation. The composition of the extracted species was also predicted for platinum, through the log D vs. log[HCl] and log D vs. log [TBP] plots obtained

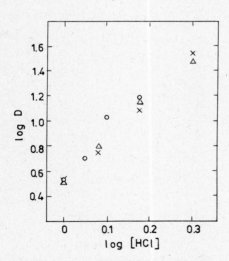

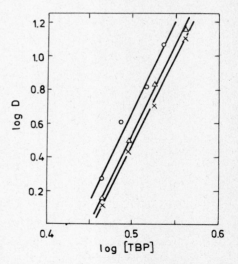

Fig. 10. Dependence of the distribution ratio of $[Fe(CN)_6]^{4-}$ on the concentration of HCl in the TBP – HCl system (from Ref. [79]).
○ – Column chromatography; Δ – Batch extraction on Kel–F; x – Liquid-liquid extraction

Fig. 11. Dependence of the distribution ratio of $[Fe(CN_6]^{4-}$ on the concentration of TBP in the TBP – HCl system (from Ref. [79]).
○ – Column chromatography; Δ – Batch extraction on Kel–F; x – Liquid-liquid extraction

from extraction chromatography, and also through the results of liquid-liquid extraction and batch extraction on Daiflon, under the same conditions as in chromatography. The results of these studies, illustrated in Fig. 8 and Fig. 9, coupled with other data obtained by the continuous-variation method, suggested that solvated $H(H_2O)_n TBP_6^+ HPtCl_6^-$ would be the principal extracted species. These studies and the magnitudes of the distribution ratio of Pt, agreed well in all three extraction techniques.

Another work dealing with the separation of $[Fe(CN)_6]^{3-}$ and $[Fe(CN)_6]^{4-}$ on a TBP column[79], may be considered an additional example. Just as anticipated through the solvent extraction curves[80], these ions behaved in the same way in the extraction column. Experiments analogous to those already described for platinum proved that the extraction mechanism for $[Fe(CN)_6]^{4-}$ in the column was identical to that of liquid-liquid extraction (Figs 10 and 11). It was suggested that the extracted species was a large, hydrated and solvated ion pair. The same results also revealed that the organic support had no effect on the extraction.

In a study of the systematic separation of numerous metals by means of the different columns[81], TBP–HCl system again showed the similarity of behaviour in solvent extraction and thin layer or column chromatography. Figure 12 shows such similarities for several metals.

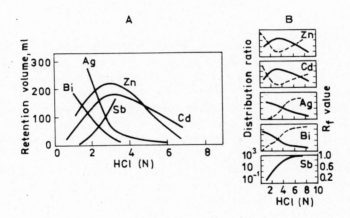

Fig. 12. Retention volumes of a TBP column and distribution ratios or R_f values as functions of the HCl concentration (from Ref. [81]).

A: Retention volume – in column extraction chromatography; B: Solid line – Ishimori's data in liquid-liquid extraction; Dotted line – Bark's data in reversed-phase thin-layer chromatography

3.5. SYNERGISTIC EXTRACTION SYSTEMS

It is sometimes possible that a combination of two extractants results in distribution ratios greater than those obtainable individually with either. This phenomenon is called the synergistic effect, and is efficiently used at present for better extraction. Although the mechanisms involved have already been clarified somewhat synergistic systems have

39

seldom been applied to practical separations. The combinations of extractants in these systems include the use of chelating agents (or acidic organophosphorus compounds) and neutral ligands, two different neutral ligands, and two different acidic ligands.

β-diketones (such as HTTA) and acidic organophosphorus compounds (HDBP and HDEHP) are the most commonly used chelating and acidic agents, respectively. The metal chleates which are produced by the these reagents give rise to adducts coordinatively saturated by reacting with electrically neutral ligands, such as neutral organophosphorus compounds (TBP and TOPO are typical ones), alcohols, ketones, amines or amides. The formation of such mixed adducts is assumed to be the reason for the enhancement of the extraction. The extraction equilibrium can again be treated with equation (6), where both the chelating agent (A) and the neutral ligand (L) are taken into account. Assuming that various kinds of adducts are formed when L is added to a metal complex MA_n, so that L can range from zero to 1, the following equation can be written for D:

$$D = [MA_n]_{org}(1 + \sum_1^1 \beta_l[L]_{org}^l)/[M^{n+}]_{aq}$$

where β is the adduct formation constant. If D_o is the value of D at $l = 0$,

$$\log D/D_o = \log \sum_1^1 \beta_l[L]_{org}^l$$

It can be seen from this relation that the increase of the distribution ratio by the synergistic effect is expressed as a function of β and of the concentration of L. In its application, therefore, factors which will affect β should be considered: the larger the basicity of L and the lower the stability of the chelate, for example, the larger the stability of the adduct will be, the effect probably being the greater, the polarity of the diluent lower.

There are very few examples of application of synergistic systems to reversed-phase extraction chromatography. A report has been published on the distribution behaviours of Ce, Sm, and Er by using paper treated with a mixture of TOPO and HTTA in various proportions, the total molarity being kept constant[33]. As Fig. 13 shows, the R_M values (and hence log D) of all three ions are greatly enhanced by the use of the HTTA–TOPO mixture as compared to those obtained with each agent independently, and became maximum at a certain composition of the mixture. This behaviour conforms closely to the typical synergistic curves found in solvent extraction[82]. The dependence of the R_f values on $[H^+]$, [HTTA] and [TOPO] was determined via plots of R_M vs. each logarithmic concentration of the above three parameters, and the slopes of the resulting straight lines were approximately 3, 3, and 2, respectively. The synergistic species was therefore considered as $M(HTTA)_3 (TOPO)_2$. This result is in accordance with that obtained in the solvent extraction of tervalent lanthanides and actinides[82]. A synergistic curve analogous to that shown in Fig. 13 was also found in the behaviours of Am, Ce and La on paper treated with combinations of HTTA and TOP[83].

Some examples of the column method include the separations of Sm–Eu and Cf–Cm by means of a mixture of HTTA with dibutyldiethyl carbamoyl phosphonate (DBDECP) [84, 85]. In Ref. [84], the distribution ratios obtained with columns of HTTA, DBDECP and HTTA + DBDECP as stationary phases were 3.1, 3.3, 106.7, respectively and these values clearly demonstrate the synergistic enhancement due to the application of the mixed

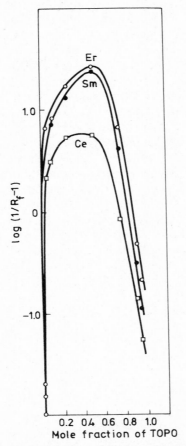

Fig. 13. Synergistic curves for Ce, Sm and Er for paper chromatography using HTTA — TOPO mixtures (from Ref. [33], by courtesy of the author)

extractant. The synergistic effects of the combinations of HDBM with TBP or TOPO were also applied to the separation of Li from alkali metals by a column method[86]. These few examples indicate that the synergistic effect takes place not only in liquid-liquid extraction, but also in reversed-phase extraction chromatography.

4. CONCLUDING REMARKS

It has been pointed out in this chapter how liquid-liquid extraction and extraction chromatography are closely correlated practically as well as theoretically. On the basis of this correlation, almost all extraction systems used in solvent extraction can be applied to extraction chromatography, and many authors have successfully solved problems concerning separation of elements that cannot be satisfactorily achieved by single-stage liquid-liquid extraction. Specifically, extraction chromatography is very effective for the separation of analogous elements, and also for the clarification of complicated extraction systems such as the extraction of Ru; to all chromatographic results an explanation may be given in connection with liquid-liquid extraction.

Either solvent extraction or extraction chromatography can be effectively utilized as a means of preliminary investigation of the other method, and at the same time a combined study of the results from the two methods makes the explanation of the extraction mechanisms easier. Because of the effectiveness of chromatography, all methods of separation used in solvent extraction can be tried in extraction chromatography as well. They include, for example, batch extraction with an extractant loaded on a solid support for rapid determination and "milking" in radiochemistry, and the systematic separation of multicomponents. A systematic scheme can even be constructed for the separation of metal ions into groups and for their mutual separations[81,87—91] on columns treated with different extractants. This is readily possible by referring to the great number of data obtained by liquid-liquid extraction in the wide variety of extraction systems reported so far and also to selectivity of extractants derived from these data, taking full advantage of the correlations between chromatography and liquid-liquid extraction. The systematic separation by extraction chromatography is, so to speak, what has been evolved from the similarities of the two methods. It will probably be further developed in the future, aiming at the establishment of more general separation schemes which will make possible the quantitative separation of each ion from mixtures of various metals of any composition. It is also probable that this process can be automatized when necessary.

REFERENCES TO CHAPTER 2

1.   A. J. P. MARTIN, R. L. M. SYNGE, Biochem. J., 35 (1941) 1358.
2.   A. I. M. KEULEMAN, Gas Chromatography, Chap. 4, Reinhold, New York, 1957.
3.   E. GLUECKAUF, Trans. Faraday Soc., 51 (1955) 34.
4.   H. B. F. DIXON, J. Chromatog., 7 (1962) 467.
4/a. I. ROSENTHAL, A. R. WEISS, V. R. USDIN, in Treatise on Analytical Chemistry (Eds. I. M. Kolthoff, P. J. Elving), Part 1. Vol. 3, Interscience, N. Y., 1961.
5.   R. CONSDEN, A. H. GORDON, A. J. P. MARTIN, Biochem. J., 38 (1944) 224.
6.   K. KIMURA, Bull. Chem. Soc. Japan, 33 (1960) 1038.
7.   K. KIMURA, Bull. Chem. Soc. Japan, 34 (1961) 63.
8.   T. ISHIMORI, K. WATANABE, E. NAKAMURA, Bull. Chem. Soc. Japan, 33 (1960) 636.
9.   T. ISHIMORI, K. KIMURA, T. FUJINO, H. MURAKAMI, J. At. Energy Soc. Japan, 4 (1962) 117.
10. T. ISHIMORI et al. JAERI 1062 (1964).
11. Y. MARCUS, A. S. KERTES, Ion Exchange and Solvent Extraction of Metal Complexes,Wiley-Interscience, New York, 1968.
11/a. P. MARKL, Extraktion und Extraktions-Chromatographie in der anorganischen Analytik, Akademische Verlagsgesellschaft, Frankfurt am Main, 1972.
12. H. M. IRVING, Ouart. Rev., 5 (1951) 200.
13. G. H. MORRISON, H. FREISER, Solvent Extraction in Analytical Chemistry, J. Wiley and Sons, New York, 1957.
14. R. M. DIAMOND, D. G. TUCK, Progress in Inorganic Chemistry (Ed. F. A. Cotton), Vol. II, Interscience, New York, 1960, p. 109.
15. Y. MARCUS, Chem. Rev., 63 (1963) 139.
16. D. F. PEPPARD, C. A A. BLOOMQUIST, E. P. HORWITZ, S. LEWEY, G. W. MASON, J. Inorg. Nucl. Chem., 32 (1970) 339.

17. I. FIDELIS, S. SIEKIERSKI, J. Inorg. Nucl. Chem., 29 (1967) 2629.
18. S. SIEKIERSKI, J. Inorg. Nucl. Chem., 32 (1970) 519.
19. I. FIDELIS, J. Inorg. Nucl. Chem., 32 (1970) 997.
20. C. K. JØRGENSEN, J. Inorg. Nucl. Chem., 32 (1970) 3127.
21. L. J. NUGENT, J. Inorg. Nucl. Chem., 32 (1970) 3485.
22. I. FIDELIS, S. SIEKIERSKI, J. Inorg. Nucl. Chem., 33 (1971) 3194.
23. S. SIEKIERSKI, I. FIDELIS, J. Inorg. Nucl. Chem., 34 (1972) 2225.
24. D. F. PEPPARD, J. R. FARIS, P. R. GRAY, G. W. MASON, J. Phys. Chem., 57 (1953) 294.
25. D. F. PEPPARD, W. J. DRISCOLL, R. J. SIRONEN, S. McCARTY, J. Inorg. Nucl. Chem., 4 (1957) 326.
26. E. HESFORD, E. E. JACKSON, H. A. C. McKAY, J. Inorg. Nucl. Chem., 9 (1959) 279.
27. G. W. MASON, S. LEWEY, D. F. PEPPARD, J. Inorg. Nucl. Chem., 26 (1964) 2271.
28. D. F. PEPPARD, G. W. MASON, S. LEWEY, J. Inorg. Nucl. Chem., 27 (1965) 2065.
29. D. F. PEPPARD, G. W. MASON, S. LEWEY, J. Inorg. Nucl. Chem., 31 (1969) 2271.
30. I. FIDELIS, S. SIEKIERSKI, J. Chromatog., 17 (1965) 542.
31. I. FIDELIS, S. SIEKIERSKI, J. Inorg. Nucl. Chem., 28 (1966) 185.
32. T. B. PIERCE, P. F. PECK, Analyst, 88 (1963) 217.
32/a. T. B. PIERCE, P. F. PECK, R. HOBBS, J. Chromatog., 12 (1963) 81.
33. N. CVJETIĆANIN, J. Chromatog., 74 (1972) 99.
34. F. ŠEBESTA, J. Radioanal. Chem., 6 (1970) 41.
35. F. ŠEBESTA, J. Radioanal. Chem., 7 (1971) 41.
36. I. AKAZA, Bull. Chem. Soc. Japan, 39 (1966) 980.
37. J. S. FRITZ, D. R. BEUERMAN, J. J. RICHARD, Talanta, 18 (1971) 1095.
38. E. CERRAI, G. GHERSINI, Analyst, 94 (1969) 599.
39. O. V. STEPANETS, Yu. V. YAKOVLEV, I.P. ALIMARIN, Zh. Anal. Khim., 25 (1970) 1906.
40. I. P. ALIMARIN, Yu. V. YAKOVLEV, O. V. STEPANETS, J. Radioanal. Chem., 11 (1972) 209.
41. F. ŠEBESTA, A. LÁZNIČKOVÁ, J. Radioanal. Chem., 11 (1972) 221.
42. I. AKAZA, Bull. Chem. Soc. Japan, 39 (1966) 971.
43. D. F. PEPPARD, G. W. MASON, G. GRIFFIN, J. Inorg. Nucl. Chem., 27 (1965) 1683.
44. C. J. HARDY, D. SCARGILL, J. Inorg. Nucl. Chem., 11 (1959) 128.
45. B. F. GREENFIELD, C. J. HARDY, AERE/R–3686 (1961).
46. D. DYRSSEN, Acta Chem. Scand., 11 (1957) 1771.
47. D. DYRSSEN, L. D. HAY, Acta Chem. Scand., 14 (1960) 1091.
48. C. F. BAES, Jr., J. Inorg. Nucl. Chem., 24 (1962) 707.
49. C. F. BAES, Jr., H. T. BAKER, J. Phys. Chem., 64 (1960) 89.
50. N. CVJETIĆANIN, J. Chromatog., 32 (1968) 384.
51. C. J. HARDY, B. F. GREENFIELD, D. SCARGILL, J. Chem. Soc., 1961, 174.
52. E. CERRAI, G. GHERSINI, J. Chromatog., 24 (1966) 383.
53. E. CERRAI, G. GHERSINI, J. Chromatog., 15 (1964) 236.
54. D. F. PEPPARD, J. Inorg. Nucl. Chem., 24 (1962) 321.
55. A. P. SAMODELOV, Radiokhimiya, 6 (1964) 286.
56. G. DUYCKAERTS, Ph. DRÈZE, A. SIMON, J. Inorg. Nucl. Chem., 13 (1960) 332.
57. D. DYRSSEN, L. D. HAY, Acta Chem. Scand., 14 (1960) 1100.
58. E. P. HORWITZ, C. A. A. BLOOMQUIST, D. J. HENDERSON, J. Inorg. Nucl. Chem., 31 (1969) 1149.
59. E. CERRAI, G. GHERSINI, M. LEDERER, M. MAZZEI, J. Chromatog., 44 (1969) 161.
60. H. HOLZAPFEL, LE VIET LAN, G. WERNER, J. Chromatog., 24 (1966) 153.
61. K. A. GAVRILOV, Talanta, 13 (1966) 471.
62. C. A. KRAUS, J. Phys. Chem., 60 (1956) 129.
63. M. BORN, Z. Physik, 1 (1920) 45.
64. U. A. Th. BRINKMAN, G. De VRIES, E. van DALEN, J. Chromatog., 22 (1966) 407.
65. C. TESTA, Anal. Chem., 34 (1962) 1556.
66. E. P. HORWITZ, L. J. SAURO, C. A. A. BLOOMQUIST, J. Inorg. Nucl. Chem., 29 (1967) 2033.

67.. E. A. HUFF, J. Chromatog., 27 (1967) 229.
68. T. B. PIERCE, W. M. HENRY, J. Chromatog., 23 (1966) 457.
69. U. A. Th. BRINKMAN, G. De VRIES, E. van BALEN, J. Chromatog., 31 (1967) 182.
70. U. A. Th. BRINKMAN, J. Chem. Ed., 49 (1972) 244.
71. U. A. Th. BRINKMAN, Progress in Separation and Purification (Eds., E. S. Perry, C. J. van Oss), Vol. 4, Wiley-Interscience, New York, 1971.
72. I. AKAZA, Bull. Chem. Soc. Japan, 39 (1966) 585.
73. I. AKAZA, Bull. Chem. Soc. Japan, 39 (1966) 465.
74. S. TRIBALAT, M. GRALL, Solvent Extraction Research (Eds. A. S. Kertes, Y. Marcus), Wiley-Interscience, London, 1969, p. 121.
75. M. KYRŠ, L. KADLECOVÁ, J. Radioanal. Chem., 1 (1968) 103.
76. I. AKAZA, T. KIBA, T. KIBA, Bull. Chem. Soc. Japan, 43 (1970) 2063.
77. I. S. BARK, G. DUNCAN, R. J. T. GRAHAM, Analyst, 92 (1967) 347.
78. E. W. BERG, W. L. SENN, Jr., Anal. Chim. Acta, 19 (1958) 12.
79. I. AKAZA, T. KIBA, M. TABA, Bull. Chem. Soc. Japan, 42 (1969) 1291.
80. T. KIBA, K. TERADA, N. KISAKA, Preprints for the Annual Meeting of the Chemical Society of Japan (April, 1962), No. 16407; (April, 1963), No. 3105.
81. I. AKAZA, T. TAJIMA, T. KIBA, Bull. Chem. Soc. Japan, 46 (1973) 1199.
82. T. V. HEALY, J. Inorg. Nucl. Chem., 19 (1961) 314.
83. N. CVJETIĆANIN, J. Chromatog., 34 (1968) 520.
84. H. F. ALY, M. A. EL—HAGGAN, Radiochem. Radioanal. Lett., 3 (1970) 249.
85. H. F. ALY, M. RAIEH, Anal. Chim. Acta, 54 (1971) 171.
86. D. A. LEE, J. Chromatog., 26 (1967) 342.
87. K. SAMSAHL, P. O. WESTER, O. LANDSTRÖM, Anal. Chem., 40 (1968) 181.
88. K. SAMSAHL, Analyst, 93 (1968) 101.
89. R. MALVANO, P. GROSSO, M. ZANARDI, Anal. Chim. Acta, 41 (1968) 251.
90. J. S. FRITZ, G. L. LATWESEN, Talanta, 17 (1970) 81.
91. R. DENIG, N. TRAUTMANN, G. HERRMANN, J. Radioanal. Chem., 6 (1970) 331.

CHAPTER 3

TECHNIQUES IN COLUMN EXTRACTION CHROMATOGRAPHY

P. MARKL, E. R. SCHMID

INTRODUCTION

INTRODUCTION

Extraction chromatography offers a number of advantages to the analyst engaged in inorganic analysis[1]:

It is the simplest and most effective way of achieving multistage extractive separations whenever multistage separation processes are needed.

In many applications, especially in radiochemistry, column separation techniques are preferable to other separation methods, since the equipment is simple and does not necessarily contain moving parts. Replacement of equipment and execution of separations can be done by remote handling techniques.

The possibility of using a wide variety of extracting agents as stationary phases tremendously enlarges the number of separations which can be achieved. In some cases, otherwise useful extracting agents form stable emulsions when used in liquid-liquid partition experiments, but these phase separation problems can be avoided if the extractant is used as stationary phase in extraction columns.

When the problem is to make efficient use of a small and limited amount of extractant to determine a maximum of partition data with a minimum of extractant loss, column extraction chromatography is the method of choice. The same holds for the problem of the determination of partition data for the components of a complex sample mixture which is only available in small amounts.

Since stationary phase parameters (type and size of support particles, type and concentration of extracting agents used in diluents, or type and degree of loading of liquid or solid extractants) and mobile phase parameters (flow rate, concentration of partitioning substances, type and concentration of complexing or salting out reagents) can — in addition to temperature — be adjusted within certain limits, extraction chromatography offers more opportunities for optimizing a separation than many other techniques.

Although many of the special advantages offered by column extraction chromatography stem from practical points of view, the working techniques used are not specific to this sort of chromatography and can largely be found in other liquid column chromatographic techniques as well.

Historically, inorganic extraction chromatography had its origin and initial period of development outside the main stream of development of chromatographic theory and techniques, but in the near future it will probably be integrated in this stream as a special case in the rapidly developing field of modern liquid chromatography[2].

1. STANDARDIZATION OF SUPPORT MATERIAL

Since support material used in inorganic column extraction chromatography differs in chemical nature, adsorptive properties and particle size, it is necessary to standardize it in order to get reproducible results.

The chemical nature of supports is discussed in Chapter 5 and procedures to eliminate adsorptive properties of support particles will be discussed in connection with support material loading (Paragraph 2.1.). This paragraph only deals with methods of standardizing particle size. Chromatographic theory shows that support particle size influences plate height by influencing eddy diffusion and mobile phase mass transfer[3]. Eddy diffusion is the result of the mobile phase flow pattern in the column, which depends only on the structure of support material packing and on the incorporated flow space. According to Giddings[4], the flow pattern has a larger influence in band broadening than any other process (with the possible exception of diffusion).

Since spherical support particles can be packed more densely, reproducibly and homogeneously than particles with irregular shape, they should be used whenever possible. Differences in particle size reduce packing density by introducing packing irregularities, so that support particles should be spheres of uniform diameter.

A reduction of particle size reduces plate heights, and smaller particles will give more efficient columns. Beyond a certain limit, however, the decrease in plate heights resulting from particle size reduction is outweighed by practical disadvantages: excessive pressure is needed to force the mobile phase through the column, and it is difficult to reach uniform packing with extremely fine material.

Standardization of particle size

Support particle diameters should be reported in mm or μm (microns).

Since particle diameters are frequently standardized by sieving, support particle diameters are also characterized by reporting mesh numbers of standardized sieves, mostly US-mesh or British BS-mesh numbers.

Table 1a shows the correlation between US-mesh numbers and the particle diameter in mm. A rough estimation is gained by using the formula

$$\text{Particle diameter in mm} = \frac{16}{\text{US-mesh number}}$$

Table 1b shows the correlation between BS-mesh numbers and the particle size in mm.

TABLE 1a

Conversion of support particle diameter from
US-mesh to millimeter

US-mesh number	Particle diameter in mm
16−20	1.2−0.85
20−50	0.85−0.29
50−100	0.29−0.15
100−200	0.15−0.08
200−400	0.08−0.04

TABLE 1b

Conversion of support particle diameter from
BS-mesh to millimeter

BS-mesh number	Particle diameter in mm	Conversion factor
10	1.5	15.0
20	0.735	14.7
30	0.485	14.55
40	0.36	14.4
50	0.286	14.3
60	0.235	14.1
70	0.200	14.0
80	0.173	13.84
100	0.136	13.6
120	0.111	13.3
140	0.093	13.02
160	0.079	12.64
180	0.069	12.42
200	0.061	12.2

Estimates are gained by using the formula

$$\text{Particle diameter in mm} = \frac{A}{BS\text{-mesh number}}$$

where A is a factor which falls linearily from 15.0 to 12.2 when going from 10 to 200 BS-mesh.

Fractionation of support particles aims at obtaining fractions with a minimum spreading of particle diameters. Since commercially available support particle mixtures are — with the exception of some specialty packings — not uniform enough for use in high resolution columns, additional grading is frequently needed.

Fractionation of support particles can be achieved by sieving, sedimentation or elutriation techniques.

Sieving: using sieves with standardized aperture size has the advantage that a large amount of particles can be fractionated in one operation. The main difficulty is to determine when the separation of particles is complete. The time needed depends on the amount of substance, on the sieving surface, and on the method of shaking. If shaking is done by hand, different operators can obtain widely differing results according to their patience and shaking efficiency.

Sedimentation techniques are based on the fact that particles with different size have different weight and therefore settle with different speed. After a short initial period of acceleration, the speed of a particle falling under the influence of gravity in a viscous medium will be constant. According to Stokes'law of hydrodynamics, the speed depends on the particle radius, the viscosity of the medium and the difference in the density of the particles and the medium.

Primitive sedimentation techniques have frequently been used to remove smaller particles or residues from broken particles contained in particle mixtures which had already been subjected to fractionation by sieving. After the elimination of a fraction containing bigger particles by sieving, the residual particle mixture is thoroughly mixed with water (or another medium of suitable density and viscosity) in a conical flask. The fraction of particles that settle after a certain time is isolated by decantation. Smaller particles, which are still suspended, are removed with the liquid.

Elutriation techniques separate particle mixtures into fractions of different mean diameter by suspending the particles in a liquid moving upward with a specific velocity. Particles smaller than a given size move upwards with the fluid, while larger particles settle down. Particle mixtures are separated in fractions by using different fluid velocities. Elutriation techniques, which need only simple and inexpensive equipment, were successfully used to fractionate particle mixtures in ion exchange[5 −7] and column extraction chromatography[8].

Figure 1 shows an apparatus used by Hamilton[5]. If a fluid flows through a V-shaped separatory funnel with constant rate, the changing diameter of the funnel leads to the formation of a gradient of linear flow velocity. Particles placed at the bottom of the funnel will rise with diminishing velocity until the upward velocity reaches the sedimen-

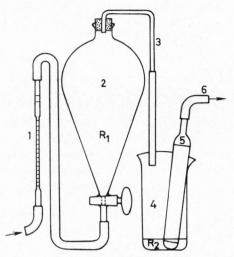

Fig. 1. Apparatus of Hamilton for the preparation of support material of homogeneous particle size.
1 – Flow meter; 2 – Separatory funnel; 3 – Overflow; 4 – Beaker; 5 – Filter; 6 – Suction (under
pressure); R_1 – Original, ungraded support material; R_2–Separated, graded support material (From
K. Dorfner, Ref: [37], with permission)

tation velocity given by the Stokes' law. The minimum linear velocity occurs at the
widest part of the funnel: particles having a sedimentation velocity greater or equal to it
will remain below or at the level of the maximum funnel diameter. Particles with sedimentation velocities smaller than the minimum velocity at the widest part, will be
carried up and swept out of the funnel. Thus the linear flow at the widest part is a limiting rate   which governs the size of particles leaving the funnel. It depends on the
input flow rate, which can be controlled by a constant volume measuring pump or
simply by supplying the flow at constant pressure from a reservoir, and adjusting it
with the funnel stopcock. The flow rate can be determined by a flowmeter and its accompanying calibration chart.

Fractions of particles isolated by this technique at constant flow rate contained
particles which differed up to 15 μm in diameters, 70–80% of the particles having
diameters within ±3 μm from the average[5].

A related method has been used by Aly and Latimer[6] for ion exchange resin particles and by Horwitz et al.[8], for extraction chromatographic column support material.
Their equipment consisted of four different size columns connected in the order of
smaller to larger diameter. A constant volume flow of a liquid with a suitable density
was maintained by a metering pump which recircled the solution through a fine porous
glass filter. Ungraded support material was placed in the smallest column and particles
of diminishing size were collected in successive columns.

Up to now, the majority of extraction chromatographic column experiments have
been executed using support particle mixtures with average diameters between 50 and
150 μm (about 100 to 350 US-mesh) and with a difference from maximum to minimum
particle diameters from about 50 to 180 μm. Smaller particles with average diameters in

the 10 to 50 μm range have been used in a number of high resolution experiments and will surely be used more frequently in the future.

2. LOADING OF THE SUPPORT MATERIAL WITH STATIONARY PHASE

Column extraction chromatography makes use of the extraction properties of a stationary phase which is more or less permanently fixed on or in the support particles.

Extraction properties of the stationary phase can be blurred by adsorption or ion exchange properties of the carrier material. In addition, extractants and diluents generally used are completely or predominantly hydrophobic, which makes it difficult to fix them on hydrophilic materials.

Since silicagel or kieselguhr are hydrophilic and show adsorption as well as ion exchange properties, they have to be deactivated and made hydrophobic before loading.

2.1. HYDROPHOBIZATION OF THE SUPPORT MATERIAL

Deactivation and hydrophobization of hydrophilic support materials is usually achieved by silanization with dimethyldichlorosilane (DMCS) (silanization by hexamethyldisilizane has also been used, but DMCS treatment is now almost exclusively adopted). DMCS converts surface silanol groups (Si$-$OH) of the material to silyl ether groups ($-$Si$-$O$-$Si$-$(CH$_3$)$_2$$-$).

Silanization methods used by the different authors differ with respect to the way the material is exposed to DMCS and the way the hydrochloric acid formed in the reaction is subsequently removed. The material is prepared for silanization by washing it with acid and water, and then dried to remove all water after the washing step. Drying temperatures and times depend on the material used; kieselguhr $-$ for instance $-$ can be dried by heating to 120 °C for 24 hours.

Dried support particles are exposed to DMCS by treatment with DMCS vapours or with dilute ($\sim$5%) solutions of the reagent in an inert diluent.

An open bottle of DMCS is placed on a dessicator containing the support particles in an open vessel with a large surface. Vacuum is applied to remove air and to saturate quickly the dessicator with DMCS vapours. Twice a day the formed hydrochloric acid and some excess DMCS vapours are pumped off and the support material is mixed. Silanization is finished after 3$-$4 days.

Alternatively, the support material can be placed in a glass column as reaction vessel, and a stream of DMCS vapours carried by dry air or nitrogen is maintained for about 2 hours. Hydrochloric acid is removed after treatment by pumping it off under vacuum, or washing with dry methanol or ethanol. Silanization in the liquid phase is exemplified by a procedure given by Smith et al.[9]: A solution of 2.5 g DMCS in 50 ml toluene is mixed with 10 g acid-washed solid support. The resulting slurry is heated to gentle boiling with continuous swirling, and allowed to stand and cool. The slurry is filtered without vacuum through a Whatman Nr. 1 filter and the support washed twice with enough

toluene to cover the particles each time. The resulting solid is similarly washed twice
with methanol, transferred to a beaker, slurried and heated to boiling. After cooling,
the support is filtered and washed twice again with methanol before being dried at 120°C
overnight.

Fragile supports, such as kieselguhr particles, have to be handled cautiously during
these operations. Vigorous mixing might fragment a fraction of the particles and thereby
lead to a loss in particle size uniformity and column permeability.

2.2. SUPPORT MATERIAL LOADING TECHNIQUES

The amount of stationary phase loaded on support particles is one of the most important primary parameters in column extraction chromatography: it determines the
capacity factor of the column and the diffusion path length in the stationary phase. Via
these secondary parameters stationary phase loading influences most other working
features. In addition, the stability of the fixation of the stationary phase on the support,
as well as column packing characteristics, depend — at least to a certain degree — on
the amount of stationary phase used.

The amount of stationary phase should be given in grams or moles per gram of dry
support material. If the stationary phase consists of an extractant dissolved and/or diluted with a diluent, weight or volume of stationary phase of a certain extractant concentration per gram of dry support material should be reported.

Techniques for loading supports should be simple, reproducible and lead to support
particles uniformly coated with a known amount of stationary phase of known composition. The technique most widely used by inorganic extraction chromatographers is
the solvent evaporation technique. Some workers also used a solvent filtration technique,
while a few others loaded prepacked columns.

Solvent evaporation technique
Dry silanized (if necessary) support particles are slurried with a known amount
of extractant dissolved in a volatile solvent. The solvent is then evaporated by gentle
stirring under a stream of air or nitrogen. If necessary, heat can be applied cautiously
until the mixture is completely dry. This technique is simple and leads to uniformly
coated particles. The amount of stationary phase can be exactly predetermined and can
easily be reproduced.

Solvent filtration technique
Dry silanized (if necessary) support particles are put into a flask and covered
with an excess amount of a dilute mixture of an extractant in a volatile solvent. Vacuum
is applied several times to the flask to remove air from the particles and ensure a more
complete wetting. After filtering off the excess solution, the support material is dried as
described for the solvent evaporation technique.

The solvent filtration technique leads to uniformly coated particles. It partly avoids
time consuming evaporation steps which are necessary in the solvent evaporation technique.
But it has disadvantages: it is impossible to determine in advance the amount of station-

ary phase fixed on the support, and the resulting amount of stationary phase can only be determined by weighing the support material before and after loading. Unless extensive standardization is done, it is also difficult to load different batches of particles reproducibly.

Loading prepacked columns

A silanized column is carefully packed by a dry packing procedure (see Paragraph 3.3) using hydrophobized support material when necessary. Then a surplus of stationary phase is filled into the column and a flow of stationary phase is maintained until all air bubbles disappear. Alternatively, support material loading and column packing can be combined using a slurry packing technique (see Paragraph 3.3). The slurry used is a mixture of support particles and a surplus of stationary phase.

The surplus stationary phase is then removed by rinsing the column at high flow rates with 10 to 15 free column volumes of the first aqueous phase used in the subsequent separation procedure.

In certain cases, loading prepacked columns or combining column packing and loading in one operation can overcome some practical problems and spare time. It is possible, for example, to minimize diluent or solvent losses. A carefully standardized procedure, combining support loading and column packing, is particularly useful when separation procedure must be optimized for a minimum time needed (see for instance Ref.[10]). Careful standardization is necessary, since loading prepacked columns or combining support loading and column packing suffer from the same disadvantages as the solvent filtration technique: it is not possible to pre-determine accurately the amount of stationary phase in the column and to load the support material within narrow limits of reproducibility. In addition, it is obviously impossible to determine the degree of loading simply by weighing the columns before and after the preparation step, and the amount of stationary phase in the column has to be determined separately by removing it from columns packed using the same packing procedure. A suitable organic solvent is used to quantitatively elute the stationary phase, whose components are then determined in the eluate.

3. BED DIMENSIONS, COLUMN TYPES AND COLUMN PACKING PROCEDURES

3.1. BED DIMENSIONS

Bed length is one of the most important parameters in extraction column chromatography: elution volumes, separation time and number of theoretical plates increase linearly with bed length, while column resolution increases with its square root. On the other hand, an increase in bed length leads to a proportional increase in pressure needed to force the mobile phase through the column.

Up to now, most extraction chromatographic experiments have been run at flow rates obtained by gravity or by small overpressures. Most columns used had bed lengths ranging between 50 and 150 mm, mainly determined by the resolution needed, and by the time which can be tolerated for analysis.

Theory in its simplified form[11] shows that bed diameter neither affects resolution nor plate height. Simplified theory is a good approximation when the tube/particle diameters ratio ranges within 5 and 50 and flow rate is near to its optimum: flow velocity optima, irrespective of column diameter, can only be found in gas chromatography; since diffusion coefficients in liquids are 10^4 to 10^5 times smaller than in gases, there is always a slight influence of bed diameter on the resolution of liquid-liquid chromatographic systems. Serious resolution losses will occur when large bed diameter columns are operated at high flow rates.

Part of the resolution losses which occur in operating large diameter columns simply stem from the fact that it is difficult to pack them uniformly. It is therefore a good practice to reduce column diameters as much as allowed by experimental limitations. A bed diameter of about 5–50 particle diameters, i. e. column internal diameters from 2.5 to 7.0 mm for 50 to 150 μm particles is a compromise frequently used.

Larger columns should only be used when capacity considerations justify a loss in resolution: for a given bed volume, that is a given total capacity column efficiency continuously improves as the bed length to bed diameter ratio is increased. Nowadays, analytical extraction chromatographic columns designed for high resolution have length diameter ratios ranging from 1:10 to 1:100.

3.2 COLUMN TYPES

The type of column to be used depends on the desired operating characteristics (capacity of the column, operating pressure, operating temperature, manual or automatic operation).

Generally columns should have minimal dead volume, since all volumes the mobile phase passes after leaving the bed contribute to extra-bed peak broadening and thereby reduce resolution.

Up to now, the majority of inorganic column extraction chromatographic experiments have been carried out manually, at room temperature and atmospheric pressure, using the simplest equipment possible.

Figure 2a shows type and bed dimensions of typical columns used for high resolution analytical separations. The bed is supported by glass or quartz wool plugs or by sintered glass plates. A layer of quartz or glass spheres, or a narrow wool plug from the same material prevents disturbances of the surface by solutions entering the bed. Flow rate can be roughly controlled by keeping the liquid in the reservoir at a suitable level. If necessary, the flow rate can be reduced and controlled by a stopcock placed after the column outlet. If slight overpressure is needed to maintain a suitable flow rate, the liquid reservoir can be closed with a ball-and-socket joint connected to some pressure source (Fig. 2b): in this case, a better flow rate control is normally obtainable.

It is absolutely necessary to prevent every part of the bed from running dry. At the cost of a partial loss of resolution, due to the increase of the dead volume, this can be ensured by making use of the principle of communicating vessels, in columns of the type shown in Fig. 2c.

Figure 2d shows a simple, versatile and convenient column set-up proposed by Snyder[12] for routine operation. The solvent reservoir and sample introduction system are kept mounted permanently. Columns can be easily changed.

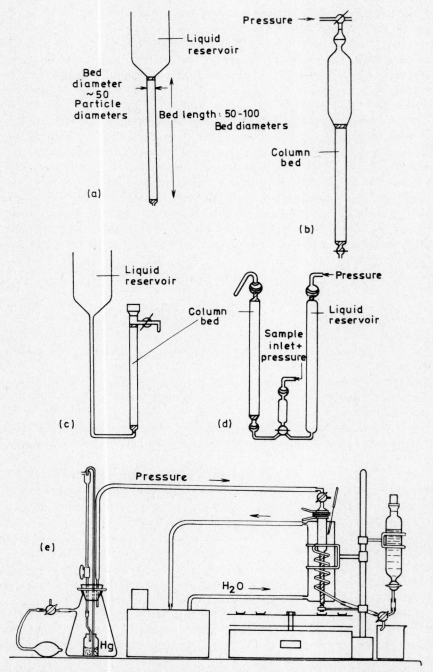

Fig. 2 (a)-(d) Column types; (e) Experimental set·up for extraction chromatographic separations at different temperatures and flow rates (From Ref.[10] by courtesy of the author)

Figure 2e shows a simple experimental set-up used by Eschrich[10] to run thermo-stated columns using slight overpressure.

Up to now, comparatively small overpressures have been used in inorganic extraction column chromatography: to the authors' knowledge, the pressures slightly higher than 1 atm used by Horwitz and Bloomquist[13,14] to run their high efficiency columns are about the maximum. If optimization of separations should make it necessary to apply higher pressures in the future, equipment will not be a problem: precision pumps, tubes, sample injection valves, fittings and columns suitable for pressures up to 35 atm and chemically inert (using only Teflon, Kel-F or glass for parts in contact with the liquid stream) are already available on the market.

3.3. COLUMN PACKING

The quality of column packing is an important parameter in extraction column chromatography: it determines the mobile phase flow pattern which has larger influence on band broadening than any other factor[4].

In a review article on high resolution liquid chromatography, Gouw and Jentoft[15] state: "In practice, the preparation of a good column necessitates a judicious combination of patience, savoir faire, and applied witchcraft".

There is no doubt that it is more difficult to pack long columns for high resolution liquid chromatography than to achieve uniform bed density in relatively short inorganic extraction chromatography columns. But the problem still remains, as documented by a statement from Cerrai and Ghersini[16]: "Column preparation is a very critical step; it is the only one where "reversed-phase art" plays an important role".

In spite of this situation, systematic investigations of column packing techniques are rare in inorganic extraction column chromatography, since it is difficult and time consuming to try to standardize all relevant parameters and to investigate their importance in column packing.

In order to prevent the formation of preferential channels with high migration rate along the walls of columns, glass columns should be silanized by exposing them to di-methyldichlorosilane for about 30 minutes, and washing them with acetone after this treatment.

Column packing procedures depend on the type, regularity and particle size of the packing materials used[17]. In general, two different types of procedures are used; namely dry packing and slurry packing procedures.

Dry packing procedures can be used to fill narrow columns with non-spherical particles having low density and particle size down to about 50 μm (e. g. silicagel or diatomaceous earth): Smaller particles of the same material usually do not give high resolution columns when this packing technique is used. On the other hand, spherical and dense particles can be dry-packed down to a particle size of about 30 μm[17].

Packing material is introduced in small amounts in a vertically held column. To ensure dense packing, either the columns are vertically tapped on the bench, or the bed is gently tamped with a glass rod after each addition of an incremental amount of packing material. After having protected the upper end of the bed by a glass or a quartz wool plug, or by a layer of glass beads, the bed is wetted by introducing the first aqueous mobile phase to be used in operating the column. Care has to be taken to remove all air bubbles from the bed. Elution with about 10 bed volumes of mobile phase should remove the last traces of air and precondition the column for the first run. Braun and Farag[18] propose a method of column packing in which the problem of the inclusion of air bubbles is overcome by evacuating dry packed columns and filling them with the mobile phase under vacuum suction (see Chapter 13, Paragraph 2.2).

Slurry packing procedures are more universally applicable than dry packing ones, and therefore have been used to pack the majority of inorganic extraction columns.

Dry packing material is suspended in a surplus of the first mobile phase to be used in column operation, and the resulting slurry is stirred until all air bubbles disappear (vacuum can be applied if necessary). Small portions of the well-stirred slurry are then poured in a vertically held column. The solution in the column is gently stirred during settling of the packing material, and after addition and settling of each incremental portion of slurry, the bed is gently tamped with a glass rod. After fixing the upper bed end by a suitable plug, column preconditioning is completed by elution with a few bed volumes of mobile phase.

Horwitz and Bloomquist[13] made a careful and remarkably detailed comparison of a dry and slurry packing technique they worked out to pack silanized graded Celite loaded with di(2-ethylhexyl) phosphoric acid into high efficiency columns. They concluded that uneven compactness of beds of 50 μm Celite particles packed leads to considerable band spreading. Their data show a definitely improved performance of columns packed by the slurry packing technique; a detailed study is available of the time stability of the bed structure and on the lifetime of such columns[14].

Since degradation of column performance with time is extremely important in assessing extraction chromatographic methods, this type of study is needed for other packing materials and column packing methods.

4. DETERMINATION OF COLUMN PARAMETERS

The knowledge of column parameters is necessary if one wants to characterize the efficiency of a column by giving the number of theoretical plates, or to determine the partition data from elution or break-through curves (see Paragraph 5).

The total column volume (defined as the volume from the beginning of the bed to column end) is the sum of the bed volume and the column dead volume.

Bed volume is the sum of the volume of support material, the volume of the stationary phase and the volume of the mobile phase, this latter parameter often being called also interstitial volume or void volume or free column volume.

Total column, bed, and support material volumes are auxiliary parameters not directly needed for the calculation of plate heights or partition data. Bed volume is used in a method to determine column dead volume from total column volume, while support material volume is useful in checking the results of other volume determinations by adding all of them up to the total column volume.

Total column volume is determined by weighing the amount of water in a column filled to the upper bed layer, and using the density of water at the experimental temperature for calculation.

Bed volume is determined by weighing the amount of water between two marks indicating bed length. Since the columns are cylindrical, bed cross-sections can be obtained by dividing bed volume by bed length.

Column dead volume is the column volume that the mobile phase has to pass after leaving the bed and before being collected or analyzed. Since column dead volume, in combination with the volumes of the tubes connecting column and detector and detector volume contribute to extra-bed peak broadening by back-mixing and diffusion, it must be kept as low as possible by a suitable column design. In many applications, skilful design makes column dead volumes negligibly small when compared with other peak-broadening parameters of the column.

In many cases, it is sufficient to have an estimate of column dead volume gained from the geometry of dead space. If necessary, it can be determined more accurately by subtracting the bed volume from the total column volume, both being determined as indicated above.

Support material volume is determined through the weight of unloaded support material used to make up the column and its density (either known or expressly determined).

Stationary phase volume (V_s) is determined by dividing the weight of stationary phase put into the column by its density at the experimental temperature, either known or expressly determined. The weight of stationary phase is calculated most often from the weight of loaded support material put into the column and the degree of loading, that is generally given in grams of the stationary phase per gram of unloaded support material (see Paragraph 2.2).

Mobile phase volume (interstitial volume, void volume, free column volume) is usually determined by measuring the elution or break-through volume of a "non-partitioning" substance which can be detected in the eluate by simple and sensitive methods. Gamma-emitting radionuclides, such as ^{137}Cs, are frequently used for this purpose.

"Non-partitioning" substances eluate or break-through the column after the passage of one mobile phase and one column dead volume (see Paragraph 5).

It is generally advisable to follow a complete elution or break-through curve, since the forms of these curves give information on the quality of the bed and of the adopted column set-up. Extraordinary high peak broadening and/or asymmetry of the elution peak and unusual flat break-through curves are normally caused by a lack in homogeneity of the support particle size and/or of column packing. In particular, possible channelling of the column can easily be detected.

Mobile phase volumes can also be determined in connection with column packing procedures:

A slurry, consisting of a weighed amount of loaded support material in water or of an aqueous solution of known density, is poured in a weighed column. The surplus solution is removed after column packing, and the column is weighed again, filled with the solution up to the upper bed layer. The difference in column weight before and after filling, subtracted of the weight of the loaded support material, gives the weight of the solution in the column; its volume is calculated through its density at the experimental temperature. Subtraction of the column dead volume leads to the volume of the mobile phase. An analogous procedure can obviously be used when the columns are packed by the dry column packing procedure.

Hornbeck[19] determined mobile phase volumes by feeding columns with a solution of a radioactive tracer in known concentration until the concentration in the eluate reached the concentration in the feed. Feeding was then interrupted, and the column was eluted with a solution (which did not strip extracted species from the stationary phase) until the tracer disappeared from the eluate. The total amount of tracer in the eluate was determined, and the resulting value, divided by the initial concentration in the feed solution, gave the sum of mobile phase and column dead volumes. The results agreed within 5% with those obtained by the weighing method.

Mobile phase flow rate influences the plate height and the migration rate of bands in the column. It is therefore an important experimental parameter which can be suitably varied to optimize column operation for a minimum time needed to achieve a certain minimal resolution, sufficient for a given separation problem.

In liquid chromatography an increase in mobile phase flow rate always leads to the increase in plate height and therefore to the reduction of the number of theoretical plates of a given bed. Since many routine separations do not demand very high resolution, columns are frequently operated with a large surplus of theoretical plates. In such cases, it is generally possible to increase flow rate until a point is reached, when no more plates can be sacrificed without an unacceptable resolution loss.

Mobile phase flow rates should be reported as linear flow velocities ($ml \cdot cm^{-2} \cdot min^{-1}$, or $cm \cdot min^{-1}$), that are obtained from the determination of the bed cross-section and of volumetric flow rates ($ml \cdot min^{-1}$).

The most accurate method for the determination of flow rate is the gravimetric method: the amount of eluate collected in an accurately known period of time is weighed, and its volume is then calculated through its density. Since this method is considerably more precise and accurate than the ability of a pump to deliver mobile phase at a constant predetermined flow rate, the gravimetric method is frequently used to calibrate pumps.

For many practical applications, it is sufficient to determine the flow rate volumetrically by collecting the eluate for a known period of time in a volumetric measuring vessel. In many cases, the volume of a drop of eluate was determined by measuring the volume of a known number of drops: the number of drops per minute indicated flow rate, the total number of drops gave the volume of eluent which had passed through the column. It must be pointed out, however, that the drop volume can appreciably vary with temperature and/or mobile phase composition. Flow measuring devices as bubble timers or flow meters, often adopted in other types of liquid column chromatography, have been rarely used up to now in inorganic extraction chromatography.

When columns are run at the limit of resolution needed for the wanted separation, it may be essential to maintain the flow rate at a preselected value. When stepwise or gradient elution are done, viscosity, density and surface tension of the mobile phase may considerably change during separation, so that flow driven by gravity or constant column inlet pressure may also vary considerably. This can be overcome by using constant volume pumps.

The capacity of a column towards a given ion for a given mobile phase composition is generally expressed as meq per ml of (bulk volume) bed. Methods for the determination of column capacities from break-through curves are discussed in Paragraph 5.2.3.

5. WORKING TECHNIQUES IN OPERATING EXTRACTION CHROMATOGRAPHIC COLUMNS

Extraction column chromatography can be used for the separation of the components of mixtures or for the determination of partition data. The purification of substances or the enrichment of traces are special cases of separations.

5.1. PRINCIPLES

In principle, extraction columns can be operated using the three main working techniques developed in chromatography: elution analysis, frontal analysis and displacement analysis. In Fig. 3 the main principles of these working techniques are schematically illustrated.

In elution analysis a single portion of the sample mixture is introduced into the column, where it forms a narrow band at the head of the bed. Columns are eluted using a solvent which is less strongly absorbed than the sample components. According to different degrees of retention caused by differences in partition equilibria and/or mass transfer kinetics, the different components of the sample migrate with the mobile phase with different speed. The different migration rates of the sample components on their way through the bed lead to a separation of the mixture: in the ideal case, completely separated bands are eluted from the columns.

In frontal analysis[20,21] the sample mixture is fed continuously into the column, the sample solution being used as mobile phase for column elution. All sample components are initially retained in the stationary phase, but different degrees of retention again lead to differences in their migration rate. The component showing the least retention will come out first, and for some time the eluate will contain this substance in the pure form. After the break-through of the second substance, a mixture of two com-

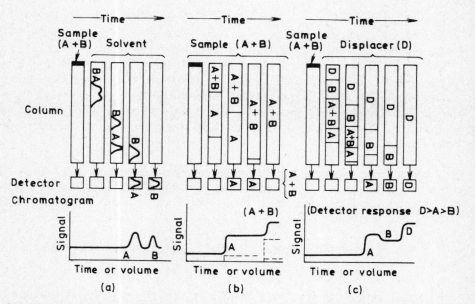

Fig. 3. Schematic diagram showing three ways of column operation. (a) Elution analysis; (b) Frontal analysis; (c) Displacement analysis (Adapted from Ref. [38] with permission)

ponents will leave the column until the third substance breaks through. The disadvantages of frontal analysis for the separation of mixtures are obvious: only a part of the least strongly retained sample component is isolated in pure form; since the height of successive steps in the frontal analysis curves cannot be directly used as a measure of the concentration of any component, the technique finds little use also for analytical purposes.

However, break-through curves of pure substances contain information on column performance (see Paragraph 5.2) and can be used for the determination of partition data.

In displacement analysis [21,22] a single portion of sample mixture is introduced into the column as in elution analysis, but — in contrast to this latter technique — the sample is not eluted with a pure solvent, but with a solution containing a substance (the displacer) which is more strongly retained on the column than any component of the sample mixture. This substance completely displaces the sample components from the stationary phase and pushes them down through the column, in the ideal case without any mixing with the sample. Displacement also occurs between sample components which show different retention: on their way through the column, the sample components are dispersed in different bands, which, in contrast to frontal analysis, contain pure substances. Since the heights of each step are characteristic of a particular substance, in principle, step height can give a basis for qualitative analysis. The length of the step is proportional to the amount of substance present.

Compared with frontal analysis, displacement analysis has the advantage of separating mixtures into bands containing pure substances. Its main disadvantage stems from the

60

fact that bands always are in immediate contact: it is therefore difficult to separate the bands completely. "Carrier displacement analysis"[21,23], a technique designed to overcome this difficulty, is very complicated to use.

Because of the limitations inherent to frontal analysis and displacement analysis, these working techniques are rarely used in column chromatography: this holds also for inorganic extraction columns.

The remainder of this chapter is therefore devoted to a discussion of elution techniques and of the use of break-through curves for the determination of partition data.

5.2. ELUTION ANALYSIS

In practice, inorganic extraction column chromatography essentially is elution analysis applied for a multistage separation of sample mixtures or for the determination of partition data.

5.2.1. SEPARATION BY ELUTION ANALYSIS

Elution analysis using extraction chromatographic columns is by far the simplest technique to make profit of the large separation potential of a large variety of extractants for multistage separations. It is therefore especially suited for the resolution of mixtures whose components have separation factors rather close to one.

Column conditioning

Column conditioning serves to equilibrate the stationary phase with the first mobile phase used in the procedure. If columns have not already been conditioned during column packing (see Paragraph 3.3 on page 55 and 56) they should be conditioned by washing them with 5 to 10 free column volumes of the first mobile phase used.

Sample application

Column chromatography theory shows that, in order to obtain highest resolution, it is necessary to introduce the minimal amount of sample dissolved in the minimal volume of solvent. Whenever possible, the solvent should have a composition identical or similar to the composition of the first mobile phase used in the separation procedure. In the ideal case the sample amount should not exceed the capacity of the first theoretical stage of the column.

In practice, the lowest amount of sample used is determined by the minimum amount of a specific component that one wishes to detect, and by the detection limits of the available analytical methods or detectors. Since the increase of the amount of sample implies the increase of danger of resolution losses caused by band broadening and overloading, the maximum sample size will be determined by the minimum resolution still acceptable for the separation.

In any case the sample should be dissolved completely, since undissolved sample components or precipitates may plug the column or cause bad tailing when they gradually dissolve in the eluents subsequently used in the separation procedure.

Considering these factors a suitably sized sample aliquot is pipetted into the column using a minimum of liquid for rinsing the pipette. An alternative possibility for applying a sample is the addition of a small portion of loaded support material soaked with the mixture to be separated.

Elution

In order to keep to a minimum the amount of stationary phase lost by dissolution into the mobile phase, all the eluting solutions must be presaturated with the extractant used as stationary phase. This can be achieved by shaking the phases in a separatory funnel or by the use of presaturation columns. "Column bleeding", that is a rather massive loss of stationary phase from a column during operation, is a symptom of insufficient fixation of the stationary phase on the support material. It can be caused by sudden changes in bed structure, temperature or composition of the stationary phase.

Whatever their origin, stationary phase losses reduce in any case the useful lifespan of a column.

In trying to achieve a separation by using a single eluent, one usually meets what Snyder named the "general elution problem": sample components which migrate through the bed almost as rapidly as the eluent front are clustered and poorly separated, while components which move very slowly need excessive elution volumes, their bands being broadened to an extent that eventually makes detection difficult[24]. All solutions to the general elution problem involve the change of band migration rates during the separation.

The migration rate of a substance is determined by its equilibrium distribution ratio, mass transfer kinetics and mobile phase velocity. Any of these parameters can therefore in principle be used to optimize separations. Since most distribution reactions used in liquid-liquid extraction show rapid mass transfer kinetics, only in rare cases considerable local non-equilibrium is observed in extraction columns operated at low mobile phase velocities. In addition to this, it is difficult to selectively influence mass transfer kinetics, since the experimental parameters which affect kinetics, as temperature for instance, influence a number of other parameters as well. In practice, therefore, attempts to solve the general elution problem by changing band migration rates during elution are limited to changing distribution ratios and mobile phase velocities.

Liquid chromatographers developed a number of operation techniques to solve the general elution problem along these lines: solvent programming (stepwise or gradient elution) [12,24—26], flow or pressure programming[27], temperature programming[27—29] and coupled column operation[26,30]. The relative merits of these techniques have been discussed by Snyder[30]. Some of them are not very effective, and others can lead to experimental difficulties.

Flow programming, which consists in a reduction of the mobile phase velocity in the beginning of a separation and in an increase in velocity during the last part of the separation procedure, transfers resolution from the last part of the chromatogram (where it is seldom needed), to the front part (where it is often needed). The technique and the equipment necessary for applying it are simple, but only a marginal increase in front end resolution is obtained, useless for samples of very broad range[26].

Temperature programming consists of a temperature change during separation. Depending on the enthalpy of the extraction reaction, temperature changes affect distribution ratios to a varying degree. However, the effect of temperature programming is slight in comparison with the effects gained by solvent programming. In addition, temperature changes can promote "column bleeding" by disturbing the equilibrium between the support particles and the stationary phase.

In general, most liquid chromatographic separations have been achieved by using solvent programming techniques. To the authors' knowledge, extraction column chromatographers engaged in inorganic analysis have not yet felt a need to apply more sophisticated techniques than stepwise or gradient elution.

In inorganic extraction column chromatography, solvent programming consists in eluting a column with an eluent containing a varying concentration of reagent which changes distribution ratios of partitioning substances. Changes in distribution ratios can be caused by salting out reagents, reagents extracting in competing extraction reactions, redox reagents and complexing reagents.

Salting out reagents and reagents which are extracted in competing extraction reactions do not specifically interact with distributed species. Salting out reagents — mostly inextractable salts not having common ions with an extractable species — change partition ratios by changing mobile phase bulk properties. Reagents which are extracted in competing reactions change distribution ratios by changing the number of extractant molecules available in the stationary phase.

Redox and complexing reagents change distribution ratios in a more specific way: they chemically react with certain species forming other species with other distribution ratios.

In order to increase the migration rate of slowly moving bands, the elution strength of the eluent can be increased stepwise (stepwise elution) or continuously (gradient elution). The principles of stepwise and gradient elution, and design principles of various gradient devices have been discussed by Snyder[24]. A discussion of Snyder and Saunders on the experimental optimization of gradient elution is also available, which might be of interest for extraction column chromatographers[25].

In general, similar results are obtainable with gradient and stepwise elution, when a sufficiently large number of steps is used. Both methods are therefore comparable in their relative separation ability.

Stepwise elution is the method of choice for the great majority of routine separations, when samples of unknown composition containing components of widely varying migration rates have to be analysed. If the separations are carried out under conditions where the components have linear partition isotherms, elution volumes will be independent of the concentration of the components and can therefore be predicted.

If the sample composition is unknown, or samples have to be analyzed under non-linear isotherm conditions, stepwise elution procedures require close attention for deciding when the eluent has to be changed. If many eluent changes are necessary, separation procedures become rather tedious. All these difficulties can be overcome by using gradient elution methods.

In most cases, gradient elution is carried out using a mixture of a weak eluent and a strong eluent mixed in a gradient device. Elution starts with the weak eluent and continues with eluent mixtures of increasing strong eluent concentrations. The form of the gradient, that is the plot of concentration of strong eluent versus eluate volume, is an important experimental parameter which can be adjusted to achieve optimum resolution for a given separation. According to Snyder[12,24], concave gradients as delivered by so-called "proportional volume" gradient devices are generally more advisable than convex gradients as delivered by "exponential type" gradient devices. Most manufacturers of chromatographic equipment offer gradient devices of varying degree of sophistication. General reviews have been given by Snyder[12,24] and Dorfner[31]. Henry [32] discussed the criteria for the selection of such devices.

5.2.2. DETERMINATION OF PARTITION DATA FROM ELUTION CURVES

If columns are operated under equilibrium conditions, elution curves provide the basis for the determination of equilibrium distribution ratios and separation coefficients.

Evaluation of partition data from elution curves has a number of advantages over their determination by liquid-liquid batch distribution experiments.

The evaluation of an elution curve containing several peaks allows obtaining partition data and separation coefficients for several species in one single operation. The method is therefore especially suited for the determination of the distribution ratios of the components of a mixture which is only available in small amounts. If a solution contains an element in different oxidation states or in form of different complexes, this can normally be seen in the elution curve which can be used to obtain the partition ratios of all species resolved. A contamination of species of one oxidation state with small amounts of species involving the same element in another oxidation state can easily be overlooked in batch distribution experiments, thereby leading to erroneous distribution data. Such contaminations are frequently revealed in elution curves by more than one peak or by a distortion of the main peak. Peak distortion or tailing are also observed in the case of kinetically labile species.

Figure 4 illustrates the method of evaluation of an elution curve for the determination of partition data and of the number of theoretical plates of a column.

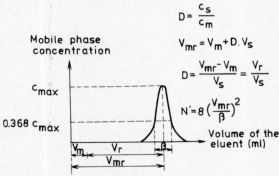

$$D = \frac{c_s}{c_m}$$

$$V_{mr} = V_m + D \cdot V_s$$

$$D = \frac{V_{mr} - V_m}{V_s} = \frac{V_r}{V_s}$$

$$N' = 8 \left(\frac{V_{mr}}{\beta}\right)^2$$

Fig. 4. Evaluation of an elution curve. V_m −Mobile phase (ml) (interstitial volume, free column volume, void volume); V_s−Stationary phase volume (ml); $V_{mr} = V_r + V_m$−Total retention volume (ml); V_r−Retention volume (ml); β- Peak width at $c = 0.368\ c_{max}$ (ml); N'−Number of theoretical plates from the centre of the sample application band to the bed end; D−Concentration distribution ratio; c_s−Concentration of the substance in the stationary phase; c_m−Concentration of the substance in the mobile phase

64

5.2.3. DETERMINATION OF PARTITION DATA AND COLUMN CAPACITY FROM BREAK-THROUGH CURVES

If columns are run under equilibrium conditions, equilibrium distribution ratios of the single substances can also be determined by evaluating break-through curves.

Since the height of the elution peaks decreases with increasing elution volumes, and elution peaks are distorted if the columns have been overloaded, the evaluation of break-through curves is advantageous for species with very high partition ratios and/or species present in large initial concentrations.

The composition of the feed solution corresponds to the aqueous phase equilibrium concentration (not to the initial concentration) of the corresponding liquid-liquid batch distribution experiments.

Figure 5 shows the evaluation of a break-through curve of a species for the determination of its distribution ratio and of the number of theoretical plates of the column.

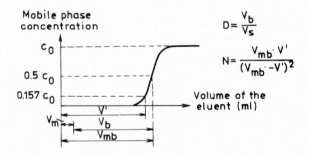

Fig. 5. Evaluation of the break-through curve.

V_m — Mobile phase volume (ml) (interstitial volume, free column volume, void volume); V_s — Stationary phase volume (ml); V_b — Break-through volume (ml); $V_{mb} = V_m + V_b$ — Total break-through volume (ml); V' — Eluent volume when the concentration of the eluent reaches $0.157\ c_0$ (ml); c_0 — Initial concentration; D — Concentration distribution ratio; N — Number of theoretical plates of the bed

Break-through curves can also be used for the determination of column capacity (meq per ml bed volume). Column capacities are usually determined by feeding a solution of a given substance in the column, until its concentration in the eluate equals the feed concentration. Then feeding is interrupted, and the bed washed with an eluent which displaces the interstitial volume without stripping the substance retained in the stationary phase. When the concentration of the substance in this eluate has fallen under the detection limit, the eluents are changed once more, and the column is eluted with an eluent which strips the substance from the stationary phase until no substance is any longer detectable in the eluate. The determination of the concentration of the substance in the collected volume of this stripping eluent gives the amount initially retained in the column.

Hornbeck[19] described a method that determines the mobile phase volume of the column, its capacity for a given species, and the distribution ratio of this species in a single operation.

6. DETECTION METHODS

Sample components leaving the columns have to be detected in the eluate. This can be done by cutting the eluate in fractions which are analyzed separately or by continuously monitoring the solute concentration using a selective or non-selective detector. Since the fraction volume or the detector volume contribute to extra-column band broadening, diminishing the overall resolution of the equipment, they should be kept as low as possible.

Up to now, most inorganic extraction columns have been operated by cutting the eluate in fractions, either manually or with a fraction collector, and analyzing the fractions with any method which is sufficiently specific, accurate, precise and sensitive. In high resolution work, especially when radiometric detection methods can be applied, every drop of the eluate has been analyzed.

The use of continuous monitors has been restricted to radiometer detectors, monitoring the α-, β-, or γ-activity of an eluent stream. If peak heights discrimination is used, α- and γ-monitors can be made specific for a given radionuclide. The problems met in using other types of detectors are strictly analogous to those met in designing detectors for other types of liquid chromatography. They are discussed in a number of articles summarizing the present state of art in the liquid chromatography detector field[15,33–36].

REFERENCES TO CHAPTER 3

1. P. MARKL, Extraktion und Extraktions-Chromatographie in der anorganischen Analytik. Vol. 13 of: Methoden der Analyse in der Chemie. F.HECHT,R.KAISER,E. PUNGOR and W. SIMON Eds. Akademische Verlagsgesellschaft, Frankfurt am Main, 1972.
2. J. J. KIRKLAND, Ed., Modern Practice of Liquid Chromatography. Wiley-Interscience, New York, London, Sidney, Toronto, 1971.
3. J. C. GIDDINGS, Dynamics of Chromatography. Part I. Principles and Theory. Marcel Dekker Inc. New York, 1965.
4. See Reference 3, Chapter 5, pp. 195–225.
5. P. B. HAMILTON, Anal. Chem., 30 (1958) 914.
6. H. F. ALY, R M. LATIMER, J. Inorg. Nucl. Chem., 29 (1967) 2041.
7. D. C. SCOTT, Anal. Biochem., 24 (1968) 292.
8. E. P. HORWITZ, C. A. A. BLOOMQUIST, D. J. HENDERSON, J. Inorg. Nucl. Chem., 31 (1969) 1149.
9. E. D. SMITH, J. M. OATHOUT, G. T. COOK, J. Chromatogr. Sci., 8 (1970) 291.
10. H. ESCHRICH, Z. Anal. Chem., 226 (1966) 100.
11. See Reference 3, p. 283.
12. L. R. SNYDER, Techniques of Liquid Column Chromatography. Chapter 5 of: Chromatography. E. HEFTMANN Ed., Reinhold Publishing Corporation. New York, 2nd ed., 1967.
13. E. P. HORWITZ, C. A. A. BLOOMQUIST, J. Inorg. Nucl. Chem., 34 (1972) 3851.
14. E. P. HORWITZ, C. A. A. BLOOMQUIST, J. Inorg. Nucl. Chem., 35 (1973) 271.
15. T. H. GOUW, R. E. JENTOFT, High Resolution Liquid Chromatography. Chapter 2 of: Guide to Modern Instrumental Analysis. T. H. GOUW Ed., Wiley-Interscience. New York, London, Sydney, Toronto 1972.
16. E. CERRAI, G. GHERSINI, Reversed Phase Chromatography in Inorganic Chemistry. In: Advances in Chromatography Vol. 9 p. 3. J. C. GIDDINGS, R. A. KELLER Eds. M. Dekker Inc., New York, 1970.
17. L. R. SNYDER, J. Chromatogr. Sci., 7 (1969) 352.

18. T. BRAUN, A. B. FARAG, Anal. Chim. Acta, 62 (1972) 476.
19. R. F HORNBECK, J. Chromatogr., 30 (1967) 447.
20. S. CLAESSON, Arkiv Kemi, Mineral. Geol., 23A (1946) 1.
21. L. HAGDAHL, Chapter 5 of: Chromatography. E. HEFTMANN Ed., Reinhold, New York, 1961.
22. L. HAGDAHL, R. J. P. WILLIAMS, A. TISELIUS, Arkiv Kemi, 4 (1952) 193.
23. A. TISELIUS, L. HAGDAHL, Acta Chem. Scand., 4 (1950) 1261.
24. L. R. SNYDER, Chromatogr. Rev., 7 (1965) 1.
25. L. R. SNYDER, D. L. SAUNDERS, J. Chromatogr. Sci., 7 (1969) 195.
26. L. R. SNYDER, The Role of Mobile Phase in Liquid Chromatography. Chapter 4 of Ref. 2.
27. R. P. W. SCOTT, J. G. LAWRENCE, J. Chromatogr. Sci., 7 (1969) 65.
28. R. J. MAGGS, J. Chromatogr. Sci., 7 (1969) 145.
29. R. P. W. SCOTT, J. G. LAWRENCE, J. Chromatogr. Sci., 8 (1970) 619.
30. L. R. SNYDER, J. Chromatogr. Sci., 8 (1970) 692.
31. K. DORFNER, Chem. Ztg., 87 (1963) 871.
32. R. A. HENRY, Chapter 2, p. 80–84 of Ref. 2.
33. S. H. BYRNE, Jr., Detectors in Liquid Chromatography. Chapter 3 of Ref. 2.
34. H. VEENING, J. Chem. Ed., 47 (1970) 549 A; 749 A.
35. J. F. K. HUBER, J. Chromatogr. Sci., 7 (1969) 172.
36. R. D. CONLON, Anal. Chem., 41 (1969) 107 A.
37. K. DORFNER, Ionenaustauscher, Verlag W. de Gruyter, Berlin, 1964.
38. G. E. BENNETT, S. DAL NOGARE, L. W. SAFRANSKI, in I. M. KOLTHOFF, P. J. ELVING, Eds., Treatise on Analytical Chemistry, Part, I, Vol. 3., Interscience, New York, 1961.

CHAPTER 4

STATIONARY PHASES IN EXTRACTION CHROMATOGRAPHY

G. GHERSINI

INTRODUCTION
1. GENERAL PERFORMANCES OF STATIONARY PHASES
   1.1. Retention
   1.2. Selectivity
   1.3. Resolution
   1.4. Capacity
   1.5. Physical stability
   1.6. Chemical stability
   1.7. Regeneration
   1.8. Repeatability and reproducibility
   1.9. Tailing effects
2. SYSTEMATIC SURVEY OF THE EXTRACTANTS USED AS STATIONARY PHASES
   2.1. Acidic extractants
         2.1.1. General behaviour
         2.1.2. Dialkylphosphoric acids, HDEHP
         2.1.3. Other acidic organophosphorous compounds
         2.1.4. Sulphonic acids
         2.1.5. Phenols
   2.2. Neutral organophosphorous extractants
         2.2.1. General behaviour
         2.2.2. Trialkylphosphates, TBP
         2.2.3. Phosphine oxides, TOPO
         2.2.4. Bifunctional neutral organophosphorous compounds
   2.3. Amines and quaternary ammonium compounds
         2.3.1. General behaviour
         2.3.2. Secondary amines
         2.3.3. Tertiary amines, TNOA
         2.3.4. Quaternary ammonium compounds, Aliquat-336
   2.4. Miscellaneous extractants
         2.4.1. Carbon-bonded oxygen donor extractants, MIBK
         2.4.2. Tri-i-octylthiophosphate
         2.4.3. Di-n-octylsulphoxide
         2.4.4. Other extractants
         2.4.5. Synergistic systems
         2.4.6. Electron-exchange columns

INTRODUCTION

The great advantage of reversed-phase extraction chromatography in inorganic chemistry relies on the large choice of stationary phases, each of them displaying peculiar selectivity characteristics so as to allow for chromatographic systems that enable simple resolution of a great number of separation problems.

The large amount of available information on extraction chromatographic columns includes significant data for almost all stationary phases used up to now, and at the same time allows a general picture to be drawn of the performances that one may expect from such chromatographic systems.

In this chapter, some general comments are given on the performance characteristics of stationary phases, and on their effect upon the overall behaviour of chromatographic columns. They are followed by a systematic survey of the different compounds used in column chromatography, where their extraction behaviours are outlined, together with their most significant features as stationary phases.

1. GENERAL PERFORMANCES OF STATIONARY PHASES

The main performance characteristics of an extraction chromatographic stationary phase may be collected within the following headings:
— retention
— selectivity
— resolution
— capacity
— physical stability
— chemical stability
— regeneration
— repeatability and reproducibility
— tailing effects.

In this section, each of these headings is separately considered, and comments are given based on the information available in the literature and on the author's experience.

1.1. RETENTION

The retention of a given metal by a reversed-phase extraction column results from its distribution between the stationary phase loaded on the support and the aqueous phase used as the eluent. The direct correlation between the chromatographic and the liquid-liquid extraction systems permits the retention ability of a column to be foreseen from the liquid-liquid extraction behaviour of the compound used as the stationary phase. In the great majority of cases, these qualitative anticipations find full agreement in the experimental results obtained.

When a quantitative correlation is attempted, complications arise from the difficulty in assessing the actual physical state of the stationary phase in the different extractant

+ support combinations, and the thermodynamic activity of the resulting non-aqueous chromatographic phase.

Let us take as an example the retention of Ca^{2+} from an aqueous acidic noncomplexing solution by a column prepared with an inert support loaded with di (2-ethylhexyl) phosphoric acid (HDEHP). The reaction believed to be responsible for the extraction of calcium is

$$Ca^{2+}_{aq} + 3 \ (HDEHP)_{2 \ org} = Ca \ (DEHP)_2 \cdot 2 \ (HDEHP)_{2 \ org} + 2H^+_{aq} \qquad (1)$$

and, assuming an ideal behaviour of the chromatographic process, the retention volume V_r of calcium can be related to thermodynamic activities of the chemical species taking part in the above equilibrium (see Chapter 2) through the equation

$$\frac{V_r - V_m}{V_s} = K \frac{\gamma_{Ca^{2+}}}{\gamma_{adduct}} \frac{\{(HDEHP)_2\}^3_{org}}{\{H^+\}^2_{aq}} \qquad (2)$$

where V_m and V_s are the volumes of the mobile and stationary phases in the column, K is the equilibrium constant of equation (1), $\gamma_{Ca^{2+}}$ and γ_{adduct} are the activity coefficients of the calcium containing species taking part in the equilibrium, and braces denote activities.

Assuming now that the above activity coefficients are roughly constant in the involved range of experimental conditions, equation (2) can be written as

$$\log (V_r - V_m) = 3 \log \{(HDEHP)_2\}_{org} - 2 \log \{H^+\}_{aq} + \log V_s + const. \qquad (3)$$

The above equations relate the chromatographic results to the thermodynamic parameters of the extraction system involved. No troubles arise for what the aqueous phase parameters are concerned, and plots of $\log (V_r - V_m)$ vs. $\log \{H^+\}_{aq}$, for calcium eluted through the very same column, give straight lines of slope -2, as expected from the coefficient of H^+ in equation (1).

When the role of the extractant in equation (3) is considered, first of all it is necessary to assess the physical meaning of what is normally referred to as "support treated with the extractant", in order to define the extractant activity $\{(HDEHP)_2\}$ and the volume V_s of the non-aqueous phase involved in the extraction process, both relevant for the resulting retention volume of the column.

Most extraction chromatographic systems involve supporting materials that are likely to be practically inert to the liquid extractant loaded on them. In such cases, the stationary phase can reasonably be assumed to consist of the extractant invading the pores and coating the particles as a more or less uniform liquid film, definitely separated from the solid material. The thermodynamic state of such films of extractant has been thoroughly discussed by Siekierski in Chapter 1. According to this author, the activity of the stationary phase can be calculated on the basis of the change of the chemical potential of the extractant due to its presence in the capillary pores of the supporting material. This implies a remarkably lower activity of the extractant, as compared to that in the analogous unsupported organic phase, but scarcely affects the actual distribution coefficients of a metal, because of the parallel effect of the support on the activity coefficient of the metal-containing adduct.

These conclusions, derived by Siekierski for supports coated with a film of undiluted extractant, can obviously be extended to systems where the extractant is dissolved in appropriate diluents. In the example of calcium extraction, the stationary phase should be the HDEHP (or HDEHP solution) actually loaded on the support, and its volume V_s should correspond to the volume of extractant, or of extractant solution. The retention volume of calcium could be easily calculated through equation (3), from the volumes V_s and V_m of the stationary phases in the column, and from the distribution coefficient of calcium in the liquid-liquid extraction system involving a HDEHP solution having the very same composition of the stationary phase. A change of the amount of HDEHP (or HDEHP solution) loaded on the support should not result in a relevant change of its activity, and should affect the retention volume of calcium merely changing the volume of the stationary phase. Therefore, plots of $\log(V_r - V_m)$ vs. log (amount of HDEHP − or HDEHP solution − present in the column), should give straight lines of slope nearly equal to one, irrespective of the number of extractant molecules involved in the extraction reaction.

The above reported approach to the nature of the stationary phase finds support on the generally observed first-power dependency of the retention volumes on the amount of extractant loaded on the support, for the great majority of column extraction chromatographic systems for which such data are available [1]. The quantitative correlation between the activities of the extractant when loaded on the support and when used as a free organic phase in liquid-liquid extraction, however, still needs to be fully confirmed by appropriate experimental data.

Among the different extractant + support combinations adopted in column extraction chromatography, some peculiar ones can hardly be assumed to consist of a liquid organic phase definitely separated from the solid supporting particles; supports that swell when put in contact with the extractant, for instance, most probably involve a more intimate interaction between the two, and are likely to give rise to a more or less homogeneous "solution" of the extractant into the supporting material. In such cases, a different approach appears to be necessary, in order to assess the physical and thermodynamic properties of the resulting stationary phases. Unfortunately, very little effort has been made in this respect, and relevant experimental data are also very scarce.

In a number of cases, the treated support is regarded as an homogeneous phase, that in its whole is considered to be the stationary phase of the system, having an extraction activity that depends on the amount of extractant present. A change of the extractant loading affects this activity, more or less proportionally, while it has hardly any effect on the volume V_s because of relatively small volume of extractant as compared to the whole phase.

If a relation is known that enables the activity of the extractant to be calculated from its amount loaded on the supporting material, $\log(V_r - V_m)$ vs. log (extractant activity) should give slopes equal to the coefficient of the extractant in the reaction responsible for retention: in the example, slope +3 should be obtained (eq. (1)). In any case, the log-log plots of $(V_r - V_m)$ vs. (amount of loaded extractant) would not necessarily imply the obtainment of straight lines, nor of slopes equal to one.

71

The very few experimental data available for extractant + support combination of this kind refer to columns made with styrene-divinylbenzene swollen in tributylphosphate (TBP) [2], and to HDEHP-treated styrene grafts on the surface of cellulose [3]. In both cases, slopes of log-log plots were obtained very near to those expected from the coefficient of the extractant in the extraction reaction, thus confirming the validity of the "homogeneous phase" approach for such systems.

Slopes differing from one were sometimes found also when making analogous correlations with paper chromatographic results [1]. However, they are sharp exceptions within a general behaviour of paper systems for which slopes close to one are found in the great majority of cases, indicating that most probably the stationary phase actually consists of a film of extractant on cellulose, and not of an homogeneous phase involving both extractant and support.

Mention must be made here to the slopes near to those expected from the extraction reactions, obtained by Cerrai and Ghersini with HDEHP-treated paper introducing an "effective concentration" of the extractant on paper, as determined by means of static distribution experiments between treated paper and aqueous solutions [4]. With regard to the present opinion of the authors, this experimentally determined quantity can hardly be identified with the actual activity of the stationary phase, so that the reasons for the apparent agreement with the expected slopes are not so clear. In any case, such results are not suitable for supporting the hypothesis of an "homogeneous" phase of paper and HDEHP.

In conclusion, the retention volumes obtainable for a given extraction chromatographic system can be quantitatively correlated to the distribution coefficients of the analogous liquid-liquid extraction system, through relations such as those presented in equations (2) and (3), and on the hypothesis of an ideal chromatographic behaviour.

In doing this, no difficulty arises as far as the aqueous phase parameters are involved. On the other hand, the assessment of the thermodynamic parameters of the stationary phase needs some hypotheses on the physical state of the support treated with the extractant.

In the great majority of cases, the stationary phase can be considered to consist of the organic extractant definitely separated from the supporting material, and filmed on it. The consequences of this are discussed in Chapter 1, and reasonably apply to chromatographic systems involving polytetrafluoroethylene, polytrifluorochloroethylene, polyethylene, kieselguhr, silica gel, and porous glass as the supporting materials, and probably also to paper chromatographic systems.

Some peculiar extractant + support combinations most probably give rise to stationary phases of completely different nature, consisting of a sort of "solution" of the extractant in the supporting material. Such combinations include supports that swell in the presence of the extractant, such as styrene-divinylbenzene copolymers and probably also vinylchloride-vinylacetate copolymers and polyurethane foams. In these cases, very little is known about the actual characteristics of the stationary phase.

As we have seen, the liquid-liquid extraction behaviour of the compound used as the stationary phase allows the retention ability of an extraction column to be foreseen, at least qualitatively. However, some cases of compounds are reported that failed to display the expected behaviour. In particular, this rather rare phenomenon was noticed with solid extractants, and it was easily explained by the impossibility for the adduct species to diffuse into the organic phase; such compounds appear to be remarkably more effective when loaded on the support together with discrete amounts of solvent.

Certain carbon-bonded oxygen solvating compounds were also found in some instances to be unexpectedly ineffective extractants in columns, and in this case the phenomenon was explained with a particularly disadvantageous phase volume ratio, due to their great mutual solubility with aqueous solutions [5]. A remarkable departure of such systems from ideal chromatographic behaviour, as suggested by Bark and co-workers [6] for laminar chromatography (see Chapter 14), may also account for the unexpected results.

Contrasting opinion exists on the possible influence of the diluent occasionally present with the extractant on the retention features of the resulting stationary phase. Diluents are known to have great effects on the distribution coefficients for certain extractants, in particular amines and quaternary ammonium compounds. Many authors more or less explicitly admit such effects on the retention features of columns, but no specific experiments have been performed to confirm this reasonable belief. Conversely, results obtained with a tertiary amine appear to exclude such effects in paper chromatography [7]: the impregnation of paper, however, normally results in the complete removal of the diluent originally present with the extractant, so that the statement can hardly be extended to columns loaded with diluted stationary phases.

Impurities of the extractant and active sites of the supporting material can sometimes interfere on the retention features of the bulk stationary phase, and are the reason for most observed unexpectedly high retentions of trace amounts of metals. Impurities can derive from the original compound employed, and an effective purification of the extractant is always advisable before use; another source of impurities can be the low chemical stability of the extractant; the acidic degradation products of TBP, for instance, are known to affect the extraction features of the bulk compound.

When present, active sites of the support may affect the retention of a column, usually acting as solvating sites for neutral species formed in the aqueous phase; the unexpected retention of the tetrachlorogallium complex by HDEHP-cellulose columns may be taken as example. Not easily definable adsorption phenomena can also take place. Conversely, retention by the support through cation exchange reactions is not likely to occur often, since the possible acidic sites generally are too weak to be effective in the strong acidic solutions normally used as the eluents.

1.2. SELECTIVITY

By selectivity of the stationary phase is meant its ability to display suitable distribution coefficients between the two chromatographic phases, so as to allow for the separation of the elements of interest. It is worth underlining that distribution coefficients are "suitable" for column separation not only when they result in an advantageous separation factor for the elements, but also when they are not so low or so high as to prevent the actual separation of elements or their recovery into acceptable volumes of eluent.

Obviously, the selectivity of a column is strictly bound to the nature of the compound

73

used as the stationary phase, and results from a suitable combination of extractant and composition of the aqueous phase used as the eluent. The expected selectivity of a given column chromatographic system, however, can be affected by several more or less predictable factors.

When a stationary phase is loaded with macroamounts of retained elements, variations of its retention features may often occur. Distribution coefficients usually are referred to the extraction of micro or even tracer amounts of metals, and it is known that the behaviour of many extractants at such metal levels can be quite different from that shown at high organic phase loading conditions. The behaviour of acidic organophosphorous compounds is a well-established example.

In additon to this, there are cases of metal-extractant complexes acting themselves as extractants and displaying distribution coefficients much greater than the original compounds (see Paragraph 2.1): bulk amounts of uranium loaded on a HDEHP column make rare earths separations difficult or even unfeasable [8,9]. These effects, found also in liquid-liquid extraction systems, are more likely to occur in chromatography, where the retained bulk element may concentrate in the upper part of the column bed.

Besides limiting cases as that just reported for rare earths, the effect of macroamounts of certain elements on the peak position of other elements (and also of their own) is rather frequent in the literature, for nearly all stationary phases. While sometimes this could result from the lower amount of available unbound stationary phase, often the extent and/or the nature of the effect is such that only actual modifications of the extractive power of the stationary phase can account for it.

When possible, it is advisable to avoid excessive loading of the stationary phase. When microamounts of a given element have to be separated from macroamounts of another one, the best solution is to choose a chromatographic system that allows for a high separation factor and retains the microcomponent while the macrocomponent is in the meantime eluted away. However, when the available systems allow only for low separation factors, greater retention of the macrocomponent is advisable, to avoid contamination of the microcomponent peak by the probable long tail of the other one.

Finally, the possible effect on selectivity must also be mentioned of impurities in the extractant and of active sites of the supporting material.

1.3. RESOLUTION

The resolution ability of a chromatographic column is generally evaluated in terms of theoretical plates. Although the number of plates is the physical property of a column directly related to its actual separative features, heights of the theoretical plates (HETP) are usually reported, since they allow for simple comparisons among column beds of different dimensions.

The HETP features of a column depend on several different factors, and their influence has been thoroughly discussed by Siekierski in Chapter 1. Two of these factors are strictly bound to the nature of the extraction system, and hence to that of the involved stationary phase: they are the organic phase diffusion coefficient of the extracted complex, and the possible slow rate of the chemical reactions involved in the distribution of the element of interest. Both these parameters, on their side, strongly depend on temperature.

Diffusion coefficients of the extracted complexes may appreciably affect HETP when the influence of the other parameters is optimized, and elutions are carried out at high flow rate and/or low temperatures. At present, very few diffusion data are available in the literature, so that deductions of the influence of organic phase diffusion on the HETP of a column must forcedly be based on viscosity data, as done by Siekierski in Chapter 1. It is interesting to note, however, that Aliquat allows for very good HETP values, in spite of its extremely high viscosity (1450 Cp at 30 °C).

Moderately slow chemical reactions in the extraction process are likely to affect the overall HETP value to a small extent, comparable to that deriving from diffusion controlled processes. Actually, the little information available on extraction kinetics often makes it hard to distinguish between the two cases. On the contrary, definitely slow extraction kinetics will be mainly responsible for HETP: however, such cases are relatively rare, and easily detectable.

It is worth pointing out that slow chemical steps may derive from processes peculiar to chromatographic operations, and generally absent in the liquid-liquid extraction practice. For instance, when a column is eluted with an aqueous phase having a different composition from the preceding eluent, slow kinetics may derive from slow attainment of equilibrium of the new metal species that might form in the aqueous phase: the double peaks obtained when eluting gallium from HDEHP columns with hydrochloric acid solutions of increasing concentrations [10] are probably due to a slow rate of formation of chlorogallium complexes.

Since both the organic phase diffusion terms and possible slow rate of chemical reactions vary with the nature of the extracted species, the HETP values obtainable with a given column depend on the nature of both the element and the aqueous phase involved. This can have great importance in theoretical considerations, but may often be overlooked from the practical view point. Actually, the contribution of the stationary phase diffusion terms to the overall HETP value is generally small, or at least its variations with the nature of the extracted species do not affect the overall HETP value to a notable extent. On their side, definitely slow chemical steps are not so frequent, and generally are easily detected when present. In most cases, therefore, the resolution ability of a given column can be meaningfully defined (as it is usually done) by means of one single HETP value, provided that it is referred to a system apparently unaffected by slow chemical steps.

Hydrolysis phenomena can also result in slow dissolution rates upon elution with a different aqueous phase.

Besides organic-phase diffusion and slow chemical kinetics, other factors that affect the HETP features of a column are the flow-rate of the eluent, the size and porosity of the supporting material, and the amount of extractant loaded on the support. The effects of the former two factors are discussed in Chapter 1.

A typical pattern of the HETP values at different extractant loadings is reported in Fig. 1, referring to 0.09 mm diameter silica gel treated with undiluted HDEHP [11]. The minimum of the curve is believed to correspond to the minimum extractant loading

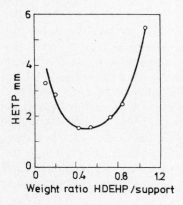

Fig. 1. The effect of extractant loading on the HETP of columns of HDEHP supported on 0.09 mm diameter silica gel (from Ref. [11], by courtesy of the authors)

sufficient to completely cover the support with an homogeneous liquid layer. Greater amounts of extractant increase the thickness of this layer, increasing HETP values because of the enhanced effect of the organic phase diffusion terms. Conversely, amounts of extractant lower than that corresponding to the minimum, result in an incomplete coverage of the supporting material, and in the subsequent worsening of HETP.

Most separations of elements having particularly low separation factors have been performed with columns carefully optimized for their amount of loaded extractant.

HETP values of a given stationary phase may vary to unforeseeable extents when the process responsible for extraction essentially changes in its nature. This could be the case of retention of acidic organophosphorous compounds from concentrated acids (see Paragraph 2.1.1.), as compared to their normal cation exchange behaviour; and of the changes in extraction mechanism deriving from high loadings of the organic phase. However, no pertinent data are available to confirm such assumptions.

Finally, the possible effect on HETP must be mentioned of appreciable solubility into the aqueous phase of the extractant and/or of the extracted species.

1.4. CAPACITY

A high capacity of a chromatographic column is important when the more retained elements to be separated are present in macroamounts. In these cases, the relatively low capacity of reversed-phase systems is considered as their main drawback, expecially when they are compared to ion exchange resins. Actually, column beds of the order of cube centimeters usually are saturated with no more than some milligrams of metals. In addition to this, loading values near to bed saturation generally are not recommended, because of possible undesirable peak broadening or serious tailing effects in the subsequent recovery of the retained element.

Obviously, the capacity of a reversed-phase column is proportional to the amount of extractant that is present in the column bed, and maximum capacity will depend on the maximum amount of the extractant that can be loaded on the supporting material, without being readily drained away by the eluting solutions. On its turn, the maximum loadable amount of extractant depends on the supporting material chosen, and also on the composition of the aqueous phases planned to be used. Significant data are collected by Katykhin in Chapter 5 (Tables 12 and 13).

Columns have been prepared for maximum capacity on the basis of empirical choice of supporting material and procedure for its impregnation. Actually, no definite information is available on the reasons for which an extractant is retained on the support. It is generally assumed that physical reasons such as the intrinsic low solubility in aqueous solutions of most extractants, their surface tension, and capillarity forces are the only responsible reasons. Pertinent investigation has been carried out only for electron-exchange systems (tetrachlorohydroquinone supported on polytrifluorochloroethylene or on siliceous material), where such hypothesis was confirmed [12].

Any interaction of the supporting material with the active groups of the extractant generally is excluded. This is most likely to be true for hydrophobic supports as siliconized ones or certain polymers, while it may not hold with other materials, such as cellulose or untreated silica gel. Actually, such interactions were suggested to explain unexpected results obtained in thin-layer chromatographic experiments done with several different extractants supported on silica gel [6].

Many capacity figures are available for extraction chromatographic systems. In normal practice, they are reported as the amount of a given ion retained on the unity of weight of treated (or untreated) support. Such figures certainly are the most meaningful ones when the capacity of columns prepared with a given support has to be assessed or optimized for a specific separation problem. However, as Katykhin points out [13], comparisons of columns prepared with different supports are better performed on the basis of capacity figures referred to the unity of bed volume of the wet column, that take into account the density of the supporting material and the degree of packing of the different beds. Obviously, capacity data obtained with a given element can hardly be correlated to those obtained with other elements, because of the relevant differences that might exist in their extraction reactions: exceptions are closely similar ions, such as lanthanides or actinides having the same oxidation state.

Columns are often prepared with the aim of obtaining a good resolution ability, and since their HETP feature apparently worsens when great amounts of extractant are loaded on the supporting material (Fig. 1), the available capacity data do not always refer to columns prepared with the maximum amount of extractant that could have been loaded on that given support.

1.5. PHYSICAL STABILITY

By physical stability of a column is meant the tendency to more or less lose upon elution the stationary phase originally loaded on the support.

Losses of extractant from the column may derive either from dissolution of it into the eluents, or from drainage by the eluent of undissolved portions of extractant, scarcely retained on the supporting material. They may result in the often undesirable presence of extractant in the eluate, and also in the variation of the characteristics of the column.

Although most extractants used as stationary phases are very slightly soluble in aqueous solutions, some of them appreciably dissolve in the eluents. In these cases, the eluent parameters, such as pH and electrolyte concentration, are often carefully chosen to keep solubility to a minimum. However, the presence in the eluate of relatively great amounts of extractant is unavoidable, and eluting solutions are generally presaturated with it to maintain the constancy of the column characteristics.

Traces of extractant are inevitably found in the eluate also when very slightly soluble compounds are used. Nevertheless, homogeneous saturated aqueous solutions of them will be unacceptable eluates only when the presence of the organic compound seriously interferes in their subsequent treatments, and practice shows that this is a rather infrequent case. On the contrary, when the very slightly soluble extractant leaks from the column because of drainage by the eluent, it will be present within the eluate as small organic droplets, or even as a well-defined heterogeneous phase: any interference phenomenon will be enhanced, and possible complexation could result in extraction and easy loss of the element of interest. In addition to this, when a second more retained element is involved, leakages of extractant could also include losses of the extracted compound, and give rise to contamination of the eluate and/or incomplete recovery of the second element in the subsequent steps.

The possible presence of organic material in the eluate is often considered a serious drawback of extraction chromatography when compared to ion-exchange resins. Actually, it must be remembered that resins also contaminate the eluate with organic degradation products, whose possible interference has sometimes led to the choice of column extraction systems. As an example, the purification of plutonium from iron before its controlled potential coulometric determination can be done with TBP columns and not with anion-exchange resins, whose degradation products poison the platinum electrode [14].

Percolation through a small column of untreated supporting material may be a simple way of freeing the aqueous solution from organic substances leaked from an extraction column. Supports treated with a normal organic solvent such as benzene can be an advantageous alternative [13].

The possible variations of the characteristics of the column, caused by dissolution or leakage of the extractant, include the lowering of its capacity and a scarce repeatability. Several stability data are available, for different extractant-support combinations, mostly referred to the check of capacity figures after reiterate elutions. They usually demonstrate the acceptable physical stability of the stationary phases suggested in the literature, the capacity loss per cycle seldom exceeding fractions of per cent of the original value.

Since nearly nothing is definitely demonstrated about the reasons for which the extract is retained on the support, the choice of a suitable combination of extractant and supporting material generally derives from empirical considerations, mostly based on the well-established fact that a material that already allowed for stable columns, very seldom results in a bad support for a new extractant. Once prepared, columns may loose by drainage a relatively great amount of extractant upon the first elutions, then maintain satisfactory physical stability in the subsequent ones.

Extractants, commonly considered as practically insoluble ones, may be appreciably soluble in particular eluting solutions; acidic compounds, for example, are much more soluble in pure water than in acid solutions, and may be completely soluble in alkaline eluents.

Possible degradation of the supporting material, when in contact with particular eluents, can also result in the loss of appreciable amounts of stationary phase: elution with very concentrated hydrochloric acid, for instance, results in the leakage of extractant from cellulose columns.

Physical stability may become a delicate problem when the stationary phase is a mixture of two or more compounds that might have different solubility features. Up to now, such cases are extremely rare, but are likely to become more frequent if synergistic systems will find greater popularity in chromatography.

1.6. CHEMICAL STABILITY

The chemical composition of a stationary phase can change either because of reactions with the chemical agents occasionally present in the eluting solutions, or because of its degradation caused by other factors, such as light, temperature and nuclear radiation. Possible unwanted redox reactions of the extractant with the element of interest can also be included among the chemical stability features of a stationary phase. All the above mentioned characteristics are very peculiar to the particular extraction system considered, and very little can be said in general terms.

Most compounds used in chromatography directly derive from previous liquid-liquid extraction experience, and generally are expected to be satisfactorily stable to most eluting solutions, simply because they have already been selected on the basis of their good stability in their early applications. Nevertheless, degradation phenomena can never be excluded, also because chromatography may require longer contact times than those adopted in liquid-liquid extraction. TBP, for example, slowly "hydrolyses" when loaded with nitric acid; it can appreciably be degraded in columns, especially when left as such after elution with concentrated acid, without a subsequent washing with a more dilute solution.

Most stationary phases are satisfactorily stable to moderately oxidizing and especially to moderately reducing solutions. They generally display some reductive properties, that may result in unwanted effects when traces of elements at easily reducible oxidation states have to be separated: trace amounts of uranium (VI), for example, are reduced to uranium (IV) by most neutral organophosphorous compounds, in the presence of light.

Most stationary phases are stable to the highest temperatures ($\sim 85\ ^{\circ}$C) adopted in extraction chromatography, and to nuclear radiation, even at appreciably high doses. Conversely, several compounds are unstable to light, where they may slowly degrade spontaneously. Degradation products usually display extraction features of their own, and may either affect the selectivity of the stationary phase or give rise to tailing phenomena.

1.7. REGENERATION

The possibility of regenerating a reversed-phase column, so as to use it for more than one separation cycle, obviously depends on the physical and chemical stability of the stationary phase to the eluents necessary for the separation cycle itself and for the subsequent regeneration steps.

The satisfactory chemical and physical stability of most stationary phases usually allows for a reasonable number of reiterate cycles, in general of the order of twenty or thirty. The most common reason for discarding a column is the excessive loss of capacity deriving from continuous small leakage or dissolution of the extractant.

Regeneration may become a difficult problem for stationary phases that display very high distribution coefficients for certain elements from most common aqueous phase compositions: this is the case of many tetravalent ions retained by acidic organophosphorous compounds. When trace amounts of strongly retained elements are involved, and their recovery is not imperative, columns may be re-used without attempting their complete regeneration, but this does not apply when macroamounts of elements are concerned. Often the recovery of such elements implies the use of rather strong chemical agents such as hydrofluoric acid, so that the possibility of regeneration of the stationary phase can be severely limited.

1.8. REPEATABILITY AND REPRODUCIBILITY

By repeatability of a column is meant the degree of accordance among the results obtained in carrying out the same separation procedure several times with the very same column Significant data are the retention volumes and the width of the peaks obtained. Repeatability can play an important role when routine separation work is performed with elements displaying low separation factors, and their appearance in the eluate cannot be easily followed by simple detection methods such as radiotracing. Again, when high separation factors are concerned, the knowledge of the degree of variation of the broadening of elution peaks can keep the eluate volumes to a minimum, with possible advantages in the subsequent chemical manipulations.

In spite of this, very few quantitative data of repeatability of columns have been reported, and all refer only to elution volumes. Generally, they are strictly related to the stability features of the columns, and satisfactory stable stationary phases may allow for per cent standard deviation of peak volumes of the order of a few per cent.

To the author's knowledge, quantitative information on reproducibility of HETP is given only by Horwitz and Bloomquist for their high speed - high efficiency columns[15], that is likely to be representative of other carefully prepared beds. Very satisfactory figures are reported, such as 0.20 ± 0.01 or 0.58 ± 0.03 mm at 7 and 26 cm·min^{-1} linear flow rates, respectively. Correspondingly, the peak retention volume variations did not exceed 1.6%. However, an abrupt worsening of the column HETP features was observed after an unforeseeable number of runs, probably imputable to the collapsing of the bed under repeated applications and releases of pressure.

Reproducibility of a column is the degree of accordance between results obtained when the same separation cycle is applied to different columns prepared in the same way. In contrast with repeatability, reproducibility represents an important information only in very particular cases, such as that of remarkably unstable columns that can be used only once.

A peculiar case are the HDEHP columns used for ^{148}Nd isolation before mass spectrometric routine analysis of nuclear fuel burn-up. They cannot be loaded with anything else but the solution to be analyzed, since the relative amounts of neodymium isotopes appear to be very easily affected by contamination by natural neodymium.

To the author's knowledge, no data are available on reproducibility of columns. In the above mentioned neodymium isolation, this lack of knowledge is overcome by the fact that quantitative recovery of the element is unnecessary: nevertheless, the portion of eluate collected for analysis is forcedly larger than that necessary if the peak position were precisely known.

Both repeatability and reproducibility obviously depend on all factors affecting the performance of the column: among them, the physical and chemical stability of the stationary phase plays a role of paramount importance.

1.9. TAILING EFFECTS

Tailing effects are definitely distinguished from the resolution ability of a column in terms of HETP, since this latter one naturally derives from the dynamics of the chromatographic process, while the former ones usually are mainly related to spurious situations interfering in the normal pattern of the descending portion of the elution curve. The main consequence of tails is the possible contamination of the more retained element by the less retained one. They can generally be accepted when separations are carried out for analytical purposes, since they usually involve very low relative amounts of elements; on the contrary, they appreciably affect the purity of the resulting products in preparative work.

Factors that can give rise to serious tailing effects are believed to include anomalous slow diffusion in the desorption of the element of interest, and the presence in the stationary phase of compounds that might display higher retention power for the element than the bulk original extractant, either for their own properties or for synergistic effects. Adsorption by active sites of the supporting material must also be mentioned.

Anomalous diffusion effects derive from the fine structure of the supporting material, where the extracted species may enter small and deep micropores invaded by the extractant, expecially when the eluted element is present in considerable amounts. When the element has to pass back into the aqueous phase, noticeable delay may occur as desorption goes on, because of the small gradients of concentratior involved. The actual existence of such diffusion effects seems confirmed by the fact that tails are more frequent with supporting materials that are known to have high microporosity. For the same supporting material, the delay of desorption of the extracted element depends on the diffusion coefficient of the extracted species into the organic phase, and the same comments apply that were already reported when dealing about resolution of columns.

Impurities of the extractant and possible active sites of the support lead to tailing effects when they retain the element more strongly than the bulk stationary phase. Contamination may result either because the second element peak falls within the tail of the first one, either because of possible abrupt release of uneluted traces of the first element upon elution with different eluents.

Very peculiar tailing phenomena often occur to microamounts of an element, when they are separated from macroamounts of a more retained one. They are generally imputed to the formation of mixed adducts or to a sort of drainage by the extracted macrocomponent of traces of the other element into the organic phase.

Tailing phenomena are frequent in reversed-phase chromatography, although often they are not detected or detectable. Nevertheless, very effective columns have also been prepared, where very high decontamination factors were obtained such as those usually wanted in radiochemical separations.

2. SYSTEMATIC SURVEY OF THE EXTRACTANTS USED AS STATIONARY PHASES

In the next paragraphs, stationary phases will be described that actually have been used in reversed-phase column extraction chromatography.

A precise classification of the different extractants was not a simple task; they could have been divided in different classes, according either to the nature of the extracted species or to the mechanism of their extraction process, but in any case complication would have arisen since most compounds actually behave in a way that cannot be definitely included within one and only one of these classes.

The relative amounts and importance of reversed-phase work with the different extractants suggested treating three main classes of compounds separately, namely acidic extractants, neutral organophosphorous compounds, and amines and quaternary ammonium salts. Extractants not included in such classes found relatively little application in chromatography, and are dealt with in a separate section, that includes also stationary phases made of a combination of extractants displaying synergistic behaviour, and the so-called "electron-exchange" columns.

Chelating compounds are a class of extractants with rather peculiar characteristics, and are the matter of a separate chapter.

Information is given here on the extraction characteristics of each class and on the mechanism that is believed to be responsible for the general extraction behaviour of the compounds included therein. When possible, information on extraction kinetics is also given, taken from the few data available in the literature. In doing this, continuous reference is made to the comprehensive work of Marcus and Kertes on solvent extraction chemistry[16]. The information here included is forcedly limited, and the reader is referred to the specialized literature for further details, in particular to several comprehensive books which have recently appeared [16—18].

The available data are also summarized on the performance of each extractant as a column stationary phase. Capacity figures are given as volumes of extractant held on the support, per volume unity of column bed. To the author's opinion, this is a less precise but more generalizable information than capacity referred to specific ions, for the reasons discussed in Paragraph 1.4. HETP figures are given as calculated according to Glueckauf, on the hypothesis of a Gaussian shape of the elution curves (i. e., they are obtained dividing by the column bed length the number of theoretical plates $N = 8(V_R/w)^2$, where V_R is the retention volume of the peak, and w is the peak width at $1/e$ times its maximum value)[19].

For the most significant compounds, details are given on their physical and chemical properties, and on their actual applications to reversed-phase chromatography; procedures are also given for their purfication and purity assay.

The viscosity and specific weight of some of the liquid extractants used in column chromatography are listed in Table 1; they were collected from usual information sources, such as handbooks or information sheets of Manufacturers. The corresponding data for more common solvents, such as carbon-bonded oxygen donor extractants, can be found in Ref.[16].

The toxicity has to be underlined of most extractant dealt with in the following paragraphs: besides very few exceptions, they can be harmful either when inhaled or upon prolonged contact with the skin.

TABLE 1

Viscosity and specific weight of liquid extractants used as stationary phases

Extractant	Viscosity, Cp $^\circ$C	Specific weight, g/ml*
HDEHP	45.6^{25}	0.97
HEHΦP	391.6^{25}	1.071
H_2MEHP	1601.3^{20}	1.096
TBP	3.32^{40}	1.072
Amberlite LA−1	72^{25}	0.84
Amberlite LA−2	18^{25}	0.83
TNOA	9.7^{20}	0.812
TLA	20.63^{25}	0.82
Alamine-336	11^{40}	0.81
Aliquat-336	1450^{40}	0.884
MIBK	0.80^{25}	0.804
TIOTP	−	0.934

*
Data refer to temperatures ranging between 20 and 25 $^\circ$C

2.1. ACIDIC EXTRACTANTS

Acidic extractants are organic compounds that have acid groups in their molecule, and exploit their extraction features mainly by exchanging the hydrogen ions of these groups for cationic forms originally present in the aqueous solution. They are often called "liquid cation exchangers", since their behaviour can be easily related to that exhibited by cation-exchange resins.

Acidic extractants used as column stationary phases include acidic organophosphorous compounds, sulphonic acids and substituted phenols. Among the various types of acidic organophosphorous compounds that have found practical application in liquid-liquid extraction, only monoalkylphosphoric, dialkylphosphoric and alkylarylphosphonic acids have been used in chromatography. Since the nomenclature of such compounds is rather confusing, their general formulae are reported in Table 2.

Most chelating agents also exchange their hydrogen ion in extraction; however, they exploit extraction features and selectivities mainly bound to their chelating ability, and will be dealt with in a separate chapter. This is not the case for acidic organophosphorous compounds, that also exploit chelating features, but whose extraction behaviour mainly relies on cation exchange reactions.

TABLE 2

Acidic organophosphorous compounds used in
reversed-phase column chromatography

Monoalkylphosphoric acids	Alkyl – O, O / P / HO, OH
Dialkylphosphoric acids	Alkyl – O, O / P / Alkyl – O, OH
Alkylarylphosphonic acids	Alkyl – O, O / P / Aryl, OH

2.1.1. GENERAL BEHAVIOUR

At fixed conditions of the aqueous phase, all acidic extractants exploit distribution coefficients that increase with the charge of the cationic species involved, as in the case of cation-exchange resins. The extent of the difference between the distribution coefficients of cations of different charge varies with the nature of the extractant considered, but in most cases is adequate for good and easy chromatographic separations. In the case of organophosphorous compounds, however, chelate or other complex formation may be superimposed to the simple cation-exchange mechanism, so that some exceptions may be found to the above mentioned rule.

Selectivity toward ions having the same charge strongly depends on the nature of the active groups of the extractants. The behaviour of sulphonic acids is very similar to that of the most popular cation-exchange resins, having the very same active groups. On the other hand, acidic organophosphorous compounds display selectivity features rather different from those of resins, thus being an interesting alternative to them not only in rare earths or actinides separations, but also in more usual analytical problems. The use of substituted phenols, that also display selectivity features that differ from those of sulphonic resins, has been generally limited to the extraction of alkali metals and alkaline earths, and little is known about the behaviour of other elements.

The mechanism of retention from dilute aqueous solutions can be referred to the equation

$$M_{aq}^{+z} + z\, HB_{org} = MB_{z,org} + zH_{aq}^{+} \tag{4}$$

where HB is the acidic extractant fixed on the support, MB_z is the adduct species whose formation is responsible for retention, and M^{+z} is the cation involved, that may be a simple metallic ion or the cationic species of charge z+ resulting from aqueous phase complexation of the element of interest.

Although the liquid-liquid extraction behaviour of most acidic extractants is rather well known, any complete description of the chromatographic process is severely limited both by the difficulties often encountered in assessing the real state of the element of interest in the aqueous solution and by the very limited knowledge of the thermodynamic properties of the stationary phase. Nevertheless, equation (4) often correctly describes the observed pH dependencies of the retention volumes of reversed-phase systems, so that a two-fold increase of the hydrogen ion concentration in the aqueous phase normally results in a decrease of the retention volumes of a factor 2, 4 or 8 for mono-, di-, or tri-valent cations, respectively.

Reversed-phase paper chromatography results with HDEHP show an abnormal behaviour of most elements when very concentrated acid solutions are used as the eluents; R_f values appear to decrease as the acid concentration is increased, in sharp contrast with a cation-exchange mechanism[1]. The very same phenomenon is most likely to occur also with columns, but only few experimental data are available. In most cases the decrease of R_f values is scarcely relevant making it hard to distinguish between an actual enhancement of distribution coefficients and the possible departure from ideality of the interactions between the two phases[6]; in other cases, however, and particularly in the case of gallium and iron(III), the extent of the phenomenon is particularly relevant, and a retention mechanism completely different from the expected one is apparent.

Distribution coefficients of iron(III), gallium and several other elements behave in a similar way in all HDEHP-mineral acid liquid-liquid extraction systems, increasing as acidity increases in the highest concentration ranges[20—23]; analogous results are reported for uranium(VI) and a few other ions extracted by di-n-butylphosphoric acid from concentrated nitric acid[24,25] and for lanthanides and actinides extracted by monoalkylphosphoric acids from concentrated hydrochloric acid[26].

The most plausible explanation of such data is based on a partial dehydration of the cation with the subsequent extraction of an electrically neutral species, formed with the anions of the acid, through a solvation mechanism involving the donor groups of the organic phase. In the case of acidic organophosphorous compounds, solvation occurs by means of their P→O groups; in other words, with aqueous concentrated acid solutions the acidic extractant behaves like neutral trialkylphosphates, known to extract by means of the above mentioned mechanism.

85

In chromatographic systems, solvation may take place by both the donor groups of the stationary phase and those that might be present in the chemical structure of the supporting material. In effects, the phenomenon is remarkably more apparent in chromatography with cellulose as the support than in liquid-liquid extraction, at least as far as acidic organophosphorous compounds are concerned. All that has been said, however, probably applies also to stationary phases made with sulphonic acids, that also display abnormal extraction of metals from highly acidic aqueous solutions[27].

2.1.2. DIALKYLPHOSPHORIC ACIDS, HDEHP

Dialkylphosphoric acids, and among them di(2-ethylhexyl) phosphoric acid (HDEHP), are by far the most popular acidic extractants in reversed-phase column chromatography. In addition to HDEHP, di-n-butyl, di-n-octyl and di-i-amyl phosphoric acids have also been used.

They are rather viscous liquids, very slightly soluble in water, and even less in acidic aqueous solutions; their solubilities obviously decrease as the length of their alkyl chains is increased. Only in the case of di-n-butylphosphoric acid (HDBP) does solubility in the eluents appreciably affect the results to be obtained and this accounts for its unpopularity in chromatography, as compared to liquid-liquid extraction.

Dialkylphosphoric acids can be prepared either by de-alkylation of the corresponding trialkylphosphates or by conventional organic synthesis techniques, and once prepared must be carefully purified of the unreacted reagents and of other reaction products that usually greatly affect their extraction properties. Synthesis and purification methods for the different compounds can be found (or are referred to) in the original articles dealing with their use in chromatography.

To the author's knowledge, among the different dialkylphosphoric acids used in cromatography, HDEHP is the only one available on the market. It is obtainable as a pale-yellow viscous liquid usually at a purity grade suitable to be used without any further purification.

As for all dialkylphosphoric acids, the corresponding monoalkylphosphoric acid is the impurity most likely to be present, that can strongly influence separations of small amounts of substances, since it generally displays higher and less selective distribution coefficients when compared to those of the bulk compound.

The presence of the monoalkylderivative may be detected by simple chromatographic procedures. Following the method worked out by Hardy and Scargill[28], down to 0.5% v/v of monoalkylderivative can be detected in HDEHP by ascending chromatography (20–30 cm long) on Whatman N. 3 paper eluted with a freshly prepared mixture of equal volumes of n-butanol, acetone and concentrated (sp.w. = 0.88 g/ml) ammonia. After drying with warm air, spots are developed by spraying 2% w/v ferric chloride in 80% ethanol, followed by 2% w/v sulphosalicylic acid in 80% ethanol. White spots are obtained on a mauve background, the monoalkylderivative being detectable behind the bulk HDEHP, at R_F values of about 0.35.

Purity can also be checked by precise potentiometric titration, with aqueous sodium

86

hydroxide, of water-ethanol (1:4 by weight) solutions of the extractant. Theoretical equivalent weight for HDEHP is 322.4 g.

The purification of commercial HDEHP can be performed in different ways; the most popular ones derive from a general procedure outlined by Stewart and Crandall[29] that takes advantage of the different solubilities of the organic acids in aqueous or alcoholic solutions. A simple and effective method[30] implies the dilution of the compound into diethylether (1:5) and two or three successive scrubs with equal volumes of ethylene glycol. After washing with water, the ethereal phase is evaporated at room temperature under reduced pressure, leaving the purified HDEHP. When the presence of pyroesters is also suspected, partition between ether and glycol must be preceded by a hydrolysis step: the commercial product and 6M hydrocloric acid (4:1) are stirred at 60 °C for about 16 hours; after several water scrubs to remove hydrochloric acid, the extractant is treated as described above.

More recently, an even simpler purification method was proposed, based on the selective precipitation of the copper-HDEHP salt[31]. Approximately 1M HDEHP in a suitable organic solvent is saturated with copper(II) either by directly adding freshly prepared $Cu(OH)_2$ or by adding NaOH to an aqueous solution of $CuSO_4$ in contact with the organic phase. Saturation is very important for effective purification. Benzene, carbon tetrachloride, cyclohexane and diethylether can be used as solvent, while high hydrocarbon mixtures are unsuitable. After removal of the aqueous phase, acetone is slowly added to the well-stirred organic solution, the copper salt of HDEHP being precipitated. The precipitate is filtered, washed with acetone, air-dried, and converted back to the acid form with dilute acid. Repeated washings with the aqueous acid may be necessary to remove all copper. Water washings remove the dissolved acid; finally, water is evaporated off at ~ 50 °C and reduced pressure.

This latter procedure is constantly gaining greater popularity; in addition to simplicity, a more effective purification is claimed.

In one case, the chromatographic features of a HDEHP column were greatly improved by loading the extractant with 1% luthetium or zirconium before the treatment of the support[32]. In this way, the monoalkylderivative impurities were saturated with these strongly extracted ions, and did not affect the column performance any more.

When suitably purified, HDEHP is a colourless liquid. Its solubility in water is 0.14 g/l, and sharply decreases in acidic aqueous solutions. As all dialkylphosphoric acids, HDEHP contains both an exchangable acidic hydrogen atom and an electron donor oxygen atom. In organic solutions, extraordinarily strong intermolecular hydrogen bonding occurs for both the unhydrated and hydrated compound. With most diluents it forms dimers, while nothing is known about its state of aggregation when fixed on a support as a stationary phase; however, it is most likely that analogous or even greater polymerization takes place.

Due to the popularity of HDEHP in spent nuclear fuel reprocessing and in rare earths separation, its mechanism of extraction of metals has been thoroughly studied, and its behaviour in liquid-liquid extraction systems is fairly well known.

When traces of metals are extracted, so that the extractant involved in adducts formation is a very low fraction of the total amount available in the system, equation (4) well represents the extraction mechanism, provided that the extractant molecules (HB) be considered as monoionizable dimers $(HDEHP)_2$. In other words, each dimer binds to the extracted metal by breaking only one of the two hydrogen bonds between the two molecules, and chelating the extracted metal by means of the oxygen atom formerly bound by the exchanged proton.

However, the number of molecules of HDEHP bound to the metal depends not only on the charge of the ionic form extracted from the aqueous solution, but also on the coordination number of the metal involved. With some metals, the number of dimers that participate to the extraction process is greater than the number of exchanged hydrogen ions, thus indicating that also non-ionized dimers bind to the extracted metal. It is the case of alkaline earths, for instance, whose extraction mechanism involves three extractant dimers, as shown by equation (1) in Paragraph 1.1.

The relative figures of the distribution coefficients for cations of different charge agree well with the already mentioned general rule for cation exchangers. However, there are several exceptions, such as beryllium and some oxyions ($UO_2{}^{2+}$, $NpO_2{}^{2+}$ and $PuO_2{}^{2+}$) whose distribution coefficients are remarkably higher than expected.

Selectivity toward cations of the same charge is greatly affected by the chelating properties of the dimeric molecules, and for strictly related metals generally increases as the ionic radius of the metal decreases, thus suggesting that unhydrated species are extracted into the organic phase. This behaviour is opposite to that shown by the conventional sulphonic cation-exchange resins.

The composition of the aqueous phase obviously plays an important effect on the distribution coefficients, but a detailed analysis of this is beyond the scope of the present chapter. As a general and obvious rule, the higher the competition of the aqueous ligand against the extractant, the lower the resulting extraction coefficient. Complexes of the metals with the aqueous ligand are often involved in the reaction responsible for extraction: provided that they keep the necessary cationic nature, they fit well into the general equation (4). As already reported, in very acidic aqueous solutions even the extraction of neutral species is likely to occur; in this case, however, the mechanism of extraction is completely different, following a reaction similar to that describing extraction by neutral trialkylphosphates (see Paragraph 2.2.1.).

When the amount of the metal to be extracted is such as to approach saturation of the extractant, things get rather more complicated and a clear scheme of the extraction process is no more available. In many cases the composition of the saturated organic phase is found to approach the theoretical stoichiometry of salts formed by neutralization of the metallic cation with anions of the monomeric extractant, thus implying a consistent departure from the mechanisms involved at low extractant loadings. A third phase often separates, either gel or solid precipitate in nature, suggesting the formation of highly polymerized species. The formation of a gel-like third phase was noticed also in reversed-phase experiments, loading a HDEHP column with rare earths[33].

Kimura has determined the extraction coefficients of a great number of trace-amount metals in the system 50% HDEHP in toluene-hydrochloric acid[34], and his results are reported in Fig. 2. Qureshi and coworkers[20,21] studied the extraction of metals by 0.75M HDEHP in cyclohexane from concentrated hydrochloric, nitric and perchloric acid solutions. Figures 3 to 5 report the distribution coefficients obtained, as calculated from the original per cent extraction data. In all systems, the departure is apparent

88

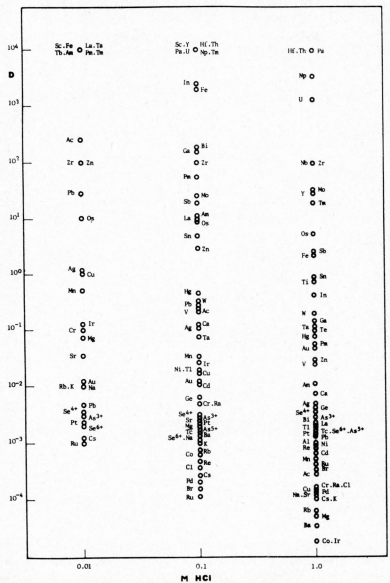

Fig. 2. Distribution coefficients of trace elements between 50% HDEHP in toluene and aqueous hydrochloric acid (from Ref.[34], by courtesy of the author)

from the expected cation-exchange mechanism in the highest acid concentration range.

Some elements such as zirconium, hafnium, and several actinides are strongly extracted by dialkylphosphoric acids at all acid concentrations in the aqueous phase, and

89

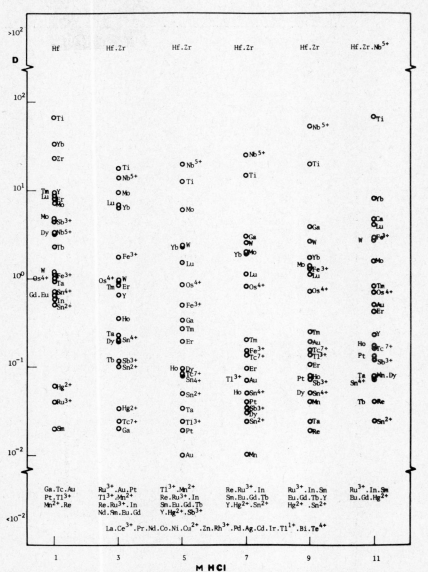

Fig. 3. Distribution coefficients of trace elements between 0.75 M HDEHP in cyclohexane and aqueous concentrated hydrochloric acid (calculated from percent distribution data of Ref. [21])

can be stripped from the organic phase only with aqueous phases containing suitable complexing agents. Obviously, the same applies to chromatography, both for the recovery of such elements or for column regeneration. For these purposes, hydrofluoric acid is often effective.

In most cases, the behaviour of dialkylphosphoric acids in chromatographic systems qualitatively agree with the corresponding liquid-liquid extraction data, and retention

90

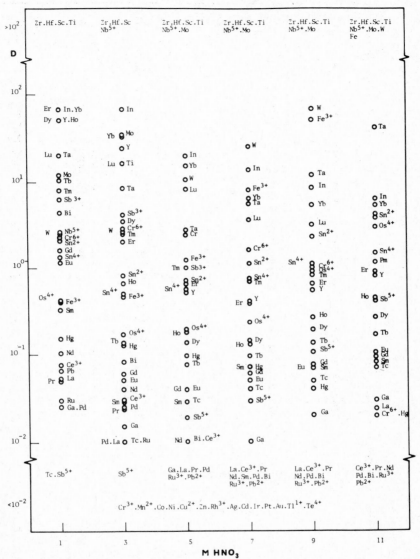

Fig. 4. Distribution coefficients of trace elements between 0.75 M HDEHP in cyclohexane and aqueous concentrated nitric acid (calculated from percent distribution data of Ref.[21])

volumes appear to depend on the aqueous phase acidity in the expected way. On the other hand, the little available information on the thermodynamic state of the stationary phase does not give easy quantitative correlations. Nevertheless, distribution coefficients for a number of metals between HDEHP and hydrochloric acid, calculated by Denig and coworkers from batch extraction data on polytrifluorochloroethylene, agree well quantitatively with column results[35].

91

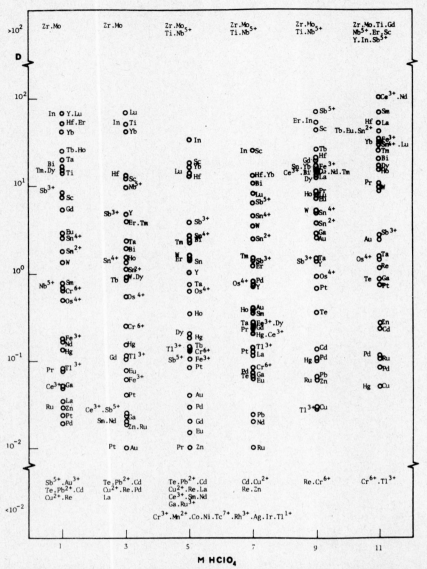

Fig. 5. Distribution coefficients of trace elements between 0.75 M HDEHP in cyclohexane and aqueous concentrated perchloric acid (calculated from percent distribution data of Ref.[21])

The kinetics of extraction usually does not appreciably affect the chromatographic results, and retention volumes are not strongly affected by the speed of elution. Conversely, desorption kinetics was proved to affect HETP values in the HDEHP–HNO$_3$–actinides system[36]. Iron(III) extracted from chloride solutions was found to follow a rather peculiar slow kinetics[1,34,37]; slow chemical steps were reported also for beryllium and aluminium[16].

Loading of the extractant with certain metals may give rise to compounds which are much stronger extractants for cations than the original HDEHP, and bulk amounts of uranium make rare earths separation difficult or even unfeasable on a HDEHP column[8,9]. Analogous effects were observed in the liquid-liquid HDEHP—HCl system, where extraction coefficients of rare-earths were remarkably enhanced by the presence of bulk amounts of zirconium[38].

Columns of HDEHP have been prepared with nearly all materials commonly used as supports. Good capacity was often obtained: data of maximum loadable extractant volume for different supporting materials are reported by Katykhin (Chapter 5, Table 12), and range from 0.24 to 0.67 ml of extractant per ml of column bed.

Very satisfactory HETP values were also obtained, expecially with hydrophobized kieselguhr or silicagel, and sintered polytetrafluoroethylene: the reported figures range from less than 0.1 to 0.35 mm (see Chapter 5, Table 14). It must be remembered that such low HETP values are obtainable with supports loaded with far less than the maximum amount of loadable extractant.

Column stability can be rather satisfactory; as an example, no more than 3% variations in peak positions were obtained with 10% HDEHP-Hyflo Supercel columns, after more than 20 runs involving a large range of nitric or hydrochloric acid concentrations[39]. Eluents containing oxidizing agents (such as persulphate, hydrogen peroxide, or bromate) did not weaken the properties of HDEHP columns.

Stability to nuclear radiation is also very good, and 15—20 micrograms of ^{252}Cf had no effect on the performance of HDEHP-kieselguhr columns[40]. However, the radiolytic decomposition of HDEHP by high fluxes of radiation accounted for the worsening of the decontamination factors obtained when separating more than 10^{10} dpm ^{244}Cm from ^{249}Bk[41].

The high speed - high efficiency HDEHP-Celite columns of Horwitz and Bloomquist deserve a particular mention[15,36]. They were prepared to allow rapid separations of transplutonium elements, and surprisingly good HETP performances were obtained even at linear flow rates of the order of 20—30 cm·min^{-1}: at 26 cm·min^{-1}, for instance, 0.65±0.02 mm HETPs were obtained over 18 runs, while retention volumes remained within 1.5%. These columns were the result of a very accurate optimization, and Refs [10] and [31] actually represent the most accurate approach to the theory underlying extraction chromatographic separations.

No data are available on the features of columns of dialkylphosphoric acids other than HDEHP, besides a good HETP value (about 0.2 mm) obtained with HDBP supported on hydrophobized Hyflo Super Cel, in spite of the relatively high water solubility of the extractant[42].

2.1.3. OTHER ACIDIC ORGANOPHOSPHOROUS COMPOUNDS

The extraction features of dialkylphosphonic acids have not been studied so much; from the available information, however, it appears that their behaviour is rather similar to that of dialkylphosphoric acids. In particular, a strong effect is reported of high organic loadings of certain metals on the distribution coefficients: the zirconium salt of di(2-ethylhexyl) phosphonic acid produces extraction coefficients for trace amounts of alkaline earths and lanthanides which are 10^4 times greater than those of the original acidic compound[43].

Rare earths separation factors by 2-ethylhexyl phenyl-phosphonic acid (HEHΦP), the only extractant of this class used in reversed-phase chromatography, are even better than those obtainable with HDEHP. Columns of HEHΦP were prepared with Hyflo Super Cel as the support, and very good stability and HETP features were obtained[44].

HEHΦP is a very viscous liquid and may be prepared from the corresponding commercial phosphonate, by hydrolysis at 80 °C for several hours, in the presence of aqueous 10% sodium hydroxide (50% in excess to the ester). The aqueous sodium salt is scrubbed with benzene; upon acidification, the product is dissolved in benzene, that is eliminated by evaporation at reduced pressure[45].

Monoalkylphosphoric acids are much more soluble in aqueous solutions than the corresponding dialkyl derivatives: this remarkably affects their applications in reversed-phase chromatography. They are diacidic compounds, having also an electron-donor oxygen atom in their active groups. As dialkyl-derivatives, they have a strong tendency to form intermolecular bonding in organic solutions, and give rise to highly polymeric species.

Little is known about their mechanism of extracting metals. While the pH dependency of the distribution coefficients agrees well with the general equation for cation exchange, only one polymeric molecule seems to supply all exchange sites necessary for binding the extracted cation at trace level. At high organic phase loadings, the available information is contradictory and confusing, suggesting either monoionization or diionization of the acid, depending on the cation involved.

2-ethylhexyl phosphoric acid (H_2MEHP) and 2,6,8-trimethylnonyl-4 phosphoric acid (H_2MDP) were used in extraction chromatography. H_2MEHP is an extremely viscous liquid, and was applied to a number of separation problems involving lanthanides, alkaline earths and several transition metals, but was found to be "a considerably less versatile and convenient stationary phase than HDEHP"[46] for such applications. Conversely, H_2MDP found a very interesting application in the separation of alkali metals[47].

H_2MDP decomposes when kept pure at room temperature, but is stable when diluted in an organic diluent such as chlorobenzene or kerosene. Stationary phases made by keeping the diluent on the support together with the extractant displayed better HETP features. Hydrophobized Hyflo Supercel was used as its support, and HETP values ranging between 0.2 and 0.4 mm were obtained. Columns could be used for many runs. Separations of alkali metals were performed with slightly acidic or neutral aqueous solutions: in the latter case, the aqueous phase was saturated with the extractant before elution to avoid appreciable dissolution of the stationary phase. Partition coefficients for alkali metals increase with the crystallographic ionic radius, thus suggesting the extraction of hydrated species, in sharp contrast with the behaviour of dialkylphosphoric acids.

Sulphonic acids are crystalline solids, relatively slightly soluble in aqueous electrolyte solutions. They are monoacidic compounds and are known to dimerize or further poly-merize in organic solutions. Their salts with the extracted metals are rather hydrophilic, and tend to form micelles in the organic layer.

Extraction takes place in good agreement with the general equation (4); however, only one aggregate appears to exchange all protons necessary for binding to the extrac-ted cation, as in the case of monoalkylphosphoric acids. Selectivity toward ions of the same charge increases as the hydrated ionic radius of the metal decreases, as in the case of sulphonic cation-exchange resins, and in contrast with the behaviour of dialkylphos-phoric acids.

Only dinonylnaphtalene sulphonic acid (HDNNS) was used in reversed-phase chromatog-raphy. It is available on the market as a $\sim$ 40% solution in n-heptane that contains various isomers of the acid, resulting from the sulphonation of dinonylnaphta-lene, and a certain amount of neutral impurities, probably starting materials and inter-mediates of synthesis.

The commercial solution was used in chromatography without any further purifica-tion, and only very recently a method has been described for its purification from the neutral components[48]. The original n-heptane solution is diluted 1:1 with absolute ethanol, and eluted on Dowex 1×2 anion-exchange resin in the OH^- form. The column is then washed with absolute ethanol until a clear eluate is obtained. HDNNS is re-covered with 1M hydrochloric acid in ethanol, and the eluate is evaporated to a solid mass at 50°C under 1 mm Hg pressure. The final product is repeatedly washed with small portions of water and then of absolute ethanol, followed each time by evaporation under vacuum. The purity grade of the resulting HDNNS is 99%, and can be checked by potentiometric titration with alcoholic NaOH: the equivalent weight is 460 when the acid is stored under diphosphorous pentoxide, otherwise two molecules of hydration water must be expected.

A method for the synthesis of HDNNS is available in the chromatographic litera-ture[49]. Its distribution coefficient between toluene and water is about 10^{-4}, and its solubility in water is expected to range around 0.4–0.5 g/1[50].

Extraction coefficients for many elements extracted from hydrochloric acid solutions have been published for 3% unpurified HDNNS solutions in a 1:1 mixture of di-ethylether-ethylacetate[51]. Paper chromatographic data are also available, for hydrochloric or nitric acids as the eluents[52].

Columns were prepared on hydrophobized silica gel[53]. Because of its relatively high solubility in aqueous solutions, the stationary phase appeared to be sufficiently stable only when prepared after an accurate hydrophobization of the support, and when eluted with solutions at least 0.1N in acid. No data are available on the HETP features of the columns.

2.1.5. PHENOLS

Substituted phenols have been successfully applied to the selective extraction of alkali metals and to their mutual separation, allowing for separation factors between cesium and rubidium higher than 20. They behave as weak acids and therefore extraction takes place only from moderately or strongly basic aqueous solutions.

The lower hydrated ionic radius the better the metals extracted. Extraction takes place by a combination of salt formation through cation exchange and of solvation by additional extractant molecules. In accordance with a cation-exchange mechanism, distribution coefficients increase as the pH of the aqueous solution is increased. At the highest pH values, however, competition occurs between the ion to be extracted and the cation of the base selected as the bulk electrolyte, and (distribution coefficients) vs. (base concentration) curves pass through a maximum. The partial substitution of the base with a salt of the same cation results in a decrease of distribution coefficients at given pH, thus confirming the suggested competitive effect.

A survey of the extraction features of 4-sec-butyl-2 (α-methylbenzyl) phenol (BAMBP) demonstrated that among 49 common cations only alkali metals and alkaline earths are more or less extracted into the organic phase, from 1 M sodium hydroxide solution containing tartrate[54]. The presence of ammonium cations, however, drastically reduces the extractability of alkali metals.

BAMBP is available on the market, and was used to test some graft copolymers studied with the aim of obtaining supports loadable with particularly great amounts of stationary phase: "Functioning columns were fabricated with adsorbed BAMBP on which it was possible to adsorb and elute milligram quantities of cesium"[55].

Columns were also prepared with polytetrafluoroethylene (Ftoroplast-40) loaded with 4-ter-butyl-2 (α-methylbenzyl) phenol (tBAMBP)[56]. The extractant was prepared by arylalkylation of 4-t-butylphenol by styrene: a nearly colourless oily liquid was obtained, with boiling point 180–185 °C. The results obtained with rubidium and cesium retained from NaOH + NaNO$_3$ solutions were in good agreement with the corresponding liquid-liquid extraction data. Cesium was also satisfactorily separated from some fission products, complexed with EDTA.

2.2. NEUTRAL ORGANOPHOSPHOROUS EXTRACTANTS

Several classes of neutral organophosphorous compounds were usefully applied to the liquid-liquid extraction of metals. Among them, only trialkyl phosphates and phosphine oxides found popularity in extraction column chromatography. Columns prepared with bifunctional compound were also reported, namely methylene bis (dialkylphosphine oxide), tetrabutyl hypophosphate, and a 1:1 mixture of tetrabutyl hypophosphate and tetrabutyl pyrophosphate. The general formulae of the compounds used in chromatography are collected in Table 3.

TBP, TOPO and dibutyl diethylcarbamoyl phosphonate have been used in association with chelating compounds to make up synergistic stationary phases, that are described in Paragraph 2.4.5. Dibutylphosphorothioic acid, whose extraction mechanism is essentially similar but whose selectivity is remarkably different from those of the above-mentioned extractants, is dealt with in Paragraph 2.4.2.

TABLE 3

Neutral organophosphorous compounds used in reversed-phase column chromatography

Trialkylphosphates

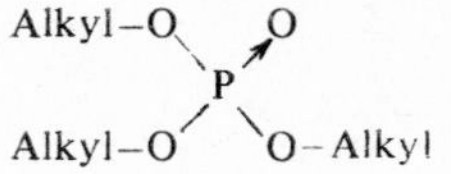

$$\begin{array}{c} \text{Alkyl}-\text{O} \\ \text{Alkyl}-\text{O} \end{array} \mathrm{P} \begin{array}{c} \nearrow \text{O} \\ \text{O}-\text{Alkyl} \end{array}$$

Trialkylphosphine oxides

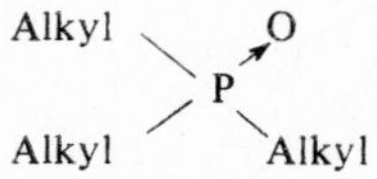

$$\begin{array}{c} \text{Alkyl} \\ \text{Alkyl} \end{array} \mathrm{P} \begin{array}{c} \nearrow \text{O} \\ \text{Alkyl} \end{array}$$

Triarylphosphine oxides

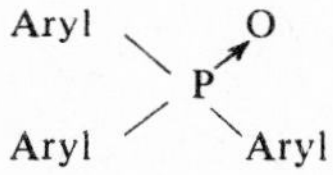

$$\begin{array}{c} \text{Aryl} \\ \text{Aryl} \end{array} \mathrm{P} \begin{array}{c} \nearrow \text{O} \\ \text{Aryl} \end{array}$$

Tetraalkylalkylene-diphosphine oxides

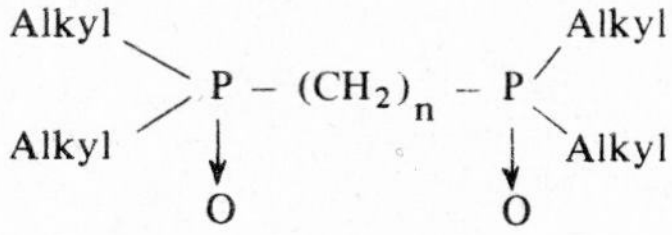

$$\begin{array}{c} \text{Alkyl} \\ \text{Alkyl} \end{array} \mathrm{P} - (\mathrm{CH_2})_n - \mathrm{P} \begin{array}{c} \text{Alkyl} \\ \text{Alkyl} \end{array}$$
$$\qquad \downarrow \qquad\qquad\qquad \downarrow$$
$$\qquad \mathrm{O} \qquad\qquad\qquad \mathrm{O}$$

Tetraalkyl pyrophosphates

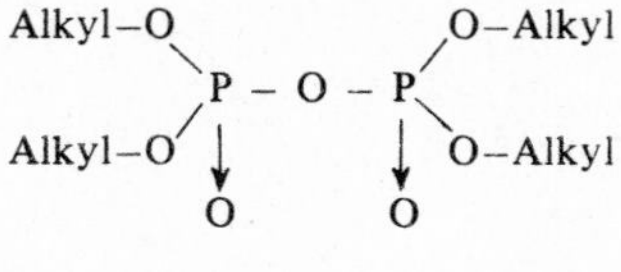

$$\begin{array}{c} \text{Alkyl}-\text{O} \\ \text{Alkyl}-\text{O} \end{array} \mathrm{P} - \mathrm{O} - \mathrm{P} \begin{array}{c} \text{O}-\text{Alkyl} \\ \text{O}-\text{Alkyl} \end{array}$$
$$\qquad\quad \downarrow \qquad\qquad \downarrow$$
$$\qquad\quad \mathrm{O} \qquad\qquad \mathrm{O}$$

Tetraalkyl hypophosphates

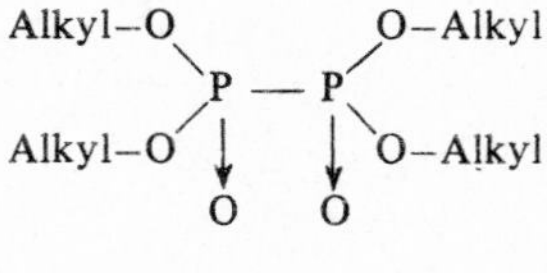

$$\begin{array}{c} \text{Alkyl}-\text{O} \\ \text{Alkyl}-\text{O} \end{array} \mathrm{P} - \mathrm{P} \begin{array}{c} \text{O}-\text{Alkyl} \\ \text{O}-\text{Alkyl} \end{array}$$
$$\qquad\quad \downarrow \qquad \downarrow$$
$$\qquad\quad \mathrm{O} \qquad \mathrm{O}$$

2.2.1. GENERAL BEHAVIOUR

Neutral organophosphorous compounds extract by substituting water in solvating neutral species originally present in the aqueous solution. Under certain conditions, extraction by an ion association mechanism may also take place. Extraction of a metal

8*

cation normally results from the competition of the extractant, water and the anion (or another ligand that might be present in the aqueous solution) in solvating the cation itself. The extracted species is a neutral complex of the cation with the anion or the anionic ligand, not necessarily the predominant metal-containing form actually present in the aqueous solution.

The role of water is very important in the overall process, because it competes in solvating the extractable species. High extraction coefficients are obtained only from solutions having high ionic strength, where activity of water is lowered and its competitive effect is therefore depressed. High ionic strength favours also the formation of unionized metal-bearing species essential for extraction: its overall effect is usually referred to as salting-out.

The effectiveness of the different electrolytes that may be used as salting-out agents mainly depends on the nature of their cation. Their relative effect generally matches with the above reported interpretation of the salting-out phenomenon, and higher extraction coefficients are obtained with smaller ionic radius and greater ionic charge of the cation involved. For example, the salting-out effect of different cations in the extraction of europium into TBP from 5–6N unacidified nitrate solutions decreases in the order Al>Mg>Zn>Li>Cu>Na, Fe, Ca>NH$_4$, distribution coefficients ranging from 270 to 2[16].

When the metal is present in great amounts, its concentration in the aqueous phase may appreciably contribute to ionic strength, and auto-salting-out may take place: this is likely to play an important role in chromatographic systems involving macroamounts of elements.

Mineral acids are relatively poor salting-out agents, since they also are easily extracted by solvation and are therefore in competition with the metal-bearing species. The great amount of mineral acids that can be extracted is a peculiar feature of all neutral organophosphorous compounds: extraction probably results from the solvation of the proton (with a mechanism similar to that described for cations) and the ion-association between the extractant-solvated proton and the anion of the acid. The same mechanism is believed to be responsible for the extraction of those metals that form oxyanions, and of transition metal haloacids.

The distribution of mineral acids depends on the nature of their anion. Together with the acid, variable but always significant amounts of water are also extracted and large changes usually occur in the volumes of the two phases. Water is believed to participate in the bonding of the acid to the extractant.

The anionic ligand present in the aqueous phase affects metal extraction both for its major or minor ability to form neutral complexes with the metal, and for the more or less extractability of the complexes formed. When more ligands are contemporarily present, mixed complexes are often responsible for the resulting distribution. Uranium, for example, is not extracted into TBP from sulphuric acid solutions, but the addition of nitrates to the system strongly enhances its distribution ratio. Conversely, the addition of sulphates to the nitrate system remarkably reduces the amount of uranium transferred into the organic phase. Both phenomena are known to occur because of the formation of more or less extractable mixed complexes with the metal, but since the extracted complex necessarily is not the predominant one in the aqueous solution, a definite description of the actual process is hardly available.

The above reported information on the mechanism of extraction by neutral organophosphorous compounds is meant only to point out the many parameters that are involved, and to stress the importance of the composition of the aqueous phase on the resulting extraction coefficients. Actually, a tremendous amount of investigation has been done on extraction by this class of substances, and a detailed discussion of the results obtained would go far beyond the scope of the present work.

Extraction is often pictured out by equations such as the following ones (written by taking TBP as representative for all types of extractants)

$$M^{+z} + zA^- + y\ TBP_{org} = MA_z\ (TBP)_{y,org}$$

or

$$MA_{z_{aq}} + yTBP_{org} = MA_z\ (TBP)_{y,org}$$

A^- being the anionic ligand present in the aqueous phase.

The latter equation perhaps better describes the solvation of actually neutral species born in the aqueous phase.

Such equations obviously are a great simplification of the complex phenomena already described, overlooking the effects of competition by water and the other neutral forms present in the aqueous phase. Nevertheless, equations such as the above ones, fitted with known aqueous ligand activities and with suitably calculated (or even experimentally determined) activities of the extractant and extracted species in the organic phase, gave the stoichiometry of many extraction processes, often confirmed by analytical data on the nature of the extracted complex.

The behaviour of the different metals cannot be described in a few words or with simple rules. Distribution coefficients usually increase as the concentration of the salting-out agent is increased. However, since the salting-out electrolyte also competes with the metal-bearing species in being solvated by the extractant, distribution coefficients often are non-monotonic functions of the electrolyte concentration. This is particularly true when extraction is carried out from strongly acidic solutions.

As for the effect of the structure of the organophosphorous compounds on their extraction features, it is a general rule that the extractive power increases with the number of direct carbon-to-phosphorous bonds. This is explainable by the more electronegative nature of alkoxy groups, as compared to alkyl groups, that decreases the availability of electrons on the phosphoryl oxygen, and thus its solvating ability. Of the two most popular neutral organophosphorous stationary phases, therefore, TOPO displays higher retention features than TBP for almost all metals. Polyfunctional compounds such as hypo-or pyro-phosphates could depart from this rule, because of possible chelating phenomena.

The extraction by neutral organophosphorous compounds was often compared to that of anion-exchange resins and of "the liquid anion exchangers" long-chain amines. Such comparisons can also be found in the chromatographic literature[57,58]. Actually, a close resemblance is apparent among the relative selectivities for the different metals especially when aqueous hydrohalic acids are involved.

As a matter of fact, a sharp distinction between the mechanism of extraction of donor-oxygen bearing neutral extractants and that of the more basic amines or resins is to some extent arbitrary, insofar as under certain conditions (such as the above mentioned extraction of acids) the former extractants are known to act through ion association as the latter ones are believed to do.

It has been suggested that TBP-solvated protons, associated in an ion-association system with chloride ions, can exchange the chloride ion for negatively charged chlorocomplexes of the metal ions, thus behaving like anion exchangers[58].

Nevertheless, a number of indirect proofs of substantial differences in the extraction mechanisms, and significant differences in the relative distribution coefficients for a number of metals, actually justify the net distinction between the two classes of extractants.

2.2.2. TRIALKYLPHOSPHATES, TBP

Tri-n-butylphosphate (TBP) was the stationary phase used by Siekierski and Kotlinska in their very first work where the application of compounds normally used as liquid-liquid extractants was extended to inorganic chromatography[59], and up to now is one of the most popular compounds used in extraction chromatography. The use of tri(2-ethylhexyl) phosphate as a stationary phase was also reported.

TBP is a colourless, high-boiling, relatively fluid liquid. It is available on the market, and is often used without any further purification. Impurities likely to be present in the commercial product are butanol and degradation products such as mono-and dibutylphosphoric acids.

The presence of acidic impurities may be detected by a simple chromatographic procedure[60]. A spot of the extractant is eluted on Whatman N. 1 paper with the water-poor phase separating from a butanol-formic acid-water mixture (10:15:20 by volume). Descending chromatograms, about 30 cm long, are recommended. Eluted spots are detected by spraying the paper with a solution containing molybdate (5 ml of 60% $HClO_4$, 10 ml of 0.1 N HCl and 25 ml of 4% ammonium molybdate, to 100 ml with water). After spraying, paper is kept at 85 °C for 7 minutes, and exposed to H_2S vapours: organophosphorous compounds are visible as blue spots on a faint buff background. R_f values of 0.10, 0.75 and 0.93 are obtained for phosphoric, monobutylphosphoric, and dibutylphosphoric acids, respectively.

An effective method for purification of TBP consists in refluxing the extractant with 0.4% sodium hydroxide (1:5 by volume), and distilling under atmospheric pressure about 200 ml of distillate. The extractant remaining in the flask is washed at least six times with equal volumes of water, and dried by heating under reduced pressure[61].

In spite of purification, thin-layer extraction chromatographic experiments have shown a significant variation of the results obtained with different batches of purified TBP, that were related to the "lack of agreement in the values for a particular ion in the supposedly same" TBP/HCl "systems (which) is evident throughout the (liquid-liquid extraction) literature."[60].

TBP is slightly soluble in aqueous solutions, its solubility decreasing as the temperature is increased. Saturated water at room temperature contains 0.4 g/1 of the extractant; solubility is even lower in neutral electrolyte solutions.

In dilute or relatively concentrated hydrochloric or nitric acids, the solubility of TBP decreases as the acid concentration is increased, reaching a minimum at 1 M HCl and 8 M HNO_3. A further increase in acid concentration gives rise to a steep increase of the solubility that eventually greatly exceeds that in pure water, expecially in the case of hydrochloric acid. Mineral acids, on their turn, are appreciably extracted by TBP: their relative extractability increases in the order $H_2SO_4 < HCl < H_3PO_4 < HNO_3 < HClO_4$.

Although TBP is considered to have a high chemical stability, it may be partially decomposed, the rate of decomposition largely depending on the nature and the amount of dissolved acids and salts. Such decomposition is normally referred to as hydrolysis, but actually is dealkylation, the main products being dibutyl and monobutyl phosphoric acids. Their presence, even at trace level, may strongly affect the behaviour of the extractant toward very low amounts of metals, that can be abnormally extracted not only because of possible higher distribution coefficients with acidic compounds, but also because of synergism between TBP and decomposition products.

Care must be taken therefore in avoiding to keep TBP stationary phases in those conditions that are known to induce dealkylation, such as the great amounts of acids that dissolve in the stationary phase when highly concentrated acid solutions are used as the eluent, the presence of extracted zirconium, or strong nuclear irradiation.

TBP appears to reduce chromium(VI) to chromium(III), and uranium(VI) to uranium(IV), the latter one only in the presence of light.

Information on the mechanism of extraction by TBP is reported in the preceding paragraph. Actually, most studies on the extraction by neutral organophosphorous compounds were performed with this extractant.

Several authors surveyed its metals extraction behaviour, collecting the distribution coefficients of many elements in extraction systems involving pure or diluted TBP and aqueous solutions having a great variety of compositions. Figures 6 to 8 report the data of Ishimori and coworkers for undiluted TBP and nitric, hydrochloric[51] and sulphuric acid[62] solutions.

Actinides (uranium(VI) in particular), zirconium and hafnium are strongly extracted from nitric or hydrochloric acid, but display rather low distribution coefficients in the sulphate systems. Metals that form oxyanions in their highest oxidation state are strongly extracted from all media, while lanthanides are not. The effect of the aqueous phase composition is particularly apparent in the case of hydrochloric acid, where protonated tetrachlorocomplexes are responsible for the very high extraction coefficients of gold, iron(III), gallium, indium and thallium(III).

Distribution coefficients in the highest acidity range tend to decrease as acid concentration is increased. This derives from the competitive effect of the acid in being extracted in place of the species containing the element of interest. When the acid is partially replaced by one of its salts, such effect is greatly lowered. In such cases, distribution of the different elements can be roughly derived from the data obtained with aqueous solutions containing the acid alone and having the same anion concentration, keeping in mind that the presence of salts may appreciably increase the distribution coefficients, but generally do not change so much their relative values for the different elements.

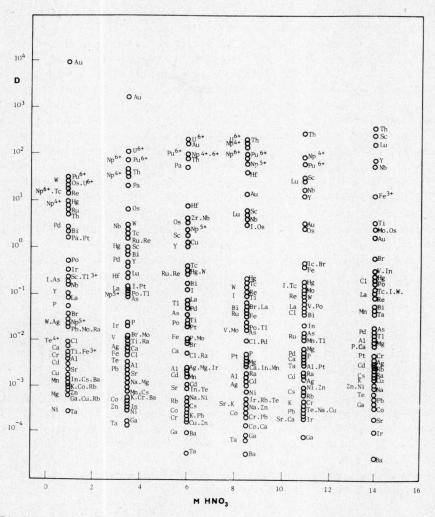

Fig. 6. Distribution coefficients of trace elements between 100% TBP and aqueous nitric acid (from Ref.[51], by courtesy of the authors)

The salt effect is greater the greater the extractibility of the acid, and as far as the data of Fig. 6 to 8 are concerned, it increases in the order sulphate < chloride < nitrate. In the case of nitrates, the substitution of acid with salt may even affect the lower concentration range of electrolyte.

TBP stationary phases behave in good accordance with liquid-liquid extraction data. At high acid concentration, however, the stationary and mobile phases gives rise to uneasily predictable volume changes in the chromatographic process, and correlation between liquid-liquid and chromatographic data often is difficult or even unfeasible. As in the case of HDEHP, columns of TBP have been prepared with almost all materials

102

commonly adopted as supports. Conversely, their HETP features were not studied as widely as in the case of acidic extractants, since most TBP separations involve elements having sufficiently high separation factors.

Nevertheless, good HETP figures were obtained from the very first use of TBP as a stationary phase[59], and further investigation suggested that neutral organophosphorous compounds display by their own nature better HETP features than acidic extractants on

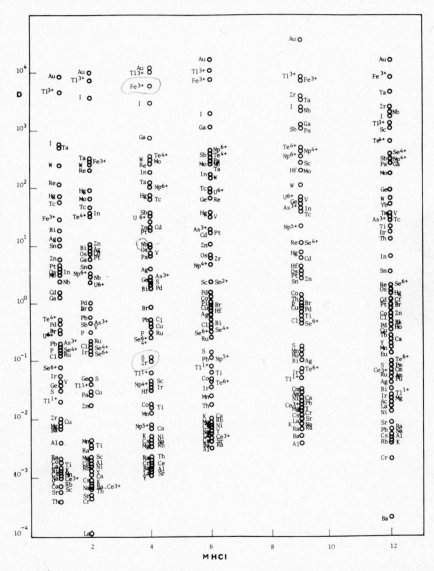

Fig. 7. Distribution coefficients of trace elements between 100% TBP and aqueous hydrochloric acid (from Ref.[51], by courtesy of the authors)

103

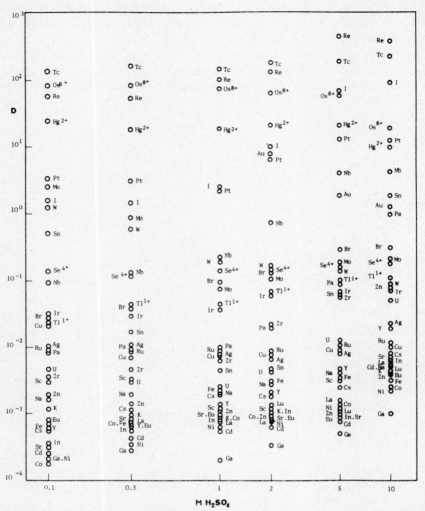

Fig. 8. Distribution coefficients of trace elements between 100% TBP and aqueous sulphuric acid (from Ref.[62], by courtesy of the authors)

siliconized Hyflo Super Cel[44]. Such hypothesis, referred to a support known for its low pore diameters, seems confirmed by the results of a very recent study on the effects of pore diameter on the HETP values of TBP and HDEHP columns[63], where it was related to the faster transfer rate of TBP molecules from the interface layer to the contiguous aqueous phase. The different rates of diffusion of the adducts into the organic phase could also explain the phenomenon.

A number of maximum capacity data are available for TBP columns, ranging from 0.28 to 0.49 ml of extractant per ml of column bed: they are collected by Katykhin in Table 12 of Chapter 5.

104

Because of the relatively appreciable solubility of the extractant, aqueous solutions are often presaturated with TBP before elution. A good hydrophobization step was reported to be essential for a satisfactory behaviour of kieselguhr columns[64]; in any case, a small loss of extractant is unavoidable, even when pre-saturated eluents are used. Using polytrifluorochloroethylene (Kel-F) as the support, remarkable amounts of the extractant were reported to be eluted away by 5.5 M nitric acid or water[65]. Nevertheless, one of such columns could be used for more than fifty uranium separations involving the above reported eluents[66].

Stability of TBP columns to chemical agents is rather good, provided that prolonged contact with too concentrated acids is avoided: concentrated nitric acid is more harmful than hydrochloric. As already mentioned, great care must be taken to avoid TBP degradation, since the resulting products can noticeably affect the retention features of the bulk compound.

TBP stationary phases appear to be stable to reducing agents, such as hydrazine sulphamate or sulphonamide, often present in the eluent for actinides separations. No damages were reported also for columns eluted with dilute hydrogen peroxide or 7 M perchloric acid. Accordingly, plutonium(III) and uranium(IV) appear to be fairly stable on TBP columns[67].

Separations involving uranium(VI) must be performed in the dark, since it is photochemically reduced by the stationary phase. Platinum is reported to be reduced on TBP columns[68]; also neptunium(VI) is reported to be slowly reduced[69].

Although TBP columns were extensively used in radiochemistry, no significant data are available on their stability to high radiation doses. It was recently reported that when solutions containing highly radioactive elements are pre-equilibrated with the extractant before feeding into the column, zirconium may unexpectedly be retained on the column itself: this is explainable with TBP radiation-induced dealkylation in the aqueous solution, followed by the dissolution of zirconium-alkylphosphoric acids complex (with or without TBP as synergistic agent) into the stationary phase[70].

Kinetic effects were noticed in the paper chromatographic behaviour of rare-earths, eluted with nitric acid[71]. No other kinetic data are available.

TBP has been used also as the stationary phase in capillary columns[72]: a detailed description of a similar system, involving the amine Amberlite LA-2, is reported in Paragraph 2.3.2.

Tri(2-ethylhexyl) phosphate and di(2-ethylhexyl) phenylphosphate were used in extraction chromatography, supported on Ftoroplast-4[73]. For the former compound, the maximum capacity of the column is available, being 0.25 g of extractant per gram of support.

2.2.3. PHOSPHINE OXIDES, TOPO

Together with TBP, although to a minor extent, also tri-n-octylphosphine oxide (TOPO) found popularity as a neutral organophosphorous stationary phase. Among other phosphine oxides, only triphenylphosphine oxide was once referred to, as an effective stationary phase on Kel-F in extracting uranium from nitrate solutions[74].

TOPO is a waxy low-melting solid, available on the market with satisfactory purity characteristics. Melting points ranging between 51 and 53 °C have been reported. All chromatographic work has been accomplished without any further purification of the commercial product. On the contrary, accurate liquid-liquid extraction work normally involves its recrystallization from cyclohexane[75].

Because of possible photo-degradation of the extractant, its solutions should be stored in opaque containers. Although phosphine oxides are far more soluble in water than the corresponding phosphates (the solubility in water of tributylphosphine oxide, for instance, is 60 g/l at 25 °C, as compared to 0.4 g/l of TBP), the relatively long alkyl chains of TOPO reduce its water solubility to 1.5 mg/l, remarkably lower than that of TBP. As for other phosphine oxides, its solubility decreases as temperature is increased.

Extraction by TOPO takes place with the general mechanism already outlined in paragraph 2.2.1. When compared to TBP, noteworthy variations are the different relative extractabilities of mineral acids and the remarkable enhancement of distribution coefficients of metals.

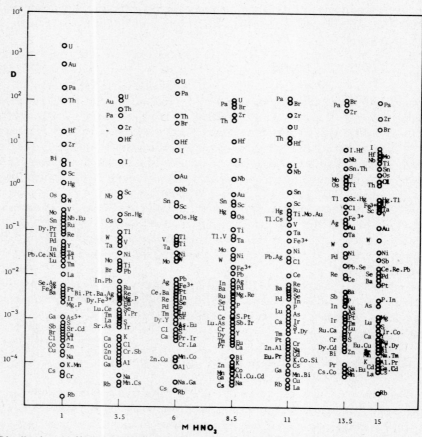

Fig. 9. Distribution coefficients of trace elements between 5% TOPO in toluene and aqueous nitric acid (from Ref. [51], by courtesy of the authors)

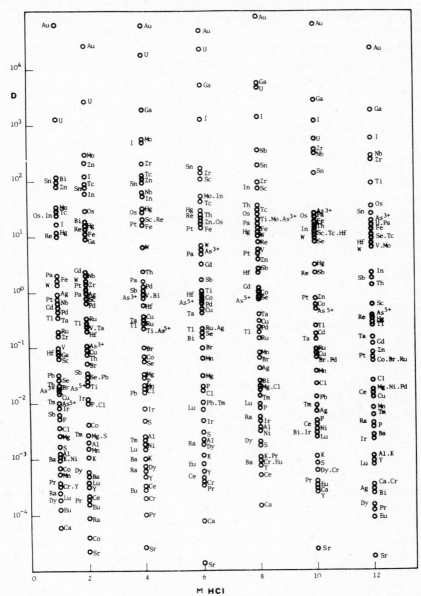

Fig. 10. Distribution coefficients of trace elements between 5% TOPO in toluene and aqueous hydrochloric acid (from Ref.[51], by courtesy of the authors)

The relative amounts of extracted acids depend on their concentration in the aqueous phase at equilibrium. For acid concentrations lower than 2 M, acid extraction increases in the order $H_2SO_4 < H_3PO_4 < HCl < HClO_4 < HNO_3$; at 6 M concentration, the order is changed in $H_3PO_4 < H_2SO_4 < HNO_3 < HCl$. Perchloric acid forms a third phase above 3 M.

107

The distribution coefficients of a number of metals are reported in Figs 9 and 10, for nitric and hydrochloric acids as the aqueous electrolytes, and for a 5% solution of TOPO in toluene[51]. In general, the distribution coefficients of metals apparently are roughly of the same order of magnitude as the corresponding ones reported in Figs 5 and 6 for 100% TBP systems, in spite of the great dilution of TOPO.

(Metal distribution coefficients) vs. (acid concentration) curves show the same patterns already described for TBP, because of the competition of the acid in being extracted into the organic phase. Again, the partial substitution of the acid with one of its salts results in an increase of metal extraction, especially in the higher range of electrolyte concentrations.

Chromatographic columns of TOPO have been prepared with many different supports such as cellulose, glass beads, polyethylene and polyfluoroethylene (Kel-F, Teflon, etc.).

To prepare a TOPO column with pre-loaded support, it is advisable to avoid complete evaporation of the diluent and stop drying when the material still smells of the diluent itself. Actually, columns prepared with pre-loaded supports seem to be uneffective if the diluent used to prepare the extractant solution is evaporated off completely. The phenomenon was reported only for TOPO-treated Kel-F columns[76], but it is likely to take place with all supports; probably it may be ascribed to the solid nature of the extactant, since analogous effects were noticed for thin-layers pre-treated with other solid compounds, Hyamines and tribenzylamine, that gave acceptable results only when diluted with decanol or similar liquid substances[77].

The first TOPO-Kel-F columns were reported to lose 30% of their original capacity after 15 consecutive cycles, during which the bed was in contact with rather concentrated chemical agents[76]. However, the batch extraction experiments currently performed by Testa with TOPO-treated Kel-F or Microtene demonstrate the good stability of the stationary phases (see Chapter 10).

Little is known about the stability of TOPO stationary phases to chemical agents and to radiation: the only information refers to experiments with stationary phases made of a mixture of TOPO and 2,3-dichloro-1,4-naphtha-hydroquinone (see Paragraph 2.4), that showed no interference effect of the extractant on the trace level redox processes performed.

Theoretical plate performances of columns were seldom reported, but published elution peaks show an overall satisfactory behaviour of TOPO to this respect. Very small HETP values are referred to by Fidelis and Siekierski[44], probably comparable to the 0.15 mm values they obtain with TBP and Hyflo Super Cel as the support.

No data are available on the maximum capacity of columns. Great amounts of TOPO were loaded on microporous polyethylene (1:1 TOPO to support weight ratio), but batch extraction results in the absorption of uranium(VI) led the authors to deduce that "a remarkable portion of TOPO was in positions of hindered accessibility, some perhaps completely inaccessible to the sorbable uranium"[78]. TOPO-Kel-F columns were reported to retain 0.33 μmoles/ml iron(III)[76].

2.2.4. BIFUNCTIONAL NEUTRAL ORGANOPHOSPHOROUS COMPOUNDS

Alkylene bis(di-alkylphosphine oxide)s as extractants were thoroughly studied at Ames Laboratories, both by liquid-liquid extraction and by reversed-phase paper chromatography, and a review is available of the results obtained[79].

Methylene bis (di-n-hexylphosphine oxide) (MHDPO) and methylene bis(di(2-ethylhexyl) phosphine oxide) (MEHDPO) were used in reversed-phase column chromatography. Several methods for their synthesis are reported, that are discussed in the above-mentioned review. The method of Richards and coll. seems to be preferable[80]. Methods for their purification are available in the chromatographic literature[81].

Both extractants are low-melting solids, less soluble in water than the corresponding monofunctional compounds: the solubility of MHDPO is reported to be 40 mg/1, and to decrease as temperature is increased. Their chemical stability appears to be rather satisfactory, only a partial degradation being reported when treated with concentrated nitric acid.

The extraction mechanism of these compounds may be connected to that of the monofunctional alkylphosphates or alkylphosphine oxides. However, their ability to form chelates may strongly affect their behaviour, both in the composition of the extracted species and in the resulting distribution coefficients. Water extraction from acidic solutions, for example, decreases as the acid concentration is increased, in contrast with the behaviour of TBP and TOPO. Distribution coefficients often are remarkably higher than those obtainable with the analogous monofunctional compounds, as in the case of uranium from perchlorate aqueous solutions, whose distribution coefficient is 10^4 times greater with MHDPO than with TOPO solutions.

Generally, perchlorate salts are best extracted, followed by nitrates and chlorides in this order; the degree of extraction decreases with decreasing ratio of charge to unhydrated radius of the cation.

O'Laughlin and Banks claim that "the chemical stability, insolubility in water and aqueous solutions of mineral acids, low melting point, and wide range of cations which are extracted make MHDPO and MEHDPO particularly useful stationary phases"[82] Several supports have been used, the best HETP being obtained with kieselguhr, that gave values of about 0.6 mm[83]. The columns appeared to be useful for separations such as that of rare earths among themselves [83] (also at milligram level), of niobium from zirconium[82], and of several trace elements from rare earth matrices[83,84].

Tetrabutyl pyrophosphate plus tetrabutyl hypophosphate were used as column stationary phase for studies on lanthanides and transuranium elements, and for the separation of a number of more common metals[85,86].

The extractants, synthesized for the purpose[86], were loaded on hydrophobized Hyflo Super Cel or Ftoroplast-4, and good HETP features were obtained with both supports. Besides reversed-phase chromatographic data, very little is known about the extraction features of the mixture of extractants.

The use of tetrabutyl hypophosphate, supported on cellulose powder, was reported for separations involving actinides and lanthanides[87].

2.3. AMINES AND QUATERNARY AMMONIUM COMPOUNDS

High molecular weight organic amines and ammonium salts are often referred to as liquid anion-exchangers, since their behaviour in the extraction of metals is noticeably similar to the absorption of metal complexes by anion-exchange resins. They found an extensive interest in the field of liquid-liquid extraction, and many of them are produced on an industrial scale, resulting in a mixture of different products usually sold and referred to with trade names hardly understandable from the chemical view-point, but nevertheless thoroughly used also in the scientific literature.

A great number of them has been screened for their behaviour towards metals in laminary reversed-phase chromatography. Conversely, a relatively limited number has been applied to column chromatography, namely the secondary amines Amberlite LA-1 and LA-2 the tertiary tri-n-octylamine, tri-i-octylamine and Alamine-336, and the quaternary Aliquat-336 and trilauryl monomethyl ammonium salts. Primary amines as column stationary phases were never reported.

A comprehensive and detailed review on the use of such compounds in extraction chromatography was recently published by Brinkman[88].

2.3.1. GENERAL BEHAVIOUR

All amines and quaternary ammonium compounds to be used in liquid-liquid extraction fulfil a number of basic requirements such as good extraction power, low solubility in aqueous solutions and sufficient chemical stability.

The solubility of amines in aqueous solutions strongly depends on their class, sharply decreasing in the order primary>secondary>tertiary. For each class, solubility is lower the longer the alkyl chains, and may reach very low values; tertiary amines with more than eight carbon atoms, for instance, dissolve in water to no more than 5 ppm. On their side, quaternary tetraalkylammonium compounds display a fairly higher solubility, and rather long alkyl chains are necessary to ensure acceptably insoluble compounds.

The solubility of amine and tetraalkylammonium salts decreases in electrolyte solutions, apparently following a rule comparable to the solubility product. The concentrations of trihexylamine hydrochloride in lithium chloride solutions, for instance, in equilibrium with its $\sim$ 0.1 M solution in benzene, appear to satisfy a constant product $[amine \cdot H^+] [Cl^-] = 10^{-6}$ in the aqueous phase[50].

The majority of amines decompose upon heating and only the lower members can be distilled without decomposition. Besides this, the stability of saturated amines and quaternary ammonium compounds is fairly good, both to oxidizing agents and to nuclear radiation. Stability to radiation, however, appears to be strongly reduced in the presence of nitric acid, and becomes lower than that of TBP. Nevertheless, amines offer the great advantage of yielding degradation products that generally do not alter the extracting properties of the bulk compound[89].

Amines react with acids to form ammonium salts. In reversed-phase work such salts are obtained by equilibrating the amine with an acidic aqueous solution: the following equilibrium takes place

$$R_3N_{org} + H^+_{aq} + A^-_{aq} = R_3NH^+A^-_{org}$$

(5)

where $R_3NH^+A^-$ is a polar ion pair with high ion-association constant in the organic phase, R_3N is a tertiary amine, taken as representative of all classes, and HA is a mineral acid, originally present in the aqueous phase.

Equilibration of the amines with aqueous acid solution usually results in the extraction of amounts of acids far exceeding those required by the stoichiometry of reaction (5); in contrast with the behaviour of oxygen donor extractants, the acids are accompanied by very little amounts of coextracted water.

The extractability of haloacids increases in the order $HF < HCl < HBr < HI$ when the extractant is in stoichiometric excess. After equivalence, things get more complicated. When concentrated acids are involved, hydrogen dihalides such as HCl_2^- are believed to be extracted. Nitric acid is extracted to a much greater extent than the stoichiometric amount, but the excess acid appears to be bound to the extractant in a rather weak way, and perhaps merely results from simple dissolution of the acid into the organic phase. The distribution of sulphuric acid is complicated by the sulphate-bisulphate equilibrium, since bisulphate has a greater affinity for the amines than sulphate.

Quaternary tetraalkylammonium compounds extract excess acid as the ammonium salts derived by the amines. The extractability of excess hydrohalic acids, however, is markedly lower.

The extraction of inorganic species from an aqueous solution into an organic phase is believed to take place by an anion-exchange equilibrium reaction such as

$$nR_3N\,H^+A^-_{\,org} + X^{n-}_{\,aq} = (R_3NH^+)_nX^{n-} + nA^-_{\,org} \qquad (6)$$

where $R_3NH^+A^-$ is the original salt of the tertiary amine R_3N, A^- and X^{n-} are the anionic forms (occasionally metal containing species) in the aqueous solution, and $(R_3NH^+)_nX^{n-}$ is the resulting amine salt, responsible for the extraction of X^{n-} species into the organic phase.

Actual extraction of a given anionic species results from the competition of the different negatively charged species present in the system in associating with the bulky cation in the organic phase. Distribution depends on the composition of the aqueous phase and on the relative values of the ion-association constants of the ion pairs that might be formed.

A general feature of amine salts and quaternary ammonium compounds is their preference for anionic metal containing species as compared to simple anions. Metals that form oxyanions or anionic complexes with an aqueous ligand easily substitute the simple anion originally bound to the extractant. However, the different amines apparently behave by supplying bulky organic cations that generally display the same selectivity features regardless of their class or chemical structure. The nature of the extractant, and other organic phase parameters such as its concentrations and nature of its diluent, will affect the thermodynamic activity of the bulky cations and of the adducts (and hence the overall effectiveness of the extraction), but will have scarce influence on the relative extractability of the different anionic species.

In other words, at a given aqueous phase composition, the different nature of the organic phase will result in higher or lower distribution coefficients for all extractable metal containing species, but will scarcely affect their relative values, so that the selectivity in

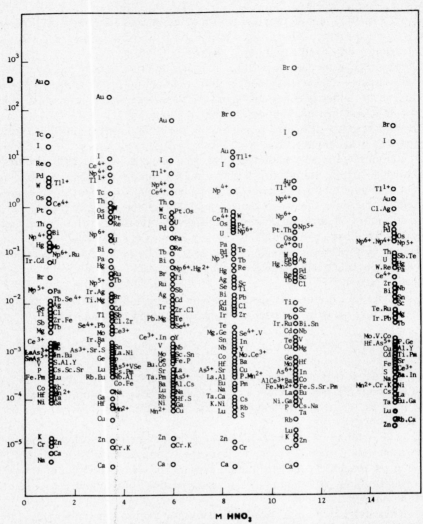

Fig. 11. Distribution coefficients of trace elements between 0.11 M TIOA in xylene and aqueous nitric acid (from Ref.[51], by courtesy of the authors)

the extraction of metals essentially depends only on the composition of the aqueous phase. The relevant parameters include the aqueous anionic ligand concentration, concentration of other complexants that may compete for the metal ions, presence of other anions that compete for the organic cation, the hydrogen ion concentration and the concentration of the metal ions themselves.

Reversed-phase processes usually correspond to extractions performed from aqueous solutions of mineral acids, and by means of ammonium salts formed with the anion of the same acid. In this case, the extraction can be easily schematized as the result of the competition for the bulky organic cation between the anionic metal species and the

112

anion of the mineral acid: an equilibrium will be established easily represented by equation (6).

As a general rule, the distribution coefficients of most of the extractable metals increase with increasing acid concentration, because of easier formation of anionic complexes. However, at high acid concentrations the increasing competition of the anion of the acid, or of other species that might arise in the solution, may result in the existence of maxima in the (distribution coefficients) vs. (acid concentration) curves.

Figures 11 to 13 report the distribution coefficients collected by Ishimori et al. for

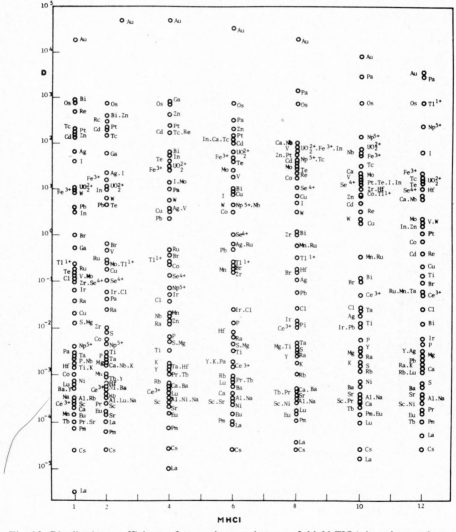

Fig. 12. Distribution coefficients of trace elements between 0.11 M TIOA in xylene and aqueous hydrochloric acid (from Ref.[62], by courtesy of the authors)

metals extracted from nitric[51], hydrochloric[62] and sulphuric acid[90] solutions by 0.11 M TIOA in xylene.

Because of the scarce complexing ability of the nitrate anion, only few metals are appreciably extracted from nitric acid (Fig. 11). (Distribution coefficient) vs. (acid concentration) curves tend to bend at the highest acid concentrations, but the effect is low, in accordance with the weak interaction of nitric acid with the extractant. Replacement

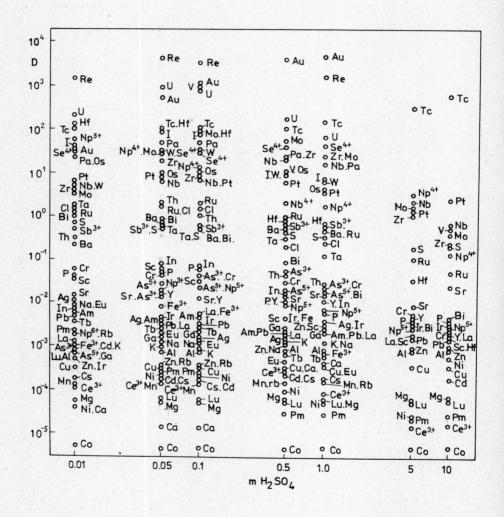

Fig. 13. Distribution coefficients of trace elements between 0.11 M TIOA in xylene and aqueous sulphuric acid (from Ref.[89], by courtesy of the authors).

of nitric acid with metal nitrates results in an increase of distribution coefficients only in the lower concentrations range. Above about 7 M overall nitrate concentration, extraction is scarcely affected by the nature of the cation in the aqueous electrolyte, but no definite explanation is available for this.

In chloride systems (Fig. 12), the highest distribution coefficients are exhibited by polyvalent metals readily forming anionic chlorocomplexes, followed by metals that form oxyanions and by bivalent transition metals. Besides oxyanions, the order of extractability is what should be expected from the tendency of metals to form anionic chlorocomplexes. However, the existence of extractable chlorometalates anions sometimes appears not necessarily to be a prerequisite for extractability, as in the case of zinc, whose predominant species in aqueous solution appears to $ZnCl^+$, and nevertheless is extracted, probably as $ZnCl_4^{-2}$ or $ZnCl_3^-$ [88].

Most (distribution ratio) vs. (hydrochloric acid concentration) curves show a maximum. Conversely, the substitution of the acid with alkali chlorides results in monotonic curves, the distribution coefficients being remarkably enhanced for the highest chloride concentrations. This behaviour is usually attributed to the competitive effect of hydrogen dichloride anions in the extraction from acid solutions, but another explanation could be the lack of dissociation of protonated chlorometalates in concentrated acids, in contrast with the complete dissociation of their salts.

Extraction from sulphuric acid solutions (Fig. 13) strongly depends on the hydrogen ion concentration in the aqueous phase. Actually, pH affects the sulphate-bisulphate equilibrium, and bisulphate is known to strongly compete with the metal sulphate complexes for the available organic cation. Distribution ratios reach a maximum value at acidities lower than 0.1 M, usually followed by a very abrupt decrease as the acid concentration is increased, in accordance with the increase in bisulphate formation. When the acid is replaced by alkali or ammonium sulphates, the initially higher distribution coefficients decrease much less with increased electrolyte concentration.

More complex aqueous phase compositions obviously can remarkably affect the extraction behaviour of the different metals. In many cases, the effect of composition on extraction is easily foreseeable, as it is the enhancement of extraction of a metal by the addition, to a non-complexing electrolyte, of a ligand that forms an anionic complex with the metal. In other cases, however, the effects cannot be easily foreseen nor explained. Zirconium, for example, is almost not extracted at all by TNOA or Aliquat-336 from nitric acid solutions, but the addition of moderate amounts of nitric acid to concentrated hydrochloric acid results in a great enhancement of zirconium extraction[91]. Conversely, the distribution coefficient of hafnium is affected in the opposite way, being remarkably depressed.

As already mentioned, the selectivity of the extractant is scarcely affected by its class and structure, so that Figs 11 to 13 actually illustrate, at given aqueous phase compositions, sequences of distribution coefficients for the different metals that roughly hold for all amines and quaternary ammonium compounds. However, the absolute values of the distribution coefficient will strongly depend on the actual extractant involved.

The relative extraction power of the different amines and quaternary ammonium compounds has been thoroughly investigated both by liquid-liquid extraction and by reversed-phase paper and thin-layer chromatography: a fair agreement was found among

115

the results obtained with the different techniques. Quoting Brinkman's review[88], "the sorption sequence of the exchangers in HCl medium is: quaternary⌣tertiary>secondary> >primary". The same sequence holds also for HNO_3, HF, HBr, HI and HSCN. "Deviations for this sequence occur only seldomly, but a more suitable division cannot easily be introduced. Tri-i-octylamine, for instance, surpasses Aliquat in the HCl system, but the latter exchanger shows higher sorption strength with SCN^- and LiCl".

In sulphate systems, the sequence appears to be reversed, and extraction strength decreases in the order primary>secondary>tertiary amines. Quaternary ammonium compounds again seem to behave roughly as tertiary amines. "It has been suggested that the reversal of the sequence is due to the fact that the metal-sulphate complexes present in the organic phase will, on the average, be more highly charged than, for example, the metal chloro anions. This necessitates the attachment of a larger number of amineH$^+$ cations in the case of the sulphate complexes, thus favouring the sorption by means of primary amines over that involving more bulky secondary and tertiary exchangers"[88].

Within a given class of extractants, the structure and length of the alkyl chains can also affect the extraction power and even modify the above reported sequences. However, little information on the effects has yet been achieved.

More detailed information on the relative retention power of amines and quaternary ammonium compounds used as stationary phases can be found in Refs.[88] and [1], where the results published by the different authors have been summarized.

Distribution coefficients of metals obtained with liquid-liquid extraction experiments have often been treated on the basis of heterogeneous equilibria such as those of equations (5) and (6), in the attempt to elucidate the chemistry underlying the extraction process. However, definitely positive results were seldom obtained, mainly because of the many complications arising from the state of the extractant and of the extracted species in the organic solution.

A peculiar characteristic of amines, tetralkylammonium compounds and their salts with acids or metal-containing anions is the very complex and unforeseeable aggregation phenomena that occur in the organic solutions, and the great effect that such phenomena have on the resulting distribution coefficients. Polymerization occurs even at concentration as low as $10^{-6}-10^{-7}$ M, and generally goes up to micelle formation. The degree of aggregation depends on the nature and chemical structure of the extractant (or its salt), on its concentration and on the nature of the diluent. In some cases, the diluent is believed to participate also in the structure of the adduct, and its overall effect can be so great and unrationalizable that it was even stated that "the amine-diluent combination rather than the amine alone should be considered as the effective extractants"[92].

In spite of these difficulties, the chemistry of several extraction processes has been fairly elucidated, often by integrating with data of maximum loading and of spectral analysis of the adducts, the results of the analysis of the dependency of the thermodynamic (experimental or more or less theoretically calculated) activities of the species that take part in the equilibria. A great number of reversed-phase paper and thin-layer chromatography data are available, due especially to the work of Brinkman, Przeszlakowski and Bark and Graham. The behaviour of several extractants for each different class has been screened for many elements, using eluents containing chlorides, bromides, iodides, fluorides, thiocyanates, nitrates, perchlorates and sulphates. In all cases, a fairly similar qualitative behaviour was found when compared to liquid-liquid extraction data[88], both for what the patterns of R_f vs. (electrolyte concentration) curves and the relative retention powers of different stationary phases are concerned. However, quantitative interpretations of the exact mechanism underlying chromatographic retention are severely limited by the uncertainties on the organic phase thermodynamic parameters, analogous to those of liquid-liquid extraction systems.

Reversed-phase columns with amines as the stationary phase are often compared to anion-exchange resins, and the overall usefulness of the amine columns is sometimes denied, because of the easier availability and greater popularity of resins. Actually, both systems consist of nearly the same cationic sites available to anionic species formed in the eluent, and although the actual chemical process involved in the distribution appears to be substantially different, in most cases the resulting selectivities and even separation factors are strictly similar.

The advantages of resins can be summarized in a remarkably higher exchange capacity per unit volume of the bed, and in a generally lower leakage of unwanted organic matter into the eluate. On the other hand, the advantages of amine columns over resins are the "tailor-cut" retention power of the column and the generally lower HETP values obtainable.

Actually the low capacity of amine columns is their greatest draw-back, and limits their application when macroamounts of retained elements are involved; when the amount of the other element to be separated is not so much, the great versatility of reversed-phase systems could suggest "reversing" the separation and choose a different stationary phase; objectively, however, this would often result in a useless complication.

On the other hand, when microamounts of elements having low separation factors have to be separated, the amine columns may be substantially more advantageous than resins, both for their generally better HETP features and because of the large choice of extractants that can make profit of small but important enhancements of the separation factor. In addition to this limiting case, the "tailor-cut" stationary phase can control the volume or the concentration of the eluent necessary to recover elements that are partially retained in most aqueous conditions. The possible leakage of extractant, on its side, often does not affect the subsequent chemical treatments of the eluate at all, and sometimes can even overcome interferences shown by the unavoidable degradation products leaking from resins.

In the opinion of the author, therefore, amine columns keep their usefulness not only in particular problems such as actinide separations, but also in other less critical analytical applications.

2.3.2. SECONDARY AMINES

Distribution coefficients of secondary amines are lower than those reported in Figures 11 and 12 for the tertiary TIOA and aqueous nitric or hydrochloric acids. Conversely they are higher than those of Figure 13, concerning extraction from sulphate solutions.

The use of Amberlite LA-1 on Kel-F has been reported in alternative to TBP for the recovery of uranium from aqueous solutions[73]. It is an unsaturated N-dodecenyl trialkylmethylamine, a viscous yellow liquid sold by Rohm and Haas. It contains 24—27 carbon

117

atoms, and its approximate molecular weight ranges between 351 and 393; 15 ppm is reported as its solubility in 1 N sulphuric acid.

Amberlite LA-2 is a saturated amine with structure analogous to that of LA-1, and is also available from Rohm and Haas. It is much less viscous, slightly more basic and remarkably less soluble in water solutions than LA-1, and its approximate molecular weight ranges between 353 and 395. As LA-1, LA-2 has been used in reversed-phase chromatography as supplied. In one case, however, it was filtered or distilled under reduced pressure before use for thin layers.

Supported on cellulose[93] or hydrophobized Hyflo Super Cel[94], LA-2 allowed for several good column separations of metals from chloride solutions.

A very peculiar application of this amine was also reported[95]. A solution of the extractant in tetrahydrofuran was forced into a polyethylene capillary having 0.5 mm internal diameter and 10 m length. A film of the extractant was stuck to the capillary walls, and was able to retain elements eluted with aqueous solutions passing through the capillary itself. Using hydrochloric acid as the eluent, microgram amounts of iron(III) were readily separated from comparable amounts of nickel or aluminium; very neat peaks were obtained. The column had total capacity of 4 μ equivalents.

The advantages of such capillary columns for extraction chromatography are discussed by Eschrich and Hansen, that applied the technique with TBP as the stationary phase[72].

2.3.3. TERTIARY AMINES, TNOA

Tertiary amines, and TNOA in particular, were among the most popular stationary phases in the early chromatography works; more recently, however, they have been gradually substitued with quaternary ammonium compounds, because of the somewhat more promising features of these latter extractants. Stationary phases have been reported made of tri-n-octylamine (TNOA), tri-i-octylamine (TIOA), methyldioctylamine, trilaurylamine (TLA) and Alamine 336.

TNOA is available on the market, and it was always used in chromatography as the technical commercial product, without any further purification. Its liquid-liquid distribution coefficients generally are slightly lower than those reported in Figs 11 to 13 for TIOA. A lower retention power of TNOA-treated paper or silica gel thin layers, as compared to that of TIOA-treated ones, was repeatedly reported by Przeszlakowski and by Brinkman for many different eluents.

Distribution coefficients of several elements were also calculated from batch extraction experiments, and showed a fair agreement with liquid-liquid extraction results[35].

Columns have been prepared supporting the extractant on a number of different materials, and in some cases very good HETP figures were obtained: for example, 0.2 mm were obtained with hydrophobized Hyflo Super Cel[96].

Capacity was satisfactory with most supports, and Ftoroplast-4 allowed for 0.21—0.25 ml of extractant per ml of column bed (see Chapter 5, Table 13). Another maximum capacity figure refers to TNOA-cellulose columns, reported to be loadable with 0.28 millimoles of iron(III) per millilitre of column bed[97].

TNOA-treated Kel-F columns were reported to be stable to 10 consecutive cycles in-

118

volving 10 M HCl, 7 M HCl + 1 M HF, 3 M HCl and 1 M HNO_3[98]. Conversely, cellulose columns were reported to be rather unstable: "cellulose was partially destroyed by 8 M HCl and ..... TNOA was taken off and appeared in the eluate. The column could only be used twice"[99]. Although this statement is perhaps exaggerated, 10–12 M hydrochloric acid actually damages cellulose and drains off appreciable amounts of extractant.

TNOA columns have been successfully applied to a great number of analytical problems; in particular, cobalt was reported to be separated from nickel with remarkably higher decontamination factors than when anion-exchange resins are used, so as to easily meet the requirements of ^{58}Co for medical use[100].

Data reported by Ishibashi and coll., who eluted iron, cobalt and nickel with hydrochloric acid on paper treated with TNOA solutions prepared with twelve different diluents[7], point out that the behaviour of TNOA stationary phases is scarcely affected by the diluent used to load the amine on the support: as already mentioned, this is probably limited to paper chromatography, and does not apply to columns.

Tri-i-octylamine (TIOA) is an isomeric mixture of branched-chain amines, which principally contains 3,5-, 4,5-, and 3,4-dimethylhexyl groups. It is available on the market, and it is normally used without any further purification. In one case, however, the commercial product was distilled at about 1.5 mmHg, and the fraction boiling between 132° and 142 °C was used for column preparation[101].

Liquid-liquid distribution coefficients obtained with TIOA from different aqueous solutions are given in Figs 11 to 13. TIOA columns have been prepared with either kieselguhr (Celite) or a poly(vinylchloride-vinylacetate) copolymer (Corvic) as the support. By means of batch extraction experiments, Pierce and Henry calculated the distribution ratios of about 30 elements between TIOA-treated Corvic and hydrochloric acid: they found "a general pattern of extraction which is similar to that found for TIOA in solution, although the actual distribution ratios are dependent upon the TIOA loading of the solid phase"[102]. Corvic columns enabled obtaining rather sharp elution peaks.

No HETP figures are available for kieselguhr columns, but it is likely that they display the usual good features that characterize this support. A rather peculiar datum is available for TIOA-treated kieselguhr, that stored in water for two years still retained an absorptive ability nearly equal to that of the freshly prepared material[103].

TIOA-kieselguhr columns are reported to reduce tracer amounts of americium(VI) regardless of the oxidants (AgO or $(NH_4)_2S_2O_8$),contemporarily put in contact with the column bed[101].

Methyldioctylamine was proved to allow for more favourable cerium(III)-berkelium(III) separation factors from nitrate solutions, as compared to TNOA; a column prepared with the former extractant supported on silicagel enabled obtaining a better separation of these two ions than an analogous column prepared with TNOA, in spite of its poorer HETP features[104].

Trilaurylamine (TLA, tridodecylamine) is a relatively viscous liquid with melting point at about 16°C and with molecular weight 522. Its solubility in water is in the order of 5mg/l. It is available on the market at 99% purity, the impurities being mono- and di-laurylamines and dodecanol. It has been used as supplied or purified by stirring for 20 min at 40 °C with "Actissil TS" activated earths (1 g per 5 ml amine), that preferentially absorb primary and secondary amines[105]. It is reported to be remarkably unstable, being easily degraded by oxygen and especially by light. Therefore, it must be kept carefully closed in dark containers.

Columns have been prepared with TLA supported on polytrifluorochloroethylene (Kel-F or Voltalef 300) or hydrophobized silica gel.

Voltalef columns displayed 3 mm HETP. As for maximum capacity, 0.18 ml of TLA+5% octanol or 0.13 ml of 0.8 M TLA in cyclohexanol were reported, per ml of column bed.

Together with the extractant, 2-octanol was loaded on Kel-F columns, to "promote wetting" by the eluent[105]. It appeared that extraction coefficients decreased as the amount of 2-octanol was increased, in accordance with liquid-liquid extraction results of the system TLA–2-octanol – dodecane[106]. The optimum conditions were found to be 19% TLA and 5% 2-octanol (w/w), as referred to the support.

Alamine-336 (tricaprylylamine) has straight alkyl chains mainly consisting of octyl and decyl groups. The total number of its carbon atoms ranges between 26 and 32, and its average molecular weight is 392. It is a relatively fluid liquid, available from General Mills with 95% tertiary amines content on the average. Alamine-336 S now is also available, which contains over 99% of tertiary amine.

It has been used in column chromatography supported on Chromosorb W, to separate nickel for copper or cobalt, using thiocyanate solutions as the eluents[107]. Some tests are also reported that demonstrated a low capacity of polystyrene-divinylbenzene beads for the extractant, and slow equilibration rates on the column[108].

2.3.4. QUATERNARY AMMONIUM COMPOUNDS, ALIQUAT-336

Quaternary tetralkylammonium compounds show very promising features in liquid-liquid extraction applications such as nuclear fuel processing and fission product recovery, because of the relatively easy control of their extraction features, their relatively low solubility in aqueous solutions and their advantageous separation factors. As already mentioned recent reversed-phase literature concerning columns loaded with strong organic bases is mainly devoted to the use of quaternary ammonium compounds as the stationary phases: Aliquat-336 and trilaurylmethyl ammonium bases have been used, both displaying very attractive chromatographic features.

Aliquat-336 is an extremely viscous liquid, obtainable as the chloride salt from General Mills. It is a methyl tri-n-alkyl ammonium salt, usually referred to as "methyltricaprylylammonium chloride", alkyl chains mainly consisting of octyl and decyl groups. The product has 27–33 carbon atoms on an average, and its average molecular weight is 475. It is supplied with a minimum purity of 88%; Aliquat-336 S is also available, with 95–98% purity grade.

The retention power of Aliquat-336 is generally lower than that of TNOA and thus of TIOA, in hydrochloric and nitric acid systems. Conversely Aliquat displays a higher sorption strength than the above mentioned amines when thiocyanate or lithium chloride solutions are used as the eluent[88].

Aliquat columns were mainly prepared using polytrifluorochloroethylene, and hydrophobized kieselguhr or silica gel as the supporting materials. Contrasting information is

120

given about column preparation. According to some authors, Aliquat-treated supports introduced into the column as a dried powder tend to bleed and lose capacity with time, whereas this does not occur when the column is filled with a slurry of the treated support[109, 110]. Conversely, other authors successfully used dry-filled columns with analogous materials as supports for the extractant [100, 111].

An accurate optimization of the chromatographic conditions can give very good performances of Aliquat-treated kieselguhr columns eluted with nitrate solutions: HETP figures as low as 0.35 mm were obtained[111,112]. Column performance was reported to be less affected by flow rate than those of other systems such as HDEHP columns, suggesting fairly fast kinetics of transfer of the adduct between the two phases[111].

Columns were satisfactorily stable to chemical agents, and were used for 15 different runs without any noticeable change in properties[113, 114]. They were also fairly stable to high alpha and neutron irradiation; one column absorbed a total dose of $\sim$ 0.5 watthours without being substantially damaged. The capacity of the above mentioned HETP optimized columns is reported to be 0.15 meq/ml for Cl[111]; remarkably higher capacity figures (2.2 meq/ml for Fe(III)) were reported for other Aliquat-Celite columns[115].

A maximum capacity figure is available also for Kel-F as the supporting material, that was loaded with 0.40 ml of 0.1 M Aliquat in chloroform per ml of bed volume[109].

A thorough investigation on the use of trilaurylmethylammonium nitrate (TLMA·NO$_3$) for the separation of actinides was performed by van Ooyen both by liquid-liquid extraction and reversed-phase chromatography. In a detailed report, simple methods are included for the synthesis and purity assay of the extractant[112]. The synthesis starts from TLA and methyl iodide, the purity assay involves the separation of TLMA·NO$_3$ from TLA·HNO$_3$ by means of an aluminium oxide column, and the subsequent separate titration with perchloric acid of the compounds dissolved in 100% acetic acid.

Columns of TLMA·NO$_3$ were prepared supported on Kel-F or hydrophobized Celite, and their features were optimized mainly for the separation of americium from curium. Very good plate heights were obtained, expecially with the latter support (0.3mm) whose selected fraction actually had smaller particle size than that of Kel-F. On the other hand, Kel-F columns allowed for more symmetrical peaks.

In general, the performance of TLMA·NO$_3$ columns in Am-Cm separations are strictly comparable to those obtained with Aliquat-336. The columns were applied also to milligram scale separations and demonstrated a very good stability to nuclear radiation.

A commercial product fairly similar to TLMA is Adogen-464, a methyl-n-alkylammonium chloride with a minimal quaternary ammonium salt content of 92%, and average molecular weight 431. It is obtainable from Archer-Daniels-Midland; although thoroughly used in laminar chromatography, its use as column stationary phase has never been reported.

2.4. MISCELLANEOUS EXTRACTANTS

In addition to the extractants falling in the already surveyed classes, a number of other stationary phases have been more or less frequently reported, and are collected in this paragraph.

Carbon-bonded oxygen donor extractants, sulphur-containing organophosphorous compounds, and alkylsulphoxides are treated under separate headings. Next, the available information is collected concerning stationary phases of which very few examples are found in the literature, namely nitrobenzene, nitrobenzene solutions of dipicrylamine, benzene, tetraphenylboron in isoamylacetate and 2,2,4-trimethylpentane-1,3-diol in chloroform + benzene solution.

The few examples of synergistic systems as stationary phases is then reported. Finally, reference is made to the so-called "electron-exchange" columns.

2.4.1. CARBON-BONDED OXYGEN DONOR EXTRACTANTS, MIBK

In spite of the great popularity of carbon-bonded oxygen donor compounds in liquid-liquid extraction, relatively few applications of them have been reported in reversed-phase chromatography. Most of the works were performed with methylisobutylketone (MIBK, Hexone) as the stationary phase, but the uses of other ketones (methylhexyl and methylheptyl ketone), ethers (diethyl, dibutyl and diisopropyl ether), esters (iso-amylacetate), alcohols (n-pentanol, decanol, cyclohexanol), and glycols (di-iso-butyl-carbitol) are also reported.

All these compounds are fluid liquids; their viscosity at 25 °C ranges between 0.22 Cp (diethyl ether) and 0.86 Cp (isoamyl acetate), the only exceptions being the slightly less fluid nitrobenzene (1.63 Cp) and cyclohexanol (4.6 Cp).

Extraction takes place through a mechanism essentially similar to that of neutral organophosphorous compounds, extracting metals by solvation of neutral species present in the aqueous solution, and extracting acids by solvation of the proton and ion-association with the anion.

The extraction of acids generally results in great volume changes of the two phases, but the phase that actually increases depends on the nature of the acid considered.

Nitric acid is extracted to a very high extent. For example, 2 M is the acid concentration in MIBK, after equilibration with an aqueous solution originally 6 M in the acid; the extraction results in a ratio of 1.2 of the equilibrium to initial volumes of the organic phase. Analogous behaviours are shown by perchloric acid and hydrofluoric acid.

Hydrochloric acid is extracted by ethers and ketones to a far less extent, its distribution coefficients ranging in the $10^{-3}-10^{-2}$ level. Equilibrium results in a great increase of the aqueous phase, and for high acid concentrations complete miscibility of the two phases is often reached. Hydrobromic, sulphuric and phosphoric acids behave as hydrochloric acid, complete miscibility however being seldom obtained.

The behaviour of esters differs from that of ethers and ketones since appreciable amounts of hydrochloric, sulphuric and phosphoric acids are extracted into the organic phase, especially from concentrated aqueous solutions.

Metal extraction normally takes place by solvation of hydrated metal-bearing species. The extractant molecules apparently are hydrogen bonded to the water molecules of the primary hydration shell of the metal. For a given system, several more or less hydrated or solvated species are believed to exist in equilibrium and a simple picture of the extraction reactions is therefore unfeasable.

Distribution coefficients of metals are strongly dependent on the electron donor properties of the oxygen in the functional group, and therefore are greater in the order esters<ketones<ethers. However, none of the extractants of these classes displays a basicity comparable to that of neutral organophosphorous compounds, and distribution coefficients seldom reach values higher than 10.

For the same class of compounds, basicity apparently decreases with increasing molecular weight. Branching of the alkyl groups also affects extraction, particularly when substitution is near to the active oxygen; steric hindrance effects are probably responsible for this.

In analogy with neutral organophosphorous compounds, high electrolyte concentrations in the aqueous phase promote extraction of metals. When appreciable concentrations are present of the element to be extracted, auto-salting-out often occurs.

Ethers, ketones and esters are particularly suitable for metal extraction from chloride solutions, and their use for this purpose in analytical chemistry goes back to the early nineteens. Protonated metal halocomplexes are strongly extracted, preferentially when single-charged: this accounts for the high distribution coefficients of many divalent or trivalent metal chloride complexes, and for the poor extraction of most metals from fluoride solutions, where high negatively charged complexes are preferentially formed. Essentially covalent metal halides are also appreciably extracted.

Distribution coefficients from nitric acid solutions are generally low. When salts partially substitute the acid, extraction may be enhanced, the effect depending on the nature of the salting-out cation in a way not easily explainable. The distribution coefficients of metals from sulphate or perchlorate solutions is even lower than from nitrate.

Both ether and ketones are especially unstable at high nitric acid loadings, as those deriving from contact even with moderately concentrated nitric acid.

As already mentioned, MIBK is the extractant of this class most frequently used in reversed-phase chromatography. It is a fluid liquid, boiling at 115.9 °C; it is available on the market both at high purity and technical grades. The technical grade commercial product is usually purified by distillation before use.

Its solubility in water is about 17 g/l; solubility is greatly enhanced in concentrated hydrochloric acid, where complete miscibility is reached for about 10 M HCl.

Distributions coefficients for the different metals from hydrochloric or nitric acid are available in the liquid-liquid extraction literature[16,18]. Besides a few exceptions, only metal extraction from chloride solutions involves distribution coefficients sufficiently high to be useful in reversed-phase work.

"Experiments with 2-octanone, 2-undecanone, n-hexyl alcohol, and MIBK show that the latter is the best extractant for iron(III) on Kel-F"[116]. In addition to Kel-F, several other supports have been adopted, and comparisons among the different column features are available: they are summarized in ref. 1. Teflon 6 appears to be one of those that gave the best results, for what both maximum capacity (0.32 ml per ml of column bed) and HETP (∿3.5 mm) are concerned[117]. The other ketones used in chromatography behave like MIBK, the main difference being their lower solubility in the eluent. In some instances, aqueous solutions were pre-equilibrated with the extractant to avoid excessive losses of the stationary phase.

Besides the enrichment and separation of iron, MIBK columns have been successfully applied to other problems, such as zirconium-hafnium separation[118]. Good agreement was found between liquid-liquid extraction and reversed-phase results.

Coupled with thenoyltrifluoroacetone (TTA), MIBK was used to prepare synergistic stationary phases, that are described in Paragraph 2.4.5.

Methylhexylketone (2-octanone) is also available on the market. Supported on Haloport F, it was successfully used for the isolation of iron from chloride solutions[119]. In one instance, Fluoropak was preferred to Haloport as the support because "it does not lump together under slight pressure"[120].

Methylheptylketone was used by Katykhin and coworkers as reported later in this section.

Diethylether supported on polytetrafluoroethylene was advantageously applied to the separation of gold and thallium(III) among themselves, and from a complex mis ure of different elements[121]. All eluting solutions were saturated with the extractant, to avoid appreciable losses of the stationary phase: very good separations are reported. A diethylether stationary phase was also used to separate zirconium from hafnium[122].

Di-isopropylether (IPE) supported on Amberlyst XAD-2, an inert cross-linked polystyrene polymer, was compared to MIBK-XAD-2 as stationary phase for the separation of gallium, indium and thallium(III) from hydrobromic acid solutions[123]. Very good separations were obtained, but the lower solubility of IPE in such aqueous media was not a sufficient advantage to prefer it to MIBK, that proved to display a higher extraction power. Nevertheless, IPE columns, together with MIBK ones, were advantageously included in a separation scheme for the analysis of multicomponent samples[124].

Several oxygen-containing extractants were screened by Katykhin and coll. for the separation of metals of the palladium and platinum groups; dibutylether, methylheptylketone and isoamylacetate were chosen for column chromatography experiments[125]. The stationary phases were supported on polytetrafluoroethylene (Ftoroplast-4), and gave satisfactory results.

On the other hand, other authors report that isoamylacetate-polytetrafluoroethylene columns were found to be ineffective in absorbing even those elements whose liquid-liquid distribution coefficients are considerably greater than unity. Similar negative features were found also for n-pentanol and were explained with a particularly disadvantageous volume ratio between organic and aqueous phases for such compounds, known for their great mutual solubility with water[5].

Di-iso-butylcarbitol (di-iso-butyldiethylene glycol) supported on politrifluorochloroethylene (Hostaflon) was successfully applied for rapid radiochemical separations[126]. Very good decontamination factors were obtained.

Good results were also reported for columns having cyclohexanol[127] and decanol [128] as the stationary phases.

2.4.2. TRI-I-OCTYLTHIOPHOSPHATE

Organic thiophosphorous compounds are not as popular extractants as the analogous oxygen-bearing compounds; however, both acidic dialkylphosphorodithioates, and neutral trialkylthiophosphates and trialkylphosphine sulphides have found several interesting applications in analytical chemistry, because of their peculiar selectivity for some metals.

A number of such compounds were also considered in reversed-phase paper chromatography, but only tri-i-octylthiophosphate (TiOTP) has been advantageously applied as a column stationary phase.

Extraction by trialkylthiophosphates is believed to take place through the very same mechanism suggested for the analogous organic phosphoryl compounds such as TBP. However, these extractants display a remarkably lower basicity as compared to that of compounds containing the $P{\rightarrow}O$ group, and are able to coordinate with "class b" acceptors, in contrast with phosphates that chiefly donate to "class a" or border region acceptors. The result of this difference is the highly selective extraction of silver and mercury(II) by TiOTP from nitric acid solutions[129].

TiOTP is available on the American market; shipping regulations, however, make it difficult to obtain it in overseas countries. It was used in column chromatography after a purification step, since the commercial product was found to contain appreciable amounts of acidic substances; purification was accomplished by diluting the reagent approximately four-fold with carbon tetrachloride, and passing the resulting solution through a column of alumina[130].

As all thiophosphates, TiOTP hydrolyses when contacted with 6–8 M nitric acid. "However, during a 5 to 10 minutes extraction period the extraction properties of TiOTP are not appreciably affected by the aqueous nitric acid, but longer contact periods should be avoided..... It is probable that the hydrolysis products prevent quantitative back extraction into the aqueous phase. Back extraction is accomplished by washing the organic phase with water, aqueous ammonia or hydrochloric acid"[129].

TiOTP columns were prepared using microporous polyethylene or Fluoropak as the supports, to avoid possible large adsorption of silver by inorganic materials such as kieselguhr. Elution peaks for Ag(I) were extremely broad and tailed at 25 °C. "At 60 °C, however, the elution bands were considerably sharper with both supports, presumably because of increased rates of diffusion of the Ag(I) species"[130]. Very good separations were obtained of micro and macro-amounts of silver from a large number of other elements.

Mercaptans are extractants that could display selective features analogous to thiophosphorous compounds. However, the only report on their use in reversed-phase chromatography refers to a t-hexadecylmercaptan-polystyrene column, unsuccessfully tested for copper extraction[131].

2.4.3. DI-N-OCTYLSULPHOXIDE

High molecular weight sulphoxides found little attention in solvent extraction work, in spite of the popularity as a solvent of their lowest homologue dimethylsulphoxide. The promising behaviour of several different sulphoxides was systematically investigated by reversed-phase paper chromatography[132], but only di-n-octylsuphoxide (DOSO) was actually used for column separations.

DOSO can be prepared by oxidizing at room temperature di-n-octylsulphide with a dilute solution of hydrogen peroxide in glacial acetic acid. After 2 hours, a large amount of water is added, and the solid sulphoxide is filtered off, washed with water, air-dried

125

and recrystallized from 95% ethanol. Melting point of the purified product was 71.5–72.5 °C.

The extraction by DOSO takes place with a mechanism very similar to that of neutral organophosphorous compounds. In hydrochloric acid media, only tin(IV) and antimony(V) are appreciably extracted from relatively dilute acid solutions, while at the highest acid concentrations several metals, such as molybdenum, titanium, zirconium, hafnium, uranium, iron, zinc, cadmium and mercury are also extracted.

In nitric acid systems, extraction takes place for fewer metals (uranium(VI), thorium, titanium, zirconium, hafnium and few others) while only uranium(VI) and gold are appreciably extracted from perchloric acid.

DOSO columns were prepared by supporting the extractant, dissolved in 1,2-dichloroethane, on a macroreticular polystyrene resin, XAD-2[133]. Eluting solutions were previously equilibrated with the organic phase. Chromatographic results were in good agreement with liquid-liquid extraction data, and satisfactory separations were obtained of uranium(VI) from several other metals.

2.4.4. OTHER EXTRACTANTS

Nitrobenzene has been used as the stationary phase for the separation of alkali metals as polyiodides.[134]. The column was prepared by loading Kel-F powder with the solvent suitably equilibrated with the iodine-iodide aqueous solution subsequently used as the eluent. Inorganic materials, such as alumina and silicagel, were found to be useless as supports, as they failed to retain any nitrobenzene even after being treated with dimethyldichlorosilane.

Eluting solutions were saturated with the organic solvent and volumes of eluent as large as 3 litres had little effect on peak widths and positions obtainable from a given column. The maximum capacity of the columns was 0.34 ml of extractant per ml of column bed; HETP ranged from 1.7 to 17 mm, linearly increasing as the amount of retained element was increased.

The behaviour of the nitrobenzene as the stationary phase proved to be in perfect agreement with previously obtained liquid-liquid extraction data[135]: retention obviously takes place because of simple physical partition of alkali polyiodided between the two phases. Good separations were performed of micro- and macro-amounts of alkali metals among themselves.

Nitrobenzene supported on Kel-F was again used together with di-picrylamine as a stationary phase to separate ^{137}Cs from the other fission products[136]. A quantitative and reproducible separation was obtained, with good (up to 10^3) decontamination factors. The column could be used for more than fifteen cycles without loosing its properties.

Several other supports were evaluated for a nitrobenzene-ammonium dipicrylaminate stationary phase[137]. Hydrophobized silica gel was found to greatly adsorb cesium by itself, no extractant being loaded, while siliconized kieselguhr (Chromosorb) gave unsatisfactory peaks on elution. Polytetrafluoroethylene (Teflon or Ftoroplast) and perfluorochloroethylene (Teflex) appeared to be suitable materials. Alkali metal separations were performed with such columns, also by frontal chromatography. Columns appeared to keep their properties after ten or more elution cycles.

Benzene supported on microporous polyethylene was successfully used to retain iodine from sulphuric acid[138]. Iodine was recovered by elution with ascorbic acid, and the overall yield of the process was about 90%. Carbon tetrachloride or carbon disulphide were also used for the same purpose, supported on Kel-F or other materials[139].

A column of sodium tetraphenylboron and isoamylacetate loaded on Kel-F was also used to separate ^{137}Cs from fission products[142]. Again an accurate and quantitative separation of cesium was obtained, but the column could be used only for one cycle, because of the appreciable solubility of the tetraphenylboron reagent in aqueous solutions.

The ability of 1,3-diols to form complexes with boric acid was exploited to obtain an interesting separation and/or enrichment of boric acid by extraction chromatography. Among several different 1,3-diols, 2,2,4-trimethyl-pentane-1,3-diol was found to give the most advantageous extraction performances, and columns were prepared treating microporous polyethylene (Microthene-710) with a 50 per cent. w/v of the diol (solid at room temperature) dissolved in a chloroform + benzene (1:1) mixture[141]. Milligram amounts of boric acid were quantitatively retained from 0.5 M NaCl + 0.01 M HCl, and recovered with alkaline solutions. No data are available on the stability of such columns.

2.4.5. SYNERGISTIC SYSTEMS

Under well-defined experimental conditions, the distribution coefficients for a number of metals, as obtained with certain combination of two extractants, are greater than one could expect from the distribution coefficients obtainable with each of the two extractants, used separately.

This phenomenon, called synergism, is believed to take place mainly because of at least one of two fundamental factors, namely the enhancement of the activity of the extractants in the organic phase (due to various kinds of possible interactions between the two extractants), and the existence in the organic phase of metal-bearing species having peculiar compositions, different from those present in the case of one extractant alone. In any case, the synergistic effect depends on the relative amounts of the two extractants, maximum extraction taking place in a short range around well-defined relative compositions of the organic phase.

A great number of synergistic systems have been found and investigated, and at present they are one of the topics most often dealt with in the solvent extraction literature. On the contrary, only a few examples of synergistic stationary phases can be found in reverse-phase chromatography, probably because they usually imply a general enhancement of the distribution coefficients of metals, but seldom give rise to attractive separation factors.

The most popular synergism is that shown by the combination of an acidic chelating compound with a neutral extractant, either a phosphorous-bonded or a carbon-bonded oxygen donor extractant. In this case, the enhancement of distribution coefficients appears to take place because of the formation of mixed complexes of the metal with the two ligands. When chelation does not saturate the coordination shell of the metal, the donor extractant may substitute the hydration water of the chelate, thus accounting for extraction of the mixed complex. When the metal is coordinatively saturated by the attached chelating anions alone, things get more complicated, and several different mechanisms have been suggested. In all cases, the less hydrophilic nature of the resulting mixed-complex accounts for its greater distribution into the organic phase.

Synergism in these systems is not a specific feature of certain particular metals, but occurs with all elements that form extractable complexes with bidentate chelating agents. The enhancement of extraction appears to be greater the greater the basicity of the neutral donor extractant (at least for organophosphorous compounds), and the weaker the chelating power of the bidentate compounds. Steric effects may also play an important role.

Akaza first prepared a synergistic stationary phase, made with a mixture of 2-thenoyl-trifluoroacetone (TTA) and methylisobutylketone (MIBK) supported on Kel-F, and successfully applied it to the separation of alkaline earths[142]. The organic mixture was previously equilibrated with a solution having the same composition of the planned eluent, before being loaded on the support. The eluting solutions, on their side, contained suitable dissolved amounts of both extractants. Temperature was controlled during the whole process, to avoid variation of the solubility of MIBK, and hence of that of TTA, in the aqueous solutions. Columns appeared to remain capable of satisfactory use after four eluting processes in a period of a week, provided that temperature was kept below 12 °C and light was excluded.

HETP were not particularly satisfactory, but sufficient to obtain acceptable separations. A satisfactory agreement with corresponding liquid-liquid and batch extraction data was found.

Aly and coll. reported the use of a stationary phase made with TTA and dibutyl diethylcarbamoyl phosphonate, supported on Celite[143,144]. A synergistic effect was apparent in the behaviour toward europium and samarium, their separation factor being also affected in a positive way. Optimization of the chromatographic parameters gave HETP lower than 0.2 mm, and separations were reported of the above-mentioned rare earths, and of curium from californium.

The combined effects of dibenzoylmethane (DBM) and TOPO (or TBP) allowed Lee to selectively extract lithium into the organic phase and to separate it from the other alkali metals on columns made with polytetrafluoroethylene (Haloport F) loaded with suitable mixtures of the two compounds[145]. Tailing effects where minimized keeping flow-rate relatively low when recovering lithium from the column.

Columns were stable to at least twenty-five runs. On a DBM—TOPO column, lithium isotopes were fractionated by the same author[146]. The single stage separation factor for ^6Li and ^7Li was determined to be 1.0024. Lithium-6 concentrated in the aqueous phase.

To the author's knowledge, no additional examples are available of reversed-phase chromatographic works where synergistic effects are recognizable, besides the studies of the system TTA-tri-n-octylphosphate[147] and TTA—TOPO[148, 149], made by Cvjetičanin with paper chromatography.

A mixture of tri(2-ethylhexyl) phosphate and HDEHP, supported on Plaskon and used for a specific analytical problem[150], did not exhibit those features which are peculiar of synergistic systems. Nothing can be said about the already mentioned columns made with tetrabutyl pyrophosphate and tetrabutyl hypophosphate, since no comparable data with the separate extractants are available.

2.4.6. ELECTRON-EXCHANGE COLUMNS

A rather peculiar and very interesting application of the reversed-phase extraction chromatography principles are the electron-exchange columns, made with suitable organic redox substances supported on inert materials.

A solid insoluble substance capable of oxidizing or reducing molecules or ions in an adjacent aqueous phase, without any addition of reagents to the aqueous solution, obviously represents a very interesting analytical tool.

Cerrai and Testa prepared columns of tetrachlorohydroquinone supported on Kel-F, following the very same technique used for the preparation of normal extraction chromatography stationary phases[151]. Tetrachlorohydroquinone (chloranil) is a crystalline solid, available on the market. It was chosen among five different insoluble redox organic compounds, being the only one that proved to undergo reversible reactions on columns at acceptable elution rates.

The obtained columns had good capacity for the organic compound ($\sim$1.7 meq/ml)and were satisfactorily stable, loosing no more than 2% of their capacity after each single oxidation or reduction cycle.

Owing to the sufficiently high redox potential of the tetrachlorohydroquinone-tetrachloroquinone couple and to its dependency on the hydrogen ion concentration in the aqueous solution, several selective oxidations or reductions were obtained, such as $Fe^{2+} \rightarrow Fe^{3+}$, $Cu^{1+} \rightarrow Cu^{2+}$, $Sn^{2+} \rightarrow Sn^{4+}$, $I^- \rightarrow I_2$, ascorbic acid $\rightarrow$ dehydroascorbic acid, and $Fe^{3+} \rightarrow Fe^{2+}$, $I_2 \rightarrow I^-$, $Ce^{4+} \rightarrow Ce^{3+}$, $V^{5+} \rightarrow V^{4+}$, $Cr^{6+} \rightarrow Cr^{3+}$.

Tetrachlorohydroquinone is white, while its oxidized form is yellow: therefore, it was possible to easily judge the extent of reaction within the columns. Columns were run in acidic or neutral media, while their use with strongly alkaline solutions was avoided, because of possible remarkable losses of the stationary phase due to some salt formation of the weakly acidic phenolic groups.

Several interesting applications were possible; in particular the use of the "memory" of the column enabled indirect determinations,so preserving the uncontaminated effluent for other purposes. The precision and accuracy of such indirect titrations was checked by Belcher and coll. and the results obtained for iron(III) reduced to iron(II) were very precise but 1% lower than expected. "These slightly low results may be due to tetrachloroquinone being slightly more soluble than the hydroquinone; possibly the effect could be overcome by incorporating a short section of untreated substrate at the bottom of the column"[12].

Belcher's columns were analogous to the above mentioned ones, but were somewhat more stable, loosing no more than $0.1-1.2\%$ of capacity per oxidation or reduction stage, after an initial $4-10\%$ loss. Alternative redox agents were examined, but they were not retained on the column, with the exception of anthraquinone, whose very low standard redox potential (0.15V), however limited its application. Different supports were also tested, and inorganic materials such as kieselguhr and Celite behaved as well as organic polymers, the capacity losses per single cycle being 0.2–0.35%.

Tetrachlorohydroquinone column were recently prepared using open-cell polyurethane foam as the supporting material[152]. Before use, "pure" grade reagent was recrystallized from chlorobenzene or by sublimation under vacuum. A negative effect of high column loadings was noticed, presumably imputable to larger crystals of the compound, and subsequent less rapid reactions. A slow reaction rate with vanadium(V) was reported, that was remarkably accelerated by performing the elution at 35 °C, where quantitative reductions of vanadium(V) were obtained even at 4 cm·min^{-1} linear flow rate. Columns could be used for over 30 reduction cycles without loosing their efficiency.

Finally, a "redox-extraction" column was very recently reported, where 2,3-dichloro-1,4-naphthahydroquinone and TOPO were loaded together on microporous polyethylene [153]: making profit of both the reducing and extracting power of the column, interesting quantitative separations were performed of several transuranium elements. The stability of this column, however, is not very satisfactory since the reducing agent easily leaks away with the eluent.

REFERENCES TO CHAPTER 4

1. E. CERRAI, G. GHERSINI, in J. C. GIDDINGS, R. A. KELLER, Eds., Advances in Chromatography, Marcel Dekker, Inc., New York, Vol. 9, p. 3, 1970.
2. H. BERANOVA, M. NOVAK, Coll. Czech. Chem. Comm., 30 (1965) 1073.
3. R. F. HORNBECK, J. Chromatog., 30 (1967) 447.
4. E. CERRAI, G. GHERSINI, J. Chromatog., 13 (1964) 211.
5. L. N. MOSKVIN, L. I. SHUR, L. G. TSARITSYNA, G. T. MARTYSH, Radiokhimiya, 9 (1967) 377.
6. L. S. BARK, G. DUNCAN, J. Chromatog., 49 (1970) 278.
7. M. ISHIBASHI, H. KOMAKI, M. DEMIZU, Mitsubishi Denki Giho, 39 (1965) 907.
8. B. TOMAZIČ, S. SIEKIERSKI J. Chromatog., 21 (1966) 98.
9. B. TOMAZIČ, Anal. Chim. Acta, 49 (1970) 57.
10. E. CERRAI, G. GHERSINI, J. Chromatog., 16 (1964) 258.
11. H. GROSSE–RUYKEN, J. BOSHOLM, J. Prakt. Chemie, 25 (1964) 79.
12. R. BELCHER, R. MAJER, G. A. H. ROBERTS, Talanta, 14 (1967) 1245.
13. G. S. KATYKHIN, Zh. Anal. Khim., 27 (1972) 849.
14. G. PHILLIPS, G. W. MILNER, in P. W. SHALLIS, Ed., Proc. SAC Conference Nottingham 1965, W. Heffer & Sons Ltd., Cambridge, p. 240, 1965.
15. E. P. HORWITZ, C. A. A. BLOOMQUIST, J. Inorg. Nucl. Chem., 35 (1973) 271.
16. Y. MARCUS, A. S. KERTES, Ion Exchange and Solvent Extraction of Metal Complexes, Wiley-Interscience, London, 1969.
17. A. K. DE, S. M. KHOPKAR, R. A. CHALMERS, Solvent Extraction of Metals, Van Norstrand Reinhold Company, Ltd., London, 1970.
18. P. MARKL, Extraktion und Extraktions-Chromatographie in der Anorganischen Analytik, Akademische Verlagsgesellshaft, Frankfurt am Main, 1972.
19. E. GLUECKAUF, Trans. Faraday Soc., 51 (1955) 34.
20. I. H. QURESHI, L. T. Mc CLENDON, P. D. La FLEUR, 156th ACS National Meeting, paper 156, New York, 1968.
21. I. H. QURESHI, L. T. Mc CLENDON, P. D. La FLEUR, Radiochimica Acta, 12 (1969) 107.
22. T. SATO, M. UEDA, J. Inorg. Nucl. Chem., 35 (1973) 1003.

23. A. I. MIKHAILICHENKO, R. M. PUMENOVA, R. V. KOTLIAZOV, Zh. Prikl. Khim., 42 (1969) 1262.
24. T. V. HEALY, J. KENNEDY, J. Inorg. Nucl. Chem., 10 (1959) 128.
25. C. J. HARDY, J. Inorg. Nucl. Chem., 21 (1961) 348.
26. D. F. PEPPARD, G.W.MASON, R.J. SIRONEN, J. Inorg. Nucl. Chem., 10 (1959) 117.
27. J. HALA, J. Inorg. Nucl. Chem., 30 (1968) 1334.
28. C. J. HARDY, D. C. SCARGILL, J. Inorg. Nucl. Chem., 10 (1959) 323.
29. D. C. STEWART, H. W. CRANDALL, J. Am. Chem. Soc., 73 (1951) 1377.
30. D. F. PEPPARD, G. W. MASON, J. L. MAIER, W. J. DRISCOLL, J. Inorg. Nucl. Chem., 4 (1957) 334.
31. J. A. PARTRIDGE, R. C. JENSEN, J. Inorg. Nucl. Chem., 31 (1969) 2587.
32. L. N. MOSKVIN, B. K. PREOBRAZHENSKII, in Radiochemical Methods for the Determination of Microelements, Izdatel'stvo Nauka, Moscow-Leningrad, p. 85, 1965.
33. T. B. PIERCE, R. S. HOBBS, J. Chromatog., 12 (1963) 74.
34. K. KIMURA, Bull. Chem. Soc. Japan, 34 (1960) 63.
35. R. DENIG, N. TRAUTMANN, G. HERRMANN, J. Radioanal. Chem., 5 (1970) 223.
36. E. P. HORWITZ, C. A. A. BLOOMQUIST, J. Inorg. Nucl. Chem., 34 (1972) 3851.
37. E. CERRAI, G. GHERSINI, in P. W. SHALLIS, Ed., Proceedings SAC Conference Nottingham 1965, W. Heffer & Sons Ltd., Cambridge, p. 462, 1965.
38. E. CERRAI, G. GHERSINI, R. TRUCCO, Proc. Int. Symposium on Analytical Chemistry, paper 58, Birmingham, 1969.
39. R. J. SOCHACKA, S. SIEKIERSKI, J. Chromatog., 16 (1964) 396.
40. J. KOOI, J. M. GANDOLFO, N. WÄTCHER, R. BODEN, R. HECQ, E. van HOOF, M. LEYNEN, Euratom Rept. EUR 2578. e, 1965.
41. R. F. OVERMAN, Anal. Chem., 43 (1971) 600.
42. B. KOTLINSKA–FILIPEK, S. SIEKIERSKI, Nukleonika, 8 (1963) 607.
43. B. WEAVER, J. Inorg. Nucl. Chem., 30 (1968) 2233.
44. I. FIDELIS, S. SIEKIERSKI, J. Chromatog., 17 (1965) 542.
45. D. F. PEPPARD, J. R. FERRARO, G. W. MASON, J. Inorg. Nucl. Chem., 12 (1959) 60.
46. S. J. LYLE, V. C. NAIR, Talanta, 16 (1969) 813.
47. W. SMUŁEK, S. SIEKIERSKI, J. Chromatog., 19 (1965) 580.
48. P. R. DANESI, R.CHIARIZIA, G. SCIBONA, J. Inorg. Nucl. Chem., 35 (1973) 3926.
49. G. WERNER, J. Chromatog., 22 (1966) 400.
50. R. CHIARIZIA, P. R. DANESI, Private communication.
51. T. ISHIMORI, E. NAKAMURA, Japan At. Energy Comm. Rept. JAERI 1047, 1963.
52. M. N. SASTRI, A.P.RAO, Z. Anal. Chem., 196 (1963) 166.
53. A. P. RAO, M. N. SASTRI, Z. Anal. Chem., 207 (1965) 409.
54. W. J. ROSS, J. C. WHITE, Anal. Chem., 36 (1964) 1998.
55. R. F. HORNBECK, USAEC Rept. UCRL – 70151, Part I, 1966; J. Chromatog., 30 (1967) 438.
56. J. RAIS, J. KŘTIL, V. CHOTIVKA, Talanta, 18 (1971) 213.
57. U. A. Th. BRINKMAN, H. VELTKAMP, J. Chromatog., 24 (1966) 489.
58. L. S. BARK, G. DUNCAN, R. J. T. GRAHAM, Analyst, 92 (1967) 347.
59. S. SIEKIERSKI, B. KOTLINSKA, Atomnaya Energ., 7 (1959) 160.
60. E. CERRAI, C. CESARANO, F. GADDA, Energia Nucleare, 4 (1957) 405.
61. T. L. DAWSON, A. R. LOWE, AERE Rept. RDBW-8048, 1953.
62. T. ISHIMORI, E. AKATSU, K. TSUKUECHI, T. KOBUNE, J. USUBA, K. KIMURA, G. ONAWA, H. UCHIYAMA, Rept. JAERI–1106,1966.
63. S. SPECHT, R. F. NOLTE, H. J. BORN, 7th Radiochemical Conference, Marianske Lazne, 1973.
64. S. SIEKIERSKI, I. FIDELIS, J. Chromatog., 4 (1960) 60.
65. A. G. HAMLIN, B. J. ROBERTS, W. LAUGHLIN, S. G. WALKER, Anal. Chem., 33 (1961) 1547.
66. A. G. HAMLIN, B. J. ROBERTS, Nature, 185 (1960) 527.
67. E. R. SCHMID, Monatsch. Chem., 101 (1970) 1856.
68. C. POHLANDT, T. W. STEELE, Talanta, 19 (1972) 839.
69. S. LIS, E. T. JÓZEFOWICS, S. SIEKIERSKI, J. Inorg. Nucl. Chem., 28 (1966) 199.

70. S. HÜBENER, A. HERMAN, 7th Radiochemical Conference, Marianske Lazne, 1973.
71. J. W. O'LAUGHLIN, C. V. BANKS, Anal. Chem., 36 (1964) 1222.
72. H. ESCHRICH, P. HANSEN, Eurochemic Rept. ETR 239, 1969.
73. T. J. HAYES, A. G. HAMLIN, Analyst, 87 (1962) 770.
74. I. ALIMARIN, T. A. BOLSHOVA, T. G. DOLZHENKO, G. M. KANDIBYNA, Zh. Analit. Khim., 26 (1971) 741.
75. B. MARTIN, D. W. OCKENCHEN, J. K. FOREMAN, J. Inorg. Nucl. Chem., 21 (1961) 96.
76. E. CERRAI, C. TESTA, J. Chromatog., 9 (1962) 216.
77. U. A. Th. BRINKMAN, G. de VRIES, E. van DALEN, J. Chromatog., 22 (1966) 407.
78. USAEC Rept. ORNL–3627, p. 182, 1964.
79. J. W. O'LAUGHLIN, in D. C. STEWART, H. A. ELION, Eds., Progress in Nuclear Energy, Series IX, Analytical Chemistry, Pergamon Press, Oxford, Vol. 6, p. 97, 1966.
80. J. J. RICHARDS, K. E. BURKE, J. W. O'LAUGHLIN, C. V. BANKS, J. Am. Chem. Soc., 83 (1961) 224.
81. J. W. O'LAUGHLIN, J. W. FERGUSON, J. J. RICHARDS, C. V. BANKS, J. Chromatog., 24 (1966) 376.
82. J. W. O'LAUGHLIN, C. V. BANKS, in D. DYRSSEN, J.-O. LILJENZIN, J. RYDBERG, Eds., Solvent Extraction Chemistry, North Holland Publ. Co., Amsterdam, p. 270, 1967.
83. C. V. BANKS, J. W. O'LAUGHLIN, J. W. FERGUSON, G. J. KAMIN, J. J. RICHARD, R. K. HANSEN, D. F. JENSEN, USAEC Rept. IS–1500, C–49, 1966.
84. J. W. O'LAUGHLIN, R. K. HANSEN, R. J. HOFER, R. Z. BACHMAN, C. V. BANKS, Symposium on Trace Characterization, Chemical and Physical, Gaithersburg, Md., Oct. 1966.
85. J. MIKULSKI, K. A. GAVRILOV, V. KNOBLOCH, Nukleonika, 10 (1965) 81.
86. J. MIKULSKI, I. STRONSKI, Nature, 207 (1965) 749.
87. Yu. S. KOROTKIN, JINR–P6–6401, 1972.
88. U. A. Th. BRINKMAN, in E. S. PERRY, C. J. VAN OSS, Eds., Progress in Separation and Purification, Wiley-Interscience N. Y. Vol. 4. p. 241, 1971.
89. F. BARONCELLI, G. M. GASPARINI, G. GROSSI, in A. S. KERTES, Y. MARCUS, Eds., Solvent Extraction Research, Wiley-Interscience, New York, p. 401, 1969.
90. T. ISHIMORI, E. AKATSU, W. CHENG, K. TSUKUECHI, T. OSAKABE, JAERI–1062, 1964.
91. F. BONFANTI, E. CERRAI, G. GHERSINI, Energia Nucleare, 14 (1967) 578.
92. C. F. COLEMAN, K. B. BROWN, J. G. MOORE, D. J. CROUSE, Ind. Engng. Chem., 50 (1958) 1756.
93. M. LESZKO, W. KARCZ, Zesz. Nauk Univ. Jagellon., Pr. Chem., 15 (1970) 183.
94. I. STRONSKI, Radiochem. Radioanal. Letters, 5 (1970) 113.
95. D. JENTZSCH, G. OESTERHELT, E. RÖDEL, H. G. ZIMMERMANN, Z. Anal. Chem., 205 (1964) 237.
96. W. SMUŁEK, Croatica Chem. Acta, 35 (1963) A12.
97. E. CERRAI, C. TESTA, J. Chromatog., 6 (1961) 443.
98. C. E. ESPAÑOL, A. M. MARAFUSCHI, J. Chromatog., 29 (1967) 311.
99. R. KREFELD, G. ROSSI, Z. HAINSKI, Microchim. Icnoanal. Acta, 1 (1965) 133.
100. W. SMUŁEK, K. ZELENAY, Proc. Anal. Conf. Application Physico-Chemical Methods in Chem. Analysis, Budapest, Vol. 2, p. 47, 1966.
101. E. K. HULET, J. Inorg. Nucl. Chem., 26 (1964) 1721.
102. T. B. PIERCE, W. M. HENRY, J. Chromatog., 23 (1966) 457.
103. K. WATANABE, 20th Conf. Chem. Soc. Japan, 1967.
104. E. G. CHUDINOV, S. V. PIROZHKOV, J. Radioanal. Chem., 10 (1972) 41.
105. M. PETIT-BROMET, CEA Rept. R–3469, 1968.
106. D. GOURISSE, A. CHESNÉ, Anal. Chim. Acta, 45 (1969) 311.
107. E. E. KAMINSKI, J. S. FRITZ, J. Chromatog., 53 (1970) 345.
108. USAEC Rept. ORNL–3314, p. 110, 1962.
109. L. RAMALEY, W. A. HOLCOMBE, Anal. Letters, 1 (1967) 143.
110. J. van OOYEN, in D. DYRSSEN, J.–O. LILJENZIN, J. RYDBERG, Eds., Solvent Extraction Chemistry, North Holland Publ. Co., Amsterdam, p. 485, 1967.

111. E. P. HORWITZ, C. A. A. BLOOMQUIST, K. A. ORLANDINI, D. J. HENDERSON, Radiochim. Acta, 8 (1967) 127.
112. J. van OOYEN, Reactor Centrum Nederland Rept. CRN–113, 1970.
113. E. P. HORWITZ, K. A. ORLANDINI, C. A. A. BLOOMQUIST, Inorg. Nucl. Chem. Letters, 2 (1966) 87.
114. E. A. HUFF, J. Chromatog., 27 (1967) 229.
115. M. El–GARHY,H.F.ALY, S. El–REEFY, Isotopenpraxis, 6 (1970) 478.
116. J. S. FRITZ, C. R. HEDRICK, Anal. Chem., 36 (1964) 1324.
117. J. S. FRITZ, G. L. LATWESEN, Talanta, 14 (1967) 529.
118. J. S. FRITZ, R. T. FRAZEE, Anal. Chem., 37 (1965) 1358.
119. J. S. FRITZ, C. E. HEDRICK, Anal. Chem., 34 (1962) 1411.
120. M. A.WADE, S. S. YAMAMURA, Anal. Chem., 36 (1964) 1861.
121. B. K. PREOBRAZHENSKII, A. V. KALYAMIN, I. MIKHALCA, Radiokhimiya, 6 (1964) 111.
122. A. GORSKI, J. MOSZCYNSKA, Chemia Analit., 9 (1964) 1071.
123. J. S. FRITZ, R. T. FRAZEE, G. L. LATWESEN, Talanta, 17 (1970) 857.
124. J. S. FRITZ, G. L. LATWESEN, Talanta, 17 (1970) 81.
125. G. S. KATYKHIN, M. K. NIKITIN, V. P. SERGEEV, S. K. KALININ, Primenienie ekstraktii i ionogo obmena v analize blagorodnykh metallov, Tsvetmetinformatsiya, p. 39, Moscow, 1967.
126. R. DENIG, N. TRAUTMANN, G. HERRMANN, Z. Anal. Chem., 216 (1966) 41.
127. D. S. DZHELEPOV, G. S. KATYKHIN, K. K. MAIDENYUK, A. I. FEOKTISTOF, Izvest. Akad. Nauk SSSR, Ser. Fiz., 26 (1962) 1030.
128. N. M. KUZMIN, E. V. EMELYANOV, S. V. MESHANKINA, V. L. SABATOVSKAYA, I. A. KUZOLEV, T. I. ZAKHAROV, Zh. Anal. Khim., 26 (1971) 282.
129. T. H. HANDLEY, Talanta, 12 (1965) 893.
130. F. NELSON, J. Chromatog., 20 (1965) 378.
131. A. L. CLINGMAN, J. R. PARRISH, J. Appl. Chem., 13 (1963) 193.
132. D. C. KENNEDY, J. S. FRITZ, Talanta, 17 (1970) 823.
133. J. S. FRITZ, D. C. KENNEDY, Talanta, 17 (1970) 837.
134. I. AKAZA, Bull. Chem. Soc. Japan, 39 (1966) 585.
135. I. AKAZA, Bull. Chem. Soc. Japan, 39 (1966) 465.
136. C. TESTA, C. CESARANO, J. Chromatog., 19 (1965) 594.
137. M. KÝRS, L. KADLECOVA, J. Radioanal. Chem., 1 (1968) 103.
138. C. TESTA, Anal. Chim. Acta, 50 (1970) 447.
139. C. TESTA, Health Physics, 12 (1966) 1768.
140. C. CESARANO, G. PUGNETTI, C. TESTA, J. Chromatog., 19 (1965) 589.
141. C. TESTA, L. STACCIOLI, Analyst, 97 (1972) 527.
142. I. AKAZA, Bull. Chem. Soc. Japan, 39 (1966) 980.
143. H. F. ALY, M. A. El–HAGGAN, Radiochem. Radioanal. Letters, 3 (1970) 249.
144. H. F. ALY, M. RAIEH, Anal. Chim. Acta, 54 (1971) 171.
145. D. A. LEE, J. Chromatog., 26 (1967) 342.
146. D. A. LEE, USAEC Rept. ORNL–4164, p. 24, 1967.
147. N. CVJETIĆANIN, J. Chromatog., 34 (1968) 520.
148. N. CVJETIĆANIN, Int. Symp. VI Chromatographie et Electrophorèse, Bruxelles, 1970, Presses Académique Européennes, Bruxelles, p. 82, 1971.
149. N. CVJETIĆANIN, J. Chromatog., 74 (1972) 99.
150. E. A. HUFF, S. J. KULPA, Anal. Chem., 38 (1966) 939.
151. E. CERRAI, C. TESTA, Anal. Chim. Acta, 28 (1963) 205.
152. T. BRAUN, A. B. FARAG, A. KLIMES-SZMIK, Anal. Chim. Acta, 64 (1973) 71.
153. A. DELLE SITE, C. TESTA, 7th Radiochemical Conference, Marianske Lazne, 1973.

CHAPTER 5

INERT SUPPORTS IN COLUMN EXTRACTION CHROMATOGRAPHY

G. S. KATYKHIN

1. REQUIREMENTS FOR SUPPORTS IN PARTITION CHROMATOGRAPHY

The supports are of great importance in partition chromatography. Their appointment is to retain the stationary phase distributed as a thin film in order to accelerate the achievement of the equilibrium state between the aqueous and organic phases.

An ideal support has to meet the following requirements:

1. To display good wettability by the stationary phase and to retain it in sufficient amounts. The fixed phase must not tear off the support with the flow of the mobile phase

134

2. To be chemically inert: it must not dissolve or swell in the stationary phase, nor dissolve or react with the mobile phase; it must not adsorb the components of the mixture to be separated.

3. To consist of particles as identical as possible (spherical ones are the best), which allow the most uniform and reproducible column packing.

4. To have a large enough surface to retain the stationary phase as a thin, even and uniform film. Porous supports generally meet this requirement, but the pore distribution has to be within a narrow range of sizes since the effect of different pore sizes is equivalent to that of different particle diameters. Too narrow pores must not be present, as the equilibrium with the liquid retained in narrow and deep pores may be very slow, leading to an additional broadening of the chromatographic peaks.

5. To allow for columns having an acceptable pressure drop as regards the mobile phase.

6. To have sufficient mechanical stability; it must not grind during column packing, impregnation of the extractant or regeneration of the support.

7. When applied for routine analyses or for preparative purposes, it must be relatively cheap or permit regeneration.

There is no ideal support for partition chromatography, so one must choose within a wide range of substances which partially meet the above-mentioned requirements. It can be noted that there is not such a strong demand in connection with the quality of supports in the extraction chromatography of inorganic substances as in the gas-liquid or liquid-liquid partition chromatography of organic substances. The reason for this is that organic partition chromatography is usually applied to separate substances with only slightly different distribution coefficients, so that all properties of the support which influence the broadening of the chromatographic peaks (i. e. the plate height), are of great significance in the separation. Conversely, extraction chromatography of inorganic substances only seldom involves the separation of elements with very similar properties. In most cases, the separation conditions can be selected so that the difference in distribution coefficients is sufficient for separation on the principle of the "sorption filter", where one element passes through the chromatographic column without any absorption while the second is strongly held on the column. For this kind of separation, one can use any support which will hold a sufficient amount of the appropriate extractant.

Notable exceptions to the above statement are the separation of the rare earths, of Am from Cm, and of isotopes of a given element (^{6}Li–^{7}Li).

2. MATERIALS USED FOR SUPPORTS

Various powdered materials are used as the supports for extraction chromatography, being specially produced for other chromatographic techniques (supports and adsorbents for gas, adsorption, gel-permeation chromatography, etc.) or available for completely different purposes (silica gels, powders of polymers)[1].

There are two large groups of supports for extraction chromatography. The first one consists of supports whose surfaces are covered by hydroxyl groups, having high surface energy (~80 dyne/cm for silica) and being very well wetted by strong polar liquids

(e. g. water). It includes kieselguhrs (diatomites, diatomaceous earths), silica gels, glasses, celluloses and aluminas, which find wide application as supports for partition chromatography when an aqueous solution is the stationary phase. In reversed-phase extraction chromatography these supports may be used without any special treatment, but the untreated supports are suitable only for the retention of a restricted number of extractants (Table 1).

TABLE 1

Extractants that can be held on hydrophilic supports

Support	Extractant
Cellulose	Tri-n-octylamine (TNOA), trioctylphosphine oxide (TOPO), di(2-ethylhexyl) phosphoric acid (HDEHP), Lix-64 (technical mixture of oximes)
Silica gel	TNOA, methyldioctylamine, trilaurylamine (TLA), tributyl phosphate (TBP)
Kieselguhr	Alamine-336, methylenebis(di-n-hexyl-phosphine oxide) (MHDPO), methylene-bis(di-2-ethylhexylphosphine oxide) (MEHDPO)
Alumina	TNOA

Furthermore, the hydroxyl groups of the supports are capable of ion exchange, and therefore display objectionable adsorptive properties.

The surfaces of these supports can be modified by treatment with organosilicon compounds (trimethylchlorosilane (TMCS), hexamethyldisilazane (HMDS), dimethyldichlorosilane (DMCS)) in order to make them hydrophobic and capable of being wetted by different organic solvents. By this treatment, the ion-exchange properties of the supports are simultaneously depressed.

When a hydroxylic support is treated with organosilicon compounds, the following reactions take place:

$$\text{(a)} \quad -Si-OH + Si(CH_3)_3Cl \longrightarrow -Si-O-\underset{\overset{|}{CH_3}}{\overset{\overset{CH_3}{|}}{Si}}-CH_3 + HCl$$

$$\text{(b)} \quad 2 -Si-OH + NH[Si(CH_3)_3]_2 \longrightarrow 2 -Si-O-\underset{\overset{|}{CH_3}}{\overset{\overset{CH_3}{|}}{Si}}-CH_3 + NH_3$$

Because of the large volume of the $(CH_3)_3Si$ group, part of OH groups disposed nearby remains unreacted.

The best siliconizing reactant is DMCS:

$$
\begin{array}{c}
-\,Si\!-\!OH \quad Cl \quad CH_3 \\
\diagdown\,\diagup \\
O \quad + \quad Si \\
\diagup\,\diagdown \\
-\,Si\!-\!OH \quad Cl \quad CH_3
\end{array}
\longrightarrow
\begin{array}{c}
-\,Si\!-\!O \quad CH_3 \\
\diagdown\,\diagup \\
O \quad Si \\
\diagup\,\diagdown \\
-\,Si\!-\!O \quad CH_3
\end{array}
+\ 2HCl
$$

Because both Si—OH groups cannot always react, the following reaction is also possible:

$$
\begin{array}{c}
O \quad\quad Cl \quad CH_3 \\
| \diagdown\,\diagup \\
-\,Si\!-\!OH \;+\; Si \\
| \diagup\,\diagdown \\
O \quad\quad Cl \quad CH_3
\end{array}
\longrightarrow
\begin{array}{c}
O \quad\quad CH_3 \\
| | \\
-\,Si\!-\!O\!-\!Si\!-\!CH_3 \;+\; HCl \\
| | \\
O \quad\quad Cl
\end{array}
$$

The very reactive Si—Cl group is removed by washing with methanol:

$$
\begin{array}{c}
CH_3 \\
| \\
-\,Si\!-\!O\!-\!Si\!-\!CH_3 \;+\; CH_3OH \\
| \\
Cl
\end{array}
\longrightarrow
\begin{array}{c}
CH_3 \\
| \\
-\,Si\!-\!O\!-\!Si\!-\!CH_3 \;+\; HCl \\
| \\
OCH_3
\end{array}
$$

It should be kept in mind, however, that the Si—OCH$_3$ groups can hydrolyse and be reconverted again to Si—OH groups when the support is used for longer periods.

In practice, a number of silylation methods are used.

The support, dried at 105–110 °C:

(a)   is kept in a desiccator over DMCS vapour for 1–7 days, then dried at 100–200 °C for from 15 min to 2 hr; it is sometimes washed with methanol or absolute ethanol before drying; the same treatment can be performed by suction of DMCS vapours through a layer of the support;

(b)   is stirred with 100% DMCS, according to 1–5 g of DMCS per 10 g of support, and dried at 120 °C for 2 hr;

(c)   is immersed in a 0.5–5% solution of DMCS in benzene, toluene, hexane, ether or ethanol; the excess of solvent is decanted, and the support is dried at 20–80 °C for 2–24 hr;

(d)   is loaded in a DMCS solution in CCl$_4$ (1:4), using 4 g of DMCS per 10 g of support, and is boiled in a flask under reflux for 3–5 hr.

The degree of the replacement of Si—OH groups can differ somewhat, depending on the method of treatment, but comparisons of the different methods have not been carried out.

In the treatment of a support, it is necessary to remember that:

(a)   if silica is treated with DMCS vapour, the entire replacement of the Si—OH groups is achieved only after 24 hr[2];

(b)   increase of the temperature increases the rate of replacement of OH groups[2];

(c)   when a support is treated with a solution of silane, the rate of the silylation reaction depends on the diluent used: for HMDS in dioxane, dibutyl ether, xylene, butyl acetate or chlorobenzene, it is slower than for pure HMDS; while it is higher for HMDS in quinoline, methyl ethyl ketone, pyridine, or   m-nitrotoluene[3].

Silylation closes the finest pores of the support so that its surface is reduced to 80–85% for silica gels[2] and to 50% for kieselguhrs[4,5] and for glass beads[5].

The second group of supports for extraction chromatography includes different organic polymers which have low surface energies (polyethylene: 33 dyne/cm; polytetrafluoroethylene (PTFE): 19 dyne/cm[6], polytrifluorochloroethylene (PFCE): 31 dyne/cm[4]), and since they are hydrophobic they are well wetted by various organic solvents.

2.1. DIATOMACEOUS EARTH (KIESELGUHR, DIATOMITES)

Diatomaceous earth is a deposit rock, resulting from the decomposition of the silica skeletons of diatoms, single-celled algae, and can be found in various parts of the world. The skeletons consist primarily of a hydrated microamorphous silica and some impurities, generally metal oxides. They are in the form of perforated particles containing holes about 1 μm in diameter. Examination of these holes under the electron microscope shows that they have a fine structure, consisting of many very small pores. Diatomaceous earth is widely used in industry for filter materials, firebricks, fillers for paper, etc. It has found application in chromatography as the adsorbent for adsorption chromatography, and as the support of the water phase in partition chromatography; for this latter purpose special brands (named "Celite") are produced. With the development of gas chromatography, diatomaceous earth became widely applied as a support in this technique too.

For the improvement of their mechanical properties, natural diatomites have been calcined with a flux of Na_2CO_3 at 1600 °C, whereby the original material fuses, and amorphous silica is converted to crystalline cristobalite. A number of supports are produced by calcination without flux (supports of the "firebrick" type), but they are not used in extraction chromatography. During calcination the surface of the support is decreased, and the size of the pores increased.

The chemical composition of kieselguhr depends on the initial deposit, and is shown for some brands in Table 2.

A great number of brands of diatomite supports are offered for gas chromatography, and they can be used in extraction chromatography after treatment with silanes. Since the treatment affects (depresses) their adsorptive properties, a number of supports already treated with silane are also available on the market. As supports for extraction chromatography, all brands are interchangeable; trade names of supports based on diatomaceous earth are shown in Table 3, together with their manufacturers.

TABLE 2

Chemical composition of diatomite supports (in %)

Supports	SiO$_2$	Al$_2$O$_3$	Fe$_2$O$_3$	CaO	MgO	K$_2$O	Na$_2$O	TiO$_2$	P$_2$O$_5$
Celite-503, 535, 545, 560[1]	89.6	4.0	1.3	0.5	0.6		3.3*	0.2	0.2
Hyflo Super Cel[1]	89.6	4.0	1.5	0.5	0.6		3.3*	0.2	0.2
Filter Cel[1]	85.8	3.8	1.2	0.5	0.6		1.1*	0.2	0.2
Celite Analytical Filter Aid[1]	97.5	1.3	0.27	0.1	0.26		0.22*	0.11	–
Celite-545[8]	89.9	3.6	1.65	1.75	1.25	0.5	2.05	0.3	–
Chromosorb W (NAW)[7]	88.9	4.0	1.6	0.6	0.6		3.6*	0.2	0.2
Chromosorb W (AW)[7]	90.0	3.6	1.4	0.4	0.5		3.2*	0.2	–
Chromosorb W(AW)[8]	91.2	4.1	1.15	0.4	0.65	0.5	2.40	0.25	–
Anakrom[10]	88.79	4.58	1.50	1.16	0.17		3.8**		–
Gas Chrom S[8]	89.3	3.5	1.20	1.25	0.40	0.15	2.70	0.2	–
Chromaton N[9]	93	3.3	0.04		0.1***		3.4*	0.01	–
Porokhrom-1[1]	97	2.1	0.05	0.74					

*Na$_2$O + K$_2$O
**Na$_2$O + K$_2$O + TiO$_2$
***CaO + MgO

TABLE 3

Supports based on diatomaceous earth

Untreated brands, and brands washed with acid and base	Brands treated with HMDS	Brands treated with DMCS	Manufacturer
Filter Cel (uncalcinated)			Johns-Manville Corp., Celite Division, USA
Celite Analytical Filter Aid (uncalcinated)			"
Hyflo Super Cel			"
Celite-503, 535, 545, 560			"
Chromosorb W (NAW,AW)	Chromosorb W– –HMDS	Chromosorb W AW–DMCS	"
Chromosorb A		Chromosorb A–ST	"
Chromosorb G (NAW,AW)		Chromosorb G AW–DMCS	"
		HP Chromosorb W AW–DMCS	"
		HP Chromosorb G AW–DMCS	"
Celite-545, 545U, 545A		Celite-545 ABS	Analabs, USA
Anakrom 545, 545A			"
Anakrom A, AB, U		Anakrom ABS, A, S, SD,Q	"
AnaPrep A, U		AnaPrep ABS	"
Gas Chrom CL,CLA,A,P,S, CLP	Gas Chrom CLH	Gas Chrom CLZ,Z,Q	Applied Science Lab.,USA

(Table 3 cont.)

Untreated brands, and brands washed with acid and base	Brands treated with HMDS	Brands treated with DMCS	Manufacturer
Diatoport W		Diatoport S	Hewlett-Packard, F&M Scientific Division, USA
		Aeropak 30 (VarAport 30)	Varian Aerograph, USA
PhaseSep W, HC, CL, CL–AW, CL–AW–BW, N,N–AW, N–AW–BW	PhaseSep N– –HMDS,CL– –HMDS	PhaseSep WI–DMCS, CL–AW–DMCS,HC– AW–DMCS,N–AW– –DMCS,CL–AW– –BW–DMCS	Phase Separations Ltd., U.K.
PhasePrep A		PhaseChrom Q	"
Supasorb	Supasorb AW–HMDS		BDH Ltd., U.K.
Embacel Kieselguhr			May and Baker Ltd.,U.K.
Diaphorit			VEB Berlin-Chemie, GDR
Chromaton N,N–AW	Chromaton N–AW–HMDS	Chromaton N–AW– –DMCS	Lachema, ČSSR
Kieselgur SK			Chemapol, ČSSR
Shimalite W			Shimadzu Seisakusho Ltd., Japan
Sferokhrom-I			Soyuzglavreaktiv, USSR
Porokhrom-I,-II,-III			"

2.2. SILICA GELS

Silica gel is a hydrated silica $SiO_2 \cdot xH_2O$. The water is present in its structure as hydroxyl groups,chemically bonded to the surface silicon atoms. Silica gel is a colourless, porous, strongly hydrophilic substance, manufactured as mechanically rigid grains.

There are three principal methods of producing silica gels:
(a)   interaction of alkali silicate with acids;
(b)   interaction of alkali silicate with easily-hydrolyzed salts;
(c)   hydrolysis of silicon halides (usually $SiCl_4$).

The condensation of SiO_2 at the stage of sol formation leads to non-porous spherical particles whose surface is covered by hydroxyl groups. The transformation of the sol to the hydrogel is accompanied by aggregation of primary particles. When the hydrogel is drying, the structural network formed by the spherical particles is kept, but its resulting size and packing type depend on the conditions of the sol–hydrogel–xerogel transformation (conditions of washing and drying of the gels, nature of the intermicellar liquid, drying temperature, etc.). The size of the globules defines the surface area, while the packing density defines the total volume and the pore diameter: their control allows formation of widely differing silica gel structures; the related methods are reported in Refs.[11] and [12].

Silica gels are commonly used in industry as adsorbents, desiccants, fillers for rubbers and plastics, carriers for catalysts, etc. They can easily be prepared directly in the laboratory[13]. Different firms produce a wide assortment of silica gels for chromatography, and a list of available brands is reported here below.

(USA: Analabs; Fisher Scientific Co.; Bio-Rad Laboratories; J. T. Baker Chemicals Co.; W. R. Grace and Co., Davison Chemical Division; Mallinckrodt Chem. Works; Italy: Carlo Erba, Divisione Chimica Industriale; U. K.: BDH Ltd.; Phase Separation Ltd.; W. and R. Balston, Ltd.; GDR: VEB Berlin-Chemie; GFR: E. Merck AG.; Macherey, Nagel und Co.; M. Woelm; Serva Feinbiochemica GmbH und Co.; Holland: Becker Delft N.V.; ČSSR: Lachema; Japan: Shimadzu Seisakusho Ltd.; USSR: Soyuzglavreaktiv.)

Silica gels produced for other purposes can also be used.

A number of brands of silica gels for gel-permeation chromatography have been offered, which have a narrow pore size distribution and consist of spherical particles; they are "Spherosil" (Pechiney-Saint-Gobain, France), "Porasil" (Waters Associates, Inc., USA), "Mercko-gel SI" (E. Merck AG, GFR), and "Silokhrom" (Soyuzglavreaktiv, USSR).

Silica gels differ from kieselguhr in having 10–100 times greater surface area and in their more uniform chemical composition (>99.9% SiO_2). There are specially refined brands of silica gels for chromatography, such as "Silikagel Merck 7754" (E. Merck AG., GFR), "Bio-Sil A" (Bio-Rad Laboratories, USA) and "Whatman Silica Gel SG 32" (W. and R. Balston, Ltd., U.K.).

Like kieselguhrs, silica gels are hydrophobized by treatment with DMCS. Silica gel pretreated with DMCS is produced by Merck AG and by the Applied Science Lab., under the trade names of "Merck Kieselgel silinisiert" and "Reversil", respectively.

2.3. GLASSES

Glasses have surfaces of the same chemical nature as silica gels, because of the presence of silanol groups (SiOH). Glass powders, used for making glass filters or glass microbeads, specially produced for gas chromatography, can be used as supports after hydrophobization. As they have no pores, the surface area of these supports is small (approximately 5–10 times less than that of kieselguhrs), and they can retain only very small amounts of extractant. Further, when smooth glass beads are used, the largest portion of extractant is accumulated at the points of contact of the beads, so that the distribution of the organic phase on the support is very uneven in thickness. In order to avoid this and to increase the surface area, the smooth surface of the glass is often etched, and "surface-etched glasses" are prepared.

Porous glasses, obtained by leaching of alkali borosilicate glasses, have more interest as supports. Variation of the chemical composition of the initial glass, and of the conditions of the preliminary thermal treatment, permits production of porous glasses which have definite, and controlled pore sizes with very narrow ranges[14]. These glasses are manufactured for gas and gel-permeation chromatography.

Trade names and manufacturers of glass supports are shown in Table 4. Some firms produce pre-hydrophobized glasses.

TABLE 4

Glass supports

Untreated brands	Brands treated with silanes	Manufacturer
Smooth brands		
Glass Beads		Minnesota Mining Manufact. Co., USA
Glass Beads		Microbeads Inc., USA
Glass Beads		Phase Separation Ltd., U.K.
Glass Beads		BDH Ltd., U. K.
Glass Beads		Carlo Erba, Divisione Chimica Industriale, Italy
Ballotini		English Glass Co. Ltd., U. K.
Glassperlen		Becker Delft N. V., Holland
Glassport M		Hewlett-Packard, F&M Scientific Division, USA
Mikroglasskugeln		VEB Berlin-Chemie, GDR
Glass Beads	Glass Beads	Applied Science Lab., USA
Anaport Glass Beads	Anaport Glass Beads	Analabs, USA
Corning Code 0202	Corning Code 0201	Corning Glass Works, USA
Glasskugeln	Glasskugeln	Serva-Entwicklungslabor, GFR
Surface-etched brands		
Corning GLC– 100	Corning GLC– 110	Corning Glass Works, USA
Porous brands		
Corning Code 7930		''
Corning Code 7935		''
Porous brands with controlled porosity		
Corning CPG 10–75, –125, –175, –240, –370, –700, –1250, – 2000*		''
Bio-Glass–200, –500, –1000, –1500, –2500*	Bio-Glass–200, –500, –1000, –1500, –2500*	Bio-Rad Laboratories, USA

*The numbers refer to the size of the pores in Å

2.4. ALUMINA

Untreated alumina can be used in extraction chromatography, as well as supports based on silica for the retention of amines[15], as with silica, hydrophobic properties are given by treatment with DMCS[16]. Alumina is produced as an adsorbent for adsorption chromatography by different firms, and a list of available brands is reported here below.

(USA: Analabs; Bio-Rad Laboratories; Fisher Scientific Co.; J. T. Baker Chemical Co.; Aluminium Company of America; Mallinckrodt Chem. Works; Italy: Carlo Erba, Divisione Chimica Industriale; U. K.: BDH Ltd.; GFR: E. Merck AG; Macherey, Nagel und Co.; M. Woelm; Japan: Shimadzu Seisakusho Ltd.; France: Pechiney-Saint-Gobain; Switzerland: Fluka AG; CAMAG; USSR: Soyuzglavreaktiv).

2.5. SURFACE-POROUS SUPPORTS

For separation of substances having closely similar properties it is important to use supports that allow small plate heights (HETP). Small support particle sizes tend to decrease the HETP, but slow down the flow rate of the mobile phase considerably. At the same time, the decrease of the particle size increases the inhomogeneity of the column packing, which in turn increases the HETP. In order to eliminate this contradiction, supports have been produced which have a solid impervious core and a thin porous layer on the surface. In this case the flow through the column is determined by the total particle size, while the HETP is determined by the size of the microparticles comprising the surface layer: as a result, the HETP is nearly independent of the flow rate of the mobile phase[17,18].

The following surface-porous supports are produced: "Zipax chromatographic support" (E. I. du Pont de Nemours and Co., Inc., USA), "Corasil I" and "Corasil II" (Waters Associates, Inc., USA). The capacity of such supports for the stationary phase is low, and they are more expensive than usual ones. In extraction chromatography they have been used only very recently[18a], but they can be of great interest for radiochemical separations where it is necessary to separate microamounts of substances quickly.

2.6. CELLULOSE

Cellulose is a natural polymer of general formula $(C_6H_{10}O_5)_n$, and is the principal component of plant cell walls. It consists of fibrous macromolecules, built up from anhydroglucose units, the 1st and 4th carbon atoms being connected to the ether bond. Each elemental unit $C_6H_{10}O_5$ contains three alcoholic hydroxyls, in positions 2, 3 and 6. Cellulose is a very hydrophilic material, and displays weak cation-exchange properties because of the presence of carboxyl groups. The exchange capacity amounts to ~0.05 meq·g^{-1}, which corresponds to one carboxyl group on each one hundred structural units.

In spite of its hydrophilic character, carefully dried cellulose can retain a number of organic extractants (even such non-polar ones as benzine[19]), but in this case its capacity for the extractant is not so large.

143

A list of different cellulose powder brands for chromatography is reported here below.

(GFR: Macherey, Nagel und Co.; Carl Schleicher und Schüll AG Feldmeilen; Serva Feinbiochemica GmbH und Co.; E. Merck AG; Dr. Theodor Schuchardt GmbH und Co., Chemische Fabrik; U. K.: W. and R. Balston, Ltd.; USA: Bio-Rad Laboratories; ČSSR: Lachema).

When microamounts of substances are involved, so that the capacity of the support is not important, and their distribution coefficients are very high, thick filter-paper disks, impregnated with the extractant, can be used as supports in the form of unusually wide columns. Microamount of copper have been isolated[20] on such filters impregnated with Lix-64 (mixture of oximes).

Cellulose can be hydrophobized to achieve a higher capacity for the extractant. This is most often done via acetylation. Acetylcellulose is produced by several firms: Macherey, Nagel und Co.; Carl Schleicher und Schüll AG Feldmeilen;M.Woelm in GFR; and VEB Berlin-Chemie in GDR. There are acetylcellulose brands with different degrees of esterification.

Acetylation lowers the resistance of cellulose to the actions of organic solvents, whereas hydrophobization by treatment with silanes results in a more resistant material. Silanized cellulose is produced by Macherey, Nagel und Co., GFR, under the trade name "Cellulosepulver MN2100 WA".

2.7. POLYTRIFLUOROCHLOROETHYLENE

Polytrifluorochloroethylene (PFCE) is a thermoplastic crystalline polymer ($-[CF_2CFCl$ with a fusion temperature of 208–210°C, and a specific weight of 2.00–2.15, depending on the degree of crystallinity. It is a fine friable white powder, consisting of small spheres with regular shape and an average diameter of 200–600 μm. The polymer particles are aggregates of hundreds or thousands of such spheres. It is used for the production of plastic goods and anticorrosive coatings. The industrial product has different trade names in the different countries: GFR — Hostaflon C2; ČSSR — Teflex; France — Voltalef; Japan — Daiflon, Polyfluoron; USA — Fluoroflex, Fluorothene, Fluoron, Halon, Genetron, Kel-F, Plascon CTFE, Surflene, Trithene; USSR — Ftoroplast-3 (Ftorlon-3).

The powder produced to make plastic goods can be used directly as a supporting material. There are also specially made PFCE supports for gas chromatography. The properties of PFCE vary depending on the conditions of production; for instance Kel-F 300 loses its capacity for TBP after thermal treatment[21]; the amount of TBP, retained by Kel-F differs strongly from batch to batch (1.5–35% of the weight of the support)[22] In contrast with supports based on silica, PFCE has a non-homogeneous pore size distribution.

The trade names of PFCE used in extraction chromatography are listed in Table 5.

TABLE 5
Supports based on polytrifluorochloroethylene

Supports	Manufacturer
Supports for gas chromatography	
Kel-F 300	Minnesota Mining Manufact. Co., USA
Kel-F 3081	Minnesota Mining Manufact. Co., USA
Kel-F 6069	Minnesota Mining Manufact. Co., USA
Anaport Kel-F 300 LD	Analabs, USA
Haloport K	Hewlett-Packard, F&M Scientific Division, USA
Industrial powder materials	
Hostaflon -C2	Farbwerke Hoechst, GFR
Daiflon M-300	Daikin Kogyo Co., Japan
Plascon CTFE -2300	Allied Chemical Corp., USA
Voltalef 300 CHR,	Soc. des Resines Fluorées, France
300 LD, Micro PL	
Ftoroplast(Ftorlon) -3, -3 M	Soyuzkhimexport, USSR

2.8. POLYTETRAFLUOROETHYLENE

Polytetrafluoroethylene (PTFE) is a thermoplastic crystalline polymer $(-[CF_2CF_2]_n-)$ with a specific weight of 2.1–2.3. It is a white, friable, lumpy powder, out of which different products are made. Two types of polymer are available depending on the method of polymerization: granular resin, consisting of porous particles with sizes ranging from 30 μm to 600 μm, and having a specific surface area of 2–4 m^2/g; and dispersion resin with an average particle size of 200–700 μm, consisting of agglomerates of a large number of small fibrous particles ranging in size from 0.05 μm to 0.5 μm, and having a specific surface area of the order of 10–12 m^2/g.

It is produced in several countries under different trade names: GFR – Hostaflon TF; U. K. – Fluon, Permaflon; France – Diaflon, Gaflon, Soreflon; Italy – Algoflon; Japan – Polyflon; USA – Enflon, Surflon, Teflon; USSR – Ftoroplast-4 (Ftorlon-4), -4D, -4M.

The polymerization powder used for making plastic goods can be used directly as support. Special supports are also produced for gas chromatography, their trade names being given in Table 6 together with their manufacturers.

Like PFCE, PTFE based supports have non-homogeneous pores. The shortcoming of several PTFE brands is that they are soft, crumpling materials and are slightly electrified by friction: these properties make the regular packing of columns difficult. Good mechanical properties and only a slight tendency to be electrified are shown by "Fluoropak-80"[23].

The mechanical properties of PTFE supports can be improved by means of thermal treatment:

11*

TABLE 6
Supports based on polytetrafluoroethylene

Supports	Manufacturer
Supports for gas chromatography	
Teflon-6	E. I. du Pont de Nemours and Co. Inc., USA
Chromosorb T*	Johns—Manville Corp., USA
Haloport F*	Hewlett-Packard, F&M Scientific Division,USA
Anaport Tee-Six*	Analabs, USA
PhaseSep T6*	Phase Separation Ltd., U. K.
Fluoropak-80	Fluorocarbon Co., USA
Teflonpulver	Becker Delft N. V., Holland
Shimalite F	Shimadzu Seisakusho Ltd., Japan
Hydeflon-Pulver	VEB Berlin-Chemie, GDR
Polykhrom-I	Soyuzglavreaktiv, USSR
Industrial powder materials	
Fluon	ICI Ltd., U.K.
Hostaflon TF	Farbwerke Hoechst, GFR
Algoflon	Montecatini, Italy
Soreflon 5A	Soc. des Resines Fluorées, France
Ftoroplast (Ftorlon)-4, -4D, -4M	Soyuzkhimexport, USSR

*Prepared from Teflon-6

    (a)   At 19—20 °C, the crystalline part of the polymer undergoes phase transition, followed by a volume reduction; the strength of the support increases and its tendency to electrification decreases. It has therefore been suggested[23] to carry out grinding and fractionating of the support at 0 °C, and to fill the column with the dry support at the same temperature.

    (b)   The mechanical properties of the support improve after sintering of the polymer at higher than 300 °C temperature. Such treatment of industrial polymer powders can easily be carried out in the laboratory. The screened polymer fraction is spread on a netted tray in a friable layer 5 mm thick, put into a muffle and sintered at 300 °C for 7—10 min. The strength of the support increases to a considerable extent[24].

A support treated as described above is produced in the USSR and is called "Polykhrom-1". However, this support still tends to crush under the excess pressures necessary to percolate the eluting solutions, and the highest flow rate that can safely be reached with "Polykhrom-1" (250—500 μm fraction) is considerably lower than that obtainable with columns prepared with PTFE (fraction <100 μm) sintered at 370—390 °C For this purpose, the screened fraction of the polymer is loaded into aluminium moulds as beds 2—3 cm in height, and sintered in a muffle for 20—30 min. The resulting material is ground to the desired grain size with a knife-mill[25].

Porous blocks of PTFE can easily be made, having only open pores. Thin plates of block porous Ftoroplast-4, soldered in polyethylene tubes are used for fast separations [26,27]; columns can be made with large porous blocks, and these are useful for preparative separations and for continuous chromatography.

2.9. POLYETHYLENE

Polyethylene is a crystalline polymer with fusion temperature 105–130 °C. Depending on the methods of synthesis, of low, high and medium pressure polyethylenes exist, which differ a little in their physical and chemical properties. Polyethylene powders are used for the production of both plastic goods and protective coatings on metal surfaces. Different brands of polyethylene have been used in extraction chromatography: Hostalen W (Farbwerke Hoechst, GFR), microporous "Mipor Granules 14CN–G" (ESB Reeves Co., USA) and "Microthene-710" (U. S. Industrial Chem. Corp., USA; Columbia Organic Chemicals, USA). BDH Ltd. (U.K.) produces polyethylene powder especially for chromatography[1].

A particular case where polyethylene has been used as support is capillary extraction chromatography, where the organic phase was retained on the inner surface of thin polyethylene tubes about 0.5–1.5 mm in bore and 2–10 m in length. Amberlite La-2[28] and TBP[29] were used as extractants.

2.10. COPOLYMERS OF POLYSTYRENE WITH DIVINYLBENZENE

Copolymers of polystyrene with divinylbenzene (DVB) can also be used as supports. They are normally produced as a base for the preparation of ion-exchange resins. In these copolymers, DVB cross-links the polystyrene chains:

147

The DVB content can vary, and determines the ability of the polymer to swell in organic solvents. In order to improve the kinetic characteristics of the ion-exchange resins, methods of synthesis of macroreticular polystyrene copolymers have been devised, since the exchange rate depends mainly on the rate of diffusion through the polymer grain.

Such copolymers are obtained by polymerization of monomers in a medium made of inert diluents, which are good solvents for monomers and do not dissolve the polymer. As a result, a three-dimensional polymer is generated, which includes cells filled with the solvent. After polymerization is complete, the solvent is removed by steam distillation or by vacuum suction, leaving the porous frame. Variation of the styrene : DVB ratio and the amount and nature of the solvent has the result that the surface area, the total porosity, and the average pore diameter of the copolymer can be changed within wide limits. Since such macroporous materials have proved to be of use for the separation of many complex mixtures in gas chromatography and for the separation of substances with different molecular weights in gel-permeation chromatography, they are produced by different firms (Table 7).

TABLE 7

Supports based on copolymers of styrene and divinylbenzene

Supports	Manufacturer
XAD-1	Rohm & Haas, Ion-Exchange Dept., USA
XAD-2	Rohm & Haas, Ion-Exchange Dept., USA
Chromosorb 101	Johns-Manville Corp., USA
Chromosorb 102 (prepared from XAD-2)	Johns-Manville Corp., USA
Bio-Beads SM-1, SM-2	Bio–Rad Laboratories, USA
Pontybond	Pontyclyn Chemical Co. Ltd., U. K.
Synachrom	Chemapol, ČSSR
V_{Nb-15}-10	Research Institute of Synthetic Resins and Laquers (Pardubice), ČSSR
Polysorb-1	Soyuzglavreaktiv, USSR

As supports for extraction chromatography, copolymers of polystyrene with DVB are noteworthy for the following properties:

1. very small sizes of the pores (for the majority of brands is lower than 0.01 μm);
2. appreciable swelling in organic solvents.

It is believed that, due to the swelling in organic solvents, copolymers of polystyrene with DVB can retain greater amounts of extractant than non-swelling supports. It is necessary to take into account, however, that as the extractant gets into the matrix of the copolymer, the extracted element will also get into it, so that slow kinetics of re-extraction are to be expected. In gas chromatography, the factor of asymmetry (i.e. the ratio of the tail peak half-width to the front peak half-width at half the peak height) is used for comparison of the influence of the supports on the broadening of the chromatographic peak owing to adsorption. Some factors of peak asymmetry in extraction chromatograms for the elution of uranyl nitrate from TBP columns with water are collected in Table 8.

TABLE 8
Factors of asymmetry for TBP columns

Support	Composition of introduced solution	Factor of asymmetry	Reference
Copolymer with 2% DVB (non-porous)	2 N NaNO$_3$(5 g/l UO$_2$(NO$_3$)$_2$)	~3	[30]
V$_{Nb-15}$-10 (macroporous)	0.1 M UO$_2$(NO$_3$)$_2$ in 5.35 M HNO$_3$	~ 1.5 – 2	[31]
Kel-F 300	50 mg U in 5–10 ml 5.5 M HNO$_3$	~1	[21]

Preobrazhenskii[32] suggested that the extraction ability of the extractant which has entered the polymer matrix will be reduced, owing to the solvation of the polymer by the extractant. References[33] and [34] report that the distribution coefficients for the extractant supported on copolymers of polystyrene with DVB are lower than those obtained in liquid-liquid extraction. To explain this, the polymer matrix was considered an inert diluent for the extractant[34]. The swelling properties of the support make it difficult to calculate the chromatographic separation relations from liquid-liquid extraction data.

2.11. OTHER SUPPORTS

Besides polystyrene—DVB, the following other polymers have been used as supports: polyurethane[35], polypropylene[36], the copolymer of polyvinyl chloride with vinyl acetate[36—39], polyvinyl chloride[36,40], polyacrylonitrile[40], polyvinylidene chloride[41], rubbers[42]. Attempts have also been made to obtain supports by etherification of sulpho-cationites[43] or by graft-polymerization under the action of nuclear radiation[44]. Organophosphorus extractants, such as TBP[45] and di-octyl pyrophosphoric acid[46], are well retained by activated charcoals.

As none of the used supports definitely excels in properties, in extraction chromatography practice siliconized kieselguhrs, silica gels and fluorocarbon polymers are applied almost equally commonly.

3. THE INFLUENCE OF THE STRUCTURE AND SIZE OF THE
SUPPORT PARTICLES ON THE EFFICIENCY OF CHROMATOGRAPHIC COLUMNS

3.1. SIZE OF PARTICLES

The influence of the support particle size on the efficiency of chromatographic columns has been studied in detail in gas chromatography. The HETP value decreases in proportion to the reduction of the size[47], and a narrow range of particle sizes is also important[48]. Analogous results were obtained for liquid partition chromatography[49,5(

The supports for gas chromatography, which have been or can be applied in extraction chromatography (Tables 3 to 7), are produced in a wide range of sizes with intervals of 10—20 mesh, that corresponds to size fractions with a 10% deviation from the average value. Conversion tables from mesh to mm sizes can be found in Chapter 3.

3.2. SURFACE AREA OF SUPPORT

The value of the specific surface area of the support affects its capacity for the extractant. The capacity for the extractant of glass powders, which have a surface area of 0.04 m^2/g, is approximately 10 times lower than that of supports that have a surface area $\geqslant 1 m^2$/g.

HETP values decrease as the amount of extractant on the support is reduced, since the thickness of the extractant layer falls, and equilibrium becomes faster; for preparative separations of elements with similar properties, the capacity for the extractant is important at the minimum HETP.

For a given film thickness, supports with larger surfaces retain greater amounts of extractant[51]. Representative data are reported in Table 9.

TABLE 9

Effect of surface area of the support on column characteristics

Support	Surface area, m^2/g	Minimum value of HETP, mm	Amount of extractant, ml/ml of total column volume
Silica gel KSK	300—500	0.1	0.20—0.27
Hyflo Super Cel	10—100	0.3	0.11

3.3. POROSITY OF SUPPORT

A support with a larger pore volume offers a lower resistance to the flow of the mobile phase, for a given amount of loaded extractant. When the average pore size decreases, the HETP increases: silica gels with average pore diameters of 70—140 Å lead to an

HETP approximately twelve times lower than that obtainable with $\leqslant 50$ Å average pore diameter material[51].

Larger amounts of extractant are retained by supports which have larger pore volumes at the minimum HETP and for equal surface areas[51].

4. PROPERTIES OF THE DIFFERENT SUPPORTS

4.1. STRUCTURE

The structural characteristics of the supports for extraction chromatography are shown in Table 10. The reported data are approximate, as surface area and porosity figures depend on the method of measurement[8], on the grain size, and on the batch of support[52]. Of the large number of brands of silica gels and kieselguhrs, the Table shows only those that have been used in extraction chromatography.

4.2. CHEMICAL RESISTANCE

Polytetrafluoro-ethylene
: Not destroyed by acids, alkalies, oxidants, organic solvents, or heating up to 250 °C.

Polytrifluoro-chloroethylene
: Not destroyed (i.e. does not change, or swells by less than 1%) by acids, alkalies or oxidants. It dissolves in benzene, toluene, xylene, CCl_4 and some other solvents at 120–200 °C. At T < 100 °C the above solvents make it swell weakly (in 7 days at 25 °C it swells by 1–4%).

Low pressure polyethylene:
: Not destroyed by acids or alkalies. Strong oxidants destroy it at T > 50 °C. Organic solvents do not interact at room temperature (in 10 days it swells by 1–7%), with increasing temperature it swells, and at T $\geqslant$ 120 °C it dissolves.

High pressure polyethylene:
: Less stable than low pressure polyethylene to the action of organic solvents (in 10 days it swells by 36% in CCl_4, 21% in $CHCl_3$, and 12% in benzene).

Polyvinyl chloride (polyvinyl acetate):
: Not destroyed by alkalies or dilute acids. Does not resist concentrated acids, ketones and esters; it swells in aromatic hydrocarbons; it melts at T $\geqslant$ 60 °C.

Copolymer of polystyrene with divinylbenzene:
: Not destroyed by acids or alkalies. It swells in many organic solvents (the degree of swelling depends on the DVB content), but resists to the action of temperature up to 250 °C.

Polyurethane:
: Destroyed by concentrated acids and alkalies. Resists the action of many organic solvents.

Acetylcellulose:
: Not destroyed by dilute acids, but is destroyed by alkalies and concentrated acids. The action of organic solvents depends on the degree of esterification. It is dissolved by ketones and esters, and is dissolved or softened by aromatic hydrocarbons. Softens at 60–120 °C.

151

TABLE 10
Structural characteristics of supports

Support	Specific weight g/cm³	Bulk density, g/ml	Packed density, g/ml	Surface area, m²/g	Average diameter of pores, Å	Total porosity, cm³/g
1	2	3	4	5	6	7
Kieselguhrs:						
Filter Cel	—	0.27[1]	—	16[53]; 30[54]	—	—
Hyflo Super Cel	—	0.328[55]0.28[1]	—	2[53]; 7.9[55]; 10[54]	—	0.0173(pores<0.08μm) [55]
Hyflo Super Cel, treated with DMCS	2.26[56]		0.442[56]; 0.37–0.43 [57]; 0.38–0.45[58––60]; 0.70–0.78[61, 62]	100[56]	—	—
Celite-503	—	0.29[1]	—	1.5[53]; 8[54]	—	—
Celite-535	—	0.29[1]	—	7[54]	—	—
Celite-545	—	0.34[66]0.432[55]; 0.30[67]	—	0.67[8];1.14[64];1.37 [65];1.55[66];2.2[67]; 6[54];7.24[55]	5000[65];60000–– 70000[63]	0.0115(pores<0.08μm) [55];1.2[65];1.46 (pores 0.06–30μm)[8]
Celite-545, treated with DMCS	—		0.364[68]0.383[69]			
Chromosorb W	2.20[71]	0.18[70]0.20[71]; 0.33[66,73]; 0.27 [72]	0.24[70]0.248[76]	0.91[8];1.0[5,70];1.2 [71,73];1.25[66];1.38 [65];1.41[64];1.5–– –1.6[72]	50000[63]65000 [65];80000–90000 [7];40000[72]	1.56(pores 0.06–30μm) [8];1.06[65];2.78 (pores 0.005–100μm) [71];2.2(pores>5μm) [71];1.8(pores<5μm)[71];1.8 0.6(pores<5μm)[71];1.8 (pores 0.018–20μm) [72]
Chromosorb W, treated with HMDS	—	0.27[67]		0.71[5];1.2[67]	50000[63]	1.54(pores 0.06–30μm) [63]
Chromosorb W(AW––DMCS)	—	0.20–0.25[74]		—	—	—
Chromosorb G	—	0.47[74]	0.40[75]	—	—	—
Chromosorb G (AW–DMCS)	—	0.52[74]	0.58[70]0.665[76]	0.5[70]	—	—
Chromosorb A	—	0.40[70]	0.48[70]	2.7[70]	—	—
Anakrom	2.33[10]	0.18–0.21[10]	0.26–0.35[10]	—	—	—

PhaseSep W,N,A,HC Gas Chrom S	2.0–2.5[77]	–	0.3–0.4[77]	1–3[77]; 0.65[8]	600-ω-100000[8]	1.72(pores 0.06–30µm) [8]
Sferokhrom-1	–	0.57[66,72];0.44[73]	–	0.94[73];1.06[66]; 1.2[72]	20000[72]	0.65(pores 0.018– –20µm)[72]
Chromaton N	2.3[9]	0.235[9]	–	1.0[9]	120000[9]	1.3–1.4[9]
Silica gels:						
Silicagel KSK N 2	2.240[51]	0.39[51]	0.455–0.478[78]	338[51]	140[51]	1.19[51]
Silicagel KSK N 2.5	2.244[51]	0.46[51]	–	376[51]	103[51]	0.971[51]
Silicagel C H	–	0.38[79]	–	400[79]	80[79]	0.97[79]
Porasil A(Spherosil XOA 400)	–	0.55[1]	0.509[76]	350–500[80];320.1 [82]	<100[81]	0.7–1.0[80]
Porasil B(Spherosil XOA 200)	–	0.55[1]	0.509[76]	125–250[80];147.1 [82]	100–200[81]	0.7–1.0[80]
Porasil C(Spherosil XOB 075)	–	0.55[1]	0.509[76]	50–100[80];83.8 [82]	200–400[81]	0.5–0.7[80]
Porasil D(Spherosil XOB 030)	–	0.55[1]	0.509[76]	25–45[80];33.3[82]	400–800[81]	0.5–0.7[80]
Porasil E(Spherosil XOB 015)	–	0.55[1]	0.509[76]	10–20[80];18.7[82]	800–1500[81]	0.5–0.7[80]
Porasil F(Spherosil XOC 005)	–	0.55[1]	0.509[76]	2–6[80];6.8[82]	>1500[81]	0.3–0.5[80]
Silokhrom-1	–		–	15–30[1]	1000–1500[1]	
Silokhrom-2	–		–	40–60[1]	600–900[1]	
Silokhrom-3	–		–	70–100[1]	400–600[1]	
Glasses:						
Glass powder "Pyrex Grade 2"	2.19[83]	–	1.25–1.29[83]	0.04[83]	–	–
Glass beads	2.7[84]	–	1.4[84];1.77[76]	0.01[76];0.2[84]; 0.36[5]	no	no
Corning Code 7935	–	–	0.8–1.0[85]	200–350[85]	30–40[85]	–
Corning GLC-100	–	–	–	0.3[76];0.41[86]	–	–
Surface-porous supports:						
Corasil I	–	–	–	7[88];7.0±1[87]	<50[88]	no
Corasil II	–	–	–	14[88]	<50[88]	no
Zipax	–	–	–	0.65[87];1[88]	1000[88]	no
Cellulose:	1.47–1.63	–	0.32[89];0.43[90]; 0.55[91]	0.35–0.65[1]	–	–

(Table 10 cont.)

1	2	3	4	5	6	7
Polytetra-fluoroethylene:						
Fluoropak-80	2.14−2.23	—	0.86[5];1.1[110]	—	—	0.4[92]
Columpak T	—	—	—	4.8[7]	—	—
Haloport-F	—	—	—	7.8[7]	—	—
Chromosorb T	2.39[93]	0.42[93];0.56[24]	0.49[93]	1.31[24];7−8[93];7.8 [7];9.3[92]	500−1500[99]	0.219[99];1.6[92]
Teflon-6	2.17[95]	—	0.73[95]	0.44[95];1.12[94]; 1.48[94];9.0[96]; 10.5[23];10.9[23]; 11.67[97];15.4±1[98]	—	0.40[95]
Soreflon 5A	2.15[100]	0.32[100];0.69[99];	—	4.0[92]	200−2000[100];	0.3[92]
Ftoroplast-4D		0.66[24]	—	1.90[24];8.9[100];	500−2000[99]	0.3[100]
Polykhrom-I	2.15[100]	0.68[100]	—	10.4[99] 5.6[100]	500−2000[100]	0.25−0.30[100]
Polytrifluorochloro-ethylene:	2.1−2.2					
Kel-F 3081	—	—	0.39[103];0.45[104];	2.2[23]	—	—
Kel-F 300	—	—	0.79[105]	1.5[106]	—	—
Voltalef 300	—	0.38−0.40[101]	0.44[102]	1.3[101];2.4[92]	—	1.2[92]
Hostaflon C2	—	—	0.56[107]	2.6[108]	—	—
Ftoroplast-3	—	0.67[24]	—	2.30[24]	—	—
Polyethylene:	0.91−0.96	—	0.15−0.44[109,110]	0.03−0.61[109]	—	—
Polyvinylchloride (polyvinylacetate):	1.35−1.45					
Corvic R 51/83		0.48	0.56	—	—	—
Copolymers of polystyrene with DVB:						
XAD-2	1.03[112]	0.29[114]	0.74[113]	290−330[112]	85−95[112]	—
Chromosorb 102	—	—	—	300−400[114]	85−95[114]	—
Polysorb-I	—	1.8[99]	—	320[99]	50−2000[99≅]	1.844[99]
V Nb−15 · 10	1.06[115]	—	—	5.0[116];7.6[115]	1200[115];1700[116]	0.216[116];0.225[115]
Copolymer with 2−8% DVB(nonporous)	—	—	—	0.1[111]	—	0.006−0.002[111]

Cellulose: Partially destroyed by concentrated acids (sooner upon heating). Easily oxidized. Concentrated solutions of some salts and organic bases (thiocyanates, triethylbenzylammonium) dissolve it, especially upon heating. Resistant to the action of organic solvents.

Siliconized kieselguhr and silica gels: Not destroyed by acids (apart from HF). Concentrated acids partially destroy the siliconized coating if their action is long[56]. Alkalies destroy the siliconized coating. Silica gels have weak oxidizing or reductive properties (depending on the medium)[117]. Resistant to all organic solvents and to temperature.

4.3. RADIATION STABILITY

The radiation stabilities of supports have not been specially studied under the specific conditions of column extraction chromatography.

In general, the radiation stability of organic polymers decreases in the following sequence: polyethylene, polyurethane > polymethylsiloxane > polytrifluorochloroethylene, acetylcellulose > cellulose > polypropylene > polytetrafluoroethylene[118].

When using polymers as support, one should take into account that irradiation reduces the resistance of polymers to the action of oxidants, and that halogen containing polymers might release free halogens and hydrohalic acids (F_2, Cl_2, HF, HCl). When Teflon powder is subjected to a 10^8 R of ^{60}Co γ-radiation, 1 mg of fluorine is released from 1 g of polymer[119]. The presence of water increases the yield of decomposition products. When Teflon or Kel-F is γ-irradiated by $23 \cdot 10^6$ rad (^{60}Co), then 0.01 meq of acidic products is released from 1 g of dry polymer, but 0.032 meq and 0.103 meq of acidic products are released from 1 g of Teflon or Kel-F, respectively, in the presence of water[120]. In Ref. [121], the following limits are suggested for the application of polymers in radiochemistry:

Polyethylene, copolymers of styrene — up to 10^9 R;
Copolymers of vinyl chloride — up to $5 \cdot 10^8$ R;
Polytetrafluoroethylene, cellulose, acetylcellulose — up to 10^8 R;
Polytrifluorochloroethylene — up to $10^7 - 10^8$ R.

The influence of radiation on the results of separations was not observed when cyclotron targets, whose radioactivity reached several hundred mCi of β- and γ-radiation, were reprocessed on columns with Ftoroplast-4 as support[26,122—124].

Supports based on silica (kieselguhrs, silica gels) have much greater resistance to all kinds of radiation (α, β, γ, fast neutrons, fission fragments)[125]. On siliconized kieselguhr, 20—30 curies of ^{242}Cm was isolated from ^{241}Am[126] and on the siliconized silica gel samples were separated having a radioactivity of 10 g-eq Ra[127].

Supports based on silica (silica gels, kieselguhrs) are weak cation exchangers. The ion-exchange ability of silica is due to the presence of OH groups on its surface. Metallic ions can substitute the hydrogen of these groups, and this may lead to strong adsorption of some elements, for example Zr[128]. The presence of impurities (Al, Fe) increases the adsorptive properties of silica, because aluminosilicic and ferrosilicic acids are stronger than silicic acid[117].

Silylation depresses the adsorptive properties of silica, but since:
1. some of the SiOH groups are not siliconized;
2. the $SiCH_3$ groups are gradually hydrolyzed to Si—OH; and
3. concentrated acids gradually destroy the hydrophobic coating;

adsorption cannot be entirely eliminated. For example, 0.2% neptunium is irreversibly adsorbed on Hyflo Super Cel treated with DMCS, when Np is isolated from irradiated uranium[60].

When applying silica-based supports, one should take into account that the separated substances may be contaminated by the desorption of elements captured by silica gel in the process of its production, or contained in kieselguhr. This is important when highly pure substances must be obtained (e.g., for careful purification of Ga from Al, see Ref.[129]), and when trace amounts of substance must be analyzed[130].

Cellulose has significant adsorptive properties; many separations on cellulose can be carried out without organic solvents[131].

Polymers have no groups capable of ion exchange, but nevertheless they adsorb different ions from aqueous solutions (Zr[132], Cs, Sr, Tl[133], Ag[133,134], Y[135], Ru[136]).Most sorption studies on polymers involve polyethylene or PTFE, the sorption on PTFE generally being significantly lower than that on polyethylene[133,134,136].

It is believed[132] that sorption of uncharged molecular complexes is most frequent. In some cases, it is so great that it is used for the isolation of elements (e. g. the adsorption of Y on Ftoroplast-4 from HF medium[135]).In the presence of water-soluble extractants, there is a great increase in the adsorption of those elements that form extractable complexes with them (e.g. thorium acetylacetonate[137,138], cesium tetraphenylborate[139]).Several different organic molecules are also strongly adsorbed from aqueous solutions onto the copolymer of polystyrene with DVB (XAD-2)[140].

The adsorption of volatile chelates of Al, Fe, Cr and Cu (acetylacetonates, trifluoroacetylacetonates) on Ftoroplast-4 has been applied to their gas-adsorption separation[141].

In the course of extraction chromatographic separations, the substantial adsorption was discovered of Am on polyethylene (up to 2%)[41] and of Zr on PTFE[143].

A comparison of the adsorption of Cs on different supports is reported in Table 11[144].

It is advantageous for routine analysis to apply columns which permit the carrying out of some separations without the loading of fresh extractant. It is noted in Ref.[56] that in this case polymer supports are more preferable than kieselguhrs because of the lower adsorptive effects.

TABLE 11
Adsorption properties of different supports

Support	Adsorption of Cs, %
Siliconized silica gel	76
Siliconized kieselguhr (Chromosorb W)	35
Polytetrafluoroethylene (Teflon)	28
Polytetrafluoroethylene (Ftoroplast-4)	31
Polytrifluorochloroethylene (Teflex)	31

4.5. CAPACITY FOR THE EXTRACTANT

The ability of supports to be wetted by organic solvents and to hold them satisfactorily depends on several factors, namely: the properties of the support surface (surface energy, and nature of groups which cover the surface), the surface tension of the extractant, and the composition of the aqueous solution.

Supports with OH groups on their surface (silica, cellulose) hold amines without hydrophobization, apparently because of their ability to exchange cations. It is interesting that untreated silica does not hold HDEHP though cellulose does.

The extractant is held on the support either as a uniform layer or as isolated droplets. The structure of the layer depends on the nature of the support surface, the size of the pores, and the amount and nature of the liquid phase[145,146].

The stability of the extractant film depends strongly on the interface tension; if it is large, the columns are stable up to high amounts of loaded extractant[147]. For example, with a thermally treated powder of Ftoroplast-4, $CHCl_3$ is driven off completely from a $5 \cdot 33$ mm bed of support by 20 ml of water saturated with $CHCl_3$; if 3 M HCl-saturated $CHCl_3$ is percolated through the same column, the $CHCl_3$ (0.11–0.12 ml/g of support) remains in the column.

With extractants which are held weakly, a dependence exists between the amount of extractant which remains on the support and the flow rate of the mobile phase[148]: $CHCl_3$-loaded columns retained 0.15, 0.25, 0.42 and 0.42 ml of stationary phase upon elution with flow rates of 5.1, 2.56, 0.51 and 0.256 $ml \cdot cm^{-2} \cdot min^{-1}$, respectively.

More viscous extractants are torn off with greater difficulty[21].

The ability of polymer supports to retain the organic phase depends appreciably on their previous treatments: the polymerization powder of Ftoroplast-4 holds $CHCl_3$ well, while the thermally treated material does not. Kel-F 300 loses its ability to hold TBP on heat treatment[21].

For different purposes, the column may be loaded with different amounts of extractant. The amount of loaded extractant is often significantly less than the maximum amount that the support can hold, to decrease the HETP when substances with similar properties have to be separated. Columns loaded with the maximum amount (or a little less to avoid droplets of extractant being drained off by the mobile phase) are preferred in the following cases:

157

TABLE 12

Capacity of supports for TBP and HDEHP

Support	Capacity for TBP				Capacity for HDEHP			
	Aqueous phase	ml/g of support	ml/ml of column	Reference	Aqueous phase	ml/g of support	ml/ml of column	Reference
Hyflo Super Cel, treated with DMCS	HNO_3	0.6	0.23	[59]	–	0.5	0.28	[58]
"	HNO_3	–	0.25	[57]	–	–	–	–
"	HNO_3	0.55–0.85*	0.24–0.38*	[56]	–	–	–	–
Kieselguhr (mark is obscure)	$HNO_3+NH_4NO_3$	1.2	–	[149]	–	–	–	–
Celite, treated with DMCS	–	–	–	–	HCl	1.0	–	[150]
Chromosorb W(AW–DMCS)	H_2O	1.27*	0.37*	–	H_2O	2.3*	0.67*	–
Chromosorb W(non-silanized)	H_2O	0.32*	–	–		Not retained		–
Silicagel KSK N 2 and 2.5, treated with DMCS	HCl	0.7	0.23	[127]	HCl	0.8	0.29	[151,152]
Silica gel(non-silanized)	Saturated solution $Al(NO_3)_3$	35–40% of weight	–	–	–	–	–	–
Porasil C	No; was defined by centrifugation	43–44% of weight	–	[153]		Not retained		[51]
Kel-F 300	HNO_3	1.33	–	[22]	HCl	0.75	0.6	[105]
Kel-F 300	HNO_3	1.0	0.33	[21]	–	–	–	–
Daiflon M-300	HCl	1.08	–	[21]	–	–	–	–
Hostaflon C2, Voltalef 300	HNO_3, HCl	0.5	–	[154]	HCl	0.7*	0.39*	[107]
Teflon	HNO_3	0.7	0.35	[155,156]	–	–	–	–
Haloport F	Air	0.288*	–	[21]	–	–	–	–
Fluon	HNO_3	0.40–0.43	0.21–0.31	[157]	–	–	–	–
Ftoroplast-4(polymerization powder)	HCl	–	0.33*	[158]	–	–	–	–

158

Ftoroplast-4(polymerization powder)	HCl	0.33*	0.29*	[160]	—	—	—	—
"	H_2O	0.62*	0.49*	—	H_2O	0.30*	0.24*	—
Thermal treated Ftoroplast-4	H_2O	0.37*	—	—	HCl	0.4*	—	[161]
Polykhrom-I	H_2O	0.50*	0.43*	—	—	—	—	—
Porous block Ftoroplast-4 (sintering with filler)	HF	0.41*	0.35*	[162]	—	—	—	—
Porous block Ftoroplast-4 (sintering without filler)	H_2O	—	0.28*	—	—	—	0.4*	[212]
Polyethylene	HNO_3	0.5	0.26	[21]	—	—	—	—
Corvic R 51/83	HNO_3	0.3*	—	[21]	$HClO_4$	0.3	—	[38]
Polyurethane foam	—	2.0	—	[35]	—	—	—	—
V$_{Nb-15}$-10	HNO_3	0.22*	—	[31]	—	—	—	—
Cellulose	—	—	—	—	HCl	0.5	0.27	[91]
Polyethylene capillary tube	HNO_3	—	0.027–0.041	[29]	—	—	—	—
Activated charcoal BAU-2	Sorption from ethanol solution	0.35*	—	[45]	—	—	—	—

*Maximum amount of extractant on the support

(a)   separation of large amounts of substances;

(b)   concentration of substances, since the volume of solution before break-through from the column is proportional to the volume of the organic phase;

(c)   routine analysis since columns with a large volume of organic phase can be used for a great number of separations without renewal of the extractant.

For the determination of the volume of extractant fixed on the support, the following methods are used:

(a)   Direct measurement of the volumes of the mobile and stationary phases in the column.

For this purpose, a known volume of extractant is fed into the column, and after the excess of extractant has been driven off by the mobile phase, the volume of recovered extractant is measured and the volume of extractant remaining in the column is calculated by difference. The volume of the mobile phase can be measured in an analogous way. In other cases, the volume of the mobile phase is measured by radiometric methods (e.g. through the elution volume of an unextractable radioactive isotope) or by titration, and the volume of the extractant is calculated by means of the bed volume of the column and the volumes of the mobile phase and the support.

(b)   After-elimination of the excess of extractant by air flow.

The excess extractant is eliminated from the support by weak vacuum suction, and the amount of extractant remaining is determined by weighing. The resulting amount of loaded extractant depends considerably on the time of suction for volatile compounds, while for non volatile extractants the results correspond to those obtained with the former method. Possible small differences are due to the absence of mobile phase which is known to influence the retention of the extractants.

(c)   Centrifugation[22].

The excess extractant is eliminated by centrifugation and the amount left on the support is determined by weighing. This method originates from ion-exchange resin practice, where it is used for the determination of the swelling of the resin and is supposed to completely eliminate the water or organic solvents wetting the resin surface. It might be expected therefore that with this method low results should be obtained. In fact, when TBP is loaded on thermally treated Ftoroplast-4 from solutions in volatile solvents and in amounts 2–3 times lower than the maximum loadable on the support, 15–20% of the organic phase is removed by the centrifugation procedure.

The amounts of extractants retained by different supports are reported in Table 12 for TBP and HDEHP, and in Table 13 for the other extractants. The figures marked with an asterisk are the largest ones reported for the extractant on the given support, while the other figures refer to loadings lower than maximum, or to doubtful cases.

It is clear from Table 12 that the capacity for TBP per unit volume of the column is practically the same for kieselguhrs and fluorocarbon polymers, while it is lower for polyethylene and polyvinyl chloride (polyvinyl acetate).

The data of Table 13 show that:

— for a given extractant only a given support has the largest capacity (for dithizone in $CHCl_3$, for example, DMCS-treated Celite-545 has greater capacity than thermally treated Ftoroplast-4 or polyethylene);

— the capacities of glass powders are markedly less than those of the other supports;

160

— the capacity values for the different polymeric supports range within narrower limits than do those for kieselguhrs;

— the thermal treatment lowers the capacities of fluorocarbon polymers, considerable differences are observed for different brands of polymer (for example MIBK is well retained by Teflon-6, but not by thermally treated Ftoroplast-4 powder[177,178,188]).

4.6. SUPPORTS AND RESOLUTION ABILITY OF COLUMNS

No specific works are available on the comparison of different supports as regards their HETP features under identical conditions. It is rather difficult to compare the HETP figures given in the literature, since these parameters depend on many factors, such as the grain size of the support, the volume of loaded extractant, the method for extractant loading, the flow rate of the mobile phase, the value of the distribution coefficient, the amount of element introduced into the column, the nature and concentration of the extractant, and the composition of the aqueous phase. In addition to this, HETPs can sometimes differ for different elements (for example, rare earths) under identical experimental conditions[58,189].

Representative HETP figures available in the literature are listed in Table 14.

Kieselguhrs and fluoropolymers give similar HETP values in most cases, though it could be expected that under identical conditions the HETP on fluoropolymers should be lower.

HETP figures given in the Table for DMCS treated Hyflo Super Cel and Plascon —2300 are roughly the same but they refer to Plascon loaded with twice the amount (by weight) of organic phase as compared to that loaded on the other support. Higher figures are found for Kel-F, but in this case the loaded organic phase is 3.3 times more (by weight) than it is for Hyflo Super Cel and since the density of packing of Kel-F is higher than that of Hyflo Super Cel, the volumes of extractant in the supports are even more different.

Reference[204] reports that the peaks obtained with Plascon are narrower than those with Chromosorb W (AW–DMCS). Thermally treated Ftoroplast-4 permits lower HETP values[191] than siliconized silica gel[49] under similar separation conditions. In practice, not only the width of the peak (i.e. HETP) is important, but also the "tail" of the peak. A comparison of some polymer supports from this point of view is given in Table 15[36].

It can be seen from the tabulated results (Table 15) that fluoropolymers need lower volumes of eluents than the other polymeric supports. Tails were also found with polystyrene — DVB copolymers[30, 31, 202].

For the separation of macroamounts of elements similar in chemical properties, an important characteristic of the support is the capacity for the extractant corresponding to the minimum HETP. Data for the comparison of this characteristic are available only for silica gels of different brands[51], but not for supports of different natures.

TABLE 13

Capacity of supports for different extractants

Support	Extractant	Aqueous phase	Capacity ml/g of support	Capacity ml/ml of column	Ref.
Hyflo Super Cel, treated with DMCS	TNOA	HCl	0.37	–	[61]
Silicagel KSK N2 and 2.5	TNOA	HCl	0.55	0.13	[129]
Cellulose	TNOA	HCl		0.12	[167]
Alumina	TNOA	HCl		0.1	[15]
Ftoroplast-4 (polymeriz.powder,mark "A")	TNOA	H_2O	0.31*	0.21*	–
Ftoroplast-4, thermal treated	TNOA	H_2O	0.29*	0.25*	[129]
Ftoroplast-4, thermal treated	TNOA	HCl	0.30	0.195	[184]
Ftoroplast-4, thermal treated	0.5 M TNOA in benzene	HCl	0.6	0.3	[173]
Kel-F	50% TNOA in xylene	HCl	1.0		
Ftoroplast-4 (polymerization powder)	0.12 M TNOA in butylbenzene	H_2SO_4	0.39*		[182]
Ftoroplast-4 (polymerization powder)	0.22 M TNOA in butylbenzene	H_2SO_4	0.46*		[182]
Glass powder "Pyrex Grade 2"	0.06 M TNOA in xylene	H_2SO_4	0.035*	0.053*	[83]
Voltalef 300	TLA +5% of octanol	HNO_3	0.40*	0.18*	[102]
Voltalef 300	0.8 N TLA in cyclohexanol	HCl	0.33*	0.13*	[101]
Chromosorb W(AW) (nonsiliconized)	Alamine-336	1.0 M KCNS+0.5 M HCl	1.14*		[166]
Tee-Six	5% Alamine-336 in toluene	HCl		0.258*	[179]
XAD-2	0.05−0.2 M Aliquat-336 in toluene	H_2SO_4		0.424*	[33]
XAD-2	0.05−0.2 M Aliquat-336 in toluene	$NaNO_3$		0.40*	[176]
Kel-F	0.1 M Aliquat-336 in $CHCl_3$		0.56		[164]
Chromosorb W(AW−DMCS)	5% Aliquat-336 in Freon-214				[46]
Activated charcoal	Di-octylpyrophosphoric acid		0.175g/g*		[170]
Microthene-710	50% HDEHP in xylene	HNO_3	0.67		
Microporous polyethylene	0.3 M HDEHP in diisopropyl-benzene		1.0	0.25	[41]
Microthene-710	0.3 M TOPO in cyclohexane	HNO_3	0.80		[170]
Kel-F	0.5 M TOPO in cyclohexane	HNO_3	1.0		[172]
Haloport-F	0.1 M TOPO + 0.1 M dibenzoyl-methane in dodecane	NH_4OH		0.17*	[180]

Support	Compound	Medium			Ref.
Ftoroplast-4 (polymeriz.powder)	Tetrabutylpyrophosphate+tetra-butylhypophosphate (1:1)	$HClO_4$	–	0.50*	[181]
Ftoroplast-4, thermal treated	Tri(2-ethylhexyl) phosphate	HBr	0.25g/g*	–	[186]
Chromosorb W(AW–DMCS)	Octane	H_2O	2.5*	0.74*	–
Ftoroplast-4 (polymeriz. powder, mark "A")	Octane	H_2O	0.40*	0.32*	–
Ftoroplast-4, thermal treated	Octane	H_2O	0.44*	0.39*	–
Chromosorb W(AW–DMCS)	Benzene	H_2O	0.70*	0.21*	–
Haloport-F	Benzene	air	0.421*	–	[157]
Ftoroplast-4 (polymerization powder)	Benzene	H_2SO_4	0.27*	–	[32]
Ftoroplast-4 (polymeriz. powder, mark "A")	Benzene	H_2O	0.46*	0.37*	–
Ftoroplast-4, thermal treated	Benzene	H_2O	0.40*	0.37*	–
Haloport-F	Toluene	air	0.425*	–	[157]
Haloport-F	o-Xylene	air	0.426*	–	[157]
Ftoroplast-4 (porous block)	Xylene	H_2O	–	0.26*	–
Chromosorb W(AW–DMCS)	CCl_4	H_2O	0.36*	0.11*	–
Ftoroplast-4 (polymeriz.powder, mark "A")	CCl_4	H_2O	0.42*	0.33*	–
Ftoroplast-4, thermal treated	CCl_4	H_2O	0.31*	0.28*	–
Microthene-710	$CHCl_3$	air	Retains badly	–	[170]
Haloport-F	$CHCl_3$	air	0.347*	–	[157]
Ftoroplast-4 (polymerization powder)	$CHCl_3$	H_2SO_4	0.22*	–	[32]
Ftoroplast-4 (polymeriz.powder, mark "A")	$CHCl_3$	H_2O	0.41*	0.31*	–
Ftoroplast-4, thermal treated	$CHCl_3$	H_2O	Is not retained		–
Ftoroplast-4, thermal treated	$CHCl_3$	HCl	0.12*	0.11*	–
Hyflo Super Cel, treated with DMCS	$CHCl_3$ or $CHCl_3$+heptane(8:2)		0.89	–	[163]
Hostalen W	$CHCl_3$+heptane (1:4)		0.80	0.295	[171]
Celite-545, treated with DMCS	Solution of dithizone in CCl_4	$HClO_4$+$NaClO_4$ (pH=4)	1.4*	0.5*	[68]
Celite-545, treated with DMCS	Solution of BPHA in $CHCl_3$	$HClO_4$+$NaClO_4$	1.4*	0.5*	[165]
Microthene-710	50% 2,2,4-trimethylpentane-1,3-diol in $CHCl_3$+C_6H_6 (1:1)	HCl	0.75	0.295	[170]
Glass powder "Pyrex Grade 2"	0.5 M HTTA in xylene	HCl	0.034*	0.043*	[83]
Ftoroplast-4 (polymerization powder)	Ether	H_2SO_4	0.31*	–	[32]
Ftoroplast-4 (porous block)	Dibutyl ether	H_2O	–	0.23*	–
Ftoroplast-4 (polymerization powder)	Isoamyl alcohol	H_2SO_4	0.41*	–	[32]
Siliconized cellulose	Pentanol, hexanol		0.5	–	[168]
Haloport-F	Hexanol	air	0.452*	–	[157]
Ftoroplast-4 (polymeriz.powder, mark "A")	Hexanol	H_2O	0.31*	0.24*	–

(Table 13 cont.)

1	2	3	4	5	6
Ftoroplast-4, thermal treated	Hexanol	H₂O	0.29*	0.26*	–
Ftoroplast-4, thermal treated	Decanol	HCl	–	0.43*	[183]
Ftoroplast-4 (porous block)	Decanol	H₂O	–	0.30*	–
Ftoroplast-4 (polymerization powder)	Cyclohexanol	H₂SO₄	0.64*	–	[32]
Haloport-F	Ethylacetate	air	0.400*	–	[157]
Haloport-F	Isoamylacetate	air	0.422*	–	[157]
Chromosorb W(AW–DMCS)	Isoamylacetate	H₂O	0.76*	0.20*	–
Ftoroplast-4 (polymeriz.powder,mark "A")	Isoamylacetate	H₂O	0.32*	0.25*	–
Ftoroplast-4, thermal treated	Isoamylacetate	H₂O	0.21*	0.18*	–
Ftoroplast-4 (polymerization powder)	Isoamylacetate	H₂SO₄	0.48*	–	[32]
Kel-F	1M tetraphenylborate-Na in isoamylacetate	0.01 M EDTA (pH=6.8)	1.0	–	[174]
Teflon-6	MIBK	HCl	–	0.316*	[177]
Tee-Six	MIBK	HCl + HF	–	0.26*	[178]
Ftoroplast-4 (polymeriz. powder, mark "A")	MIBK	H₂O	0.27*	0.21*	–
Ftoroplast-4, thermal treated	MIBK	H₂O	0.03*	–	–
Haloport-F	2-octanon	air	0.420*	–	[157]
Haloport-F	Di-isobutylketone	air	0.379*	–	[157]
Ftoroplast-4 (polymerization powder)	Cyclohexanone	H₂SO₄	0.74*	–	[32]
Kel-F	0.2–1.48 M HTTA in MIBK	CH₃COOH+ +CH₃COONa	0.91*	0.36*	[103]
Haloport-F	Acetylacetone	air	0.401*	–	[157]
Haloport-F	Nitrobenzene	air	0.079*	–	[157]
Kel-F	Polyiodides in nitrobenzene	NH₄I + I₂	0.75*	0.34*	[104]
Kel-F	Dipicrylamine in nitrobenzene	0.01 M EDTA (pH = 7)	–	–	–
Teflex	0.01 M NH₄-dipicrylaminate in nitrobenzene		0.67	–	[175]
Acetylcellulose	Saturated solution of dithizone in CCl₄+CHCl₃ (1:1)	(NH₄)₂CO₃	0.55*	–	[144]
Ftoroplast-4, thermal treated	Zn diethyldithiocarbamate in CHCl₃	CH₃COONa+ +HCl (pH=5)	Till 2	–	[169]
V_Nb-15-10	2.5·10⁻³ M solution of dithizone in CCl₄	H₂O	–	0.33	[185]
		HCl	0.18	–	[187]

*Maximum amount of extractant on the support

164

TABLE 14
HETP for different supports

Support	Size of particles μm	Extractant	Elements	HETP, mm	Ref.
Hyflo Super Cel, treated with DMCS	15	HDEHP	Rare earth	0.33−0.35	[58]
"	15	HDEHP	"	0.32−6.0	[189]
Alumina, treated with DMCS	150−420	HDEHP	"	0.43	[16]
Cellulose	−	HDEHP	"	1.0−1.1	[190]
Silicagel KSK, treated with DMCS	15	HDEHP	"	0.07−0.10	[51]
Corvic R 51/83	100−150	HDEHP	"	1.3−1.6	[38]
Thermal treated Ftoroplast-4	30−50	50% HDEHP in heptan	"	0.12−0.25	[191]
Silicagel VEB Feinchemie Eisenach, treated with DMCS	50	0.81 M HDEHP in toluene	"	0.35	[49]
"	80−100	50% HDEHP in toluene	Rare earth (macroamount)	1.1	[192]
"	120−160	HDEHP	"	3.0	[192]
Silicagel KSK, treated with DMCS	40−71	HDEHP	"	2.9−3.7	[193]
Celite-545, treated with DMCS	50	HDEHP	Cf	0.32	[69]
Hyflo Super Cel, treated with DMCS	15	Di-n-buthylphosphoric acid	Rare earth	0.15−0.34	[194]
"	15	2-ethylhexyl phosphonic acid	"	0.11	[195]
"	80	TBP	"	0.28	[59]
"	74	MHDPO	"	0.42−0.95	[196]
"	74	MEHDPO	"	1.0	[196]
Kel-F	74	MHDPO	"	0.98−2.4	[196]
Plascon 2300	74	MEHDPO	"	0.63−0.94	[196]
Hyflo Super Cel, (nontreated)	74	MEHDPO	"	0.77−2.10	[196]
Hyflo Super Cel, treated with DMCS	6−16	TNOA	Co	0.17	[197]
"	6−16	TNOA	Fe	0.2−0.4	[61]
V_{Nb-15}-10	−	TNOA	U	0.90	[198]
Voltalef 300	200−400	TLA	Cu	3.25	[101]
Celite-545, treated with DMCS	37−74	Aliquat-336	Cf, Am	0.35−0.52	[199,200]
"	50−150	Aliquat-336	Am	0.40	[201]
Kel-F	420−590	Aliquat-336	F,Cl,Br,I	1.0	[176]
Kel-F	420−590	Aliquat-336 in $CHCl_3$	"	2.0	[176]
V_{Nb-15}-10	−	TBP	U	0.57	[202]
Copolymer of polystyrene with DVB	150−300	TBP	U (macroamount)	1.56	[30]
Kel-F	75−150	TBP	"	3.4	[106]
Voltalef 300 LD	−	TBP	U	0.17−3.94	[208]
Hyflo Super Cel, treated with DMCS	10−40	TBP	Np	0.24−0.86	[56]
"	−	TBP	U	0.29−0.61	[208]
Polyurethane foam	−	TBP	Pd	1.7	[35]
Teflon-6	90−100	MIBK	Hg	3.5	[177]
Tee-Six	90−100	MIBK	Nb	0.9	[178]
Kel-F	170−350	HTTA in MIBK	Ca	4.5	[103]
Kel-F	170−350	Polyiodides in nitrobenzene	Rb	1.6	[104]
Hyflo Super Cel, treated with DMCS	6−16	Monododecylphosphoric acid	Alkali metals	0.2−0.4	[62]
Teflex, Ftoroplast-4	40−60	Dipycrylamine in nitrobenzene	Rb, Cs	0.5−2.0	[144]
Haloport-F	−	TOPO + DBM in dodecane	Li	2.8	[203]

TABLE 15
Comparison of tailing effects for different polymeric supports

Support	Volume of eluent for 100% elution of element, ml			
	Ni	Co	Fe	Ni, Co, Fe
Kel-F (polytrifluorochloroethylene)	4	4	6	14
Algoflon (polytetrafluoroethylene)	5	5	5	15
Moplen (polypropylene)	5	5	6	16
Microthene (polyethylene)	6	7	8	21
Corvic (polyvinyl chloride polyvinyl acetate)	7	6	8	21
Vipla (polyvinyl chloride)	7	8	9	24

5. LOADING OF THE STATIONARY PHASE ON THE SUPPORTS. REGENERATION OF SUPPORTS

The methods for loading the support with the stationary phase vary depending on whether one wishes to load the maximum amount of extractant or not. The following procedures may be followed:

(a) The desired amount of pure extractant is added drop by drop to a given amount of support, and the mixture is stirred for a long time in order to attain a uniform redistribution of the extractant on the solid material.

(b) The support is treated as described above, but with a solution of the extractant in a volatile diluent. The amount of diluent is chosen so as to cover the whole support, and is gradually evaporated under continuous stirring. In this way, the extractant distributed much more evenly on the support than when loaded without diluent; as a result, HETP values lower by a factor of two can be obtained[51]. A device has been described for loading the extractant on large amounts of support (up to 2 kg)[205].

(c) A solution of the extractant in a volatile solvent passes through the chromatographic column containing the support, and the solvent is removed with an air stream. The method can be applied only to non-volatile extractants (TBP, HDEHP, TNOA). The HETP values obtainable are comparable to those of the second method[206].

(d) The easiest way of loading the maximum amount of extractant on the support consists in percolating the extractant through a layer of dry support, or filling the column with a suspension of the support in the extractant. The excess extractant is displaced by subsequent elution with an aqueous phase.

(e) The support, wetted with the extractant, is mixed with an aqueous phase and the excess extractant is removed by centrifugation. The suspension is then poured into the column[161].

The prices of supports based on fluoropolymers and of high-quality siliconized kieselguhrs are rather high, especially for supports of narrow size range (about 20–30 $ per 100 g). This is why it is advisable to use the supports repeatedly for routine analysis and when large columns have to be used.

166

Fluoropolymers can easily be regenerated by successive washing on a Buchner funnel with acid (HCl, HNO$_3$), water, and volatile organic solvents (ether, ethanol or acetone) to remove water. The support is dried with an air stream, and is ready to be used again[103,104,209]. Supports based on kieselguhrs can be regenerated in the same way, but it must be kept in mind that the particles of those brands that have low mechanical strengths are easily crushed in the regeneration process. Chromosorb W is believed to withstand no more than 1 or 2 regenerations, but Chromosorb G up to 20 regenerations[207]. Supports can also be regenerated by applying the above listed sequence of washing directly to the chromatographic column.

When the same extractant is involved in both the old and the new column, it is sufficient to wash the column with pure extractant, so that at the same time the fresh extractant is loaded on the support. In some works, the pure extractant is used for the elution of the last component of the separated mixture, and the recovery of the element with the regeneration of the support can be combined [162,177,210]. Both polymers and kieselguhrs can be regenerated in this way.

6. SUPPORTS FOR PREPARATIVE AND CONTINUOUS CHROMATOGRAPHY

In the making of large diameter columns for preparative chromatography, it is difficult to obtain a homogeneous filling of the support; this leads to an irregular flow of the solution through the column bed because of wall effects, and channelling, and the HETP for large columns is generally higher than for narrow ones. As time goes on, the column bed may gradually compact itself, changing the parameters of the column.

All these troubles can be avoided by using blocks of porous Ftoroplast-4 as the column bed.

For production of whole porous Ftoroplast-4 blocks, the following procedure can be followed[211]:

The polymerization powder suspended in ethanol (about three times the volume of powder) is passed several times through a colloidal mill; a suitable amount of finely powdered mineral salt (usually NH$_4$Cl, (NH$_4$)$_2$CO$_3$ or NaCl) is then added to the mixture, which is then again passed through the mill ten times, so as to obtain a satisfactory milling and mixing of Ftoroplast-4 and salt.

The resulting mixture is filtered, dried, pressed into tablets and sintered at 380 °C for 5–7 hours. During sintering the ammonium salts volatilize. Non-volatile salts are leached out with water.

The material made in this way has a uniform rigid structure with open pores, it differs from columns made with powdered supports in having more uniform properties throughout its whole volume, that do not change during use. The porosity of the material depends on the amount of salt added for its preparation; volatile salts give rise to less uniform material; non-volatile ones allow the control of the porosity and pore size, but necessitate the water leaching step, which is rather time consuming, especially for large samples.

However, the rigid porous Ftoroplast-4 blocks obtained as above have a notable defect: it is difficult to press them close to the walls of the glass tubes normally used in chromatography, that frequently have irregular profiles and appreciably different internal diameters along their length.

Blocks of porous Ftoroplast-4 can be soldered into tubes made of thermoplastic polymers, but in this case it is difficult to produce long columns. To avoid wall effects, the clearance between the porous block and the wall of the tube can be filled with molten paraffin[162].

Preobrazhenskii[212] proposed the use of blocks of elastic porous Ftoroplast-4 that overcome such complications. A simple procedure was worked out for the preparation of such blocks, that can easily be applied in any laboratory.

Polymerization powder of Ftoroplast-4 is loosened in a coffee mill, and sintered at $380\pm10\,^{\circ}C$ for 20—30 min so as to obtain a layer about 10 mm thick. The mass is again ground in the coffee mill, and is sieved to homogeneous size fractions. The fractions are sintered once more at $380\pm10\,^{\circ}C$ for 30 min, in 20—30 mm layers. The resulting material is fairly elastic, and retains sufficient mechanical strength to be easily worked with the lathe. This permits preparation of Ftoroplast-4 blocks with a diameter a little larger than the bore of the column. Thanks to its elasticity, the block can be forced into the glass column, where it fits pressed close to the walls.

Columns prepared with elastic porous Ftoroplast-4 blocks have a homogeneous structure and ensure the uniform flow of the mobile phase, without wall or other effects. The blocks can be cut out with special cylindrical knives resembling cork drills. Glass tubes with a 10 mm bore need a knife 10.3—10.4 mm in internal diameter, and 100 mm columns a 100.9—101.0 mm knife.

The low density of the blocks of elastic porous Ftoroplast-4 (0.45—0.49 g/cm^3 compared to 0.8—1.1 g/cm^3 for rigid blocks) ensure a high permeability for aqueous solutions; this allows work at high flow rates of the mobile phase without the need for overpressure. The columns have a high efficiency, their HETP features being comparable to those obtained with columns packed with Ftoroplast-4 powder of the same grain size, as that used in making the blocks.

Analogous porous blocks can be obtained from other polymers, but PTFE is to be preferred for its exceptional chemical resistance, that ensures the constancy of its characteristics after prolonged use in contact with any reagent.

The use of blocks of porous Ftoroplast-4 has led to the successful solution of an important problem in preparative chromatography: the realization of a continuous chromatographic process[213—216]. For this purpose, a ring of porous Ftoroplast-4 is put between horizontally restricting walls. The moving eluent feeder is in contact with the inner and the moving collector of the separated fractions with the outer side of the ring. They are made of monolithic Ftoroplast-4 the hydrophobicity of which preserves the tightness of the moving contacts toward water solutions. With the rotation of the ring, which is the support of the stationary phase, each of the components of the separated mixture moves in the layer of support, along a trajectory determined by the speed of rotation, the flow rate of the mobile phase and the distribution coefficients. When these values are constant, the form and the length of the trajectory of any component are constant, which means that the distance from the window of the feeder of the starting solution up to the window of the corresponding receiver is constant.

Powdered supports cannot be used for such application, since the shape of the surface and the size of the layer of support would change with time, and the sealing of the contacts of the feeder and the collector with the ring surface could not be ensured.

In addition, the use of whole porous blocks allows working with any number of eluents of different composition which is not possible with powdered supports.

REFERENCES TO CHAPTER 5

1. A. A. LUR'E, Adsorbents and Chromatographic Supports (in Russian), Khimiya Press, Moscow, 1972.
2. I. E. NEUMARK, I. B. SLINYAKOVA, Ukr. Khim. Zh., 27 (1961) 196.
3. H. ROTZSCHE, in Aspects in Gas Chromatography (H. G. STRUPPE, Ed.), Akademie-Verlag, Berlin, p. 120, 1971.
4. J. J. KIRKLAND, in Gas Chromatography (L. FOWLER, Ed.), Academic Press, New York, pp. 77, 102, 1963.
5. D. T. SAWYER, J. K. BARR, Anal. Chem., 34 (1962) 1518.
6. F. FARRE-RIUS, J. HENNIKER, G. GUIOCHON, Nature, 196 (1962) 63.
7. D. M. OTTENSTEIN, J. Gas Chromatog., 1, N4, 11 (1963).; in Advances in Chromatography Vol. 3, Marcel Dekker, New York, p. 137, 1966.
8. G. BLANDENET, J. P. ROBIN, J. Gas Chromatog., 2 (1964) 225.
9. Lachema, Brno Materials for Chromatography.
10. Analabs. Chromatography. Chemicals and Accessories. Biochemicals, 1972.
11. I. E. NEUMARK, in Gas Chromatography (Proceedings of First All-Union Conference) (in Russian), USSR Academy of Science Press, Moscow, p. 81, 1960.
12. N. V. AKSHINSKAYA, A. V. KISELEV, Yu. S. NIKITIN, Zh. Fiz. Khim., 37 (1963) 927.
13. A. GORDON, A. MARTIN, R. SYNGE, Biochem. J., 37 (1943) 79.
14. S. P. ZHDANOV, Dokl. Akad. Nauk SSSR, 82 (1952) 281.
15. A. ABRAO, Publ. IEA No. 241, 1971.
16. A. A. POLI, CNEN−RT/CHI (70) 32, 1970.
17. J. J. KIRKLAND, Anal. Chem., 41 (1969) 218.
18. J. J. KIRKLAND, Anal. Chem., 43, N12, 36A-48A (1971).
18a. S. SPECHT, R. F. NOLTE, H.− J. BORN, 7th Radiochemical Conference, Mariánské Lázne, p.18, 1973.
19. I. M. HAIS, K. MACEK, Papirova chromatografie, Nakladatelstvi Československe Akademie Ved, Praha, p. 224, 1959.
20. E. CERRAI, G. GHERSINI, Analyst, 94 (1969) 599.
21. A. G. HAMLIN, B. J. ROBERTS, W. LOUGHLIN, S. G. WALKER, Anal. Chem., 33 (1961) 1547.
22. C. POHLANDT, T. W. STEELE, Talanta, 19 (1972) 839.
23. J. J. KIRKLAND, Anal. Chem., 35 (1963) 2003.
24. V. V. BRAZHNIKOV, L. I. MOSEVA, K. I. SAKODYNSKI, Gazov. Khromatogr., No.5., Niitekhim, Moscow, p. 62, 1967.
25. B. K. PREOBRAZHENSKII, A. V. KALYAMIN, O. M. LILOVA, L. N. MOSKVIN, B. S. USIKOV, Radiokhimiya, 10 (1968) 375.
26. G. S. KATYKHIN, M. K. NIKITIN, Programme and Abstracts of Reports, 19th Conference on Nuclear Spectroscopy, Part I, Nauka Press, Leningrad, p. 267, 1969.
27. G. S. KATYKHIN, M. K. NIKITIN, Radiokhimiya, 10 (1968) 474.
28. D. JENTZSCH, G. OESTERHELT, E. RÖDEL, H.−G. ZIMMERMANN, Z. anal. Chem., 205 (1964) 237.
29. H. ESCHRICH, P. HANSEN, ETR−239, 1969.
30. H. SMALL, J. Inorg. Nucl. Chem., 18 (1961) 232.
31. H. BERANOVA, M. NOVAK, Coll. Czech. Chem. Comm., 30 (1965) 1073.
32. B. K. PREOBRAZHENSKII, G. S. KATYKHIN, Radiokhimiya, 4 (1962) 536.

33. J. S. FRITZ, J. J. TOPPING, Talanta, 18 (1971) 865.
34. H. BERANOVA, M. TEJNECKY, Coll. Czech. Chem. Comm., 32 (1967) 437.
35. T. BRAUN, A. B. FARAG, Anal. Chim. Acta, 61 (1972) 265.
36. C. TESTA, CNEN Report RT/PROT (65) 33, 1965.
37. T. B. PIERCE, P. F. PECK, Nature, 194 (1962) 84.
38. T. B. PIERCE, R. S. HOBBS, J. Chromatog., 12 (1963) 74.
39. T. B. PIERCE, W. M. HENRY, J. Chromatog., 23 (1966) 457.
40. R. BELCHER, J. R. MAJER, G. A. H. ROBERTS, Talanta, 14 (1967) 1245.
41. US At. Energy Comm. Rept., ORNL–3627, p. 183, 1964.
42. G. L. STAROBINETS, E. R. MAZOVKO, N. A. POLYAK, Vestn. Akad. Nauk BSSR, Ser. Khim. Nauk, No. 1, (1969) 15.
43. A. GORSKI, J. MOSZCZYNSKA, Chem. Anal. (Pol.), 9 (1964) 1071.
44. R. F. HORNBECK, J. Chromatog., 30 (1967) 438.
45. TSZEN SYAN–FU,I. A. KUZIN,V. P. TAUSHKANOV, Zh. Prikl. Khim., 36 (1963) 703.
46. L. A. McCLAINE, P. NOBLE, E. P. BULLWINKEL, J. Phys. Chem., 62 (1958) 299.
47. S. D. NOGARE, J. CHIU, Anal. Chem., 34 (1962) 890.
48. J. BOHEMEN, J. H. PURNELL, in Gas Chromatography-1958 (D.H. DESTY, Ed.,), Academic Press, New York, p. 6, 1958.
49. H. GROSSE–RUYKEN, J. BOSHOLM, J. Prakt. Chemie, 25 (1964) 79.
50. I. HALASZ, M. NAEFE, Anal. Chem., 44 (1972) 76.
51. E. HERRMANN, J. Chromatog., 38 (1968) 498.
52. N. C. SAHA, J. C. GIDDINGS, Anal. Chem. 37 (1965) 822.
53. J. M. BOBBITT, A. E. SCHWARTING, R. J. GRITTER, Introduction to Chromatography, Reinhold Book Corp., New York, 1968.
54. A. B. CUMMINS, Ind. Eng. Chem., 34 (1942) 403.
55. M. KUBINOVA, Chem. Listy, 53 (1959) 850.
56. H. ESCHRICH, Z. anal. Chem., 226 (1967) 100.
57. H. ESCHRICH, Kjeller Report No. 11, 1961.
58. R. J. SOCHACKA, S. SIEKIERSKI, J. Chromatog., 16 (1964) 376.
59. S. SIEKIERSKI, I. FIDELIS, J. Chromatog., 4 (1960) 60.
60. S. LIS, E. T. JOSEFOWICZ, S. SIEKIERSKI, J. Inorg. Nucl. Chem., 28 (1966) 199.
61. W. SMUŁEK, Nukleonika, 11 (1966) 635.
62. W. SMUŁEK, S. SIEKIERSKI, J. Chromatog., 19 (1965) 580.
63. G. BLANDENET, J. P. ROBIN, J. Gas Chromatog., 4 (1966) 288.
64. L. S. ETTRE, J. Chromatog., 4 (1960) 166.
65. M. KREJČI, Coll. Czech. Chem. Comm., 32 (1967) 1152.
66. V. P. PAKHOMOV, I. A. MUSAEV, V. G. BEREZKIN, P. I. SANIN, Neftekhimiya, 7 (1967) 631.
67. K. V. ALEKSEEVA, I. A. STREL'NIKOVA, Gazov. Khromatogr., No.7, Niitekhim, Moscow, p. 70, 1967.
68. F. ŠEBESTA, J. Radioanal. Chem., 6 (1970) 41.
69. E. P. HORWITZ, C. A. A. BLOOMQUIST, D. J. HENDERSON, J. Inorg. Nucl. Chem., 31 (1969) 1149.
70. Johns-Manville Corp. Bulletin FF–133A.
71. W. J. BAKER, E.H.LEE,R.F. WALL, in Gas Chromatography (H. J. NOEBELS,R.F.WALL,N. BRENNER, Eds.), Academic Press, New York, p. 21, 1961.
72. N. I. BRYZGALOVA, T. B. GAVRILOVA, A. V. KISELEV, T. D. KHOKHLOVA, Neftekhimiya, 8 (1968) 915.
73. V. G. BEREZKIN, V. P. PAKHOMOV, V. R. ALISHOEV, Gazov. Khromatogr., No. II, Niitekhim, Moscow, p. 8, 1969.
74. A. BEVENUE, J. N. OGATA, H. BECKMANN, J. Chromatog., 35 (1968) 17.
75. B. F. RIDER, J. P. PETERSON, C. P. RUIZ, Trans. Amer. Nucl. Soc., 7 (1964) 350.
76. G. GUIOCHON, Chromatographia, 4 (1971) 404.
77. Phase Separations Ltd., Queensferry Flintshire, Gas Chromatography. Materials and Accessories.

78. E. HERRMANN, H. GROSSE–RUYKEN, N. A. LEBEDEV, V. A. KHALKIN, Radiokhimiya, 6 (1964) 756.
79. J. PITRA, Chem. Listy, 56 (1962) 495.
80. C. L. GUILLEMIN, M. LePAGE, A. J. deVRIES, J. Chromatog. Sci., 9 (1971) 470.
81. M. LePAGE, R. BEAU, A. J. deVRIES, J. Polymer Sci., C8 (1968) 119.
82. M. F. BURKE, D. G. ACKERMAN, Anal. Chem., 43 (1971) 573.
83. D. C. PERRICOS, J. A. THOMASSEN, Kjeller Report No. 83, 1964.
84. J. TRANCHANT, Manuel Pratique de Chromatographie en Phase Gazeuse, Masson et Cie, Paris, p. 52, 1964.
85. Corning Chromatography Products.
86. B. L. KARGER, K. CONROE, H. ENGHELHARDT, J. Chromatog. Sci., 8 (1970) 242.
87. R. E. MAJORS, J. Chromatog. Sci., 8 (1970) 338.
88. D. J.KENNEDY, J. H. KNOX, J. Chromatog. Sci., 10 (1972) 549.
89. D. MAPPER, J. R. FRYER, Analyst, 87 (1962) 297.
90. E. CERRAI, C. TESTA, C. TRIULZI, Energia Nucl.,9 (1962) 193.
91. S. J. LYLE, V. C. NAIR, Talanta, 16 (1969) 813.
92. R. AUBEAU, G. BLANDENET, F. LECCIA, Chromatographia, 5 (1972) 240.
93. Johns-Manville Corp. Bulletin FF–124.
94. R. STACZEWSKI, J. JANÁK, Chem. Anal. (Pol.), 7 (1962) 1057.
95. R. STACZEWSKI, J. JANÁK, Chem. Anal. (Pol.), 7 (1962) 1073.
96. J. J. CHESSICK, F. H. HEALY, A. C. ZETTLEMOYER, J. Phys. Chem., 60 (1956) 1345.
97. D. GRAHAM, J. Phys. Chem., 66 (1962) 1815.
98. J. W.WHALEN,W.H.WADE, J. J. PORTER, J. Colloid. Interface Sci., 24 (1967) 379.
99. V. V. BRAZHNIKOV, L. I. MOSEVA, K. I. SAKODYNSKY, V. A. SOKOL'SKII, Gazov. Khromatogr., No. 9, Niitekhim, Moscow, p. 80, 1969.
100. K.I.SAKODYNSKY,L. PANINA, Chromatographia, 4 (1971) 113.
101. M. PETIT–BROMET, Rapport CEA–R–3469, 1968.
102. D. GOURISSE, A. CHESNE, Rapport CEA–R–3245, 1967.
103. I. AKAZA, Bull. Chem. Soc. Japan, 39 (1966) 980.
104. I. AKAZA, Bull. Chem. Soc. Japan, 39 (1966) 585.
105. E. CERRAI, C. TESTA, J. Inorg. Nucl. Chem., 25 (1963) 1045.
106. T. J. HAYES, A. G. HAMLIN, Analyst, 87 (1962) 770.
107. M. T. RICCATO, G. HERRMANN, Radiochim. Acta, 14 (1970) 107.
108. N. TRAUTMANN, R. DENIG, G. HERRMANN, Radiochim. Acta, 11 (1969) 168.
109. G. T. BENDER, C. E. MELOAN, J. Chromatog., 45 (1969) 220.
110. F. NELSON, J. Chromatog., 20 (1965) 378.
111. A. A. TAGER, M. V. TSILIPOTKINA, A. B. PASHKOV, E. B. MAKOVSKAYA, E. I. LYUSTGARTEN, M. I. ITKINA, Plast. Massy, No. 3 (1966) 23.
112. S. J. HOPKINS, Manufact. Chemist, 38, N2, 51 (1967).
113. H. L. BRADLOW, Steroids, 11 (1967) 265.
114. Johns-Manville Corp. Bulletin FF–157B.
115. R. BOROVSKY, S. KRAIS, J. MALINSKI, J. SEIDL, Chem. Průmysl., 13 (1963) 446.
116. H. BERANOVA, M. TEJNECKY, Report UJV–2557-Ch., 1971.
117. S. I. KOL'TSOV, V. B. ALESKOVSKII, Silicagel, its Structure and Chemical Properties (in Russian), Goskhimizdat Press, Leningrad, 1963.
118. A. J. SWALLOW, Radiation Chemistry of Organic Compounds, Pergamon Press, London, p. 282, 1960.
119. J. C. BRESEE, J. R. FLANARY, J. H. GOODE, C. D. WATSON, J. C. WATSON, Nucleonics,14, N9, 75 (1956).
120. I. STRONSKI, Radiochim. Acta, 13 (1970) 25.
121. I. F. BENNETT, Chem. Engng., 62 (1955) 226.
122. L. N. MOSKVIN, L. G. TSARITSINA, Radiokhimiya, 12 (1970) 187.
123. G. S. KATYKHIN, M. K. NIKITIN, Programme and Abstracts of Reports, 15th Conference on Nuclear Spectroscopy, Nauka Press, Leningrad, p. 72, 1965.

124. L. N. MOSKVIN, L. G. TSARITSINA, Atomnaya Energ., 24 (1968) 383.
125. J. KOOI, Radiochim. Acta, 5 (1966) 91.
126. E. P. HORWITZ, C. A. A. BLOOMQUIST, H. E. GRIFFIN, US At. Energy Comm. Rept. ANL-7569, 1969.
127. M. N. SHULEPNIKOV, G. I. SHMANENKOVA, V. P. SHCHELKOVA, Yu. V. YAKOVLEV, Zh. Analit. Khim., 28 (1973) 608.
128. J. RAIS, M. KÝRŠ, R. CALETKA, J. Inorg. Nucl. Chem., 27 (1965) 463.
129. G. I. SHMANENKOVA, L. M. MIKHEEVA, G. I. ELFIMOVA, Zh. Analit. Khim., 27 (1972) 2161.
130. T. B. PIERCE, P. F. PECK, AERE-R-4969, 1965.
131. M. LEDERER, C. MAJANI, Chromatogr. Rev., 12 (1970) 239.
132. I. E. STARIK, I. A. SKULSKII, V. N. SHCHEBETKOVSKII, Radiokhimiya, 3 (1961) 428.
133. I. E. STARIK, V. N. SHCHEBETKOVSKII, I. A. SKULSKII, Radiokhimiya, 4 (1962) 393.
134. F. K. WEST, P. W. WEST, F. A. IDDINGS, Anal. Chim. Acta, 37 (1967) 112.
135. V. I. LEVIN, L. N. KURCHATOVA, A. B. MALININ, Radiokhimiya, 12 (1970) 529.
136. J. PELKIČ, J. BAR, Jaderna Energia, 12 (1966) 126.
137. J. RYDBERG, B. RYDBERG, Svensk. Kem. Tidskr., 64 (1952) 200.
138. B. EDROTH, Acta Chem. Scand., 23 (1969) 2636.
139. I. A. SKULSKII, V. V. GLAZUNOV, Radiokhimiya, 7 (1965) 703.
140. A. K. BURNHAM, G. V. CALDER, J. S. FRITZ, G. A. JUNK, H. J. SVEC, R. WILLIS, Anal. Chem., 44 (1972) 139.
141. V. I. MISHIN, S. L. DOBYCHIN, Zh. Prikl. Khim., 43 (1970) 1584.
142. V. V. GLAZUNOV, I. A. SKULSKII, Radiokhimiya, 8 (1966) 604.
143. F. L. MOORE, A. JURRIAANSE, Anal. Chem., 39 (1967) 733.
144. M. KÝRŠ, L. KADLECOVA, J. Radioanal. Chem., 1 (1968) 103.
145. A. G. POPOV, Zh. Prikl. Khim., 41 (1968) 127.
146. V. G. BEREZKIN, V. M. FATEYEVA, Chromatographia, 4 (1971) 19.
147. G. A. HOWARD, A. J. P. MARTIN, Biochem. J., 46 (1950) 532.
148. I. P. ALIMARIN, Yu. V. YAKOVLEV, O. V. STEPANETS, J. Radioanal. Chem., 11 (1972) 209.
149. R. H. A. CRAWLEY, Nature, 197 (1963) 377.
150. H. F. ALY, M. A. EL-HAGGAN, Mikrochim. Acta, No. 1 (1971) 4.
151. E. HERRMANN, G. PFREPPER, D. CHRISTOV, Radiochim. Acta, 7 (1967) 10.
152. M. G. ZEMSKOVA, N. A. LEBEDEV, SH. G. MELAMED, O. F. SAUNKIN, G. V. SUKHOV, V. A. KHALKIN, E. HERRMANN, G. I. SHMANENKOVA, Zavod. Lab., 33 (1967) 667.
153. G. W. HEUNISH, Anal. Chim. Acta, 45 (1969) 133.
154. I. AKAZA, T. KIBA, T. KIBA, Bull. Chem. Soc. Japan, 43 (1970) 2063.
155. E. R. SCHMID, Monatsh. Chem., 101 (1970) 1330.
156. R. DENIG, N. TRAUTMANN, G. HERRMANN, J. Radioanal. Chem., 6 (1970) 57.
157. J. S. FRITZ, C. E. HEDRICK, Anal. Chem., 34 (1962) 1411.
158. MA HUI-CHANG, NEE CHE-MING, LIANG SHU-CHUAN, Scientia Sinica, 14 (1965) 1176; Acta Chim. Sinica, 30 (1964) 388.
159. A. V. KALYAMIN, Radiokhimiya, 5 (1963) 748.
160. I. P. ALIMARIN, T. A. BOLSHOVA, Zh. Analit. Khim., 21 (1966) 411.
161. V. I. LEVIN, L. N. KURCHATOVA, A. B. MALININ, Radiokhimiya, 11 (1969) 210.
162. G. P. GORYANSKAYA, B. Ya. KAPLAN, I. V. KOVALIK, Yu. I. MERISOV, M. G. NAZAROVA, Zh. Analit. Khim., 27 (1972) 1498.
163. J. SJÖVALL, Acta Physiol. Scand., 29 (1953) 232.
164. J. F. WEISS, A. D. KELMERS, Biochemistry, 6 (1967) 2507.
165. F. ŠEBESTA, A. LAZNIČOVA, J. Radioanal. Chem., 11 (1972) 221.
166. E. E. KAMINSKI, J. S. FRITZ, J. Chromatog., 53 (1970) 345.
167. E. CERRAI, C. TESTA, J. Chromatog., 6 (1961) 443.
168. K.-G. WAHLUND, K. GRÖNINGSSON, Acta Pharm. Suecica, 7 (1970) 615.
169. T. B. PIERCE, P. F. PECK, Analyst, 86 (1961) 580.

170. C. TESTA, L. STACCIOLI, Analyst, 97 (1972) 527.
171. O. WISS, U. GLOOR, Z. Physiol. Chem., 310 (1958) 260.
172. C. TESTA, Radiological Health and Safety in Mining and Milling of Nuclear Materials, Vol. 2, IAEA, Vienna, p. 489, 1964.
173. C. E. ESPAÑOL, A. M. MARAFUSCHI, J. Chromatog., 29 (1967) 311.
174. C. CESARANO, G. PIGNETTI, C. TESTA, J. Chromatog., 19 (1965) 589.
175. C. TESTA, C. CESARANO, J. Chromatog., 19 (1965) 594.
176. L. RAMALEY, W. A. HOLCOMBE, Anal. Letters, 1 (1967) 143.
177. J. S. FRITZ, C. L. LATWESEN, Talanta, 14 (1967) 529.
178. J. S. FRITZ, L. H. DAHMER, Anal. Chem., 40 (1968) 20.
179. J. S. FRITZ, R. K. GILLETTE, Anal. Chem., 40 (1968) 1777.
180. D. A. LEE, J. Chromatog., 26 (1967) 342.
181. J. MIKULSKI, K. A. GAVRILOV, V. KNOBLOCH, Nukleonika, 10 (1965) 81.
182. A. S. NAZAROV, B. V. GROMOV, Trans. Moscow Chemical Techn. Inst. Mendeleev, 47 (1964) 159.
183. N. M. KUZMIN, A. V. EMELIANOV, S. V. MESHCHANKINA, V. I. SABATOVSKAYA, I. A. KUZOVLEV, T. I. ZAKHAROVA, Zh. Analit. Khim., 26 (1971) 282.
184. I. P..ALIMARIN, I. M. GIBALO, N. A. MYMRIK, Vestnik Mosk. Univ. Khim., No. 4 (1971) 457.
185. O. V. STEPANETS, Yu. V. YAKOVLEV, I. P. ALIMARIN, Zh. Analit. Khim., 25 (1970) 1906.
186. I. P. ALIMARIN, T. A. BOLSHOVA, T. G. DOLZHENKO, G. M. KANDYBINA, Zh. Analit. Khim., 26 (1971) 741.
187. V. SPÉVÁČKOVA, M. KŘIVANEK, Radiochem. Radioanal. Letters, 3 (1970) 63.
188. L. N. MOSKVIN, L. I. SHUR, L. G. TSARITSINA, G. T. MARTISH, Radiokhimiya, 9 (1967) 377.
189. S. SIEKIERSKI, R. J. SOCHACKA, J. Chromatog., 16 (1964) 385.
190. E. CERRAI, C. TESTA, C. TRIULZI, Energia Nucl., 9 (1962) 377.
191. B. K. PREOBRAZHENSKII, L. N. MOSKVIN, G. S. KATYKHIN, Radiokhimiya, 10 (1968) 373.
192. H. GROSSE–RUYKEN, J. BOSHOLM, G. REINHARDT, Isotopenpraxis, 1 (1965) 124.
193. A. I. MIKHAILICHENKO, R. M. PIMENOVA, Zh. Prikl. Khim., 42 (1969) 1010.
194. B. KOTLINSKA–FILIPEK, S. SIEKIERSKI, Nukleonika, 8 (1963) 607.
195. I. FIDELIS, S. SIEKIERSKI, J. Chromatog., 17 (1965) 542.
196. J. W. O'LAUGHLIN, D. F. JENSEN, J. Chromatog., 32 (1968) 567.
197. W. SMUŁEK, K. ZELENAY, Rept. Inst. Badan Jadrow. PAN, No. 677, 1965.
198. H. BERANOVA, M. TEJNECKY, Coll. Czech. Chem. Comm., 37 (1972) 3579.
199. E. P. HORWITZ, K. A. ORLANDINI, C. A. A. BLOOMQUIST, Inorg. Nucl. Chem. Letters, 2 (1966) 87.
200. E. P. HORWITZ, L. I. SAURO, C. A. A. BLOOMQUIST, J. Inorg. Nucl. Chem., 29 (1967) 2033.
201. E. P. HORWITZ, C. A. A. BLOOMQUIST, K. A. ORLANDINI, D. J. HENDERSON, Radiochim. Acta, 8 (1967) 127.
202. M. BERANOVA, M. TEJNECKY, Report UJV–2567–Ch., 1971.
203. D. A. LEE, Isotope Effects in Chemical Processes (Advances in Chemistry, Ser. 89), Washington, D. C. p. 57, 1969.
204. B. Z. EGAN, J. E. CATON, A. D. KELMERS, Biochemistry, 10 (1971) 1890.
205. C. W. HANCHER, A. D. RYON, Biotechnol. Bioengng., 12 (1970) 1099.
206. N. HAYASHI, S. INOUE, K. TAKEUCHI, Japan Analyst, 19 (1970) 907.
207. A. D. KELMERS, H. O. WEEREN, J. F. WEISS, R. L. PEARSON, M.P. STULBERG, G. D. NOVELLI, Methods in Enzymology, Vol. 20, Part C. p. 9, 1971.
208. H. ESCHRICH, ETR–249, 1969.
209. S. S. YAMAMURA, Anal. Chem., 36 (1964) 1858.
210. J. S. FRITZ, R. T. FRAZEE, G. T. LATWESEN, Talanta, 17 (1970) 857.

211. D. D. CHEGODAEV, Z. K. NAUMOVA, Ts. S. DUNAEVSKAYA, Fluoroplasts, Goskhimizdat Press, Leningrad, 1960.
212. B. K. PREOBRAZHENSKII, L. N. MOSKVIN, A. V. KALYAMIN, O. M. LILOVA, B. S. USIKOV, Radiokhimiya, 10. (1968) 377.
213. L. N. MOSKVIN, L. G. TSARITSINA, Radiokhimiya, 10 (1968) 740.
214. L. N. MOSKVIN, L. G. TSARITSINA, Radiokhimiya, 12 (1970) 730.
215. L. N. MOSKVIN, L. G. TSARITSINA, Radiokhimiya, 12 (1970) 737.
216. L. N. MOSKVIN, L. G. TSARITSINA, Radiokhimiya, 14 (1972) 588.

CHAPTER 6

EXTRACTION CHROMATOGRAPHY OF METALLIC
AND NON-METALLIC IONS

I. STRONSKI

INTRODUCTION

INTRODUCTION

Numerous reviews have been recently published on rapid methods of separation of chemical elements and especially of radioisotopes[14,16,49–51,58,111,112]. Real success has been achieved by applying gas-phase methods which make possible the preparation of radionuclides of short half-lives of the order of tenths of a second[15,46,50].

Extraction chromatography is of importance among the various methods used for the separation of chemical elements. The copious literature on the subject, presented in the form of bibliographies[33,34] and reviews[20–22,56,97,101,105] illustrates the versatile applicaton of this method in the field of the separation of radioisotopes and of other species.

The comprehensive bibliography of Zolotov et al.[13] concerning extraction includes also extraction chromatography which is, however, "diffused" in the extensive literature cited. Some papers on extraction chromatography are discussed in the monograph of Korkisch[57].

Paper and thin-layer chromatography, apart from column chromatography[56,101], have been discussed by several authors[18,20,22,82,100]. Monographs on extraction with high-molecular-weight amines[70] and with TOPO[109,110], as well as compilations of extraction data for inorganic compounds[52], should also be cited.

1. RADIOCHEMICAL SEPARATIONS AND PRODUCTION OF CARRIER-FREE RADIOISOTOPES. THE SEPARATION OF MOTHER-DAUGHTER PAIRS

Siekierski was the first to apply in chromatography the selectivity of reagents normally used in liquid-liquid extraction and the separation of zirconium and niobium in the TBP–HNO_3 system is reported in his first work[90]. The stationary phase consisted of silicone-coated Hyflo Super Cel of ca. 0.08 mm grain size on which TBP was loaded in the amount of 0.6 ml per gram of the support. Using a small column, 5 mm in diameter and 50 mm in length, this author separated ^{95}Zr and carrier-free ^{95}Nb, i.e. chemical elements whose atomic numbers are 40 and 41, respectively. The results are presented in Fig. 1.

TBP was also applied as the stationary phase by Herrmann et al.[29] for the separation of the same radioisotopes, 6 N and 2.5 N HCl being used as the eluent.

The problem of separation of the ^{95}Zr and ^{95}Nb isotopes was investigated by several authors, among others by Caletka[19] who impregnated Whatman No.1 paper with BPHA and TTA, and also adsorbed the extracting agents on Teflon powder of 3 μm grain size. This author obtained the radiochemically pure ^{95}Zr isotope using a mobile phase which consisted of HCl added with a small amount of $NH_4 F$ or oxalic acid, or of HNO_3 added with $H_2 O_2$.

Niobium and zirconium have also been separated using TTA on Ftoroplast-4 powder, niobium being eluted with 6 N HCl and zirconium with 2.5 N HF[81].

In order to elute zirconium first, Pierce et al. used TiOA and eluted both elements with $H_2 SO_4$[79].

The problem of the preparation of carrier-free radioisotopes was the subject of numerous papers. Among them, the preparation of the following radionuclides from reactor, cyclotron, and other targets should be mentioned: ^{54}Mn from a Fe target, $^{58,60}Co$ from a Ni target, ^{199}Au from a Pt target, ^{184}Re from a W target, ^{206}Bi from a Pb target; and from targets irradiated with high-energetic protons of the order of hundreds of MeV, such as, e. g., $^{72,73}Sc$, $^{206-210}Po$, ^{225}Ac, and ^{225}Ra; or carrier-free radioisotopes derived from mother isotopes, such as: ^{90}Y, ^{99m}Tc, ^{125m}Te, ^{212}Pb, ^{212}Bi, etc.

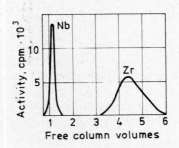

Fig. 1. Separation of niobium and zirconium[90].

Silicone-coated Hyflo Super Cel, ϕ=0.08 mm, 60% TBP–4.6 N HNO_3 +0.3% H_2O_2. Flow rate:0.12 ml/min Column size : ϕ 5 mm, length = 50 mm; Concentaration – Nb and Zr were carrier-free. (By courtesy of the authors)

The ^{54}Mn radioisotope was obtained from a spectrally pure iron target bombarded with thermal neutrons. It was eluted from a TNOA loaded silica-gel column by means of 4 N HCl, whereas Fe was recovered with 0.2 N HNO$_3$. This method allows micro-, as well as macro-amounts of both metals[92] to be separated.

The separation of the ^{56}Mn isotope from potassium permanganate bombarded with thermal neutrons and enriched by the Szilard-Chalmers method has also been reported[113]. For this purpose, HDEHP loaded on Ftoroplast-4 and HCl as eluent were used, the Mn^{4+} and MnO$_4^-$ ions being separated. The enrichment coefficient amounted to 10^4.

Smułek et al.[94] obtained the ^{58,60}Co isotopes from a Ni or NiO target bombarded with thermal neutrons. This author used TNOA and HCl, and eluted Ni by means of 8 N HCl, whereas the Co isotopes were recovered with water or diluted HCl.

Siekierski et al.[37], in an investigation on the behaviour of arsenic at various degrees of oxidation, obtained a carrier-free isotope of ^{77}As of high purity from a germanium target bombarded with thermal neutrons. The system used was TBP and HCl, and allowed for the separation of tracer amounts of arsenic from macro amounts of germanium.

The separation of arsenic and selenium was described by Moskvin[71] who also used a TBP–HCl system.

TOPO was used in the column method, on a bed of Teflon, by Stronski et al.[102], who eluted ^{90}Sr with 0.5 N HNO$_3$ and then the carrier-free ^{90}Y with 9.0 N HNO$_3$.

Both radioisotopes have also been separated in the system HDEHP–HClO$_4$[64], by applying paper, as well as column chromatography, cellulose powder being used in the latter case.

An analogous system was used for the determination of ^{90}Y by Testa et al.[106,107] in biological materials.

In the case of molybdenum bombarded with thermal neutrons, the radioactive ^{99}Mo isotope is formed, which yields the ^{99m}Tc radionuclide. Both isotopes can be separated in the system HDEHP–1 N HCl, Mo being retained on the column, whereas the carrier-free ^{99m}Tc isotope is eluted. Studies on the separation of these isotopes were carried out by Herrmann et al.[29], who used a column, 300 mm in height and 4 mm in diameter, filled with Hostaflon C2.

The carrier-free radioisotope ^{113m}In was separated from the mother ^{113}Sn nuclide, both in the anionic form, by Stronski[99], who used for this purpose a Kel-F column on which Amberlite LA–2 was adsorbed, and eluted both isotopes with 10 N HCl.

HDEHP was used for the separation of the carrier-free ^{125m}Te isotope from the mother quinquevalent radioactive ^{125}Sb isotope. Tellurium was eluted by means of 9 N HCl added with KClO$_3$. The extracting agent was loaded on Hostaflon C2, and the separation process was carried out at a raised temperature[29].

Telluric acid of natural isotope compositon, bombarded with 14 MeV neutrons yields a variety of antimony isotopes which were separated by means of a rapid extraction chromatographic procedure using DIBC adsorbed on Hostaflon C2 and 9 N HCl added with ClO$_3^-$ ions. The presence was established of the ^{126}Sb, ^{128}Sb, and ^{130}Sb isotopes, of 19, 10, and 6 min. half-lives, respectively. The results were obtained 4 min. after irradiation, the γ spectra measurements being carried out in 3 min[28,29].

Antimony and tellurium have also been separated in the system TBPP-HCl, as well as in the (TBPP + TBHP)–HCl system[67]. The separation of Sb(V) and Sn(IV) was carried out on a column in the system MIBK– 8 N HCl[43].

13*

The ^{184}Re radioisotope, which served for investigations on internal conversion of electrons, was separated in the system cyclohexanol–H_2SO_4 or HCl, from a W target irradiated with 13.6 MeV deuterons[30]. A 70·10 mm Ftoroplast-4 column was used, and the flow rate of the eluents was 0.5 ml·min^{-1}. Rhenium, which was "adsorbed" on the column, was re-extracted with water.

The carrier-free ^{199}Au isotope is prepared according to the reaction:

$$^{198}\text{Pt (n, }\gamma)^{199}\text{Pt} \xrightarrow[\text{31 min}]{\beta^-} {}^{199}\text{Au} \xrightarrow[\text{3.2 d}]{\beta^-} {}^{199}\text{Hg (stable)}$$

A spectroscopically pure sample of Pt was bombarded by a stream of thermal neutrons and, after dissolution of the target in aqua regia, washed out by means of 7 N HCl from the siliconized silica gel bed on which TBP was loaded. The ^{199}Au nuclide was recovered by means of 14 N HNO_3 [38]. The carrier-free ^{199}Au isotope was also obtained using a cellulose column loaded with 32% HDEHP, Pt being eluted with 6 N HCl saturated with Cl_2 and Au with 2 N HCl[64].

Much attention has been paid to the problem of the separation of Pb and Bi. The carrier-free ^{206}Bi isotope is formed upon irradiation of natural Pb with deuterons. Cerrai and Testa separated the two metals in the system TOPO–HCl on a cellulose powder column[23]. Stronski et al. achieved the same separation on a Teflon powder column in the system TOPO–HNO_3, Pb being eluted with 0.1 N HNO_3 and Bi with 10 N HNO_3[102]. This latter paper describes the separation of ^{212}Pb and ^{212}Bi from the disintegration products of thoron gathered on an Au plate. The two radioisotopes have also been separated in the TBP–HCl system[96], and in the HDEHP–HCl system[103].

The thiourea complexes of both metals have recently been separated in the system TBP–$HClO_4$[104].

Several papers describe the preparation of the ^{233}Pa isotope from Th irradiated with thermal neutrons. The isotope is formed according to the reaction:

$$^{232}\text{Th (n, }\gamma)^{223}\text{Th} \xrightarrow[\text{26.6 min}]{\beta^-} {}^{233}\text{Pa} \xrightarrow[\text{27.4 d}]{\beta^-} {}^{233}\text{U}$$

By applying the TBP–HCl system it is possible to separate Th from carrier-free Pa, which is washed out by means of 5% oxalic acid[36]. The same system was used by Herrmann et al.[29], who washed out Pa with 3 N HCl. The same authors obtained the ^{234m}Pa isotope from uranium as UX_2 in 5–10 sec by means of DIBC[28].

Thorium and Pa may be separated from U in the TNOA–HCl system[68]. The Aliquat 336–HCl system was used for the separation of Th, Pa, and U[98]. In this case, the extracting agent was adsorbed on Kel-F powder; Th, which does not form complexes with Aliquat-336, passed through the column when eluted with a 10 N HCl, whereas Pa was eluted using 2 N HCl.

The ^{234}Th isotope, i. e. UX_1, which is always present in uranium, was obtained by applying the TBP–HCl system. The extracting agent was in this case absorbed on Ftoroplast-4 of grain size ranging between 100 and 200 μm. Thorium was eluted first, by means of 6 N HCl, and then uranium by means of 0.1 N HNO_3[88].

The end of this paragraph is devoted to the presentation of some analytical problems connected with the preparation of various radioisotopes arising from the action of high energy protons, their energy amounting to 660 MeV.

At Dubna (USSR), a method has been developed for the preparation of selenium isotopes from a strontium target[71] and from yttrium and strontium targets in the form of $SrCO_3$[74]. Carrier-free isotopes of $^{72,73}Se$ can be obtained by using the TBP–HBr system on Ftoroplast-4, and eluting the column with water saturated with Cl_2. The selenium halides, however, are not a convenient material for the separation, on account of their volatility and the ease of passing to elementary selenium.

Khalkin et al. used a Teflon bed loaded with TBP in order to separate polonium isotopes from a bismuth oxide target[83]. The $^{206-210}Po$ isotopes were retained from 6 N HCl and eluted with 2 N HF; Bi was not extracted in such conditions, and passed through the column.

The radiochemically pure isotopes ^{225}Ac and ^{225}Ra were prepared from an irradiated Th target by applying the same extractant as the stationary phase. The target was dissolved in concentrated HNO_3 added with HF, and thorium, as well as other nuclides formed in this process (such as Pa, Zr, Hf, and Nb), were retained on the column, whereas the eluate contained the isotopes of actinium, radium, alkaline elements, alkaline earth, rare earth elements, and others. When this eluate was next introduced on a column loaded with HDEHP, Ac was eluted with 0.2 N HNO_3 and the Ra isotope with 4 N ammonium acetate[73].

The $MsTh_1$ (i.e. ^{228}Ra) undergoes a β^- transformation to $MsTh_2$ (i.e. ^{228}Ac) and both Ra and Ac radioisotopes may be easily separated using a HDEHP–HCl system[103]. Figure 2 illustrates the results of the separation of the two isotopes.

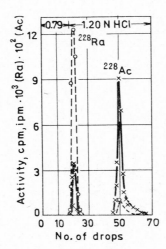

Fig. 2. Separation of radium and actinium [103].
Kel-F, 200–350 mesh, 0.15 M HDEHP–HCl. Flow rate : 0.05 ml • 0.2 cm^{-2}•min^{-1}; Column size: ϕ 5 mm, length = 55–60 mm; Concentration: Ra and Ac were carrier-free

179

2. THE SEPARATION OF ALKALI AND ALKALINE EARTH METALS

In the previous paragraph a review was presented on the separation of various chemical elements with atomic numbers differing by unity — a situation which is most interesting for radiochemists. The separations of elements belonging to a given vertical column of the Periodic Table, i. e. of elements exhibiting similarities in their chemical properties, will now be reviewed.

Prior to comments on the separations of alkaline metals and of alkaline earths, reference should be made to the attempts to separate stable ^6Li and ^7Li isotopes with the aim of obtaining samples enriched in ^6Li. Lee applied Teflon powder column on which a DBM + TOPO mixture was loaded[61]. Using a LiOH solution as the mobile phase, ^6Li was found to be enriched in the aqueous phase. The separation coefficient, however, amounts only to 1.003. Such a method of separation of lithium isotopes, therefore, has no practical value, expecially when compared with electromagnetic methods. It can be successfully applied however to studies on complexes of stable isotopes soluble in organic solvents.

The same author loaded on the column a mixture of 0.1 M DBM and 0.1 M TOPO, or 0.18 M DBM in 50% TBP in dodecane, for the separation of lithium from other alkaline elements: 3.2 N NH_4OH, 1.6 N NH_4OH, 0.1 N KOH, and 0.6 N HCl were used as the eluents[60]. Sodium was eluted by means of 0.1 N KOH and lithium, by means of 0.6 N HCl. The separation of lithium and sodium is difficult on account of the low value of their separation factor.

Akaza studied the problem of separation of the alkali metals in the form of polyiodides[1 By applying Kel-F on which nitrobenzene, iodine, and ammonium iodide was absorbed, K and Rb were separated by eluting the column with 1 N HCl, while Na, K, Rb, and Cs were separated by washing out Na, K, and Rb with 1 N HCl and Cs with 6 N HCl. Using water as the eluent, Li was separated from Na.

The best results in separating the ions of alkaline elements were obtained by Smułek and Siekierski[93], who applied the systems H_2MDP and 0.04 N $LiNO_3$ + 0.01 N HNO_3, or 0.04 N LiCl + 0.01 N HCl, or water. A chromatogram illustrating their results is presented in Fig. 3.

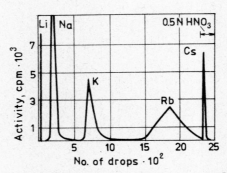

Fig. 3. Separation of alkaline elements[93].

Silicone-coated Hyflo Super Cel, ϕ = 0.08 mm, H_2MDP-H_2O. Flow rate : 0.3 – 0.5 ml · cm^{-2} · min^{-1}; Column size : ϕ = 3 mm, length = 100 mm; Concentration: Na = K = Rb = 0.1 mg, Cs was carrier-free. (By courtesy of the authors)

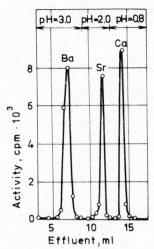

Fig. 4. Separation of barium, strontium, and calcium[53].
Silicone-coated Hyflo Super Cel, ϕ 32–58 μm, 13% HDEHP – 0.5 M NaNO$_3$ of varying pH. Flow rate:
1 ml/15–20 min; Column size : ϕ 3 mm, length 130 mm; Concentration : Ba = 0.01 mg, Ca = 0.02 mg,
Sr was carrier-free. (By courtesy of the author)

The separation of Cs and Rb has been studied by applying a nitrobenzene solution of dipicrylamine, loaded on various neutral carriers, and 0.15 M ammonium carbonate as the eluent[59], as well as by using t-BAMPB loaded on Ftoroplast-40 and 0.5 to 1 N NaOH as the mobile phase[84].

The only paper on the separation of Ca and Sc is that of Siekierski et al., who separated Ca from 5 mg of Sc in the TBP–HCl system by eluting both elements with 6 N HCl[91].

A mixture of alkaline earth elements and yttrium was separated in the MHDPO–HNO$_3$ system, and the influence of other metal ions on the results of separation of Y and Ca was studied by applying the column technique[78].

Lieser et al. studied the separation of Ca and Sr applying HDEHP and 0.03 N HCl (for Sr) and 1 N HCl (for Ca) [63]. The separation is complete when the amounts of Ca and Sr ions are equal. When the ratio between the two elements is 1 to 1000, then 0.1% of Ca ions is present in the Sr fraction, or vice versa.

Barium, Sr, Ca, and Mg have been separated as complexes with TTA by applying a 0.15 M solution of TTA in MIBK absorbed on Kel-F, and eluting with 0.5 N ammonium acetate (pH ranged between 5.5 and 6.5)[2].

By using HDEHP, and NaNO$_3$ solutions of different pH values as the eluent, fractions of milligram of Ca and Ba and carrier-free Sr ions were separated. Figure 4 presents the chromatogram of these ions[53].

181

3. THE SEPARATION OF I B AND II B SUB-GROUP ELEMENTS

These elements constitute the so-called copper and zinc sub-groups, and the problem of their separation has engaged considerable attention. The chromatographic behaviour of Cu, Ag, Zn, Cd, and Hg have been determined using paper, as well as column chromatography, in the systems comprising neutral organophosphorous compounds, e. g. TBP, TOPO, MHDPO, and inorganic acids, such as HCl, HNO_3 and $HClO_4$[77].

Nelson developed a method of separation of Ag^+ from several other metal ions, e. g. alkaline metals and some di-, tri-, and tetravalent elements, in the system 8 N HNO_3-TiOTI the organic phase being adsorbed on Fluoropak of 20—80 US. mesh[76].

Gold can be easily separated from a variety of metal ions due to the differences in the distribution coefficients in the TBP—HCl system. Preobrazhenskii et al. separated Au from composite mixtures of elements by applying a diethyl ether-HCl system[80].

Apart from the most frequently mentioned organic phosphorous compounds used in the extraction of the copper and zinc sub-group elements, particular reference should be given to the reports where high molecular weight amines were used as extracting agents. These agents were applied for the first time by Cerrai and Testa[20,21,105] for extraction chromatographic purposes, and since then they have been repeatedly used as stationary phases.

Amberlite LA-2 and HCl were used for the separation of Zn and Cd, the extracting agent being adsorbed on cellulose powder[62].

4. THE SEPARATION OF ALUMINIUM GROUP METALS

The mixture of Al, Ga, In, Tl was separated for the first time by Cerrai and Ghersini [35] in the system HDEHP-HCl. The authors have observed a double peak of Ga due to slow chlorocomplexation, presumably stripped through a $GaCl_x$ formation (see Fig. 2., Chapter 14). Later the problem of separation of Ga, In, and Tl was undertaken by Fritz et al., who applied a column with polystyrene powder (20—50 mesh) on which IPE and MIBK were loaded[40]. The separation of these elements was achieved from HBr solutions: the technique developed by these authors enabled analytical separations of Pb-In alloys to be carried out. The result of separation of the trivalent Ga, In, and Tl ions is illustrated in Fig. 5.

Alimarin et al. achieved the separation of Ga^{3+} from zinc using TBP[5] and of gallium from indium[17]. The authors also succeeded in concentrating the Ga^{3+} ion from solutions of various chlorides, such as Li, K, Na, Mg, Ca, Zn, and Cd, and from hydrochloric acid[10]. Furthermore, Alimarin separated microamounts of aluminium from gallium in the system NH_4SCN—HCl—TBP[11] and investigated the influence of temperature on the extraction of various metal salts, including gallium[32]. The application of TEHP and DEHΦP in the chromatographic extraction of In and Ga was also studied, the ratio of the two metals being $1:1.10^4$[7]. Again the same author and his co-workers [6—9] applied TBP for the enrichment of several elements, especially gallium, and investigated the influence of large amounts of chlorides on the process of enrichment of trace amounts of Ga and other metal salts from thiocyanate solutions[31].

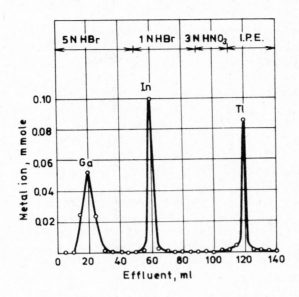

Fig. 5. Separation of gallium, indium, and thallium[40].
Amberlyst XAD-2 = polystyrene polymer, 80–100 mesh, IPE – HBr. Flow rate : 1.0 – 1.5 ml·min^{-1};
Column size : ϕ = 10 mm, length = 120 mm; Concentration : Ga = In = Tl = 100 μmole. (By courtesy
of the authors)

Activation analysis was applied for the determination of Cr, Mn, Co, Ni, Cu, and Zn
impurities in gallium arsenide, the various elements being separated in the TNOA–HCl
and TBP–HCl systems[89].

5. THE SEPARATION OF IV B SUB-GROUP ELEMENTS

The analysts have shown much interest in the tetravalent ions of Ti, Zr, and Hf. Hafnium and Zr were separated in the TBP–HNO$_3$–NH$_4$NO$_3$ system[27]; Sastri et al. achieved the separation of 4.0 mg of Ti from 4.0 mg of Zr by applying HDNNS, the column
being eluted by 0.5 N and 9 N HCl[86].
A TBP–HCl system was used for the separation of Zr and Hf, TBP being absorbed on
a Celite support[108]. Zr and Hf were also separated in the system MIBK–(NH$_4$SCN +
+ (NH$_4$)$_2$SO$_4$) [39].
The elements of the IV B sub-group may be separated also by means of liquid anion-
exchangers of the type of primary, secondary, and tertiary high molecular weight amines
or ammonium compounds, such as Aliquat–336. Cerrai and Testa separated 9.7 mg of
Zr from 0.3 mg of Hf using a TNOA–(8 N HCl + 5% conc. HNO$_3$) system on a cellulose
powder column[24].

6. THE SEPARATION OF CHROMIUM SUB-GROUP ELEMENTS

Only a few papers describe the separation of elements of the VI B group with TBP or other organophosphorous compounds as the extracting agents. Much more work has been done on the separation of these elements by means of high molecular weight amines

Chromium ions at two oxidation states, namely Cr(III) and Cr(VI), were separated on a cellulose column in the 0.1 N TOPO—HCl system[23], and Sastri et al.[86] separated $[Cr(H_2O)_5]^{3+}$ and $[Cr(H_2O)_4Cl_2]^+$ in the system HDNNS—HCl. Tungsten has been separated from other elements in the TBP—HCl system[95]; W and Mo were separated in the same system, tartaric acid being added to HCl[72].

A method of determination of Mo in steel has been developed, advantage being taken of the fact that when MIBK is used, a variety of metal ions can be separated from Mo by means of 6 N HCl[42].

TNOA was also applied in studies of "sorption" and of separation, among others, of Mo and W; it was also applied to the isolation of these elements from Zr and Nb[47] .

The chromium group elements were separated by Stronski, who applied Aliquat-336 and HCl, the extracting agent being loaded in amounts of 5% by weight with respect to the Kel-F support[98]. The chromatogram thus obtained is shown in Fig. 6.

Fritz et al. applied Aliquat-336 and HCl in paper and column chromatography for the separation of W and Mo, the two ions being eluted with 0.3 M H_2SO_4 and 4 M H_2SO_4 added with 0.5% H_2O_2[45].

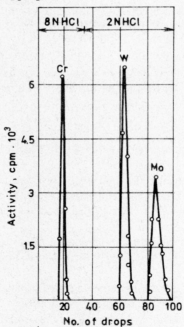

Fig. 6. Separation of hexavalent chromium, tungsten, and molybdenum ions[98]. Kel-F, 250—325 mesh, about 0.1 M Aliquat-336—HCl. Flow rate : 0.05 ml · 0.2 cm^{-2}·min^{-1}; Column size : ϕ =5 mm, length = 60 mm; Concentration : 2 · 10^{-6} — 2 · 10^{-4} M

7. THE SEPARATION OF IRON SUB-GROUP ELEMENTS

This paragraph comprises the VIII B sub-group elements, and a description of the separation of the two different oxidation states of iron, Fe(II) and Fe(III), is given at first. The first paper on this subject was published by Cerrai and Testa, who separated the two ions on a cellulose powder column in the TOPO—HCl system, 2 N HCl and 0.5 M H_2SO_4 being used for elution of Fe(II) and Fe(III), respectively[23]. The same system was applied for the separation of Ni, Co and Fe in amounts of 5 mg each[25]. The results are presented in Fig. 7.

Akaza et al. carried out the separation of $Fe(CN)_6^{3-}$ and $Fe(CN)_6^{4-}$ ions, the optimal conditions for the separation of both complexes being attained when 3 N HCl and water, respectively were used for their elution[4].

Divalent metal ions, including Co, have been separated from trivalent Fe in the TBP—HCl system, the latter element being eluted with 0.5 N HCl[66]; a similar system was used by Preobrazhenskii et al.[81], who separated Co and Fe(III).

Fritz et al. described the separation of Fe(III) from various elements in the system 2-octanone-HCl[41]. Mikulski et al. separated various divalent metal ions from Fe(III) in the TBPP—HCl system[67].

O'Laughlin et al. used neutral organophosphorous compounds, for separation purposes such as TBP, TOPO, or MHDPO, absorbed on filter paper or on Kel-F columns[77].

A separation scheme was also elaborated for the separation of multicomponent samples. For this purpose, IPE, MIBK, and TOPO were absorbed on Teflon powder columns, and the compounds washed out by appropriate eluents[44].

High molecular weight amines were frequently applied to separate iron sub-group elements. Cellulose powder columns impregnated with TNOA served for the separation

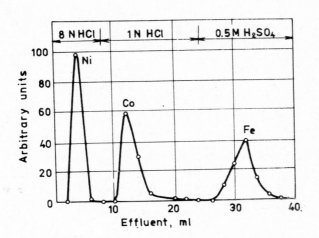

Fig. 7. Separation of iron sub-group elements[25].
Kel-F, low density type, TOPO—HCl, H_2SO_4. Flow rate : 0.5 ml·min^{-1}; Column size : ϕ = 8.6 mm; Concentration : Ni = Co = Fe = 5 mg. (By courtesy of the authors)

of Ni (8 N HCl), Co (3 N HCl), and Fe, which was eluted with 0.2 N HNO$_3$[24]. Teflon loaded with TNOA served for the separation of Ni and Co[65].

Fritz et al. applied TNOA for the preparation of pure (Co-free) Ni from thiocyanate solutions[55], whereas Aliquat-336 was applied by Sastri et al. for the separation on a Kel-F column of 5.0 mg Ni, 5.0 mg Co, and 4 mg Fe(III) which was eluted with 2 N HNO$_3$[87].

Aliquat-336 supported on cellulose powder was used in thin-layer, as well as column chromatography for the separation of Fe(III) and Co, using HCl as the eluent[75].

8. THE SEPARATION OF NOBLE METALS

The literature on the separation of noble metals belonging to the VIII B sub-group is somewhat scanty. Apart from the papers describing the preparation of carrier-free ^{199}Au from Pt, there are only a few works concerned with the separation of noble metals. Osmium was separated from W in the TBP—HCl system[96]. Akaza et al. separated a mixture of Au, Pt, and Pd in amounts of 0.5 mg each, the two latter elements being eluted with 3.5 N HCl, and Au with HNO$_3$[3].

Palladium(II), Pt(IV), and Au(III) were separated with TOPO loaded on cellulose powder, using mineral acids as the mobile phase[23]. The separation of these ions, in amounts of 3 mg each, is illustrated in Fig. 8.

Palladium was recovered from nuclear wastes by Colvin, who used the Aliquat-336—HCl system[26].

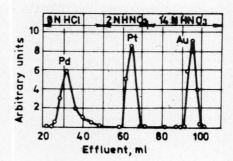

Fig. 8. Separation of palladium, platinum, and gold[23].
Whatman Cellulose powder, TOPO—HCl, HNO$_3$. Flow rate : 0.20 ml·min^{-1}; Column size : 13—16 mm; Concentration : Pd = Pt = Au = 3 mg. (By courtesy of the authors)

9. THE SEPARATION OF THE HALIDE IONS

As in the case of noble metals, the literature on the separation of the halogen ions is rather poor. Halogens may be separated only by means of high molecular weight amines.

Ramaley et al. applied Aliquat-336 and separated F⁻, Cl⁻, Br⁻, and I⁻ on a column, 0.1 N sodium acetate and $NaNO_3$ of various concentrations being used as the eluents[85]. In Fig. 9 the chromatograms of these halides are presented.

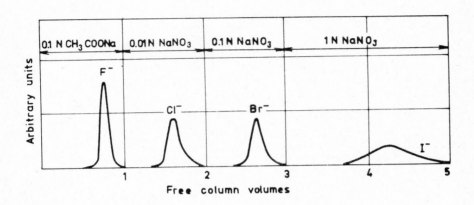

Fig. 9. Separation of halides[85].

Kel- F, 30—40 mesh, 0.1 M Aliquat-336 —acetate and nitrate of sodium. Flow rate : 1.0 — 1.5 ml·min⁻¹; Column size: ϕ 15 mm, length = 400 mm. Concentration : milligram quantities. (By courtesy of the authors)

10. THE SEPARATION OF OTHER IONS

It is easy to notice that the elements belonging to the IVth, Vth, and VIth groups, i. e. Ge, Sn and Pb, Sb and Bi, V, Nb and Ta, as well as Se and Te, have not yet been discussed here. The literature on the separation of these elements is not rich and concerns rather the isolation of one of these elements, or the determination of their R_f value. For example, Minczewski et al., in their studies on the analysis of molybdenite applied the TBP—HCl system for the determination of Nb in the presence of large amounts of Mo[69].

O'Laughlin et al. applied the MHDPO—HF system for the separation and determination of Nb and Zr in zirconium-niobium alloys[54].

On a cellulose powder column on which TOPO was absorbed, 3 mg of As(III) and 3 mg of Sb(III) were separated, the ions being eluted with 1N HCl and 5 N HNO_3, respectively[23].

Some of these elements were separated using high molecular weight amines. Alimarin et al. separated Nb from Ti on a column in the TNOA—HCl system[12]. Nb was separated from other elements by Gibalo et al.[47,48].

187

Aliquat-366 was used for the separation of Re and Ta, the former anion being eluted with NH_4OH (1:1) and the latter one with acetone which, however, washed out the extracting agent as well[98].

As for As and Sb, which are the most frequently studied among the elements of the Vth group, all three chromatographic techniques, (i.e. column, thin-layer, and paper) were applied for their separation.

The results presented in this Chapter, together with those obtained with actinides (Chapter 7), lanthanides (Chapter 8) and fission products (Chapter 9), indicate the versatility of extraction chromatography, which provides the means for the separation of almost all metal and non-metal ions.

REFERENCES TO CHAPTER 6

1. I. AKAZA, Bull. Chem. Soc. Japan, 39 (1966) 585.
2. I. AKAZA, Bull. Chem. Soc. Japan, 39 (1966) 980.
3. I. AKAZA, T. KIBA, T. KIBA, Bull. Chem. Soc. Japan, 43 (1970) 2063.
4. I. AKAZA, T. KIBA, M. TABA, Bull. Chem. Soc. Japan, 43 (1969) 1291.
5. I. P. ALIMARIN, T. A. BOLSHOVA, Zh. Analit. Khim., 21 (1966) 411.
6. I. P. ALIMARIN, T. A. BOLSHOVA, Z. Chem., 8 (1968) 411.
7. I. P. ALIMARIN, T. A. BOLSHOVA, T. G. DOLZHENKO, G. M. KANDYBINA, Zh. Analit. Khim., 26 (1971) 741.
8. I. P. ALIMARIN, T. A. BOLSHOVA, N. I. ERSHOVA, A. D. KISELEVA, Vestnik Moskovsk. Univ., ser. Khim., 4 (1967) 55.
9. I. P., ALIMARIN, T. A. BOLSHOVA, N. I. ERSHOVA, M. B. POLINSKAYA, Zh. Analit. Khim., 24 (1969) 26.
10. I. P, ALIMARIN, N. I. ERSHOVA, T. A. BOLSHOVA, Vestnik Moskovsk. Univ., ser.Khim., 4 (1967) 51.
11. I. P. ALIMARIN, N. I. ERSHOVA, T. A. BOLSHOVA, Vestnik Moskovsk. Univ., ser. Khim. 6 (1969) 79.
12. I. P. ALIMARIN, I. M. GIBALO, N. A. MYMRIK, Vestnik Moskovsk. Univ., ser. Khim., 4 (1971) 457.
13. V. V. BAGREEV, Yu. A. ZOLOTOV, I. A. KYRILINA, G. T. KALININA, Extraction of Inorganic Compounds, 2 Vol., Nauka, Moscow, p. 327, 335, 1971.
14. R. A. BAILEY, Rept NAS–NS–3106, Washington, 1962.
15. B. BAYAR, N. G. ZAITSEVA, A. F. NOVGORODOV, Rept JINR, P6–5955, Dubna p. 18, 1971.
16. D. BETTERIDGE, Chem. Zvesti, 21 (1967) 629.
17. T. A. BOLSHOVA, I. P. ALIMARIN, A. S. LITVINCHEVA, Vestnik Moskovsk. Univ., ser. Khim., 6 (1966) 59.
18. U. A. Th. BRINKMAN, in Progress in Separation and Purification, Eds. E. S. PERRY, C. J. VAN OSS, Vol. 4, J. Wiley-Interscience, New York, p. 241, 1971.
19. R. CALETKA,Chem. Listy, 62 (1968) 669.
20. E. CERRAI, Chromatogr. Rev., 6 (1964) 129.
21. E. CERRAI, Rept CISE–103, Milan p. 46, 1966.
22. E. CERRAI, G. GHERSINI, Adv. Chromatogr. 9 (1970) 3.
23. E. CERRAI, C. TESTA, Energia Nucl. 8 (1961) 510.
24. E. CERRAI, C. TESTA, J. Chromatogr., 6 (1961) 443.
25. E. CERRAI, C. TESTA, J. Chromatogr., 9 (1962) 216.
26. C. A. COLVIN, Rept ARN–1346, Richland, p. 10. 1969.
27. R. H. A. CRAWLEY, Nature, 197 (1963) 377.

28. R. DENIG, N. TRAUTMANN, G. HERRMANN, Z. anal. Chem., 216 (1966) 41.
29. R. DENIG, N. TRAUTMANN, G. HERRMANN, J. Radioanal. Chem., 6 (1970) 57.
30. B. S. DZHELEPOV, G. S. KATYKHIN, V. K. MAIDANYUK, A. I. FEOKTISTOV, Izv. Ak. Nauk USSR, ser. Fiz., 26 (1962) 1030.
31. N. I. ERSHOVA, T. A. BOLSHOVA, I. P. ALIMARIN, Zh. Analit. Khim., 26 (1971) 243.
32. N. I. ERSHOVA, T. A. BOLSHOVA, I. P. ALIMARIN, L. I. MOROZOVA, Vestnik Moskovsk. Univ., ser. Khim., 3 (1971) 348.
33. H. ESCHRICH, W. DRENT, Rept ETR–211, p. 99, Mol 1956.
34. H. ESCHRICH, W. DRENT, Rept ETR–271. p. 131, Mol 1971.
35. E. CERRAI, G. GHERSINI, J. Chromatogr. 16 (1964) 253; 18 (1965) 124.
36. I. FIDELIS, R. GWÓŹDŹ, S. SIEKIERSKI, Nukleonika, 8 (1963) 245.
37. I. FIDELIS, R. GWÓŹDŹ, S. SIEKIERSKI, Nukleonika, 8 (1963) 319.
38. I. FIDELIS, R. GWÓŹDŹ, S. SIEKIERSKI. Nukleonika, 8 (1963) 327.
39. J. S. FRITZ, R. T. FRAZEE, Anal. Chem., 37 (1965) 1358.
39a. E. CERRAI, G. GHERSINI, J. Chromatog., 16 (1964) 258; 18 (1965) 124.
40. J. S. FRITZ, R. T. FRAZEE, G. L. LATWESEN, Talanta, 17, (1970) 857.
41. J. S. FRITZ, C. E. HEDRICK, Anal. Chem., 34 (1962) 1411.
42. J. S. FRITZ, C. E. HEDRICK, Anal. Chem., 36 (1964) 1324.
43. J. S. FRITZ, G. L. LATWESEN, Talanta, 14 (1967) 529.
44. J. S. FRITZ, G. L. LATWESEN, Talanta, 17 (1970) 81.
45. J. S. FTITZ, J. J. TOPPING, Talanta, 18, (1971) 865.
46. M. GASIOR, H. I. LIZUREJ, H. NIEWODNICZAŃSKI, A. W. POTEMPA, Rept INP No. 584/PL. p. 16; Kraków 1967 Nukleonika, 13 (1968) 635; Rept AEC-tr-6931, p. 8; Sci. Techn. Aero Space, Repts, 8, 46.
47. I. M. GIBALO, N. A. MYMRIK, Zh. Analit. Khim., 25 (1970) 1744.
48. I. M. GIBALO, N. A. MYMRIK, Zh. Analit. Khim., 26 (1971) 918.
49. G. HERRMANN, Arkiv. Fys., 36 (1967) 111.
50. G. HERRMANN, Rapid Radiochemical Separation, Rev. Paper Presented at the Radiochemical Seminar, Dubna (USSR), p. 38, 1970.
51. G. HERRMANN, H. O. DENSCHLAG, Ann. Rev. Nucl. Sci., 19 (1969) 1.
52. T. ISHIMORI, et al., Rept. JAERI 1047(1), p.36, Tokai-mura 1963; kept JAERI 1062(2), p. 28; Tokai-mura 1964; Rept JAERI 1106(3), p. 31, Tokai-mura 1966.
53. H. JASKÓLSKA, Chem. Anal. (Warsaw) 14 (1969) 285.
54. G. J. KAMIN, J. W. O'LAUGHLIN, C. V. BANKS, J. Chromatogr., 31. (1967) 292.
55. E. E., KAMINSKI, J. S. FRITZ, J. Chromatogr. 53 (1970) 345.
56. G. S. KATYKHIN, Zh. Anal. Khim., 20 (1965) 615; 27, (1972) 849.
57. J. KORKISCH, Modern Methods for the Separation of Rarer Metal Ions, Pergamon Press, Oxford, p. 620, 1969.
58. Y. KUSAKA, W. W. MEINKE, Rept NAS–NS–3104, Washington, 1961.
59. M. KÝRŠ, L. KADLECOVÁ, J. Radioanal. Chem., 1 (1968) 103.
60. A. D.. LEE, J. Chromatogr., 26 (1967) 342.
61. A. D. LEE, Adv. Chem., ser., No. 89 (1969) 57.
62. M. LESZKO, W. KARCZ, Schedae Sci. Univ. Jagell., Schedae Chem., 15 (1970) 183.
63. K. H. LIESER, H. BERNARD, Z. anal. Chem., 219 (1966) 401.
64. S. J. LYLE, V. C. NAIR, Talanta, 16 (1969) 813.
65. J. MIKULSKI, Rept INP No 412/C, Kraków, p. 12, 1965; Nukleonika, 11 (1966) 57.
66. J. MIKULSKI, I. STROŃSKI, Nukleonika, 8 (1961) 295.
67. J. MIKULSKI, I. STROŃSKI, Nature, 207 (1965) 749.
68. J. MIKULSKI, I. STROŃSKI, J. Chromatogr., 17 (1965) 197.
69. J. MINCZEWSKI, C. RÓŻYCKI, Chem. Anal., 10 (1965) 95.
70. F. L. MOORE, Rept NAS–NS–3101, Washington, 1960.
71. L. N., MOSKVIN, Radiokhimiya, 6 (1964) 110.
72. L. N. MOSKVIN, L. I. SHUR, L. G. TSARITSYNA, G. T. MARTYSH, Radiokhimiya, 9 (1967) 377.
73. L. N. MOSKVIN, L. G. TSARITSYNA, At. Energiya, 24 (1968) 383.

74. L. N. MOSKVIN, L. G. TSARITSYNA, Radiokhimiya, 12 (1970) 187.
75. R. W. MURAY, R. J. PASSARELLI, Anal. Chem., 39 (1967) 282.
76. F. NELSON, J. Chromatogr., 20 (1965) 378.
77. J. W. O'LAUGHLIN, C. V. BANKS, Anal. Chem., 36 (1964) 1222.
78. J. W. O'LAUGHLIN, G. J. KAMIN, D. L. BERNER, C. V. BANKS, Anal. Chem., 36 (1964) 2110.
79. T. B. PIERCE, W. M. HENRY, J. Chromatogr., 23 (1966) 457.
80. B. K. PREOBRAZHENSKII, A. V. KALYAMIN, I. MIKHALCHA, Radiokhimiya, 6 (1964) 111.
81. B. K. PREOBRAZHENSKII, G. S. KATYKHIN, Radiokhimiya, 4 (1962) 536.
82. S. PRZESZLAKOWSKI, Habilitation Thesis, Medical School, p. 92, Lublin 1970; Chromatogr. Rev., 15, (1971) 29.
83. Kh., RAICHEV, V. KHALKIN, Radiokhimiya, 12 (1970) 778.
84. J. RAIS, J. KRTIL, V. CHOTIVKA, Talanta, 186 (1971) 213.
85. L. RAMALEY, W. A. HOLCOMBE, Anal. Letters, 1 (1967) 143.
86. A. P. RAO, M. N. SASTRI, Z. anal. Chem., 207 (1965) 409.
87. M. N. SASTRI, A. P. RAO, A. R. K. SARMA, Indian J. Chem., 4 (1966) 287.
88. B. I. SHESTAKOV, I. A. SHESTAKOVA, L. N. MOSKVIN, Radiokhimiya, 11 (1969) 471.
89. M. N. SHULEPNIKOV, G. I. SHMANENKOVA, Yu. V. YAKOVLEV, N. N. DOGATKIN, Zh. Analit. Khim., 26 (1971) 1167.
90. S. SIEKIERSKI, B. KOTLIŃSKA, At. Energiya, 7 (1959) 160.
91. S. SIEKIERSKI, R. J. SOCHACKA, Rept INR No 262/V, p. 3, Warszawa, 1961.
92. W. SMUŁEK, Nukleonika, 11 (1966) 635.
93. W. SMUŁEK, S. SIEKIERSKI, J. Chromatogr. 19 (1965) 580.
94. W. SMUŁEK, K. ZELENAY, Rept INR No 677/V, XIII/C, p. 12, Warszawa, 1965.
95. I. STROŃSKI, Kernenergie, 8 (1965) 175.
96. I. STROŃSKI, Radiochim. Acta, 6 (1966) 163.
97. I. STROŃSKI, Öster. Chem. Ztg., 68 (1967) 5.
98. I. STROŃSKI, Radiochem. Radioanal. Letters, 1 (1969) 191.
99. I. STROŃSKI, Radiochem. Radioanal. Letters, 5 (1970) 113.
100. I., STROŃSKI, Wiad. Chem., 26 (1972) 63.
101. I., STROŃSKI, Rept INP No 803/C, p. 41, Kraków, 1972.
102. I. STROŃSKI, M. BITTNER, J. KRUK, Nukleonika, 11 (1965) 47.
103. I. STROŃSKI, J. KEMMER, N. KAUBISCH, Z. Naturforsch., 23 b (1968) 137.
104. Z. ŠULCEK, V. SIXTA, Coll. Czech. Chem. Commun., 36, (1971) 1561.
105. C. TESTA, Rept RT(PROT) (65) 33, Rome, p. 27. 1965.
106. C. TESTA, Anal. Chim. Acta, 50 (1970) 447.
107. C. TESTA, G. SANTORI, Energia Nucl., 17, (1970) 320.
108. K. UENO, M. HOSHI, Bull. Chem. Soc. Japan, 39 (1966) 2183.
109. J. C. WHITE, ASTM Spec. Techn. Res., No. 238, p.54, Philadelphia, 1958.
110. J. C., WHITE, W. J. ROSS, Rept NAS–NS–3102, p. 56, Washington, 1961.
111. N. G. ZAITSEVA, Rept JINR, P–1292, p. 41, Dubna, 1963.
112. N. G. ZAITSEVA, Rept JINR, 6-3596 Dubna, p. 26, 1967.
113. Yu. G. ZHUKOVSKII, G. S. KATYKHIN, A. L. MARTYNOV, M. K. NIKITIN, Radiokhimiya, 10 (1968) 252.

CHAPTER 7

EXTRACTION CHROMATOGRAPHY OF ACTINIDES

W. MÜLLER

1. INTRODUCTION

Most separations involving actinide elements belong to one of the following examples:
— separation of actinides from large amounts of mineral compounds and structure material (nuclear fuel matrix, cladding);
— separation of actinides from fission products (except lanthanides);
— separation of actinides from lanthanides;
— separation of actinides from each other.
Whereas in general the problems of the first two categories can be easily solved, the

latter examples involve elements belonging to "f"-families which display similar chemical behaviour. When the ions coexist in different oxidation states, they can be separated by oxidation-reduction cycles. The separation of actinide-lanthanide ions present in the same oxidation state depends on small differences of ionic radius and complexing behaviour. Small separation factors must, therefore, be increased by proper choice of complexing media and repetition of the elementary separation step. Precipitation, ion exchange and solvent extraction are well established methods for the separation and/or purification of actinide ions[1].

The application of these methods in the presence of radionuclides, with high specific activities, may be disturbed by radiolysis produced gassing and changes in the acidity of an aqueous solution and of the oxidation states. Radiation damage of ion-exchange resins and solvent extraction agents is also a serious problem. Large-scale operations have additional difficulties concerning phase separation after precipitation or extraction. Removal of the decay heat, shielding, and telemanipulation requirements must be considered. Some of these difficulties have been circumvented by the introduction of improved or new techniques: e. g. high-pressure ion exchange, high-speed extraction and extraction chromatography.

Extraction chromatography combines the versatility of solvent extraction systems with the ease of operation of ion-exchange column chromatography. The low capacity of extraction chromatography columns, as compared to ion exchangers, reduces radiolysis problems and facilitates the dissipation of decay heat from the highly active nuclides.

Extraction chromatography has proved appropriate not only for the separation of trace quantities of actinides (e. g. for analytical purposes) but is increasingly being used for the isolation and/or purification of macro amounts of transplutonium nuclides (americium, curium). It has found applications in such distant fields as the processing of actinide waste solutions and the search for superheavy elements. Large-scale industrial applications however, are limited because of the low capacity and of the depletion of extractant on the solid phase by solubility losses.

Solvent extraction systems offer experimental guidelines for the choice of an appropriate extraction chromatographic system. The following review on extraction chromatography of actinides emphasizes the important usage of neutral (TBP), basic (substituted ammonium salts) and acidic (HDEHP) extractants and their application to separations of actinides in the di- to hexavalent oxidation state. Actinides with oxidation numbers VII (Pu, Np) and I (Md) have not been separated by extraction chromatography.

In preparing this review the author found useful information in comprehensive articles published recently[1–7].

2. LIQUID-LIQUID EXTRACTION OF ACTINIDES

Liquid-liquid extraction has often been classified according to the nature of the extracting agent:

Class of extracting agent	Examples	Species in the organic phase at equilibrium
I Ketones, ethers, alcohols	MIBK, R_2O, DBC	"Onium compounds"
II Neutral (organophosphorous) compounds	TBP, TOPO	
III Basic (substituted ammonium) compounds	TOAHX, TLMAX	Ion pairs (liquid anion exchange)
IV Acidic (organophosphorous) compounds	HDEHP	Solvated metal salts (liquid cation exchange)
V β-Diketones	HTTA	Chelates

In general, the distribution coefficients (D) of a given metal ion increase in the order I to V. The capacities of the organic phase for this ion decrease in the same order. Variables that influence the solvent extraction equilibria are:
— structure of the extracting agent (length of the aliphatic chain),
— concentration and polarity of the diluent (which is not inert but takes part in the extraction equilibria),
— composition of the aqueous phase (ligand concentration, pH-value).

The transfer mechanism of an ion between two immiscible phases is beginning to be understood. However, there are a number of approaches used to determine the composition of the "extracted species" in the organic phase at equilibrium. Direct determination of this composition can be accomplished by the isolation of the complex from the organic phase, or by saturation of the organic phase with the ion to be extracted. Indirect evidence, relying only on the variation of distribution coefficients as a function of the concentration of the extractant, may be misleading because of association or dissociation phenomena in the organic phase. The composition of the extracted compound as determined by "slope analysis" can be supported by absorption spectroscopy of the species in the organic phase. However, it is not necessarily correlated to the composition or concentration of the predominant species in the aqueous phase.

2.1. ACTINIDE EXTRACTION BY KETONES, ETHERS, ALCOHOLS

The extraction of U(VI) from nitrate solutions by diethyl ether was one of the first applications of solvent extraction. The composition of the complex in the organic phase varies with the composition of the aqueous phase. In the presence of dilute nitrate solutions, the ether molecules form hydrogen bridges to the crystal water molecules, while uranium extraction from concentrated salt or acid phases results in the formation of onium compound ion pairs, i. e. $M(VI)O_2(NO_3)_3^- \cdot HOR_2^+$. Dibutylcarbitol (Butex = $= C_4H_9O(C_2H_4OC_2H_4)OC_4H_9$) is also hydrogen bound to U(VI):

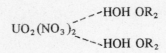

$$UO_2(NO_3)_2 \underset{\text{---HOH } OR_2}{\overset{\text{---HOH } OR_2}{<}}$$

Tetravalent actinides (Th^{4+}, Pu^{4+}) are fairly well extracted by Butex, but not by diethyl ether.

2.2. ACTINIDE EXTRACTION BY NEUTRAL
(ORGANOPHOSPHOROUS) COMPOUNDS

Representative extractants of class II are trialkylphosphates $(RO)_3PO$ (e. g. TBP) or trialkylphosphine oxides R_3PO (e. g. TOPO). They are used for actinide extraction mostly from nitric acid. The solubility of TBP is 0.38 g/l of water and 0.16 g/l of 5 M HNO_3. In the complexes of the general formula $MX_n \cdot S_m$ (m = number of molecules of extractant S associated with the metal salt MX_n) present in the organic phase, the organic compound is bound directly to the central metal ion, thereby completing the coordination sphere. Typical compositions are: $UO_2(NO_3)_2 \cdot (TBP)_2$; $Pu(NO_3)_4 \cdot (TBP)_2$ $Am(NO_3)_3 \cdot (TBP)_3$.

The stability of the complexes is inversely proportional to the phosphorous—oxygen vibration frequency; in general, phosphine oxide complexes ($\nu_{PO} < 1200$ cm^{-1}) are more stable than the corresponding neutral phosphate compounds ($\nu_{PO} > 1250$ cm^{-1}). Depending on the water and acid activities, the organic extractant may form 1:1 complexes with water or mineral acid (TBP·H_2O; TBP·HNO_3) which may compete with the formation of the metal complex.

For different nitric acid concentrations, for example, two different mechanisms have been proposed to describe the extraction of Am(III) ions. At low nitric acid concentration in the aqueous phase, the extraction equilibrium is described by the equation: $3[TBP\cdot H_2O]_{org} + [Am(NO_3)_3]_{aq} = [(TBP)_3 Am(NO_3)_3]_{org} + 3[H_2O]_{aq}$. With increasing HNO_3 concentration, this equilibrium is disturbed by competing coextraction of the acid:

$$[TBP\cdot H_2O]_{org} + [HNO_3]_{aq} = [TBP\cdot HNO_3]_{org} + [H_2O]_{aq}$$

which reduces the extraction of americium. The extraction of americium increases again as the nitric acid concentration increases (> 7 M) according to the equation:

$$m[TBP\cdot HNO_3]_{org} + [Am(NO_3)_{3+m}^{m-}]_{aq} = [(TBPH)_m^+ Am(NO_3)_{3+m}^{m-}]_{org} + m[NO_3^-]_{aq}$$

The extraction coefficients decrease in the order $Pu^{4+} > UO_2^{2+} > Am^{3+}$.

In the presence of radionuclides, hydrolysis of the neutral esters is enhanced, and the resulting acid esters are stronger extractants than the neutral compounds. Metal extraction with acid esters can be so high that separation of different ions becomes difficult.

194

2.3. ACTINIDE EXTRACTION BY BASIC (SUBSTITUTED AMMONIUM) COMPOUNDS

Substituted ammonium salts $R_{4-n}H_nNX$ are formed by reaction between amines with acids or alkylhalides. The substituted alkylammonium salts (R_4NX) exist predominantly as ion pairs $R_4N^+ \cdot X^-$. Depending on the nature and the concentration of the ammonium ion, the acid strength and the polarity of the diluent, these ion pairs dissociate or associate. In the presence of aqueous solutions, the alkylammonium salts are hydrolyzed; their solubility in water is low ($< 10^{-5}$ M).

As in the case of the TBP systems, metal ion extraction competes with extraction of water and acid. Radiolysis enhances secondary reactions of the substituted ammonium salts with diluents, mineral acids or oxygen. The reactions with nitrous acid produced by radiolysis cause degradation products which change the extraction characteristics of a given system. Nitrous acid reacts with, e. g. TLA and forms acid, nitrile, aldehyde, alcohol, nitro compounds, dilauryl formamide and secondary amine. However, nitrous acid can be removed by the addition of hydrazine or sulphamic acid to the aqueous phase.

In the organic phase the extracted metal forms the central ion of a metallate anion. Hence, extraction by substituted ammonium salts is often referred to as "liquid anion exchange".

For any given oxidation state, there is only a limited number of useful aqueous phases for the efficient extraction of actinide ions:

VI: Extraction of MO_2^{2+} ions by tertiary ammonium salts passes through maxima in the order sulphuric acid ($<$ 1 M), nitric acid (6–8 M) and hydrochloric acid ($>$ 10 M).

V: Pentavalent Pa is extracted from sulphate solutions by primary alkylammonium salts. From fluoride and chloride solutions, Pa is extracted by tertiary and quaternary salts. Extraction from nitric acid passes through a maximum near 6 M acid concentration. Extraction of Np(V) is almost 2 orders of magnitude lower.

IV: Tetravalent ions are extracted from sulphuric, nitric or hydrochloric acid. At low acidities ($<$ 1 M), best extraction is achieved from sulphuric acid with secondary ammonium salts. From concentrated acid solutions, M(IV) ions are better extracted by tertiary and quaternary ammonium compounds. Extraction from nitric acid passes through a maximum between 2 and 6 M, depending on the length of the aliphatic chain. This maximum is caused by the coextraction of nitric acid. Extraction of tetravalent ions decreases in the order: Pu $>$ Np $>$ U $>$ Th ($>$ Pa).

III: Trivalent actinides (transplutonium elements) are predominantly extracted from concentrated chloride, nitrate or thiocyanate solutions which are slightly acidic. Trivalent actinides can be separated from trivalent lanthanides by extraction from 11–13 M LiCl, 0.01 M HCl, with separation factors approaching 10^2. Actinide-lanthanide group separation is also possible in thiocyanate, but not in nitrate media. The interactinide separation is more difficult because the separation factors are small.

The following typical complex compositions have been postulated based on slope analysis, absorption spectroscopy, molecular weight determinations in the organic phase, etc.:

$$(R_2NH_2)_2\,PuO_2Cl_2\,;\,(RNH_3)\,PaO(SO_4)_2\,;\quad(R_3NH)_2\,Th(NO_3)_6\,;$$

$$R_4N\,Am(SCN)_4\cdot R_4NSCN.$$

2.4. ACTINIDE EXTRACTION BY ACIDIC (ORGANOPHOSPHOROUS) COMPOUNDS

The most representative (and used) compounds of this type are di(2-ethylhexyl) phosphoric acid (HDEHP) and 2-ethylhexyl phenyl-phosphonic acid (HEHΦP). A novel, highly efficient agent for the extraction of heavy actinides is (di)octylpyrophospha (OPPA)[8], for which extraction coefficients of 10^2 to 10^7 for all nitric or hydrochloric acid concentrations have been observed.

In organic diluents the HA type monoacids (HDEHP and HEHΦP) form dimers. The extraction of metal ions into the organic phase is described as cation exchange by the following equation:

$$[M^{m^+}]_{aq} + m[(HA)_2]_{org} = [M(HA_2)_m]_{org} + m[H^+]_{aq} \quad ([H^+]=0.1-2\ N)$$

Metal ion extraction by acidic phosphorous compounds (HDEHP) can be enhanced by the addition of neutral phosphorous compounds (TBP, TOPO).

Extraction by HDEHP has been used successfully for the isolation of actinides and their separation from each other. Separation factors for the extraction of adjacent trivalent ions from 1 M hydrochloric or nitric acid are given in Table 1.

TABLE 1

Separation factors of adjacent trivalent actinides
in the system HDEHP–1 M HCl

Z	Element	S.F.
95	Am	
		1.25
96	Cm	
		9.0
97	Bk	
		2.6
98	Cf	
		1.0
99	Es	
		2.1
100	Fm	
		4.2
101	Md	

These values show the possibility of interactinide separation between the transcurium elements, and Am + Cm. Trivalent actinides and lanthanides, which are both extracted

196

by HDEHP from 1 M lactic acid, can be separated with a separation factor of approximately 60 by the addition of 0.05 M diethylenetriaminopentaacetate (DTPA): such conditions are generally referred to as the TALSPEAK-process[9,10].

The variations of the HDEHP (and HEHΦP) distribution coefficients of both trivalent lanthanides and the corresponding actinides as a function of Z show a remarkable analogy[11]: the points of a log D vs. Z plot can be grouped into four tetrads with the gadolinium or curium point being common to the second and third tetrads. The first and second tetrads intersect for the lanthanides between $Z = 60 - 61$. The third and fourth tetrads intersect between $Z = 67 - 68$ for the lanthanides, and between 99 and 100 for the actinides. This tetrad effect is not restricted to extraction systems, but can also be observed in the variation of other properties of f-elements; such regularities have been studied in detail and described as "double-double effect" by Siekierski and Fidelis[12—15].

2.5. ACTINIDE EXTRACTION BY β-DIKETONES

The extraction of metal ions by 2-thenoyl-trifluoroacetone (HTTA) is considered as cation exchange with a (monomeric) monoacid HA present in its enol form:

$$\underset{S}{\boxed{}} - \underset{\underset{O}{\|}}{C} - CH = \underset{\underset{OH}{|}}{C} - CF_3$$

$$[M^{m+}]_{aq} + m[HA]_{org} = [MA_m]_{org} + m[H^+]_{aq}$$

The equilibrium depends on the acid activity and the formal charge of the metal ion. With decreasing charge of the metal ion, the pH-value has to be increased in order to achieve complete extraction.

Examples of the separation of some actinides (An) are:

Ac^{3+}/Ra^{2+} separation at pH = 5.4

Pu^{4+}/An^{3+} separation from 0.5 M HNO_3

Ce^{4+}/An^{3+} separation from 0.5 M HNO_3, $Cr_2O_7^{2-}$

3. EXTRACTION CHROMATOGRAPHY OF ACTINIDES

3.1. GENERAL REMARKS

The technique of extraction chromatography is based on the versatility of solvent extraction systems and the multistage case of operation of chromatography. Specific separation problems may be solved by variation of the organic phase composition. The liquid phase (extractant with or without diluent) is sorbed on a stationary inert support (kieselguhr, glass powder, paper,...). This stationary phase can be contacted with the mobile (aqueous) phase in a single step (slurry) process or in one of the multistage techniques using column or thin-layer chromatography. The ions loaded on the stationary phase are desorbed by passing an eluent. Because of the low capacity of the stationary phases, thin-layer or paper chromatography of actinides is limited to small scale, mostly analytical separations; on the other side, batch and column extraction chromatography have found successful application in preparative and purification procedures. The resistance to radiation damage and the simplicity of phase transfer and phase separation make extraction chromatography especially useful for the separation and preparation chemistry of the radioactive actinide ions.

3.2. FACTORS INFLUENCING THE CHROMATOGRAPHIC SEPARATION OF ACTINIDES

The efficiency of a chromatographic column is usually characterized by the height of a theoretical plate HETP, expressed in mm[1]:

$$HETP \ (mm) = \frac{L \cdot W^2}{8 \cdot V_r^2}$$

with L = length of the column bed;
      V = volume of eluent to peak;
      W = peak width at 1/e of the peak height.

Some variables affecting the efficiency of extraction chromatographic systems have initially been studied with lanthanides eluted from HDEHP columns. The nature and the particle size of the support materials, the loading of the extractant, the temperature and rate of the elution affect the HETP.

The following results may be considered as generally valid for actinides:

The diffusion of the metal ion–HDEHP complex from the interior of the particle to the interface is considered to be the rate-determining step. The diffusion rate can be increased by
— raising the temperature and
— using small and uniform support particles which are not saturated with the extractant

For a given particle size of the support, the HETP value can be minimized by reducing the elution rate and by increasing the temperature. However, temperature also affect separation factors. Silica gel and kieselguhr have a higher capacity than fluorinated or chlorinated polymer (Kel-F, Teflon, Corvic) supports.

The factors affecting performance and band spreading of high efficiency extraction chromatographic columns for actinide separations have been studied in detail by Horwitz and Bloomquist[16]. High efficiency columns were prepared by a slurry packing method rather than by tamping the dry graded material. The HDEHP–Celite column material was wetted by the aqueous phase to form a slurry, and compacted by application of gas (nitrogen) pressure. This technique results in uniform bed density and eluent flow,

198

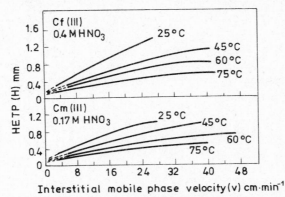

Fig.1. Height of a single plate (HETP) versus mobile phase velocity (v) as a function of temperature.
Column: HDEHP–Celite (8.82%), 3x50 mm (from Ref.[16], by courtesy of the authors)

even for separations involving high levels of alpha activity and possibility of local radiation damage.

HETP vs. flow rate (H vs. v) curves with slurry packed columns level off at high mobile phase velocities, and decrease in magnitude and slope with decreasing particle size and increasing temperature (Fig. 1). Coarser Celite fractions gave a more uniform bed structure than the smaller particle size fractions.

In the case of Cf(III), band spreading at low temperature was explained by a slow extraction rate. At higher temperatures band spreading was caused by irregular flow. Diffusion in the stationary layer was not found to be a major cause of band spreading. Differences in column behaviour for different tripositive transplutonium ions have been observed[16,17], but their interpretation is not universally accepted[1].

In a recent publication, Horwitz and Bloomquist describe high speed–high efficiency separations of transplutonium ions[18]. As a measure of the degree of separation of two ions (1 and 2), the resolution function R_s is used, which is defined as

$$R_s = \frac{(V_R)_2 - (V_R)_1}{2(\sigma_2 + \sigma_1)}$$

with V_R = volume of eluent to peak;
σ = standard deviation of the Gaussian elution curve.

If the (dimensionless) resolution function R_s is equal to 1.0, cross contamination is about 2%. For cross contaminations smaller than $< 0.1\%$, R_s has to be greater than 1.5.

R_s can be related to the three fundamental column parameters: separation factor β, number of theoretical plates N, and capacity factor k', by

$$R_s = \frac{1}{4}\left(1 - \frac{1}{\beta}\right)(N_2)^{\frac{1}{2}}\frac{k'_2}{1 + k'_1}$$

$$N = \frac{L}{HETP}$$

199

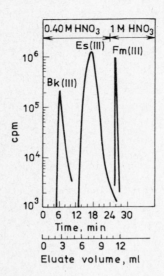

Fig. 2. High speed separation of Bk(III), Es(III) and Fm(III).
Column: HDEHP–Celite (8.82%), 0.63x100 mm; Flow rate: 6.2 ml·cm^{-2}·min^{-1}; 60°C (from Ref.[18],
by courtesy of the authors)

The validity of the resolution function was checked for the Bk, Cf separation as a function of column length and diameter. R_s increased with temperature; increasing the elution rate, R_s decreased and then became constant at high flow rate.

High speed Cm--Cf separations (5 cm column, 8.82% HDEHP, 35 μm Celite, 75 °C; interstitial linear flow rate: 29 cm·min^{-1}), although carried out at 16 times the flow rate used previously[17], gave the same purity of each fraction. Despite the 6-fold increase of the elution rate, in the high speed separation of Bk, Es, Fm (Fig. 2) the same purity of the fraction was obtained as before[19]. Despite the low separation factor (1.3) for Am, Cm, these elements were separated by elution with 0.17 M HNO$_3$ on a 8.82% HDEHP, 25 μm Celite, at 1.6 cm·min^{-1} linear flow rate and 75°C: a 10 cm column (0.064 cm^2) gave more than 1000 plates for the Cm elution.

The slurry packed Celite columns tended to suddenly shrink by 3–6% in length after a certain amount of usage. Prior to shrinkage, the plate heights remained constant. Smaller Celite particles form a less stable bed structure than the larger ones.

3.3. EXTRACTION CHROMATOGRAPHIC SYSTEMS FOR ACTINIDES

The extractants that have been mostly used in reversed-phase partition chromatography of actinides are TBP, tertiary and quaternary ammonium salts, and HDEHP. Although some extractants of Class I are rather volatile and water-soluble, viscous extractants like diisobutylcarbitol sorbed on Hostaflon found special applications, e. g. for rapid separations of short-lived protactinium isotopes[20,21]. Extraction chromatography with HTTA (Class V) in xylene has been proposed for Bk, Ce separation[22] or the isolation of

Np[23—26]. Extractants that found only limited use in extraction chromatography of actinides are dioctylsulphoxide[27], amine and arsine oxides[28], 1 to 1 mixtures of tetrabutylpyrophosphate and tetrabutylhypophosphate[29], tetrabutylhypophosphate [30—32], or dinonyl naphthalene sulphonic acid[33].

4. EXAMPLES OF ACTINIDE SEPARATIONS BY EXTRACTION CHROMATOGRAPHY

4.1. EXTRACTION WITH DIISOBUTYLCARBITOL (DIBC)

Protactinium(V) is extracted from 9 M HCl into DIBC with a distribution coefficient of 8000. Loaded on Hostaflon (5 ml DIBC on 10 g of Hostaflon) as a stationary phase, DIBC can be used for rapid separations of Pa(V) from uranium solutions[20,21]. Even at linear flow rates of 2—3 ml·sec^{-1}, more than 90% of Pa can be retained in a thin layer of the stationary phase. The technique has been applied to the separation of 70 sec ^{234m}Pa (UX$_2$) from natural uranium[20]. Protactinium isotopes produced by 15 MeV neutron bombardment of natural uranium were efficiently isolated by the same procedure from the mixture of fission products and actinides U, Np and Th[21]; reducing agents were required to reduce Sb. Protactinium could be recovered from the DIBC — Hostaflon layer by eluting with 12 M HCl + 0.5 M HF.

4.2. EXTRACTION WITH NEUTRAL (ORGANOPHOSPHOROUS) COMPOUNDS

Most applications involve the isolation and separation of uranium, neptunium and plutonium from nitric and hydrochloric acid solutions. The elution orders depend on the oxidation states, and are different for the different elements. The elution orders are: U(IV) > U(VI) ; Np(V) > Np(VI) > Np(IV) ; Pu(III) > Pu(VI) > Pu(IV).

4.2.1. ISOLATION OR SEPARATION OF U(VI) BY TRIBUTYLPHOSPHATE (TBP)

Extraction chromatography with TBP has been applied to the isolation of uranium with regard to its determination[34,35]. It was also used to extract the uranium before impurity analysis[36] and for fission product analysis[37]. On the basis of experience gained with the selective extraction of microgram to gram quantities of U(VI) from wastes[34], Hayes and Hamlin[35] used the same technique for the precise determination of uranium in complex alloys. The uranium is extracted from nitric or hydrochloric acid; the impurities or alloying constituents can then be determined in the effluent. Uranium is eluted with water and can be determined with high precision.

Kel-F powder proved to be a satisfactory support material when aqueous phases containing fluoride ions are used as eluent. The Kel-F (100—200 mesh) powder was loaded with equal weights of TBP. In analogy to the corresponding solvent extraction systems, the distribution coefficients of U(VI) were found to pass through a maximum at 5.5 M HNO_3. After equilibration with 5.5 M HNO_3 (the column contains $TBP \cdot HNO_3$), saturation of the column with uranyl ions corresponds to the formation of $UO_2(NO_3)_2 \cdot 2TBP$. Columns with 1.5 g of the stationary phase were used for the extraction of less than 50 mg of uranium, whereas 12.5 g columns were applied to uranium quantities in the 100 mg to 1 g range. For the separation of uranium from zirconium (U-Zircaloy), 1.3 −1.5 M HF has to be added to the 5.5 M HNO_3; excess hydrofluoric acid is complexed by the addition of boric acid or aluminium nitrate. Uranium can be separated from titanium by extraction from a 5 M $HCl-HNO_3$ mixture.

Huff removed the matrix constituents of thorium-uranium and plutonium-thorium-uranium alloys for trace impurity analysis[36], by a combination of anion exchange and extraction chromatography on the same column. The tetravalent actinides (Th, Pu) are sorbed on Dowex 1×8 anion exchanger from 8 M HNO_3, whereafter U(VI) is retained on a TBP−Kel-F bed.

Beranová and Novák used a styrene divinylbenzene copolymer as support material for TBP[37]. Although the maximum extraction for uranium was observed from 6 M HNO_3, the best separation from fission products was achieved by 5−5.5 M HNO_3. After removal of the fission products (Ru, Zr-Nb by 5 M HNO_3; Cs, Sr by 5.35 M HNO_3), uranium is eluted with water.

Phosphate or pyrophosphate ions interfere with usual methods for the analysis of uranium. Uranium can be quantitatively separated from phosphoric acid on a TBP−silica gel column. Phosphate is eluted with concentrated aluminium nitrate solution before the elution of uranium with water[38].

4.2.2. INTERACTINIDE SEPARATIONS

All actinide ions which can be oxidized to the hexavalent state are easily extracted by TBP phases.

Eschrich used TBP loaded on silanized Hyflo Super Cel to separate U, Np, Pu (and Am)[39]. U, Np and Pu were oxidized to the hexavalent state with $Cr_2O_7^{2-}$ in 1−2 M HNO_3. Am(III), Cr(VI) and Pu(VI) can then be eluted with 0.8−2 M HNO_3, whereas Np(VI) has to be reduced to Np(V) by hydrazine, in order to achieve separation from U(VI). Separation of Pu(III), Np(IV) and U(VI) is obtained by the elution with 0.7 M HNO_3, if ferrous sulphamate has been added to the (1−2 M HNO_3) feed solution.

To reduce Pu(IV) to Pu(III) in PUREX process solutions, U(IV) has been proposed as an efficient reducing agent. After reduction, such solutions contain Pu(III), excess U(IV) and U(VI). These ions can be separated for analysis by TBP extraction chromatography[40]. In agreement with the higher extraction of U(VI) in comparison to U(IV) in TBP solvent extraction systems, U(IV) is eluted before U(VI) from TBP extraction chromatographic columns[41,42]. Pu(III) ions (and colloidal plutonium) are known to pass through TBP chromatographic columns without adsorption[43].

Schmid has determined the distribution coefficient of U(IV) between TBP and HNO_3 from the separation factor of the couple U(IV)/U(VI)[42]: this separation factor could

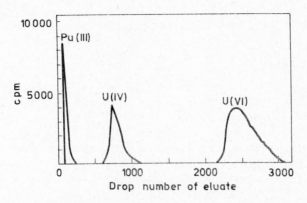

Fig. 3. Separation of Pu(III), U(IV) and U(VI).
Column: TBP–Voltalef (1:2), 5x235 mm; Flow rate: 1 ml·cm^{-2}·min^{-1} (from Ref.[40], by courtesy of the author)

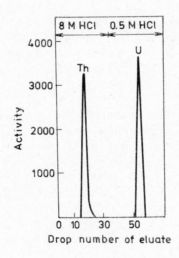

Fig. 4. Separation of Th and U.
Column: TBP–Hyflo Super Cel, 2x140 mm (from Ref.[44], by courtesy of the author)

be obtained with better precision and reproducibility from the retention volumes on a chromatographic column than by liquid-liquid extraction. Distribution coefficients of U(IV) reach a maximum ($> 10^2$) at a concentration of 8–9 M HNO$_3$.

For his detailed study of the extraction chromatographic behaviour of U(IV), Schmid used Voltalef as the support material in a 2:1 weight ratio to TBP[40]. No difference was observed between columns prepared with cyclohexane suspensions or by mechanical mixing of the support and the extractant. In general, solutions of 4.8 M HNO_3 −0.1 M hydrazine were used for loading and for eluting (1 ml·cm^{-2}·min^{-1}) (Fig. 3). After the elution of blue Pu(III) and green U(IV), yellow U(VI) was washed from the column with water. The high acidity of the eluent is required to prevent overlapping of the Pu(III) and U(IV) bands. The separation can be accelerated: as soon as the blue Pu(III) is washed out by 4.8 M HNO_3, elution of the green U(IV) is continued with 2 M HNO_3, after which yellow U(VI) is desorbed with water. During the isolation of U(VI) with 5.5 M HNO_3 on TBP−Kel-F column[34], interference by Th(IV), Ce(IV) and Pu(IV) was observed. In contrast to the reduction of Ce(IV), the reduction of Pu(IV) by hydroxylamine hydrochloride in HNO_3 is slow. Substitution of 6.5 M HCl for HNO_3 as the eluent resulted in a quantitative separation of Pu(III) and U(VI).

Using Hyflo Super Cel columns, the TBP−HCl system proved to be efficient also for the separation of Th(IV) (and Tb(III) or Hf(IV), respectively) from U(VI)[44] (Fig. 4); Th(IV) and Hf(IV) were eluted with 8M HCl. After removal of Tb(III) with concentrated HCl, U(VI) was desorbed from the column with 0.5 M HCl. On the same type of column Th(IV) can be separated from Pa(V) by elution with 3 M HCl. Protactinium can be recovered with 0.5 M HCl or 5% oxalic acid[45].

Similar methods can be used for the isolation of actinide nuclides for tracer studies, such as the separation of ^{234}Th (UX$_1$) from natural uranium[46] or the separation of ^{233}Pa (27 d) after neutron irradiation of ^{232}Th[47]. The mixture of ^{234}Th-U (ratio: 1.5 · 10^{-11}) was sorbed from 6 M HNO_3 on a TBP−Teflon column[47]. Thorium was completely eluted with 6 M HCl, uranium with 0.1 M HNO_3. The distribution coefficients of Th(IV) and Pa(V) between TBP−Hostaflon and 9 M HCl are 8 and 2500, respectively; consequently, Th can be eluted with 9 M HCl, after which Pa is washed from the column by 3 M HCl[47].

The transuranium elements Pu and Np often exist in aqueous nitric acid solution in more than one oxidation state. Aqueous processing of irradiated (natural or enriched) uranium or of irradiated neptunium (e. g. for the production of ^{238}Pu) is frequently based on TBP extraction from nitric acid solutions, and requires the knowledge of the oxidation states of Pu and Np in the aqueous (feed and product) phases. Besides polymeric or colloidal species, the following ionic forms may coexist: Pu(III), Pu(IV), Pu(VI); Np(IV), Np(V), Np(VI).

Bonnevie-Svendsen and Martini used Celite as support material for the TBP separation of plutonium species in reprocessing feed solutions[48]. The elution order with 0.5−3 M HNO_3 is Pu(III), Pu(IV) and Pu(VI). In the presence of excess U(VI) and in the day-light the fraction of "unextractable" plutonium was surprisingly large: Pu(III) was found to be formed on the column, by reduction of Pu(IV) with photochemically produced U(IV). In the absence of light, however, uranium has no influence on the results.

Eschrich separated Np(IV), Np(V) and Np(VI) on kieselguhr-supported TBP columns[49]. The elution sequence was Np(V), Np(IV), Np(VI) up to 4 M nitric acid concentrations (4−50°C). The best separations were obtained at room temperature, by elution with 0.5−2 M HNO_3 at a rate of 1 ml·cm^{-2} min^{-1}. Colloidal Np is not extractable, and is eluted with Np(V). In the presence of U(VI), the separation of the neptunium species has to be carried out in the dark, to avoid the photoinduced reduction of U(VI) to U(IV) which reduces Np(VI) to Np(V).

Possible applications of the extraction chromatographic technique are:
— the determination of the radiochemical purity of neptunium samples;
— the preparation of HNO_3 solutions of Np in only one oxidation state;
— the determination of the extractability of Np in processing solutions.

Radiochemically pure Np has been isolated from irradiated uranium[50]. In the presence of dichromate ions, U(VI), Np(VI) and Th(IV) are retained on the TBP—kieselguhr column, whereas most of the fission products are eluted with nitric acid. Subsequently, Np(V) is selectively eluted from the column with 0.2 M N_2H_4 in 5.5 M HNO_3.

The separation of trivalent transplutonium elements by extraction chromatography on TBP columns is possible only in a narrow range of concentrated solutions of nitric and hydrochloric acid (12 to 14.5 M HNO_3 or 12 to 12.5 M HCl)[51,52]. At lower concentrations, the distribution coefficient of Cf is smaller than unity; at higher acid concentration, the organic extractant is attacked by nitric acid or hydrogen chloride gas. The elution sequence Am, Cm, Cf, Fm, Md is the same in both media, but the separation factors are slightly greater in hydrochloric acid: Fm—Cf and Md—Fm separation factors respectively are 1.15 and 1.1 in 14.5 M HNO_3, but 1.45 and 1.4 in 12.3 M HCl.

The system tri-n-octylphosphine oxide (TOPO)—mineral acid has been used for chromatographic separation of trace quantities of U(VI) and Th(IV). Dietrich, Caylor and Johnson have sorbed U(VI) from partially digested urine on glass microbeads loaded with TOPO[53]:TOPO together with the uranium were eluted with alcohol. The same technique was applied in an improved semiautomatic apparatus for the analysis of uranium in urine samples[54]. The isolation of thorium and uranium from biological materials and their separation from each other were studied in a series of papers by Testa and co-workers[55-57]. The separation of these actinide ions is based on their different extraction into TOPO from sulphuric acid [55]. After decomposition (mineralization) of the urine in H_2O_2 and HNO_3, Th and U are sorbed from 4 M HNO_3 on 100—170 U.S.-mesh Microthene (microporous polyethylene) loaded with TOPO. Th(IV) is eluted with 0.3 M H_2SO_4, U(VI) with 1 M HF[56,57]. Batchwise extraction with Microthene supporting TOPO is especially useful for routine urine analyses because of its greater simplicity and speed: the mineralized urine solution is stirred with the solid extractant after which the slurry is transferred into a chromatographic column for elution. A detailed description of these techniques is reported in Chapter 10.

In a similar way, Th and U were separated by Strónski, Bittner and Kruk[58] (Fig. 5). After elution of the thorium with H_2SO_4, the uranium was desorbed with 9 M $HClO_4$ from a TOPO—Teflon column. Trivalent lanthanides can be separated from uranium by elution with 6 M HNO_3.

Fritz and Latwesen report a separation scheme for the analysis of multicomponent samples, which combines different extraction and ion-exchange chromatographic columns[59]. They used a TOPO column for the separation of Ti, Sc, Th, Zr and Hf: Ti(IV) and Sc(III) are eluted with 5 M HNO_3, Th(IV) is then desorbed with 12 M HCl; Zr(IV) and Hf(IV) are washed from the column with 1 M HCl.

Dioctylsulphoxide (DOSO) has been used as an alternative to TBP or TOPO as an extractant for U(VI)[27,60]. Encouraged by the almost selective retention of U(VI) on paper impregnated with DOSO[60], the authors applied column chromatography (with perchloric acid as the eluent) to separate uranium from most other metals. The support

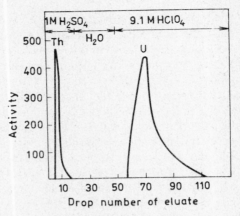

Fig. 5. Separation of Th and U.

Column: TOPO–Teflon (24%) (from Ref.[58], by courtesy of the authors)

was XAD–2 impregnated with 0.5 M DOSO in 1,2-dichloroethane. Quantitative sorption of U(VI) was obtained with 1 M LiClO$_4$ at pH values higher than 1.65. Only Au(III), Pd(II) and possibly Hg(II) are retained as strongly as U(VI); Th(IV), Zr(IV) and the trivalent lanthanides are not extracted. The composition of the extractable U(VI) complex is $[UO_2(DOSO)_4]^{2+}[ClO_4^-]_2$.

Although tetrabutylpyrophosphate (TBPP) and tetrabutylhypophosphate (TBHP) form complexes with metal salts similar to those obtained with TBP or TOPO, they found only limited use in the separation chemistry of actinide elements.

Mikulski and Strónski used a 1:1 mixture of TBPP and TBHP for the separation of several metal ions by extraction chromatography[61]: lanthanides could be eluted·with 5 M HNO$_3$, while U(VI) was retained on the column. Korotkin separated tri- to hexa-valent actinide ions by paper and thin-layer chromatography with TBHP using HNO$_3$ or HClO$_4$ as mobile phases[30]. On the basis of this study, a separation of lanthanide and actinide ions on a TBHP–cellulose column was developed: the trivalent actinides and lan-thanides are first eluted by 8 M HNO$_3$, followed by Th(IV), U(VI), Pu(IV) and Np(V). Zr and Hf are retained on the column and can be recovered with 8 M HClO$_4$.

TBHP sorbed on paper as a stationary phase was also used in the study of the hydro-lysis of transuranium ions in perchlorate or nitrate media: U(VI), Pu(VI)[32]; Am(III)[62] Cm(III)[63].

4.3. EXTRACTION WITH BASIC (SUBSTITUTED AMMONIUM) COMPOUNDS

Reversed-phase extraction chromatography with substituted ammonium salts has been applied to:
— separations of tetra-, penta- and hexavalent actinides from each other, or from triva-lent ions by tertiary or secondary ammonium salts;
— separations of trivalent actinides and lanthanides by tertiary or quaternary ammonium salts.

Most of the separations of tri- to hexavalent actinides have been carried out in chloride or nitrate solutions. Sulfate media have been used for U(VI); thiocyanate media for transplutonium element separations.

4.3.1. SEPARATION OF TETRA- TO HEXAVALENT ACTINIDES

Cellulose powder treated with tri-n-octyl-amine (TNOA) was found as useful as treated chromatographic paper for the separation of U(VI), Th(IV), Zr(IV), Ti(IV) and La(III)[64,65]. In hydrochloric acid, Th is not extracted, whereas Zr is retained at 10 M, and U(VI) almost over the complete range of acidity. Therefore, TNOA is preequilibrated and loaded with HCl; Th(VI) is removed by washing with 10 M HCl, Zr with 6 M HCl, and U(VI) with 0.05 M HNO_3 (Fig. 6). If zirconium is the major compound (Zr ores, Zircon), it should be washed from the column, whereas the trace elements (Th, U) are

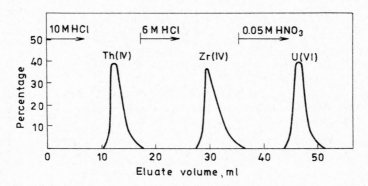

Fig. 6. Separation of Th (1 mg), Zr (10 mg) and U (5 mg).
Column: TNOA–Cellulose (0.045 M), 10x150 mm; Flow rate: 0.25 ml·cm^{-2}·min^{-1} (from Ref.[65],
by courtesy of the authors)

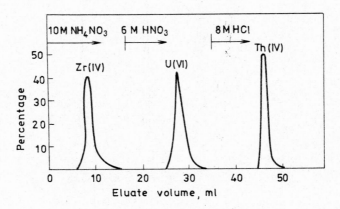

Fig. 7. Separation of Zr (10 mg), U (5 mg) and Th (1 mg).
Column: TNOA–Cellulose (0.045 M), 10x150 mm; Flow rate: 0.25 ml·cm^{-2}·min^{-1} (from Ref.[65],
by courtesy of the author)

207

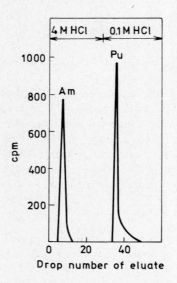

Fig. 8. Separation of Am(III) and Pu(VI).
Column: TNOA·HCl–Teflon, 2.8×80 mm (from Ref.[68], by courtesy of the author)

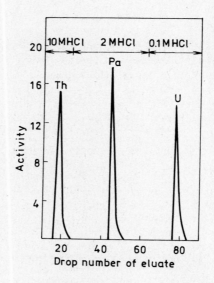

Fig. 9. Separation of Th, Pa and U.
Column: Aliquat-336 – Kel-F (0.1 M), 5×60 mm; Flow rate: 0.01 ml·cm^{-2}·min^{-1} (from Ref[69], by courtes
of the author)

retained: such a separation is also possible, by using NH_4NO_3: the feed solution is 2 M HNO_3 saturated with NH_4NO_3, zirconium is removed with 10 M NH_4NO_3, then uranium with 6 M HNO_3 and thorium with 8 M HCl (Fig. 7). In a similar way, U and Th are separated from Ti and Fe (ilmenite) or from La.

TNOA loaded on Teflon powder was used by Mikulski and Strónski to separate the mixture of Th and Pa from U, by elution with 6 M HCl[66]. Zr, Nb and Mo are separated from uranium on a column of Corvic and tri-iso-octylamine (TIOA)[67].

After elution of Tb, Hf and Am(III) with 3—4M HCl, U(VI) or Pu(VI) are washed from a TNOA·HCl—Teflon column with 0.1 M HCl[68] (Fig. 8). Strónski separated Th(IV), Pa(V) and U(VI) from each other on Aliquat-336—Kel-F by elution with 10, 2, and 0.1 M HCl, respectively (Fig. 9)[69].

Actinides, rare earths and some other fission products can be separated in nitrate media by TNOA supported on paper or cellulose powder[70—72]. The relative stability of the nitrate complexes of thorium and of the rare earths allows for their separation from each other and from uranium by elution with nitric acid or nitrate solution[70,71]. Thorium was separated on (0.25 M) TNOA-treated paper from Ce, Nd, and Gd, by elution with 3 M $LiNO_3$[71]. Knoch, Muju and Lahr were able to separate tri- to hexavalent actinides on TNOA impregnated paper with nitric acid[72]: in 1 M HNO_3, the R_f values increase (extraction decreases) in the order Pu(IV), Th(IV), U(VI), Np(V), Pa(V), Am(III), Cm(III). At higher acidities, Np(V) disproportionates to Np(IV) and Np(VI), which decreases its R_f value. With $LiNO_3$ as eluting agent, this disproportionation does not occur, and the elution sequence Am(III), Pa(V), Np(V), U(VI), Th(IV), Pu(IV) is observed over a wide range of concentrations; the R_f results obtained are reported in Table 2.

TABLE 2

R_f values of actinides on paper treated with 0.1 M TNOA
(from Ref.[72])

	2 M HNO_3	8 M HNO_3	0.5 M $LiNO_3$ + 0.5 M HNO_3
Pu(IV)	0.03	0.03	0
Th(IV)	0.18	0.24	0.20
U(VI)	0.68	0.60	0.70
Np(V)	0.86	0.20*	0.85
Pa(V)	1	1	
Am-Cm(III)	1	1	1

*Disproportionation

The identification and quantitative determination of different oxidation states in Np and Pu solutions were achieved by elution with NH_4NO_3. The elution order is Np(V), Np(VI), Np(IV) and Pu(III), Pu(VI), Pu(IV)[72]. With 2 M NH_4NO_3 + 0.5 M HNO_3 as the eluent, the R_f values for Np(V), (VI) and (IV) are 0.9, 0.26 and 0, respectively.

Analysis of a Np stock solution in 6 M HNO_3 + 0.02 M $NaNO_2$ by extraction chromatography revealed the presence of Np(V), Np(VI) and Np(IV) in a mass ratio of approximately 8 : 8 : 3. R_f values for ionic plutonium species are 0.9 for Pu(III), 0.25 for Pu(VI) and 0 for Pu(IV), with 4 M NH_4NO_3 as the eluent.

For the isolation and analysis of Np(IV) in concentrated solutions of (irradiated) uranium, Gourisse and Chesné applied loaded trilaurylamine nitrate (TLA · HNO_3) Kel-F columns[73]. In the presence of a large excess of uranium, the actinides Th, Pa, Np and Pu are loaded on the column from a 2 M HNO_3 + 0.1 M Fe(II) sulphamate solution, in which plutonium is reduced to the trivalent state. After washing the column with 1 M HNO_3 containing Fe(II), Np(IV) is eluted with a mixture of sulphuric and nitric acids. The method is selective for Np(IV), and has been successfully applied for analytical purposes to solutions with U : Np ratios greater than 10^{10}, and for the preparation of radiochemically pure ^{239}Np.

Recycled uranium from processing plants contains Np and Pu. In order to isolate Np(IV) and Pu(IV) simultaneously by extraction chromatography prior to their analysis, the oxidation-reduction potential of the aqueous solution has to be adjusted by addition of a Fe(III)-Fe(II) mixture[74]. Traces of Pu and Np, contained in up to 200 mg of U, are quantitatively retained on 40x4 mm (19%) TLA·HNO_3—Kel-F columns. The mixture of the actinides is loaded on the column from 2 ml of 2 M HNO_3 containing 200 mg of U, to which 15 μl of 1 M Fe(II) sulphamate and 0.7 ml of 2 M Fe(III) nitrate are added[73]. To remove uranium, the column is washed with 3 ml of 1 M HNO_3, 0.5 ml of 7 M HNO_3 and 1 ml of 7 M HCl. Np and Pu are eluted by 1 ml of 0.5 M H_2SO_4 + + 0.18 M HNO_3. As an alternative to the use of TBP, extraction chromatography with (TNOAH)$_2$SO$_4$ on Kel-F has been applied for routine analysis of uranium residues[34]. Pu(III) was separated by washing with 0.1 M H_2SO_4, while U(VI) was eluted with 1.5 M HNO_3. Reversed-phase thin-layer chromatography (silica gel impregnated with the secondary ammonium salt Amberlite LA-2 sulphate) was used for the separation of lanthanides, Th(IV), U(VI) and Zr by elution with 0.5 or 1 M H_2SO_4[75]. The behaviour of Th(IV) and U(VI) on thin layers of Alamine-336S oxide and trioctylarsine oxide on silica gel, with chloride solutions as eluents, was similar to that expected from solvent extraction[28]. Th was not retained, U(VI) was extractable at HCl molarities $<$ 6 or (acidified) LiCl solutions up to 10 M by amine oxide. Arsine oxide extracts U(VI) from chloride solutions (M $>$ 4), whereas Th is unextractable up to 10 M HCl or LiCl.

4.3.2. SEPARATIONS OF TRIVALENT ACTINIDES
(TRANSPLUTONIUM ELEMENTS)

4.3.2.1. REMOVAL OF PU(IV) FROM TRIVALENT ACTINIDES

The isolation or purification of transplutonium elements frequently requires the removal of Pu(IV). The processing of irradiated transuranium elements includes, in general, the separation of matrix material, fission products, plutonium (neptunium) and transplutonium elements. The irradiation of ^{239}Pu, ^{237}Np or ^{241}Am results in the formation of plutonium isotope mixtures: $^{240-242}$Pu, ^{238,239,236}Pu and ^{238,242}Pu, respectively. The

210

separation of the plutonium fraction by anion exchange is disturbed by radiolysis and by the presence of fission products. In order to remove plutonium, Zaitsev and co-workers substituted anion exchange with extraction chromatography with TNOA·HNO$_3$ on silica gel[76].

Increasing amounts of ^{244}Cm are becoming available for the investigation of the solid state chemistry of this element. Even after short storage, ^{244}Cm (T$_{1/2}\approx$ 18 y) contains appreciable amounts of ^{240}Pu which may disturb the study of curium. Removal of plutonium from gram quantities of curium by anion exchange is incomplete because of radiolysis; extraction chromatography with Aliquat-336 nitrate on Celite proved to be an efficient separation method, and no Pu could be detected in curium samples purified in this manner[77]. Multigram quantities of ^{241}Am were purified on a 1 l Aliquat-336 nitrate–Celite column. The constituents of stainless steel are washed from the column with 8 M LiNO$_3$ in the presence of KBrO$_3$; Am can be eluted by 1 M HNO$_3$, while Pu(IV) and Ce(IV) are retained[78].

4.3.2.2. SEPARATION OF TRIVALENT ACTINIDES AND LANTHANIDES

Trivalent actinides (transplutonium elements) can be separated from trivalent lanthanides by solvent extraction with substituted tertiary and quaternary ammonium salts from concentrated (lithium) chloride and (ammonium) thiocyanate solutions. In nitrate solutions, the actinide-lanthanide separation factors are low.

Moore has studied the separation of transplutonium and lanthanide ions with Aliquat-336 thiocyanate by solvent extraction from a 0.1 M H$_2$SO$_4$ + 0.5 M NH$_4$SCN solution[79]. The highest separation factor observed was 9800 for Cf–La. Separation factors of Am–Ce are 17 with cyclohexane, and 73 with benzene as the diluent[80]. Huff used Aliquat-336 thiocyanate supported on Plascon (polytrifluorochloroethylene) for the separation of lanthanides from americium[81]. Lanthanides were eluted with 0.6 M NH$_4$SCN, after which Am was stripped from the column with 0.1 M NH$_4$SCN. The distribution coefficients of the trivalent ions varied with the 2nd power of the ammonium thiocyanate concentration in the aqueous phase. They were directly proportional to the Aliquat-336-SCN concentration in the stationary phase, and inversely proportional to acidity. The same elements were separated on Celite supported columns impregnated with Aliquat-336-SC (Fig. 10)[82]. Ascending paper chromatography (Whatman No.1) developed with 0.2 M NH$_4$SCN + 0.5 M H$_2$SO$_4$, gave retention factors of 0.35 for Am and 0.81 for Eu.

In an attempt to check experiments on the possible synthesis of element 112 (Eka-Hg), proton bombarded W and U samples were analyzed by a combination of liquid-liquid extraction and extraction chromatographic methods. Lanthanides were separated from the transplutonium fraction by column chromatography with Aliquat-336 thiocyanate[83].

The separation factors between trivalent actinides and lanthanides as determined by solvent extraction into tertiary or quaternary ammonium nitrate solutions from aqueous nitrate phases are low[84-86]. Strónski succeeded in separating Am from its homologue Eu by extraction chromatography with 0.1 M Aliquat-336 nitrate on Kel-F and elution with 4.2 M LiNO$_3$ (Fig. 11)[87]. The separation factor (2.5) is in agreement with Van Ooyen's results[85,86].

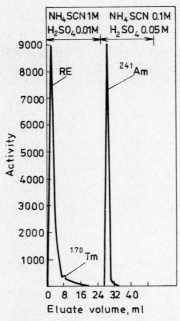

Fig. 10. Separation of rare earths (RE) from ^{241}Am.
Column: Aliquat-336-SCN—Celite (0.3 M), 5.4×100 mm; Flow rate: 0.78 ml·cm^{-2}·min^{-1} (from Ref.[82], by courtesy of the authors)

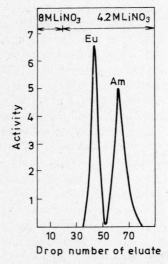

Fig. 11. Separation of Eu and Am.
Column: Aliquat-336-NO$_3$—Kel-F (0.1 M), 5×60 mm; Flow rate: 0.01 ml·cm^{-2}·min^{-1} (from Ref.[87], by courtesy of the author)

4.3.2.3. SEPARATION OF TRIVALENT ACTINIDES

Table 3 contains separation factors of trivalent actinide ions with respect to Cm, the least extractable transplutonium ion; these values have been determined by liquid-liquid extraction[84-86].

TABLE 3

Separation factors of trivalent actinides and europium relative to curium

Metal	Separation factors		
	*	**	***
Es	3.1	2.6−3.0	
Cf	2.7−3.0	1.8−2.1	1.2−2.1
Bk			1.1−1.9
Am	1.7−2.0	2.1−2.4	2.6−3.2
Eu	1.6−1.7	1.1−1.5	1.0--1.3

* from Ref.[84]: 0.59 M Alamine-336 nitrate in diisopropylbenzene. Aqueous phases: 6 M $LiNO_3$, 2.6 M $Mg(NO_3)_2$, 3.7 M $Ca(NO_3)_2$, 2 M $Al(NO_3)_3$.
** from Ref.[84]: 0.39 M Aliquat-336 nitrate in xylene. Aqueous phases: 4 M $LiNO_3$, 2.6 M $Mg(NO_3)_2$, 7 M $Ca(NO_3)_2$, 2 M $Al(NO_3)_3$.
*** from Refs [85,86]: 0.1 M trilaurylmethylammonium nitrate (TLMA·NO_3) in xylene. Aqueous phases: 3.7 M $LiNO_3$.

The separation factors are practically independent of the nature of the alkylammonium salts. Whereas a comparison of Al, Mg, Ca and Li did not reveal any significant influence of the nature of the cation in the aqueous phase[84], extraction with different alkaline nitrates[88] showed highest values of the separation factors (with respect to Cm), when the transplutonium ions were extracted by tri- or tetraoctylammonium nitrates from $NaNO_3$ or NH_4NO_3 solutions. The separation factor for Bk−Cm exceeds 5 for the extraction from $NaNO_3$ with TNOA·HNO_3−o-xylene[88]. With $LiNO_3$ as the aqueous phase, the separation factor for Am−Cm is sufficiently high to permit separation by extraction chromatography; at 10 °C it is about 3.5[85,86].

Knoch and Lahr[89] separated Am and Cm on paper impregnated with TNOA·HNO_3 by elution with 3 M $LiNO_3$ + 0.02 M HNO_3. The R_f values were 0.25 and 0.38 for Am and Cm, respectively. Separation factors of 2.5−3.2 for the separation of tracer quantities of Am and Cm were observed on kieselguhr columns loaded with quaternary ammonium nitrate (Aliquat-336 nitrate[90] (Fig. 12), trilaurylmethylammonium nitrate (TLMANO_3)[85,86,91]). The Am−Cm mixture was loaded on the columns from 8 M $LiNO_3$; eluents were 3.5 M $LiNO_3$ or 1.2 M $Al(NO_3)_3$ solutions.

Also milligramme quantities of Am and Cm could be separated by quaternary ammonium nitrate with a separation factor of 2.8 (Fig. 13)[92]. Because of the low capacity of the column (volume capacity: 0.21 mM·ml^{-1}), the purity of the separated elements decreased with increasing quantities. Assuming that 1:1 complexes ($R_4N^+Me(NO_3)_4^-$) are formed, almost 8% of the extractant is bound to Am or Cm. Whereas Cm, which is eluted first, contained only a few ppm of Am, the Am fraction contained up to 7% Cm[92]. After the separation of Cm with $LiNO_3$, the Am fraction can be eluted with

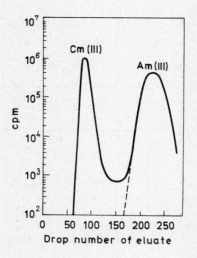

Fig. 12. Separation of ^{244}Cm and ^{241}Am (tracer) with 3.5 M LiNO$_3$–0.01 M HNO$_3$.
Column: Aliquat-336–Kieselguhr (0.14 M), 2.8×100 mm; Flow rate: 1.1 ml·cm^{-2}·min^{-1} (from Ref.[90], by courtesy of the authors)

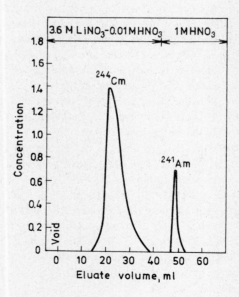

Fig. 13. Separation of ^{244}Cm (11.2 mg) and ^{241}Am (1.3 mg) with 3.5 M LiNO$_3$–0.01 M HNO$_3$.
Column: Aliquat-336–Kieselguhr (0.21 M), 8×140 mm; Flow rate: 1 ml·cm^{-2}·min^{-1} (from Ref.[92], by courtesy of the authors)

214

HNO_3. Repetition of the extraction chromatographic separation of milligramme quantities resulted in ^{241}Am fractions with only 200 ppm of ^{244}Cm[93].

Extraction chromatography is extremely useful for the isolation and preparation of Cm samples of high purity: after neutron irradiation of ^{241}Am, the mixture of ^{241}Am and ^{242}Cm (30 Ci) was extracted from $Al(NO_3)_3$ solution into Aliquat-336 nitrate in xylene[94]. Most of the americium was precipitated as $K_3AmO_2(CO_3)_2$ by oxidation in 4 M K_2CO_3. The Cm fraction was decontaminated from Am impurities by extraction chromatography with Aliquat-336 nitrate on Celite.

The same technique was used for the purification of ^{244}Cm isolated from a mixture of 6 g of ^{244}Cm and 5 g of ^{243}Am oxides[78,95]. After precipitation of the bulk of the americium by ozone oxidation in K_2CO_3, the Cm fraction was further purified on a 1 l (20%) Aliquat-336 nitrate—Celite column. BrO_3^- was added to the $LiNO_3$ eluent, and allowed for the simultaneous removal of Pu(IV), Ce(IV), Pb, Ni and Fe.

Despite the low separation factor of 1.46 for Es—Cf[96], which is unaffected by temperature changes, these elements could successfully be separated by extraction chromatography with Aliquat-336 nitrate: with 4.8 M $LiNO_3$ + 0.05 M HNO_3 as the eluent, a decontamination factor of Es from Cf higher than 100 was obtained.

In complete analogy to simple liquid-liquid extraction, the tertiary ammonium chloride Alamine-336 sorbed on Celite-360 could be used to separate Cf—Cm by elution with 11 M LiCl + 0.02 M HCl[97].

4.4. EXTRACTION WITH ACIDIC (ORGANOPHOSPHOROUS) COMPOUNDS

Many separations with this class of extractants (HDEHP is the most commonly used one) can be carried out from dilute mineral acid media. As in solvent extraction systems, synergism is achieved by addition of neutral extractants (TBP, TOPO).

In the field of actinides, the most interesting applications of HDEHP extraction chromatography are found in interactinide separations. The presence of complexing agents permits trivalent actinide-lanthanide group separations, in analogy to the TALSPEAK process[9,10].

4.4.1. SEPARATION OF TETRA- TO HEXAVALENT ACTINIDES

Since U(VI) is known to be extracted into HDEHP, even from concentrated mineral acids, its separation from fission nuclides in synthetic mixtures was attempted on kieselguhr supported columns by elution with increasing concentrations of nitric acid[98,99]. Uranium is stripped from the column with concentrated hydrochloric acid, while zirconium is retained. Uranium can also be separated from fission products on HDEHP—Hostaflon columns with hydrochloric acid. This extraction system could be applied to the rapid separation of protactinium (^{234m}Pa) from uranium[20]. Uranium could also be separated from Th by elution with concentrated hydrochloric acid: thorium was desorbed with 5 M HF[100] or 6 M H_3PO_4[101].

The use of concentrated hydrochloric acid as an eluting agent allows for the separation of U(VI) and U(IV) on the same HDEHP column[101]. After elution of U(VI), U(IV) is

215

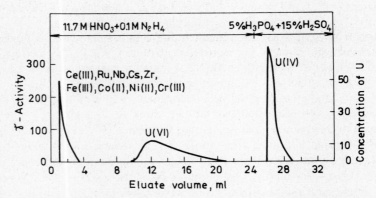

Fig. 14. Separation of fission and corrosion products from U(VI) and U(IV).
Column: HDEHP—Voltalef (30% HDEHP in xylene:support = 1 : 2), 4×60 mm; Flow rate: 0.8 ml·cm^{-2}·min
(from Ref.[102], by courtesy of the authors)

oxidized on the column with 15% H_2O_2 and can be recovered with hydrochloric acid.

U(IV) has been proposed as a reducing agent for plutonium in processing solutions. It is prepared by reduction of U(VI), and is slowly reoxidized in nitric acid. In an attempt to determine small quantities of U(VI) in U(IV), and to find a rapid purification method for U(IV) stock solutions, Schmid and Pfannhauser used HDEHP supported on Voltalef 300 columns for extraction chromatography[102]. Whereas TBP columns retain U(VI) better than U(IV), and are, therefore, used for the removal of U(IV) from U(VI) solutic HDEHP extracts U(IV) better than U(VI). Mixtures of both uranium species could be separated with 11.7 M HNO_3 + 0.1 M hydrazine. Plutonium and fission and corrosion products, were removed together with most of the U(VI) from a 30% HDEHP column by 2 M HNO_3 + 0.1 M hydrazine. The rest of U(VI) and U(IV) were separated by means of 11.3 M HNO_3 + hydrazine and U(IV) was desorbed from the column with 15% H_2SO_4 + 5% H_3PO_4 (Fig. 14). Impurities in Pu—U—Zr alloys were separated from the matrix elements on a column impregnated with a mixture of (5%) HDEHP and tri(2-ethylhexyl) phosphate by elution with 8 M HNO_3[103].

To separate Am(III) and Cm(III) from U(VI) or Pu(VI), Strónski used HDEHP on Kel-F with hydrochloric acid as the eluent (Fig. 15)[104].

Hulet describes a separation of Am(VI) and Bk(IV) from Cm(III) on HDEHP—silica columns[105]. In the presence of $(NH_4)_2S_2O_8$, Cm(III) is washed from the 0.2 M HDEHP columns by 0.1 M HNO_3; the bulk of americium is desorbed with 1 M HNO_3. Cm is decontaminated from Am by a factor of 10—14, whereas the decontamination factor for Cm in Am is 10^3. Bk(IV) is totally extracted from 10 M HNO_3 + 0.2 M $KBrO_3$ into 0.2—1 M HDEHP; it is eluted with 9 M HNO_3 + 1 M H_2O_2.

A method for the rapid separation of Am(V) is proposed by Moore[106]. Americium is oxidized by peroxidisulphate in dilute HNO_3, and retained on HDEHP—Teflon columns. This method is especially useful for the purification and determination of ^{243}Am in the presence of ^{244}Cm.

Overman used HDEHP for the rapid separation of Bk(IV) from rare earth fission products and trivalent actinides[107]. After anion exchange in the presence of bromate (8 M HNO_3 + 0.2 M $NaBrO_3$) to remove most of the cerium, the eluate containing lanthanides and transplutonium nuclides is passed through HDEHP supported on diatomaceous earth. The trivalent lanthanides and actinides are eluted with 7 M HNO_3 + + 0.02 M $NaBrO_3$, residual cerium is removed with 0.15 M HNO_3 + 10% H_2O_2, whereas berkelium is recovered by elution with 4 M HNO_3.

The oxidation-reduction separation ("OKVIBEK" process)[76] of berkelium from cerium is based on complex formation with polyaminoacetate as in the TALSPEAK procedure. The mixture of transplutonium and fission product ions is heated for 30 minutes with 0.3 M $KBrO_3$ in 6 M HNO_3 to oxidize Bk and Ce. Bk(IV) and Ce(IV) are extracted into HDEHP; Bk is reextracted into lactic acid–DTPA containing a reductant (e. g. ascorbic acid). From this aqueous phase, Bk can be recovered by the TOPEX process, i. e. the extraction into a benzene solution of TOPO.

Extraction chromatography with the OKVIBEK process was applied to the separation of Bk from irradiated Pu and Cm targets. In the case of the Pu targets, Bk was separated after removal of Pu and Zr. From Cm targets, however, Bk was separated immediately after dissolution of the irradiated sample, and isolated in this way the ^{249}Bk contained less than 10^{-10} ^{252}Cf.

A combination of several chromatographic methods was applied to the isolation of Bk, Cf and Es from irradiated Pu, Am and Cm[76]. From the feed solution (4 M HNO_3 + + 1 M $Al(NO_3)_3$) Pu was sorbed on TNOA·HNO_3–silica gel. Bk was separated by the OKVIBEK process. After the separation of Bk, the solution was made 0.1 M in HNO_3 (8 M total nitrate concentration) and passed on a silica gel column impregnated with diisoamylester methylphosphonic acid, which retains trivalent actinides and lanthanides, while Al is removed. Group separation of actinides and lanthanides is carried out by TALSPEAK and TOPEX processes. Subsequently, the transplutonium elements are separated into the two fractions Am–Cm and Cf–Es on a HDEHP column, by elution with

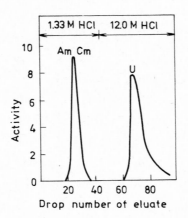

Fig. 15. Separation of trivalent actinides from uranium(VI).
Column: HDEHP–Kel-F (5%), 5x60 mm; Flow rate: 0.006 ml·cm^{-2}·min^{-1} (from Ref.[104], by courtesy of the author)

0.5 M HNO_3. Es and Cf are separated on TNOA·HNO_3 by sorption from 8 M $LiNO_3$ (pH = 2.5–3) and elution with 6 M $NaNO_3$.

4.4.2. SEPARATION OF TRIVALENT ACTINIDES

Kooi and co-workers[108-111] were among the first to use extraction chromatography with HDEHP for the separation of individual trivalent transplutonium elements. During the processing of highly irradiated ^{241}Am targets, the transplutonium nuclides were separated from each other by HDEHP supported on kieselguhr[108,109]. Microgramme amounts of Bk and Cf were isolated from several hundred milligrammes of Am and Cm by elution with 0.5 M HCl. Tracer amounts of Bk were separated from milligrammes of Ce on the same column by elution with 0.3 M HCl at 87 °C. Elution order is Am (Cm), Bk, Cf.

De Wet and Crouch proposed the use of HDEHP on kieselguhr for the separation of Am and Cm after their isolation from highly irradiated fuel[112]. The best conditions for the chromatographic separations of Cf, Fm and Md were selected by Gavrilov et al.[113] on the basis of a thorough study of the liquid-liquid extraction of these elements from hydrochloric acid. Using 14% HDEHP on kieselguhr and a temperature of 64 °C, the optimum acid concentration for Cf–Fm separation was found to be 0.7–0.9 M; for Fm–Md separations, an acid concentration of 0.9–1.0 M was used.

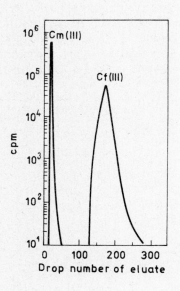

Fig. 16. Separation of Cm(III) and Cf(III).
Column: HDEHP–Celite (8.82%), 3×100 mm; Flow rate: 1.3 ml·cm^{-2}·min^{-1}; 60 °C (from Ref.[17], by courtesy of the authors)

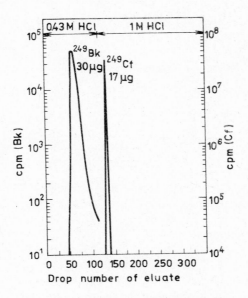

Fig. 17. Separation of ^{249}Bk and ^{249}Cf (milking of Bk).
Column: HDEHP–Celite (8.82%), 3x100 mm; Flow rate: 1 ml·cm^{-2}·min^{-1}; 45 °C (from Ref.[17], by courtesy of the authors)

Moore and Jurriaanse used dilute nitric acid as the eluent for the rapid separation of Cf–Cm and Bk–Ce on small Teflon columns[114]. The absence of chloride and operation at room temperature make this system suitable for glove box or hot cell processing.

In a series of publications[16-19, 115] Horwitz and coworkers studied in detail HDEHP column performance, separation factors and distribution coefficients of elements Am through Md, and the influence thereupon of mineral acid concentration and temperature. Figures 16 and 17 are examples of the very neat separations obtained by these authors.

Table 4 summarizes useful separation factors of trivalent transplutonium nuclides.

The results of tracer studies were applied to larger quantities, e. g. for the separation of ^{248}Cm from ^{252}Cf (Fig. 18)[116,117], or for the separation of Md from helium-bombarded Es targets[118]. Microgramme quantities of ^{248}Cm were separated from hundred microgrammes of ^{252}Cf with a decontamination factor of $10^8 - 10^9$ in a single run on HDEHP–Celite[116].

The Cf–Cm separations necessary to determine the half-lives of several Cf and Cm isotopes were carried out on HDEHP with HCl as eluent[119].

From the stability constants of the complexes between trivalent actinides and hydrocarboxylic acids it can be concluded that Cm, Bk, Cf and Es may be separated on HDEHP by elution with methyllactic acid[120].

Extraction chromatography with HDEHP has found several interesting applications in theoretical and applied chemistry of transplutonium elements. Mendelevium was separated

219

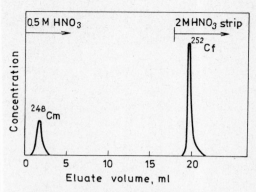

Fig. 18. Separation of ^{248}Cm from ^{252}Cf.
Column: HDEHP-Bioglas (400 mg/g), 6×600 mm; Flow rate: 3 ml·cm^{-2}·min^{-1} (from Ref.[117], by courtesy of the authors)

TABLE 4

Separation factors for trivalent transplutonium
ions on HDEHP columns (from Ref.[18])

Temp. °C	Cm-Am	Bk-Cm	Cf-Bk	Es-Cf	Fm-Es	Md-Fm
			HNO$_3$ elutriant			
25	1.18	8.7	2.94			(5.1)
45	1.22	8.5	2.82	0.989	2.24	4.7
60	1.24	8.3	2.70	1.02	2.20	4.4
75	1.26	7.9	2.57	2.03	2.13	4.2
			HCl elutriant			
25	1.20	11	2.90			
45	1.22	9.9	2.81			
60	1.26	9.4	2.67	0.993	2.04	4.0
75	1.28	8.8	2.58			3.7

Temp. °C	Cf-Cm	Es-Bk	Fm-Cf	Md-Es
		HNO$_3$ elutriant		
25	26			
45	24	2.79	2.22	11
60	22	2.75	2.24	9.7
75	20	2.65	2.19	9.0
		HCl elutriant		
25	32			
45	28			
60	25	2.65	2.03	8.2
75	23			

from Es(III) and Fm(III)[121], and its dipositive nature shown by its behaviour on a HDEHP column in the presence of reducing agents (Zn amalgam or 0.6 M Cr^{2+} solution). In order to determine the oxidation potential of No(II)/No(III) + e^-, distinction between No(II) and No(III) was made by column partition from dilute acid solutions containing different oxidants[122]. Nobelium(II) was contained in the eluent, No(III) was retained on the column, and the oxidation potential was determined from the per cent distribution as a function of the potential of the oxidizing agent, with only a few thousand atoms of the element.

The transplutonium nuclides isolated from proton-bombarded W or U targets were separated into 3 fractions: Am, Cm; Bk, Cf, Es, Fm; and Md, Lr by column chromatography with nitric acid[83]. New encapsulation techniques for the fabrication of ^{252}Cf neutron sources are based on the sorption of the neutron emitter on HDEHP (or HEHΦP)—silica in a quartz capillary. After loading and washing with 0.1 M HNO_3, the capillary is dried; after decomposition of the organic phase, the capillary is melted into a sphere of fused silica which completely encloses californium[123].

In order to overcome the drawbacks of the use of highly corrosive (lithium) chloride solutions necessary for separating lanthanide fission products from trivalent transplutonium elements (anion exchange or solvent extraction by tertiary ammonium salts), TALSPEAK [9,10], an alternative process for actinide-lanthanide group separations, has been developed at Oak Ridge National Laboratory. It is based on the simultaneous extraction of lanthanides and actinides by HDEHP from carboxylic acids, and on their separation by addition of diethylenetriaminopentaacetate (DTPA) as complexing agent.

In general, the trivalent ions are extracted from 1 M lactic acid; in the presence of 0.05 M DTPA, lanthanides and actinides form complexes of different stability. Actinides are reextracted into the aqueous phase, while lanthanides remain in the organic phase. At pH = 2.8, the separation factors for the group separation are about 60.

Although some lanthanides can be separated from trivalent actinides on HDEHP columns by elution with dilute nitric acid[114,124-126], more efficient group separations are obtained by lactic acid—DTPA as the eluents: Marsh used HDEHP in diisopropylbenzene on Chromosorb for the separation of trivalent actinides and lanthanides in spent nuclear fuel[127]. A rapid separation of lanthanides and actinides which minimizes radiation damage of the stationary phase is achieved by "lanthanide filtration" from 1 M lactic acid + 0.1 M DTPA, at pH = 2.4 on HDEHP supported on Bioglas[128].

The techniques used in different research establishments in the Soviet Union have been summarized in a paper submitted to the 4th U.N. Conference on the Peaceful Uses of Atomic Energy, Geneva, 1971[76]; whereas solvent extraction of actinides and lanthanides by HDEHP is carried out from citric acid solutions, lactic acid—DTPA solutions are used as mobile phases in extraction chromatography. The transplutonium elements are recovered from the eluent (lactic acid, DTPA) by extraction with TOPO after addition of $Al(NO_3)_3$.

Extraction chromatography on HDEHP columns could be applied to the separation and purification of multigramme quantities of americium and curium. The purification of large amounts of ^{241}Am by ion exchange in hydrochloric acid and thiocyanate media is affected by radiolysis and corrosion[78]. A TALSPEAK purification of 15 g of ^{241}Am on a 2 l column allowed the removal of lanthanide impurities. For the separation and purifica-

221

tion of the components of a mixture of ^{243}Am and ^{244}Cm oxides (5 g each), precipitation, ion-exchange and extraction chromatographic techniques were combined. The bulk of the ^{243}Am was precipitated as $K_3AmO_2(CO_3)_2$ by ozone oxidation in K_2CO_3. The remaining ^{243}Am was removed from the Cm fraction on a quaternary ammonium nitrate column. After irradiation of lactic acid and adjustment of the pH value to 2.8, the $LiNO_3$ eluent containing the Cm was immediately passed over a 1 l HDEHP column. Curium was separated from rare earths by elution with lactic acid—DTPA. Finally, Cm was isolated from the aqueous eluent and concentrated on Dowex 50 cation-exchange resin[95].

4.5 EXTRACTION WITH 2-THENOYLTRIFLUOROACETONE (HTTA)

Berkelium(IV) is selectively extracted from nitric acid solutions into HTTA—xylene in the presence of $Cr_2O_7^{2-}$ as oxidant[22]; the HTTA—HNO_3 system has therefore been suggested for the difficult separation of Bk(IV) and Ce(IV) by extraction chromatography. After removal of the lanthanide, Bk can be desorbed from the column by 10 M HNO_3.

In order to determine uranium in uraniferous coal, HTTA—xylene on glass powder was used as stationary phase for the quantitative and selective isolation of ^{239}Np after neutron irradiation of the coal[25]. After dissolution of the irradiated sample in nitric acid, neptunium was reduced to the tetravalent state by hydroxylamine. Np(IV) was sorbed from dilute HCl-hydroxylamine and desorbed for counting by 6 M HCl.

A similar technique was used for the separation of Np from fission and activation products in the determination of microgramme and submicrogramme quantities of uranium[20 The valency of the neptunium was adjusted by addition of Fe(II) and hydroxylammonium chloride. After washing with 1 M HCl, the neptunium was stripped with 10 M nitric acid.

Curium could be separated from Cf on a Celite column loaded with equal amounts of HTTA and dibutyldiethylcarbamoyl phosphonate, with a separation factor for Cf—Cm of approximately 4[129].

*

The author wishes to thank R. D. Baybarz, Oak Ridge National Laboratory, for his critical remarks and suggestions.

REFERENCES TO CHAPTER 7

1. E.K. HULET, D. D. BODÉ, in Lanthanides and Actinides, MTP International Review of Science, Inorganic Chemistry, Series One, Vol 7. K. W. BAGNALL, Edt. Butterworth and Co. Ltd., London, University Park Press, Baltimore p. 1., 1972.
2. H. ESCHRICH, W. DRENT, ETR—271, 1971.
3. G. S. KATYKHIN, Zh. Anal. Khim., 27 (1972) 849.
4. P. MARKL, Extraktion und Extraktions-Chromatographie in der Anorganischen Analytik, Akademische Verlagsgesellschaft, Frankfurt am Main, 1972.
5. E. CERRAI, G. GHERSINI, in Advances in Chromatography, J. C. GIDDINGS, R. A. KELLER, Eds., Vol. 9, M. Dekker Inc., New York, p. 3, 1970.

6. A. K. DE, S. M. KHOPKAR, R. A. CHALMERS, Solvent Extraction of Metals, Van Nostrand Reinhold Co. Ltd., London, 1970.
7. Y. MARCUS, A. S. KERTES, Ion Exchange and Solvent Extraction of Metal Complexes, Wiley-Interscience, London, 1969.
8. E. K. HULET, J. E. EVANS, R. QUONG, B. J. QUALHEIM, ACS Symposium Macroscopic Studies of the Actinides, San Francisco, 1968.
9. B. WEAVER, F. A. KAPPELMANN, ORNL-3559, 1964.
10. R. E. LEUZE, R. D. BAYBARZ, F. A. KAPPELMANN, B. WEAVER, in Solvent Extraction Chemistry of Metals, H. A. C. McKAY, T. V. HEALY, I. L. JENKINS, A. NAYLOR, Eds., McMillan, London, p. 423, 1965.
11. D. F. PEPPARD, C. A. A. BLOOMQUIST, E. P. HORWITZ, S. LEWEY, G. W. MASON, J. Inorg. Nucl. Chem., 32 (1970) 339.
12. I. FIDELIS, S. SIEKIERSKI, J. Inorg. Nucl. Chem., 28 (1966) 185.
13. S. SIEKIERSKI, J. Inorg. Nucl. Chem., 32 (1970) 519.
14. I. FIDELIS, Bull. Acad. Pol. Sci. Sér. Sci. Chim., 18 (1970) 681.
15. S. SIEKIERSKI, I. FIDELIS, J. Inorg. Nucl. Chem., 34 (1972) 2225.
16. E. P. HORWITZ, C. A. A. BLOOMQUIST, J. Inorg. Nucl. Chem., 34 (1972) 3851.
17. E. P. HORWITZ, C. A. A. BLOOMQUIST, D. J. HENDERSON, D. E. NELSON, J. Inorg. Nucl. Chem., 31 (1969) 3255.
18. E. P. HORWITZ, C. A. A. BLOOMQUIST, J. Inorg. Nucl. Chem., 35 (1973) 271.
19. E. P. HORWITZ, C. A. A. BLOOMQUIST, D. J. HENDERSON, J. Inorg. Nucl. Chem., 31 (1969) 1149.
20. R. DENIG, N. TRAUTMANN, G. HERMANN, Z. Anal. Chem., 216 (1966) 41.
21. N. TRAUTMANN, R. DENIG, G. HERMANN, Radiochim. Acta, 11 (1969) 168.
22. F. L. MOORE, Anal.Chem., 38 (1966) 1872.
23. J. A. THOMASSEN, H. H. WINDSOR, Kjeller Report KR–44, 1963.
24. D. C. PERRICOS, J. A. THOMASSEN, Kjeller Report KR–83, 1964.
25. D. C. PERRICOS, E. P. BELKAS, Talanta, 16 (1969) 745.
26. H. WERNER, S. Al-MURAB, M. STOEPPLER, Radiochem. Radioanal. Letters, 13 (1973) 1.
27. J. S. FRITZ, D. C. KENNEDY, Talanta, 17 (1970) 837.
28. H. R. LEENE, G. DE VRIES, U. A. Th. BRINKMAN, J. Chromatog., 57 (1971) 173.
29. J. MIKULSKI, K. A. GAVRILOV, V. KNOBLOCH, Nukleonika, 9 (1964) 785.
30. Yu. S. KOROTKIN, Radiokhimiya, 13 (1971) 137.
31. Yu. S. KOROTKIN, JINR-P6-6401, Dubna, 1972; KFK-tr-409.
32. Yu. S. KOROTKIN, JINR-P6-6402, Dubna, 1972; KFK-tr-410.
33. A. P. RAO, M. S. SASTRI, Z. anal. Chem., 207 (1965) 409.
34. A. G. HAMLIN, B. J. ROBERTS, W. LOUGHLIN, S. G. WALKER, Anal. Chem., 33 (1961) 1547.
35. T. J. HAYES, A. G. HAMLIN, Analyst, 87 (1962) 770.
36. E. A. HUFF, Anal. Chem., 37 (1965) 533.
37. H. BERANOVÁ, M. NOVÁK, Coll. Czech. Chem. Commun., 30 (1965) 1073.
38. G. W. HEUNISCH, Anal. Chim. Acta, 45 (1969) 133.
39. H. ESCHRICH, Kjeller Report KR-11, 1961.
40. E. R. SCHMID, Microchim. Acta, 1970, 301.   301.
41. H. ESCHRICH, ETR-249, 1969.
42. E. R. SCHMID, Mh. Chemie, 101 (1970) 1856.
43. R. GWÓŹDŹ, S. SIEKIERSKI, Polnische Akademie der Wissenschaften, Institut für Kernforschung, Report 168/v, 1960.
44. I. STRÓNSKI, Kernenergie, 8 (1965) 175.
45. I. FIDELIS, R. GWÓŹDŹ, S. SIEKIERSKI, Nukleonika, 8 (1963) 245.
46. B. I. SHESTAKOV, I. A. SHESTAKOVA, L. N. MOSKVIN, Radiokhimiya, 11 (1969) 471.
47. R. DENIG, N. TRAUTMANN, G. HERMANN, J. Radioanal. Chem., 6 (1970) 57.
48. M. BONNEVIE-SVENDSEN, V. MARTINI, Kjeller Report KR-81, 1966.
49. H. ESCHRICH, Z. anal. Chem., 226 (1967) 100.

50. S. LIS, E. T. JÓZEFOWICZ, S. SIEKIERSKI, J. Inorg. Nucl. Chem., 28 (1966) 199.
51. M. TAUBE, R. GWÓŹDŹ, K. A. GAVRILOV, J. MALI, I. BRANDSTETR, WAN TUN-SEN, Nukleonika, 7 (1962) 479.
52. I. BRANDSTETR, TUNG-SENG WANG, K. A. GAVRILOV, E. GVUDZ, Ya. MALY, M. TAUBE, JINR-P-1075, 1962; UCRL-trans-922.
53. W. C. DIETRICH, J. D. CAYLOR, E. E. JOHNSON, TID-7606, p. 195, 1960.
54. T. C. WHITSON, T. KWASNOSKI, TID-7629, p. 220, 1961.
55. C. TESTA, D. DE ROSA, A. SALVATORI, RT/PROT (68), 6, 1968.
56. C. TESTA, RT/PROT (69), 44, 1969.
57. C. TESTA, Anal. Chim. Acta, 50 (1970) 447.
58. I. STRÓNSKI, M. BITTNER, J. KRUK, Nukleonika,11 (1966) 47.
59. J. S. FRITZ, G. L. LATWESEN, Talanta, 17 (1970) 81.
60. D. C. KENNEDY, J. S. FRITZ, Talanta, 17 (1970) 823.
61. J. MIKULSKI, I. STRÓNSKI, Nature, 207 (1965) 749.
62. Yu. S. KOROTKIN, JINR-P6-6403, 1972; KFK-tr-411.
63. Yu. S. KOROTKIN, JINR-P6-6404, 1972; KFK-tr-412.
64. C. TESTA, J. Chromatog., 5 (1961) 236.
65. E. CERRAI, C. TESTA, J. Chromatog., 6 (1961) 443.
66. J. MIKULSKI, I. STRÓNSKI, J. Chromatog., 17 (1965) 197.
67. T. B. PIERCE, W. M. HENRY, J. Chromatog., 23 (1966) 457.
68. J. MIKULSKI, Nukleonika, 11 (1966) 57.
69. I. STRÓNSKI, Radiochem. Radioanal. Letters, 1 (1969) 191.
70. E. CERRAI, C. TESTA, J. Chromatog., 5 (1961) 442.
71. C. TESTA, Anal. Chem., 34 (1962) 1556.
72. W. KNOCH, B. MUJU, H. LAHR, J. Chromatog., 20 (1965) 122.
73. D. GOURISSE, A. CHESNÉ, Anal.Chim. Acta, 45 (1969) 311.
74. D. GOURISSE, A. CHESNÉ, Anal. Chim. Acta, 45 (1969) 321.
75. T. SHIMIZU, R. ISHIKURA, J. Chromatog., 56 (1971) 95.
76. A. A. ZAITSEV, E. A. KARELIN, V. N. KOSYAKOV, I. A. LEBEDEV, B. F. MYASOEDOV, V. M. NIKOLAEV, E. G. CHUDINOV, I. K. SHVETSOV, 4th Intern. U. N. Conf. on the Peaceful Uses of Atomic Energy, Geneva, Switzerland, 49/P/689, 1971.
77. K. BUIJS, J. REUL, Anal. Chim. Acta, 62 (1973) 304.
78. K. BUIJS, F. MAINO, W. MÜLLER, J. REUL, J. Cl. TOUSSAINT, Symposium on the Chemistry of Transuranium Elements, Moscow, 1972.
79. F. L. MOORE, Anal. Chem., 36 (1964) 2158.
80. P. Th. GERONTOPULOS, L. RIGALI, P. G. BARBANO, Radiochim.Acta, 4 (1965) 75.
81. E. A. HUFF, J. Chromatog., 27 (1967) 229.
82. P. G. BARBANO, L. RIGALI, J. Chromatog., 29 (1967) 309.
83. E. P. HORWITZ, 3rd Intern. Symposium on the Transuranium Elements, Argonne National Laboratory, 1971.
84. E. P. HORWITZ, C. A. A. BLOOMQUIST, L. J. SAURO, D. J. HENDERSON, J. Inorg. Nucl. Chem., 28 (1966) 2313.
85. J. Van OOYEN, RCN-113, 1970.
86. J. Van OOYEN, Thesis, Amsterdam, 1970.
87. I. STRÓNSKI, Chromatographia, 2 (1969) 285.
88. E. G. CHUDINOV, S. V. PIROZHKOV, IAE-1904, Moscow, 1969.
89. W. KNOCH, H. LAHR, Radiochim. Acta, 4 (1965) 114.
90. E. P. HORWITZ, K. A. ORLANDINI, C. A. A. BLOOMQUIST, Inorg. Nucl. Chem. Letters, 2 (1966) 87.
91. J. Van OOYEN in Solvent Extraction Chemistry, D. DYRSSEN, J. O. LILJENZIN, J. RYDBERG, Eds., North Holland Publ. Co., Amsterdam, p. 485, 1967.
92. E. P. HORWITZ, C. A. A. BLOOMQUIST, K. A. ORLANDINI, D. J. HENDERSON, Radiochim. Acta, 8 (1967) 127.

224

93. W. MÜLLER, Atompraxis, 15 (1969) 35.
94. E. P. HORWITZ, C. A. A. BLOOMQUIST, H. E. GRIFFIN, ANL-7569, 1969.
95. K. BUIJS, F. MAINO, W. MÜLLER, J. REUL, J. Cl. TOUSSAINT, Angew. Chem., (Intern. Ed.) 10 (1970) 727.
96. E. P. HORWITZ, L. J. SAURO, C. A. A. BLOOMQUIST, J. Inorg. Nucl. Chem., 29 (1967) 2033.
97. H. F. ALI, A. A. ABDEL-RASSOUL, Z. anorg. allg. Chem., 387 (1972) 252.
98. B. TOMAŽIĆ, S. SIEKIERSKI, J. Chromatog., 21 (1966) 98.
99. W. SMULEK, K. MOSZYŃSKA, Nukleonika, 14 (1969) 1123.
100. G. S. KATYKHIN, M. K. NIKITIN, Radiokhimiya, 10 (1968) 474.
101. V. A. LUGININ, I. A. TSERKOVNITSKAYA, Radiokhimiya, 12 (1970) 898.
102. E. R. SCHMID, W. PFANNHAUSER, Mikrochim. Acta, 1971, 250.
103. E. A. HUFF, S. J. KULPA, Anal. Chem., 38 (1966) 939.
104. I. STRÓNSKI, Radiochim. Acta, 13 (1970) 25.
105. E. K. HULET, J. Inorg. Nucl. Chem., 26 (1964) 1721.
106. F. L. MOORE, Anal. Chem., 40 (1968) 2130.
107. R. F. OVERMAN, Anal. Chem., 43 (1971) 600.
108. J. KOOI, R. BODEN, Radiochim. Acta, 3 (1964) 226.
109. J. KOOI, R. BODEN, J. WIJKSTRA, J. Inorg. Nucl. Chem., 26 (1964) 2300.
110. J. KOOI, J. M. GANDOLFO, N. WAECHTER, J. WIJKSTRA, EUR-2578 e, 1965.
111. J. KOOI, Radiochim. Acta, 5 (1966) 91.
112. W. J. DE WET, E. A. C. CROUCH, J. Inorg. Nucl. Chem., 27 (1965) 1735.
113. K. A. GAVRILOV, E. GVUDZ, J. STARÝ, WANG TUNG SENG, Talanta, 13 (1966) 471.
114. F. L. MOORE, A. JURRIAANSE, Anal. Chem., 39 (1967) 733.
115. E. P. HORWITZ, C. A. A. BLOOMQUIST, Inorg. Nucl. Chem. Letters, 5 (1969) 753.
116. E. P. HORWITZ, C. A. A. BLOOMQUIST, J. A. BUZZELL, H. W. HARVEY, ANL-7546, 1969.
117. R. D. BAYBARZ, J. B. KNAUER, Radiochim. Acta, 19 (1973) 30.
118. P. R. FIELDS, I. AHMAD, R. F. BARNES, R. K. SJOBLOM, E. P. HORWITZ, Nucl. Phys., A 154 (1970) 407.
119. D. N. METTA, H. DIAMOND, F. R. KELLY, J. Inorg. Nucl. Chem., 31 (1969) 1245.
120. H. F. ALY, R. M. LATIMER, Radiochim. Acta, 14 (1970) 27.
121. E. K. HULET, R. W. LOUGHEED, J. D. BRADY, R. E. STONE, M. S. COOPS, Science, 158 (1967) 486.
122. R. J. SILVA, T. SIKKELAND, M. NURMIA, A. GHIORSO, E. K. HULET, J. Inorg. Nucl. Chem., 31 (1969) 3405.
123. R. D. BAYBARZ, J. B. KNAUER, J. R. PETERSON, Nuclear Technology, 11 (1971) 609.
124. K. WANATABE, J. Nucl.Sci. Technol., 2 (1965) 45.
125. K. WANATABE, J. Nucl. Sci. Technol., 2 (1965) 112.
126. B. A. LOVERIDGE, M. S. GORDON, J. P. WEAVER, AERE-R 4747, 1964.
127. S. F. MARSH, Anal. Chem., 39 (1967) 641.
128. S. A. ALI, S. H. EBERLE, Inorg. Nucl. Chem. Letters, 7 (1971) 153.
129. H. F. ALY, M. RAIEH, Anal. Chim. Acta, 54 (1971) 171.

CHAPTER 8

EXTRACTION CHROMATOGRAPHY OF LANTHANIDES

S. SIEKIERSKI, I. FIDELIS

INTRODUCTION

The development of extraction chromatography as a method for the separation of inorganic ions is closely connected with the problem of the separation of lanthanides. Because of the similarity of their chemical properties, lanthanides represent a difficult analytical problem, and the ability to separate them can be considered as a test for any new method aimed at the separation of ions: no wonder therefore that extraction chromatography was generally accepted as a separation method only after the first successful separation of lanthanides was performed by this method[1,2]. The importance of the lanthanide separation problem for extraction chromatography is also due to the fact that the majority of practical and theoretical aspects of the method have been studied on the example of their separation. On the other hand, lanthanide chemistry owes much to the method of extraction chromatography. For example, the important effect concerning the changes in some properties of lanthanide and actinide ions as a function of their atomic number, at first called "the regularities", was discovered during the separation of lanthanides by extraction chromatography[3]*. Nevertheless, many additional potential applications of extraction chromatography to the study of lanthanide chemistry, particularly in the field of the kinetics of complex formation and extraction, are still unexplored.

Any practical application of extraction chromatography to the separation of lanthanides requires the knowledge of some basic facts concerning the extraction of these elements by the different extractants, and particularly the knowledge of the magnitude of the separation factors for neighbouring lanthanides in the different systems. These problems will therefore be discussed prior to the review of the separations achieved.

1. EXTRACTION OF LANTHANIDES

Lanthanides can and actually have been extracted by each of the three main extraction mechanism: formation of chelate complexes with acidic extractants, solvation of salts, and formation of ion pairs. Each of these systems has also been applied to their extraction chromatographic separation. The various systems and the various extractants differ in their ability to separate lanthanides by extraction chromatography; the main cause of these differences are thermodynamic factors, although kinetic factors can also be of some importance. In the following paragraphs, the thermodynamic factors will be discussed briefly.

In the separation of lanthanides (or of other elements of similar properties) two quantities related to the thermodynamic properties of the system are of great importance: the distribution coefficient D and the separation factor β. Therefore, the following discussion will be devoted to the problem of defining which step or steps in the extraction process are mainly responsible for the magnitude of the distribution coefficient and for the differences in the distribution coefficients of the different lanthanides.

1.1. EXTRACTION OF LANTHANIDES BY CHELATE FORMATION WITH ACIDIC ORGANOPHOSPHOROUS EXTRACTANTS

A classical example of this extraction mechanism is offered by the extraction with di(2-ethylhexyl) phosphoric acid (HDEHP). According to Peppard et al.[9], the extraction of a lanthanide ion into a toluene solution of HDEHP can be represented by the following equation:

$$Ln_{aq}^{+3} + 3(HDEHP)_{2,org} \rightleftharpoons Ln[H(DEHP)_2]_{3,org} + 3H_{aq}^+ \qquad (1)$$

---

*This effect was also colled "tetrad effect"[4], however, the name "double-double effect" introduced during the continuation of the original studies made by the present authors[5-8], appears to be more justified.

227

As pointed out in Chapter 1, it is convenient to divide the overall extraction process into two steps, i. e. the formation of the extractable species in the aqueous phase, and its partition between the two phases. The following equilibria can be written:

$$Ln_{aq}^{+3} + 3(HA)_{2,aq} \rightleftharpoons Ln(HA_2)_{3,aq} + 3H_{aq}^{+} \tag{2}$$

$$Ln(HA_2)_{3,aq} \rightleftharpoons Ln(HA_2)_{3,org} \tag{3}$$

where HA is an acidic organophosphorous extractant. The experimentally measured distribution coefficient is proportional to the product of the equilibrium constants for reaction (2) and (3):

$$D = A \cdot K_c \cdot P \tag{4}$$

where K_c is the stability constant of the complex, P is its partition coefficient, and A is a proportionality constant that takes into account the partition of the extractant dimer between the two phases ($HA_{2,org} \rightleftharpoons HA_{2,aq}$), and its monomerization and acidic dissociation in the aqueous phase.

The greater is the stability constant of the complex $Ln(HA_2)_3$, the more shifted to the right is the equilibrium represented by equation (2), and the greater therefore is the concentration of the extractable species in the aqueous phase and as a consequence the greater is the distribution coefficient. No experimental data are available on stability constants in water of lanthanide complexes with HDEHP or similar extractants. However, one can guess that these complexes are rather strong, from the fact that the distribution coefficient (which is proportional to stability constant of the complex) increases very rapidly with Z. This is because experimental data on the stability constants of other lanthanide complexes in the aqueous phase generally indicate that the greater the stability constant the faster it increases with Z.

As for the effect of the partition coefficients of the extractable complexes on the distribution coefficients of the different lanthanides, it has been shown in Chapter 1 that the partition coefficient of a complex increases with its molar volume and decreases with increasing hydration, both the molar volume and hydration depending on the extractant.

In the case of acidic organophosphorous extractants, the distribution coefficient of lanthanides substantially increases with Z. From equation (4) it appears that the reason for this may be either the changes in the stability constant with Z, or in the partition coefficient of the extractable complex, or both. The fact that the double-double effect exists in the free energy of extraction as well as in the free energy of complex formation in the aqueous phase (see Paragraph 2.1) is a proof that the differences in the extractability of lanthanides are mainly caused by the complex formation step of the extraction process.

Whether the complexes of the different lanthanides with the same acidic organophosphorous extractant differ in their composition and structure (which would influence their partition) seems to be an open question. It is possible that they differ in the number of water molecules which enter the coordination shell of the lanthanide cation to satisfy its demand for a high coordination number. Anyway, it seems that in most cases

228

the partition coefficient of the extractable complex does not change significantly with Z.

Extraction of lanthanides with HDEHP can be described by equation (1) provided the concentrations of both mineral acid and lanthanide salt are low. In the region of moderate or high acid concentrations (depending on the lanthanide), the distribution coefficient starts to increase with the concentration of the acid, and the composition of the extracted complex changes to $Ln(NO_3)_3 \cdot 3HDEHP$[10-12] (in the case of nitric acid): the mechanism of extraction is then probably the same as with TBP[11,12]. At high lanthanide salt concentrations, the formation of a polymeric species in the organic phase was observed, the composition of this polymer was found to be $Ln(DEHP)_3$[13]. The formation of this polymeric species at high Ln/HDEHP ratios makes difficult the separation of macroamounts of lanthanides by extraction chromatography with HDEHP as the stationary phase.

1.2. EXTRACTION OF LANTHANIDES BY SOLVATION OF SALTS

An acidic organophosphorous extractant performs three functions: it neutralizes the charge on the cation, it removes all or most water molecules from the coordination shell of the cation, and it increases the molar volume of the metal-containing species. In the extraction by solvation of salts, these functions are separated, and the role of the organic extractant consists in the removal of a part of water molecules and in the increase of the molar volume of the extracted salt, whereas the charge on the cation is neutralized by inorganic anions. The extraction mechanism can be represented as follows:

$$Ln_{aq}^{+3} + 3X_{aq}^- + nB_{aq} \rightleftharpoons LnX_3B_{n,aq} \tag{5}$$

$$LnX_3B_{n,aq} \rightleftharpoons LnX_3B_{n,org} \tag{6}$$

where X^- is an inorganic anion (NO_3^-, Cl^-, SCN^-, ClO_4^-) and B is a neutral (organophosphorous) extractant. Obviously also the partition of the extractant between the two phases ($B_{org} \rightleftharpoons B_{aq}$) must be taken into account.

The distribution coefficient depends on both steps of the extraction process, that is complex formation and partition. Because of the complex formation step (equation (5)), the distribution coefficient increases with the concentration of both ligands (X^- and B) and with their increasing ability to form complexes with the lanthanide ions. In practice, the dependence of the distribution coefficient on the concentration of the acid (HX) is quite complicated. In the case of tributylphosphate (TBP) as the extractant, a maximum and minimum can be observed on the curves logD vs. c_{HNO_3} for light lanthanides[14]. At high acid concentrations, D increases with the concentration of the acid, but even at the highest obtainable concentration of HNO_3 and with undiluted TBP, the distribution coefficients of the lightest lanthanides are so low that their separation by extraction chromatography presents some difficulties[1]. A complicated acid dependance exists also in the case of extraction with bifunctional neutral organophosphorous extractants[15]. Since the dependence of ΔG° of extraction on Z generally is less steep for neutral than for acidic extractants, it can be inferred that the mixed complexes (LnX_3B_n) are weaker than those with acidic organophosphorous extractants.

Because of the partition step (equation (6)), the distribution coefficient should incrase with the molar volume of the extractant and with decreasing hydration of the extracted complex.

The changes in the distribution coefficient with Z depend on the concentration of the mineral acid. In the case of TBP as the extractant, the distribution coefficient increases with Z at high HNO_3 concentrations, while it decreases for the heavier lanthanides with more dilute acid[14]. The reversal of the extraction order of lanthanides at low acid concentrations was also observed with methylenebis(di-n-hexylphosphine oxide) (MHDPO)[15], the "anomalous" behaviour of lanthanides at low HNO_3 concentrations can probably be explained by assuming that outer-sphere complexes with NO_3^- ions are formed at low electrolyte concentrations[16]. The changes of the distribution coefficient with Z observed in the case of neutral organophosphorous extractants are, in the majority of cases, caused also by the complex formation step: this conclusion follows from the fact that the double-double effect is observed in the extraction with TBP at high HNO_3 concentration (see Paragraph 2).

1.3. EXTRACTION OF LANTHANIDES IN THE FORM OF ION PAIRS

The essential steps in the extraction of lanthanides in the form of ion pairs can be represented by the following equilibria:

$$Ln_{aq}^{+3} + nX_{aq}^- \rightleftharpoons LnX_{n,aq}^{3-n} \quad n > 3 \tag{7}$$

$$(n-3)R_{aq}^+ + LnX_{n,aq}^{3-n} \rightleftharpoons [R_{n-3}LnX_n]_{org} \tag{8}$$

where X^- is an inorganic anion, and R^+ is a large organic cation, most frequently an ammonium cation, distributed between the two phases in the form of the salt RX. In the first step, an anionic complex is formed which is then transferred into the organic phase together with an equivalent amount of cations R^+. As it was pointed out in Chapter 1, in this case the main driving force for extraction is the ability of the large organic cations to increase the structure of water. Since the electrostatic part of the standard free energy of transfer of an ion from water into an organic solvent is positive and increases with the charge on the ion, one would expect n to be equal to 4 or at maximum 5. This conclusion is supported by experimental evidence in the case of both lanthanides[17] and actinides[18]. Besides inorganic anions, several water molecules are probably present in the first coordination shell of the lanthanide cation to satisfy its requirements for a high coordination number.

As in the case of the extraction systems previously discussed, the magnitude of the distribution coefficient depends on both complex formation and partition. The more shifted to the right is the equilibrium represented by equation (7), the greater is the distribution coefficient. The distribution coefficient increases also with the molar volume of the cation and its hydrophobic properties: therefore, quaternary ammonium salts are better extractants than tertiary, and tertiary are better than secondary.

The way in which the distribution coefficients change with the atomic number of the lanthanide depends on the nature of the complexing anion. In the case of thiocyanates, extraction increases with Z[19,20], while it decreases in the case of nitrates[21] and changes only very slightly in the case of chlorides[22]. It has been suggested that with weak complexing agents, such as nitrates and chlorides, lanthanides retain their coordination shell of water molecules[16]. Since the hydration of lanthanides increases with Z, the electrostatic interaction between the hydrated cation and the anions decreases. As a result of the reversed order in the complex formation ability, the reversed order of extraction is observed. The situation is similar to that probably occurring in the extraction of lanthanides with TBP from dilute HNO_3 solutions.

The influence of the anion on the order of extraction, and the fact that the double-double effect can be observed in the case of thiocyanates, suggest that in the case of ion pair extraction too the complex formation step is mainly responsible for the changes in the distribution coefficient with Z.

1.4. GENERAL TRENDS IN COMPLEX FORMATION AND EXTRACTION OF LANTHANIDES

In the previous paragraphs, the two steps of the extraction process (complex formation and partition) have been analyzed from the viewpoint of the differences in the extractability of the individual lanthanides. It has been pointed out that these differences are most probably caused by the differences in the stability constants of the complexes formed by consecutive lanthanides with a given extracting agent.

On the basis of many well-established instances of predominantly ionic bonding in lanthanide compounds, it could be expected that the stability of lanthanide complexes should increase with decreasing cation radius, i. e. increasing atomic number. The examination of the stability constants data for a number of complexing agents shows that although there are ligands (EDTA, NTA, DCTA, PDTA, α-hydroxy acids and others) for which a monotonic increase with Z is observed in the stability constants, there are also ligands (such as DTPA, HEDTA, EGTA or EEDTA) for which stability does not increase monotonically or even does not change at all for lanthanides heavier than gadolinium. In the case of weak organic complexing ligands (mostly monodentate), and in the case of most inorganic complexing anions, the stability constants do not increase at least significantly along the whole series, and in some cases even decrease with increasing Z. If the typical ligands for lanthanide extraction are taken into account, the situation is similar. For the most popular acidic organophosphorous extractants, HDEHP and 2-ethylhexyl phenyl-phosphonic acid (HEHΦP), the monotonic increasing of stability with Z is observed, but such a trend is not a general one. In the case of TBP and MHDPO, for instance, the general trend depends on the acid concentration in the aqueous phase.

The reasons for the various trends in stability constants for the different ligands are not fully understood. Factors that appear to play an important role are the nature of the complexes that are formed, (outer-or inner-sphere) and possible changes in the hydration number of lanthanide ions, changes in the number of water molecules in the complex, and in the arrangement of a polydentate ligand around the ion.

The changes in the extraction (or complex formation) ability observed from La to Lu for a given system are usually represented by the plot of logD (or logK) vs. Z. The gen-

231

eral trends in these plots show the order of elution of consecutive elements in the series, while the slopes demonstrate how rapidly the D (or K) values change with Z, and at the same time are a measure of the differences in the extractability or complex forma-tion ability for a given system. Such slopes, however, derived over the whole lanthanide series, do not provide an exact information on the differences existing between partic-ular neighbours in the series. Thus, for instance, the straight line plots of slope 2.5 and 2.8 found by liquid-liquid extraction with HDEHP and HEHФP, respectively[9,23], could suggest the same differences in extractability between consecutive lanthanides, but, as it will be shown in paragraph 2.5, this is not true.

The differences in extractability between two neighbouring lanthanides are expressed by means of the separation factor β, defined as the ratio of the distribution coefficients of the two elements ($\beta = D_{Z+1}/D_Z$). The experimentally measured separation factor can be presented as a product of two factors: $\beta = \beta_c \cdot \beta_p$, where β_c is the ratio of stability constants ($\beta_c = K_{Z+1}/K_Z$) and β_p the ratio of the partition coefficients ($\beta_p = P_{Z+1}/P_Z$) of the complexes of the two lanthanides. It is assumed that in the case of neighbouring lan-thanides the ratio of the partition coefficients of their complexes is, in most cases, near to unity ($\beta_p \approx 1$).

Since the value of β is of fundamental importance for the separation of lanthanides, the changes of β with the consecutive lanthanide pairs will be discussed in more detail in the following paragraphs.

1.5. DIFFERENCES IN EXTRACTION AND COMPLEX FORMATION BETWEEN INDIVIDUAL LANTHANIDES

From the very beginning of ion exchange chromatography practice, it was known that some pairs of neighbouring lanthanides, as well as actinides, were more difficult to be separated than others[24—26]. This was particularly apparent for the Eu-Gd and Am-Cm pairs, and was related to the half-filled shell effect[27—30].

The special position of the f^7 (and also of the f^0 and f^{14}) configuration was supported by a number of data, and accepted as an experimental fact. This state of knowledge lasted for a comparatively long time, probably because the most popular separation meth-ods in use at that time, liquid-liquid extraction and ion-exchange chromatography, were incapable, at their stage of development, of furthering the knowledge on the behaviour of individual lanthanides and actinides. However, it should be mentioned that "an appre-ciable difference between lanthanides of even and odd atomic number" was reported by Hesford et al. in 1959, while investigating the extraction of lanthanide nitrates with TBP[14]. The existence of the odd-even effect seemed to be supported by other data for lanthanides and actinides, and was claimed by other authors[17,31—33]. As far as the extraction of lanthanides with acidic organophosphorous compounds is concerned, the odd-even Z effect was found by Peppard et al.[32,33]. In 1964[32] these authors sum-marized that although they had previously reported[9] the plot of logD vs. Z as a straight line of slope 2.5 for HDEHP, now they decided that in the case of HDEHP and di[p-(1,1,3,3-tetramethylbutyl)phenyl] phosphoric acid their data represent the odd-even Z effect.

At the same time, however, a new effect was revealed during the separation of lanthanides by extraction chromatography with HEHΦP as the stationary phase, and was presented as "Regularities in Stability Constants" by Fidelis and Siekierski[3]. The effect of regularities consists in a definite sequence of four minima and four maxima in the separation factor along the lanthanide series; it was introduced as a general pattern of behaviour for some series of lanthanide complexes, and was demonstrated with plots of β values. After this, Peppard et al. decided that their data for the extraction of lanthanides with acidic organophosphorous compounds follow the pattern of regularities revealed for HEHΦP. It has to be emphasized that Peppard et al.[4] introduced a new name, the "tetrad effect", for the effect of regularities, and presented it as a plot of logD vs. Z, instead of the plot of β vs. Z that was used in the original paper.

Irrespective of which relation is plotted in order to illustrate the effect — the original β vs. Z, or $\Delta G°$ vs. Z[34-36], or log D vs. Z (used by Peppard et al.), or even unit cell volumes vs. Z, as it was done for lanthanide[37] and actinide[7] compounds — the essence of the effect consists in some stabilization of the two pairs of electron configurations f^3-f^4 and f^{10}-f^{11}, in addition to the stabilization of the $f°$, f^7 and f^{14} configurations. From the phenomenological viewpoint, these stabilizations result in the main division of the whole group of f-electron elements ($f°$-f^{14}) into two subgroups, $f°$-f^7 and f^7-f^{14}, and in the further internal division of each of the two subgroups by the f^3-f^4 and f^{10}-f^{11} pair, respectively. Therefore, the name "double-double effect" was introduced[5,6] as more justified and at present it is used with increasing frequency[7,8,38-42].

2. THE DOUBLE-DOUBLE EFFECT AND ITS SIGNIFICANCE FOR THE LANTHANIDE AND ACTINIDE SEPARATION

The essence of the double-double effect, as presented in paragraph 1.5, consists of some stabilization of certain f-electron configurations in lanthanides and actinides. This stabilization is important for many properties of these elements, in particular for those of significance for the separation of elements of the same series among themselves. The stabilization of certain electron configurations influences the ability of complex formation along the lanthanide series in a definite way. It has been shown in the previous paragraphs that complex formation is mainly responsible for the separation of these very similar elements; therefore, a more detailed discussion of the double-double effect in the free energies and enthalpies of complex formation by the f-electron elements appears to be advisable.

2.1. THE DOUBLE-DOUBLE EFFECT IN THE FREE ENERGIES OF EXTRACTION AND COMPLEX FORMATION

The double-double effect was originally presented[3] as a regular pattern of changes in the separation factor as function of the lanthanide pair, as shown in Fig. 1 a, for

HEHФP, and consists of an identical sequence of two maxima and two minima in each of the two subgroups of lanthanides, La to Gd and Gd to Lu. This method of presentation very clearly shows the periodical character of the effect, and the analogy between the f^n and f^{n+7} configurations; at the same time it is also of practical importance, since

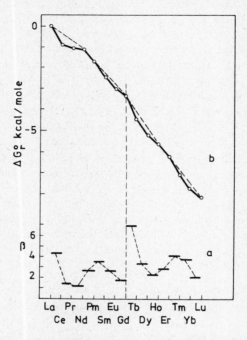

Fig. 1. The regularities or double-double effect in the case of HEHФP. a — Changes in the separation factor; b — Changes in the standard free energy of extraction (relative to lanthanum)[34].

it provides direct information on the changes in the separation factor along the lanthanide series.

Because of the relation: $-RT \ln\beta = \Delta G^{\circ}_{Z+1} - \Delta G^{\circ}_{Z}$, the pattern of regularities in β implies a definite shape of the ΔG° vs Z plot as shown in Fig. 1b[34]. As can be seen in this Figure, two main branches are formed (La to Gd and Gd to Lu), each of them subdivided into two segments in an analogous way. This method of presentation directly shows the special position of the f^0, f^7, f^{14} and f^3-f^4, f^{10}-f^{11} configurations.

The existence of the effect in the free energies was demonstrated for several extractants and complexing agents, and a systematic analysis of the published data on free energies of complex formation and extraction has fully confirmed its general character[35]. However, other factors, such as changes in the coordination number of the lanthanide ions along the series, may often modify ΔG to a much larger extent than the double-double effect, and this could partially account for the sceptical position expressed by Rowlands[43] in regard to its generality. As for its magnitude, the effect is generally more distinct, the stronger the complexes formed by a given ligand are and the more steep the plot of ΔG° vs. Z is.

234

According to the pattern of regularities in the separation factor (Fig. 1a) high separation factors are generally obtained for the La-Ce, Pm - Sm, Gd-Tb and Er-Tm pairs, and low ones for the Pr-Nd, Eu-Gd, Dy-Ho and Yb-Lu. Within each of these two sets, however, the relative values of the separation factors depend on the ligand. In the set having high separation factors, the highest value is that for either the La-Ce or the Gd-Tb pair, while the Pm-Sm and Er-Tm pairs generally display lower separation factors. In the set having low separation factors, the lowest value is usually found for the Pr-Nd pair (in the case of water-immiscible extractants), while the Eu-Gd or Dy-Ho pair show the lowest ratios of stability constants with water soluble ligands.

2.2 THE DOUBLE-DOUBLE EFFECT IN SOME OTHER LANTHANIDE PROPERTIES

Another lanthanide property for which the double-double effect has been found is the enthalpy of extraction[8,40]; in this case, however, the effect is more difficult to be observed than in the case of the free energy, and generally does not appear in the full range, because of the generally non-monotonic trend in ΔH° with Z. Anyhow, it has been established that the effect exists in that region of lanthanides where the enthalpy is mainly responsible for the increase in complex stability with Z[36,44] (Fig. 2); the significance of the double-double effect in enthalpies for the separation of lanthanides will be discussed in Paragraph 3.

Provided that the ΔG° (or ΔH°) of complex formation decreases monotonically with Z, it can be assumed that the changes in ΔG° reflect to some extent the lanthanide contraction. The existence of the double-double effect in ΔG° suggests therefore the existence of this effect in the lanthanide radii; that was confirmed by a careful examination of the difference $r_Z - r_{Z+1}$ (which is a formal analogue of the separation factor β), and on the experimental data on the unit cell volumes of many lanthanide[35,37] and actinide[7] compounds.

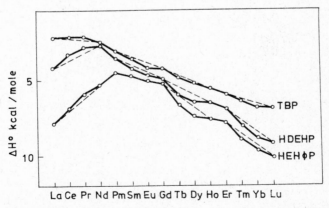

Fig. 2. Changes in the enthalpy of lanthanide extraction as a function of Z, for TBP, HDEHP, HEHΦP[8]

2.3. THE DOUBLE-DOUBLE EFFECT IN ACTINIDES

The existence of the double-double effect for actinides(III) in the Am-Md region was verified by Siekierski[35] on the example of several complexing and extracting agents, and was confirmed by Peppard et al.[45] and by Harmon et al.[46]. The changes in β or in ΔG° as a function of the number of actinide f-electrons exactly follow the same pattern exhibited by lanthanides, and have the same importance for their separation. It appears that the effect is even more distinct in actinides than in lanthanides, probably because of a lesser shielding of 5f-orbitals. The region in which the effect for actinides(III) is observable is limited by the fact that the lightest actinides do not have a sufficiently stable +3 oxidation state, and that no sufficient data are available for the heaviest actinides.

The effect exists also in unit cell volumes of actinide (III, IV, and VI) compounds [7,47]. In the case of higher oxidation states, the pattern of regularity is shifted by one or three places towards greater atomic numbers.

2.4. ON THE THEORY OF THE DOUBLE-DOUBLE EFFECT

At present, the theory of the double-double effect is not yet fully developed. The parallelism between the free ion ground-terms (in regard to the L quantum number) and the pattern of the double-double effect, demonstrated by Fidelis[5], shows that the effect is an inherent property of f-electrons. According to Jørgensen[48,49], the stabilization of the f^7 configuration results from the changes in the E^1 Racah parameter as a function of Z, and the stabilization of the f^3-f^4 and f^{10}-f^{11} configurations from the changes in the E^3 parameter.

As far as thermodynamic functions of complex formation are concerned, it should be noted that the experimentally observed magnitude of the effect depends on the difference between its value in the complex and in the hydrated ion; it depends therefore on the relative positions of the ligand and of water in the nepheloauxetic series. An attempt to explain the double-double effect using the crystal field model has also been made[50].

2.5. COMPARISON OF SEPARATION FACTORS FOR EXTRACTANTS
MOST FREQUENTLY USED IN EXTRACTION CHROMATOGRAPHY

Because of the double-double effect, a comparison of different extractants cannot be solely made on the basis of the mean separation factor. A better method consists in comparing the separation factors for the most difficult separable parts, i. e. Pr-Nd, Eu-Gd, Dy-Ho and Yb-Lu. Separation factors for these four pairs obtained with several extractants used in extraction chromatography are listed in Table 1, together with mean (geometric) separation factors and with the separation factors for the most easily separable pairs (La-Ce and Gd-Tb). From the data listed in the Table and those referring to the extraction with amine nitrates[21], the following conclusions can be drawn:

TABLE 1
Separation factors for lanthanides with different extractants

Separation factor	Extractant				Aliquat--336-SCN
	HDEHP	HEHΦP	TBP	MHDPO	
β_{Pr}^{Nd}	1.44	1.19	1.24	1.08	1.28
β_{Eu}^{Gd}	1.57	1.71	1.41	1.32	1.0
β_{Dy}^{Ho}	2.07	2.19	1.25	1.36	1.25
β_{Yb}^{Lu}	1.92	1.99	1.22	0.88	1.60
β_{La}^{Ce}	4.78	4.31	2.00	–	1.50
β_{Gd}^{Tb}	5.66	6.96	2.19	1.93	1.67
β_{La}^{Gd}	2.61	2.48	1.65	1.41	1.35
β_{Gd}^{Lu}	3.19	3.55	1.48	1.34	1.61
β_{La}^{Lu}	2.90	3.02	1.57	1.37	1.47

Data for HDEHP, HEHΦP and TBP were obtained with liquid-liquid extraction[36,53,54]; data for MHDPO and Aliquat-336-SCN from column extraction chromatography[55,56].

1. The acidic extractants (HDEHP and HEHΦP) are definitely better than both neutral and basic ones (except HEHΦP in the case of the Pr-Nd pair). The differences between the various extractants are smaller for some of the most difficultly separable pairs, e. g. for Pr-Nd and Eu-Gd, than for the easily separable pairs. However, even small differences are of extreme importance when pairs separable with difficulty are concerned.
2. HDEHP, HEHΦP and Aliquat-336-SCN exhibit better selectivity for heavy lanthanides than for light ones, in contrast with the behaviour of other types of extractants and of most complexing agents. For light lanthanides HDEHP is a better extractant than HEHΦP, but the reverse is true for heavy ones.
3. Separation factors obtained with Aliquat-336-SCN are rather low, except for the few heaviest lanthanides. As pointed out in Paragraph 1.3, amine nitrates and chlorides behave differently from thiocyanates and most other complexing agents, and display separation factors that are generally lower than unity (i. e. the lighter lanthanide is

better extracted than the heavier one). For instance, the following separation factors were reported by Stroński[51] for Aliquat-336-NO$_3$ as the stationary phase: $\beta_{Eu}^{Gd} = 0.58$; $\beta_{Ho}^{Er} = 0.75$ and $\beta_{Tm}^{Lu} = 0.20$. These separation factors differ sufficiently from unity to allow for the separation by extraction chromatography with reversed order to elution[51] Actually, the separation factor of the Eu-Gd pair in this system is one of the most favourable, its reciprocal (β_{Gd}^{Eu}) being equal to 1.7; an analogous situation exists for the Am-Cm pair for which separation factors (β_{Cm}^{Am}) lying in the range from 0.33[52] to 0.41 [19] were reported for the same extraction system.

3. THE ENTHALPIES OF LANTHANIDE EXTRACTION AND THE INFLUENCE OF TEMPERATURE ON THE SEPARATION FACTOR

Since most extraction chromatographic separations of lanthanides and actinides have been carried out at raised temperatures, a short survey of the existing data on the enthalpies of lanthanide extraction and on the influence of temperature on their separation factor appears to be advisable.

Assuming that the composition of both phases is constant and that in the expression for the thermodynamic extraction constant the term containing the activity coefficients of all involved species does not appreciably change with temperature, the following relations can be obtained:

$$\frac{\Delta \ln D}{\Delta \ 1/T} = \frac{-\Delta H^\circ}{R}$$

and

$$\frac{\Delta \ln \beta}{\Delta \ 1/T} = \frac{-\Delta \ (\Delta H^\circ)}{R} \tag{9}$$

where ΔH° is the enthalpy of extraction, and $\Delta(\Delta H^\circ) = \Delta H_{Z+1}^\circ - \Delta H_Z^\circ$ is the difference in the enthalpies of extraction for two neighbouring lanthanides. In deriving the above relations, it is also assumed that no appreciable amounts of lanthanide complexes with the extractant exist in the aqueous phase and that only one complex is present in the organic phase. From equation (9) it follows that the separation factor increases with temperature when $\Delta(\Delta H^\circ)$ is positive, and decreases when it is negative.

The are only very few data on the enthalpies of lanthanide extraction and on the influence of temperature on the separation factors, especially under conditions such as those normally adopted in extraction chromatography (i. e. direct column experiments, or liquid-liquid extraction with the undiluted extractant). Most of the authors studying the effect of temperature on the separation of lanthanides devoted their attention first of all to dynamic factors which are more sensitive towards temperature changes than thermodynamic ones. Cerrai, Testa and Triulzi[57] found that the separation factor of the Eu-Gd pair, eluted with HCl on a HDEHP-treated cellulose powder column, decreases from 1.65 to 1.46 as temperature increases from 23 °C to 75 °C. Aly and Raieh[58] determined the enthalpies of extraction of Eu and Sm with a mixture of dibutyldiethylcarbamoyl phosphonate and TTA, supported on Celite, and found that the enthalpy of extraction of Sm is much more negative than that of Eu, so that the separation factor β_{Sm}^{Eu} decreases very rapidly with increasing temperature; the same authors also noticed that the enthalpies and the separation factor depend on the mineral acid ($HClO_4$, HNO_3, HCl) used as the eluting agent. The enthalpies of the extraction of actinides with HDEHP under extraction chromatography conditions were determined by Horwitz et al.[59-61], who found that the separation factors of the Cm-Bk, Bk-Cf, Es-Fm and Fm-Md pairs decrease with increasing temperature, whereas that for the Am-Cm pair increases. According to Horwitz's results, the enthalpies of actinide extraction with HDEHP diluted with n-heptane are more negative by about 2.5 kcal/mole than those with undiluted HDEHP: this result makes the transfer possible of the data on lanthanides obtained by liquid-liquid extraction to the conditions of extraction chromatography.

A systematic study on the enthalpies of extraction of all lanthanides from HNO_3 solutions into HDEHP and HEHΦP (both diluted with heptane) and into undiluted TBP was carried out by Fidelis[36,53,54], and the results obtained are shown in Fig. 2. On the basis of these results, the following conclusions can be made:

1. The enthalpies of extraction are negative for all three extraction systems. This statement should apply also to undiluted HDEHP (and probably to HEHΦP), provided that the difference in the extraction of lanthanides with undiluted and heptane-diluted HDEHP is the same as that found for actinides.
2. For both HDEHP and HEHΦP, the enthalpies pass through a maximum (at Nd and Pm, respectively). In the case of TBP, there is a plateau in the La-Nd region.
3. Starting from Pm, the double-double effect is observed on the ΔH° vs. Z curves. This is easily explainable, since in the Pm-Lu region the enthalpy term is almost entirely responsible for the decrease in the free energy of extraction with Z[44]. On the left branch of each curve, one segment of the double-double effect also seems to appear, its trend being opposite to that observed on the right branch.

From the viewpoint of the extraction chromatographic separation of lanthanides, the differences in the enthalpies for neighbouring elements are much more important than the values for the individual elements, since they determine the changes of β with temperature. The differences in enthalpies should be practically independent of the presence of the diluent, so that liquid-liquid extraction data on the changes in β with temperature can be used to predict the behaviour of chromatographic columns. Figures 3 and 4 report the separation factors for neighbouring lanthanides as a function of temperature, in the system HDEHP in heptane – HNO_3[53,54]. On the basis of these results, the following conclusions can be made:

1. Since the enthalpy of extraction increases with Z from La to Nd, the separation factors in this region increase with temperature. The increase is significant only for the La-Ce pair, that has high enough separation factor because of the double-double effect. The change observed for the most difficultly separable pair (Pr-Nd) is negligible, $\beta^{50°}/\beta^{10°}$ being equal to 1.04.
2. In the region from Nd to Lu, the increase in temperature lowers the separation factors for all pairs. Because of the double-double effect in the enthalpies, the lowering is significant for pairs which have high separation factors, and is very small for difficultly separable pairs. In the case of Eu-Gd, $\beta^{50°}/\beta^{10°} = 0.96$.

239

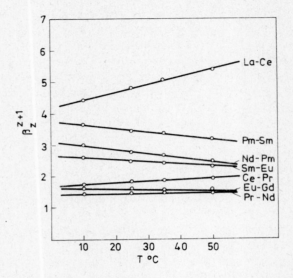

Fig. 3. Changes in the separation factor for neighbouring lanthanides as a function of temperature. Liquid-liquid extraction with 0.04 M HDEHP in heptane from 0.04 M HNO$_3$. La to Gd subgroup[54]

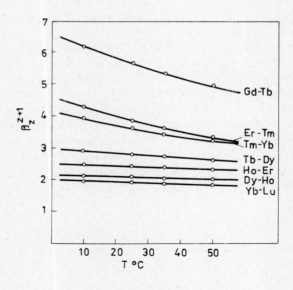

Fig. 4. Changes in the separation factor for neighbouring lanthanides as a function of temperature. Liquid-liquid extraction with 0.04 M HDEHP in heptane from 0.2 M HNO$_3$. Gd to Lu subgroup[54]

Since the influence of temperature on the extraction with HEHΦP and TBP is very similar to that observed with HDEHP, one can conclude that very little improvement in the separation factor can be achieved by changing the temperature, at least for these three extractants.

It seems that no data exist on the influence of temperature on the separation factors, for lanthanides extracted with amines, but it is possible that the effect is rather small in this case too: actually, Horwitz, Sauro and Bloomquist[62] reported a 12% decrease in β for the Cf-Es pair (an analogue of the Dy-Ho pair) in the Aliquat-NO$_3$ system, for temperature increasing from 24 to 80 °C.

4. SURVEY OF THE LANTHANIDE SEPARATIONS

4.1. SEPARATION OF LANTHANIDES FROM EACH OTHER

The separation of lanthanides by extraction chromatography has been achieved by means of extractants representing all main classes of extraction systems. In all cases elution technique was used, for both tracer and macroamount separations. In order to separate all the lanthanides in a single run, the gradient elution method was also applied. Most often separations were performed at elevated temperatures on columns whose bed length varied between some centimeters and several tens of centimeters, depending on column efficiency and the amount of lanthanides to be separated.

4.1.1. SEPARATION WITH ACIDIC ORGANOPHOSPHOROUS EXTRACTANTS

HDEHP is the extractant most often used for lanthanide separations. The first attempt to separate lanthanides by extraction chromatography with HOEHP is that of Winchester[63] who used HDEHP supported on aluminium oxide. The same author later preferred [64,65] silanized kieselguhr as the supporting material, obtaining HETP values as low as ~0.16 mm (Pr peak) that enabled him to separate tracer amounts of all lanthanides on a 6 cm long column using gradient elution with HCl at about 65 °C (Fig. 5): the resulting peaks generally were well separated, although some overlapping could be observed in the case of the most difficult separable pairs Nd-Pm and Yb-Lu.

Pierce and Peck[66,67] used Corvic (a vinyl chloride-vinyl acetate copolymer) as the support for HDEHP. The performance of such columns was not very satisfactory, since HETP was as high as ~1 mm. Nevertheless, all lanthanides were separated using a 60 cm long column and gradient elution with HClO$_4$ at 60 °C. In accordance with the regularities in the separation factors, some difficulties were encountered in the separation of Ce-Pr-Nd, Eu-Gd, Dy-Ho and Yb-Lu.

At the same time, Cerrai, Testa and Triulzi[57,68] used cellulose powder as a support for HDEHP. The performance of such columns was rather poor, since HETP was about 6 mm at room temperature, decreasing to about 1.5 mm at 75 °C. In spite of this, a tracer level separation of La-Ce-Nd-Gd-Tb-Tm was achieved on a 25 cm long column at room temperature, using HCl as the eluting agent. On the same column, milligram amounts

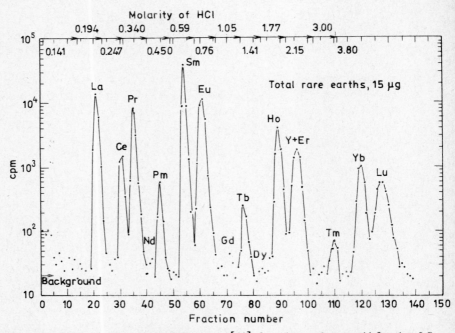

Fig. 5. Separation of lanthanides on a HDEHP column[64]. 6 cm long column, void fraction 0.7; Elution temperature: 60 to 70 °C. (By courtesy of the author)

of La, Ce, Nd and Eu were satisfactorily separated. Later, Cerrai and Testa[69] switched to Kel-F as support, and obtained much more efficient columns than the previous ones, so that the separation of almost all lanthanides was achieved by gradient elution with HCl at 85 °C. The length of the bed was 14 cm; again as expected from the double-doub effect, the separation was not complete for the Eu-Gd, Dy-Ho and Yb-Lu pairs.

Sochacka and Siekierski[70] separated lanthanides by elution with HNO_3 at room temperature, obtaining a satisfactory resolution also for the Eu-Gd, Dy-Ho and Yb-Lu. The Pr-Nd pair was almost completely separated with HCl as the eluting agent. The columns were 10 cm long and contained hydrophobized Hyflo Supercel as the support for HDEHP. The HETP was about 0.33 mm; recently, however, much lower HETP values have been obtained by the present authors (Fig. 6).

Grosse-Ruyken and Bosholm[71] performed a very neat separation of all lanthanides, including the most difficult separable pairs[72], after a thorough examination of the factors influencing the value of HETP of columns containing hydrophobized silica gel as the support for a HDEHP-toluene mixture. The length of the bed was 24 cm, and columns were operated at 40 °C; HETP was ~ 0.35 mm. The same authors achieved also successful milligram level separations[73-75].

Watanabe[76,77] separated several light lanthanides at tracer and milligram levels. Moskvin and Preobrazhenski[78] reported a neat separation of La-Ce-Pm-Eu on HDEHP-Ftoroplast columns at 90 °C. Stroński[79], using short (6.5 cm) columns containing Kel-F as the supporting material, obtained good separation of the Eu-Gd, Eu-Tb, Ho-Er, Tm-Lu

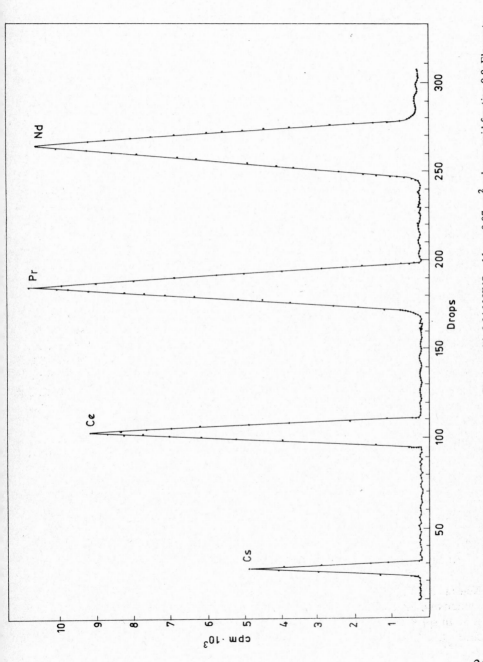

Fig. 6. Separation of Ce-Pr-Nd on a HDEHP column. Elution with 0.21 M HNO_3. 11 cm•0.07 cm^2 column, void fraction 0.8; Flow rate: 0.5 ml•cm^{-2}•min^{-1}. Room temperature

and Yb-Lu pairs. Riccato and Herrmann[80] performed a very fast separation of Eu from Tb. Hornbeck[81] used HDEHP supported on a 12% surface graft of styrene-Carbowax on Kel-F powder, and separated 5 mg of each Ce, Eu, Y, Tm and Lu. However, the column performance was not satisfactory.

Most of the above-mentioned works refer to the separation of tracer or milligram amounts of lanthanides. Attempts were also made to separate gram amounts of lanthanides among themselves, or to separate microamount of one lanthanide from a macroamount of another one. The technique used in such cases still was single elution, the dimensions of the bed being significantly increased in order to avoid the formation of poly mers in the organic phase. Herrmann et al.[82] were able to separate more or less quantitatively tracer amounts of Ce (100%), Gd (76%) and Tb (40%) from 2 g of erbium: the column used was 25 mm in diameter and 420 mm long, and contained 100 g of hydrophobized silica gel and 60 ml of HDEHP. Shmanenkova et al. [83] separated microamounts of both lighter and heavier lanthanides from 1 g amounts of either La, Nd, Sm, Eu, Gd, Ho and Er. The yield of the separation of the microcomponent was found to be above 95%, except for the lanthanide immediately preceding the element that was present in macroamount, for which the yield ranged between 50 and 80%; 7% to 20% of the total column capacity was utilized. Several micro-macro separations were also performed by Mikhailichenko and Pimenova[84]. In Oak Ridge[85] gram amounts of Ce and Eu were separated each from each other on columns (5 cm in diameter and 50 cm long) containir HDEHP supported on Chromosorb P. The separation factor for the two lanthanides was only about 2.7, showing the great influence of the amounts of elements involved on their separation factor.

Among other acidic organophosphorous extractants, only HEHΦP has been sufficiently investigated. Fidelis and Siekierski[86] separated lanthanides using 9 cm long columns with hydrophobized Hyflo Supercel as the support and HCl or HNO_3 as the eluents. Because of the generally more favourable separation factors than in the case of HDEHP, and because of the low (about 0.2 mm) HETP values, very satisfactory separations were obtained even for the pairs separable with most difficulty. Separation of several lanthanid from 10 mg of irradiated Ho_2O_3 was performed by Hunt[87], using again HEHΦP as the stationary phase.

Dibutylphosphoric acid supported on hydrophobized Hyflo Supercel was applied by Kotlińska-Filipek and Siekierski[88] to the separation of tracer amounts of Pm, Eu and Tb from macroamounts of Er.

In the case of all acidic organophosphorous extractants the order of elution was the "normal" one, i. e. each heavier lanthanide was eluted after the lighter one. The mineral acid concentrations varied from several tenths of a mole to several moles per litre.

4.1.2. SEPARATION WITH NEUTRAL ORGANOPHOSPHOROUS EXTRACTANTS

Neutral organophosphorous extractants have been less extensively used than acidic extractants in the column separation of lanthanides, because of their lower separation factors. In spite of this, several very satisfactory separations of lanthanides have been performed.

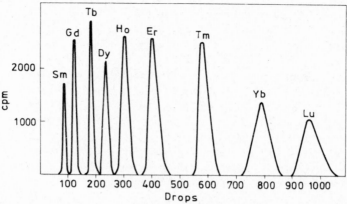

Fig. 7. Separation of some lanthanides on a TBP column[2]. Elution with 12.3 M $HNO_3 \cdot 11$ cm $\cdot 0.07$ cm^2 column, void fraction about 0.7; Flow rate: 0.4 ml$\cdot$cm$^{-2}\cdot$min^{-1}. Room temperature

Fidelis and Siekierski[1,2] used TBP supported on hydrophobized Hyflo Supercel as the stationary phase. The columns were 11 cm long and were operated at room temperature. Almost all lanthanides were separated eluting with HNO_3 and HCl solutions, the latter one being used only in the case of the heaviest lanthanides: an example of the achieved separations is shown in Fig. 7. Because of the low extraction coefficients for the lightest lanthanides, the Ce-Pr and Pr-Nd pairs could not be separated even from very concentrated HNO_3. Very satisfactory HETP values were obtained, in some cases as low as 0.1 mm. TBP supported on silica gel was also used by Brandstetr et al.[89] in the separation of the Tb-Tm, Y-Tm and Y-Ho pairs.

O'Laughlin and Jensen[55] performed an extensive study on the separation of lanthanides by extraction chromatography using bifunctional phosphine oxides, namely methylenebis(di-n-hexylphosphine oxide) and methylenebis di-(2-ethylhexylphosphine oxide). Kel-F, Plaskon and Hyflo Supercel were used as the supporting materials, and HNO_3 as the eluting agent. Although the value of HETP was rather high, most often ranging around 1 mm, a satisfactory separation of milligram amounts of Gd, Tb, Dy and Ho on a 21 cm long column was reported.

Mikulski, Gavrilov and Knobloch[90] applied a 1:1 mixture of tetrabutylpyrophosphate and tetrabutylhypophosphate supported on kieselguhr. In addition to some actinide-lanthanide separations, Eu-Tb-Tm have been separated on a 15 cm long column.

In accordance with the general extraction mechanism of lanthanides, all separations were performed using concentrated mineral acids as the eluting agents. The order of elution was the same as that obtained with acidic extractants.

4.1.3. SEPARATION WITH AMINES

The most thoroughly studied extractant of this class was Aliquat-336-SCN (methyltricaprylylammonium thiocyanate). Huff[56] used Plaskon as the support for it, and eluted lanthanides with NH_4SCN solutions of varying concentration on a 14.7 cm long column. The HETP was rather high, being ~ 3 mm (as calculated from the Yb peak); in

245

spite of this, complete milligram level separation was obtained for La-Eu, and an almost complete one for Tb-Yb-Lu; the separation of Eu-Ho-Er-Tm was rather poor. The order of elution was the "normal" one, that is the lighter lanthanide was eluted before the heavier one.

The same extractant supported on hydrophobized Celite was used by Barbano and Rigali[91] who separated tracer amounts of several lanthanides on a 10 cm long column. A satisfactory separation was obtained for the La-Ce, Ce-Pm and Eu-Tm pairs, while that of Pm from Eu was not so good.

Separations obtained with Aliquat-336-SCN are less satisfactory than those obtained either with HDEHP or TBP. One reason may be the lower separation factors, but the most important one appears to be the rather poor performance of the columns used, not necessarily caused by the properties of the extractant.

Aliquat-336-NO_3 supported on Kel-F was applied by Stroński[51] to separate the Sm-Pm, Er-Ho and Lu-Yb pairs using $LiNO_3$ as the eluting agent. The HETP was about 0.55 mm. In this system, a heavier lanthanide is eluted before a lighter one.

Yb, Nd and La (20 mg each) were separated by Testa[92] using TNOA-treated cellulose and eluting with $LiNO_3$ solutions of various concentrations.

4.2. SEPARATION OF LANTHANIDES FROM OTHER ELEMENTS

Because of the same charge and similar radii of the ions involved, the most difficult separation is that of lanthanides from actinides in the +3 oxidation state. According to Peterson and Cunningham[93], the radius of Am (based on cubic sesquioxides) is intermediate between the radii of Nd and Pm, whereas the radius of Cf is equal to that of Eu. A comparison of radii based on hexagonal trichlorides[94] shows that the radius of Am is equal to that of Pm, and that the radius of Cf is equal to that of Gd. Therefore, the separation of the whole group of lanthanides from transplutonic elements appears to be unfeasible with extraction systems for which the distribution coefficient is solely determined by the crystallographic radius of the ion.

According to the extraction chromatographic data of Watanabe, the distribution coefficient of Am(III) in the system HDEHP-HNO_3 is almost the same as that of Pr[76]. Kooi[95] found that the distribution coefficient of Cf(III) in the system HDEHP-HCl is almost the same as that of Eu; the same conclusion follows from the results of Gavrilov et al.[96]. According to Horwitz et al.[60], the distribution coefficient of Eu in the HDEHP-HCl system is intermediate between those of Bk(III) and Cf(III). From these data it follows that in the HDEHP-mineral acid systems the peak positions of transplutonic elements are near to those predicted from their radii, so that only two or three among the lightest lanthanides and the heaviest ones beginning from Tb can be separated from the Am-Fm group. Obviously, several possibilities exist of separating a single lanthanide from a transplutonic element.

Several separations of lanthanides from transplutonic elements have been performed by the extraction chromatography method using HDEHP as extractant. Watanabe[76,77] separated Am from Pm and from 5 mg of La; Stroński[79] reported the separation of Am and Cm from Eu, whereas Gavrilov et al.[96] performed the separation of Tb from bot

Fm and Md, and of Eu from Fm. Kooi and Boden[97], and also Moore and Jurriaanse [98], separated Ce from Bk; Pluym[99] separated Nd from Am by elution with 0.2 N HNO_3 + 1 M NH_4SCN. An interesting separation of the La-Eu group from actinides was reported by Marsh[100]; this group separation was possible owing to the preferential extraction of lanthanides by HDEHP from a mixture of 1 M lactic acid and 0.05 M sodium diethylenetriaminopentaacetate, used as the eluting agent. Essentially the same method was later applied by Ali and Eberle[101] to the separation of Am from Ce, Eu and Tb.

In the case of the TBP-mineral acid systems, the situation is very similar to that existing in the case of HDEHP. Liquid-liquid extraction experiments[102] demonstrate that the distribution coefficient of Am(III) from concentrated HNO_3 into 100% TBP is intermediate between those of Nd and Pm, and that the distribution coefficient of Bk(III) is almost identical with that of Eu. Only few lanthanides-actinides separations were performed with neutral organophosphorous compounds as the stationary phase. Brandstetr et al.[89] reported the separation of Eu, Ho, Y and Tm from Md(III) using TBP columns and HNO_3 or HCl as eluting agents. TBP was also applied by Ziv et al.[103] to the separation of Ac from La; Mikulski et al.[90] used a mixture of tetrabutyl hypophosphate and tetrabutylpyrophosphate to the separation of Eu from CM(III); TOPO was used by Watanabe[76] to separate Am(III) from La and Eu.

In the case of ammonium salts as extractants, the possibility of separating lanthanides from actinides on the +3 oxidation state depends on the anion. In nitrate systems (at least with Aliquat-336), the distribution coefficient of lanthanides decreases with increasing Z, whereas that of actinides increases[21,52] except for the position of Am. Because of these trends, the distribution coefficients decrease in the sequence Ce > Am > Pm > Cf > Bk > Eu > Cm > Tm. This makes a group separation of lanthanides from actinides impossible, but does not preclude the separation of various pairs; Horwitz et al.[104], for instance, showed that La, Ce and Pr can be separated from Cm(III) in the system Aliquat-336-acidified $LiNO_3$.

In the case of Aliquat-336-SCN, the distribution coefficients of both the lanthanides and actinides[19,20,56] increase with decreasing radius, except for the Am-Cm pair. In spite of this, the distribution coefficients of transplutonic elements are greater than those of lanthanides, the separation factor of Am from Lu being near to 3[56]. Owing to this difference, very satisfactory separations of lanthanides from Am(III) were performed by Huff[56] and by Barbano and Rigali[91] using NH_4SCN as the eluting agent.

Because of similar properties, the separation of yttrium from certain lanthanides also presents some difficulties. In the TBP-HNO_3 system, Y appears between Er and Tm[2], whereas in the HDEHP-diluted HCl system it is eluted together with Er[64]; except for the separation from Er, Tm and — to some extent — Ho, the separation of Y from other lanthanides is quite easy. Grosse-Ruyken and Bosholm[73] separated Y from Dy and Tb in the HDEHP-HCl system; in the same system, Tb, Dy and Ho (the latter only partially) were separated from 1 g of Y by Zemskova et al.[105]. Michelsen and Smutz[106] separated Ho from Y in the system HDEHP—2.1 M HCl; the same authors reported the separation of Er from Y in the system HDEHP-concentrated HCl, Er being eluted the first[107]. The latter separation was possible because of the different extraction mechanism with HDEHP at high acidities.

The separation of lanthanides from +1, +2 and +4 cations can be performed in nearly all extraction systems because of the great differences in the distribution coefficients involved. Such differences make also the separation possible of those lanthanides which can appear on +2 (Eu) or +4 (Ce) oxidation states from the other lanthanides[108-111]. Also the separation of lanthanides from the AnO_2^{+2} ions (An staying for U, Np or Pu) is easily obtained with HDEHP. According to Tomažič[112-113], however, the separation of lanthanides from U(VI) at high U(VI) to HDEHP ratios is incomplete because of the incorporation of lanthanides by the insoluble U(IV)-HDEHP polymer which forms in such conditions.

Several papers have been devoted to the separation of lanthanides from uranium fission products[114-119], most often by means of HDEHP columns. In these separations, extraction chromatography is usually combined with either ion-exchange chromatography or liquid-liquid extraction.

5. RESOLUTION ABILITY OF COLUMNS USED FOR THE SEPARATION OF LANTHANIDES

The attainment of a satisfactory resolution ability of columns is a problem of paramount importance in the separation of lanthanides among themselves, because of the very low separation factors for some pairs; actually, most of the work devoted to the resolution ability of columns was carried out on the example of the separation of lanthanides or actinides. A short survey of the results pertaining to the separation of lanthanides appears to be necessary, also because the resolution of columns is not an independent notion, but depends – at least in part – on the properties of the eluted substance.

HETP values ranging from 0.1 to several millimeters have been obtained for columns prepared with the very same stationary phase, but supported on different materials. The main reason for the observed differences between the results obtained with the various supports seems to be the difference in their particle size. The role of the chemical nature of the support has not been sufficiently investigated, under conditions of the same particle size and porosity; the best results were obtained with kieselguhr and silica gel, but incidentally particles of particularly small size have been used with both these types of material. The role of the specific surface of the supporting material was also insufficiently elucidated: very low HETP values (0.1 mm) have been obtained with both Hyflo Supercel and silica gel that appreciably differ in their specific surface area. Supports having greater specific surface are able to retain greater volumes of extracting agents. To this respect, silica gel is better than either Hyflo Supercel or Celite; also Kel-F and similar materials appear to be able to retain substantial amounts of organic phase. The property of retaining great amounts of stationary phase is of crucial importance in the separation of macroamount of lanthanides, but is unessential in tracer-level separations. According to Hermann[120], the pore radius does not affect HETP as long as it does not fall below ~ 35 Å, this value being somewhat greater than the assumed radius of the $Ln[H(DEHP)_2]_3$ molecule.

According to Horwitz and Bloomquist[121], the slurry technique for column preparation gives better results than dry tamping. Nevertheless, HETP values as low as 0.1 mm at room temperature have been obtained using the dry tamping method. Hydrophilic support materials can be satisfactorily silanized with dimethyldichlorosilene, either in vapour or liquid form. The best method of loading the support with the extractant is mixing the support material with a solution of the extractant in a low-boiling solvent, which is next evaporated away.

The chemical properties of the extractant do not appear to substantially influence the column resolution ability for lanthanides. Satisfactory columns have been obtained with TBP, HDEHP, HEHΦP and in some cases also with amines as the stationary phase. Some differences should, however, exist, at least at high flow rates, because of different viscosities and possible chemical effects, but their study is difficult because of the many factors that are involved (see Chapter 1). Relatively much work has been devoted to the influence of the amount of the extractant on column resolutions[70,71,86,120,121]. At low extractant loadings, the HETP value is low and constant, provided that the support has been loaded by the above-cited evaporation method, and that the amount of the lanthanide is negligible. With a further increase in the amount of the extractant, HETP values start to increase. This occurs for extractant to support ratio that substantially depends on the type of the support, and that is higher for silica gel than for kieselguhr. It seems also that greater amounts of TBP than of HDEHP can be retained on the support (at least in the case of Hyflo Supercel) without increasing HETP.

The effect of flow rate on the separation of lanthanides or actinides has also been investigated in several papers[59,60,71,121-123]. In general, HETP appreciably increases with flow rate. The actual shape of HETP vs. flow rate curves depends on many factors, the most important being particle size, pore diameter, specific surface and loading with the extractant; according to the results of Ref.[121], it also depends on the method of column preparation, and is less steep in the case of the slurry technique.

The HETP of a given column decreases as the operating temperature is increased; the magnitude of the effect depends on several factors, e. g. particle size and flow rate, and so far it has not been sufficiently investigated under comparable conditions. Since the increase in temperature is the easiest way of improving the column resolution power, most of the lanthanide separations have been performed at raised temperatures. However, with a well-prepared 10 cm long column all lanthanides can be satisfactorily separated even at room temperature, at least when HDEHP, HEHΦP or TBP are used as the stationary phase.

Under certain conditions, slow chemical steps are almost exclusively responsible for the observed value of HETP. Siekierski and Sochacka[123] found that the plate heights of HDEHP columns at low temperatures depend on the lanthanide atomic number and on the mineral acid used for elution. In the case of lanthanides lighter than Gd, plate heights are low and almost equal, irrespective of the acid (HNO_3, $HClO_4$, HCl or H_2SO_4) used as the eluent. However, when heavy lanthanides are concerned, the plate heights increase for lanthanides between Tb and Er, and remain constant and very high values for those heavier than Er: such effect was observed with HCl and H_2SO_4, but was almost absent with HNO_3 and $HClO_4$. For heavy lanthanides and for HCl as the eluting agent, HETP was found to be very sensitive to flow-rate and temperature; it decreased to the low value observed with HNO_3 when the temperature was increased to 75 °C. The great influence of flow-rate and temperature is a proof that the high values of HETP are caused by a slow chemical step. The dependence of the plate heights on the eluting mineral acid was also observed by Mikhailichenko and Pimenova[84], who found that in the case of Tb eluted on a HDEHP column the plate height is higher for HCl than for

HNO$_3$. However, it should be pointed out that other authors separating lanthanides in the system HDEHP-mineral acid did not report such an effect. This, in part, might be due to the fact that most separations have been performed at elevated temperatures, when the effect is much less relevant; nevertheless, it appears that additional causes must exist for the differences between the results of the various authors. This problem needs further investigation.

According to the explanation proposed in Ref.[123], the differences in the HETP values obtained with the different acids are caused by the differences in the kinetic lability of the inner sphere or outer-sphere complexes probably formed by lanthanides with Cl$^-$ and SO$_4^{-2}$, or NO$_3^-$ and ClO$_4^-$ ions, respectively. However, this explanation contrasts with the fact that no influence of the nature of the acid on plate height was found for HEHΦP as the stationary phase[86].

A similar effect to that observed for heavy lanthanides eluted on HDEHP columns was reported by Aly and Raieh[58] for columns having a mixture of TTA and dibutyl-diethylcarbamoyl phosphonate as the stationary phase. They found that the plate height of europium depends on the eluting acid, increasing in the seequence: $H_{HClO_4} < H_{HNO_3} < H_{HCl}$. According to their explanation, the effect is caused by the different degree of hydration of the anions. It should also be noted that an increase in plate height with increasing atomic number of the actinide was found by Horwitz et al.[60] in the HDEHP-HNO$_3$ system.

Obviously, the role of a slow chemical step in the mass transfer between the stationary and mobile phases in extraction chromatography of lanthanides and actinides still deserves further study.

REFERENCES TO CHAPTER 8

1. S. SIEKIERSKI, I. FIDELIS, J. Chromatog., 4 (1960) 60.
2. I. FIDELIS, S. SIEKIERSKI, J. Chromatog., 5 (1961) 161.
3. I. FIDELIS, S. SIEKIERSKI, J. Inorg. Nucl. Chem., 28 (1966) 185.
4. D. F. PEPPARD, G. W. MASON, S. LEWEY, J. Inorg. Nucl. Chem., 31 (1969) 2271.
5. I. FIDELIS, Bull. Acad. Polon. Sci. Ser. Sci. Chim., 18 (1970) 681.
6. I. FIDELIS, S. SIEKIERSKI, J. Inorg. Nucl. Chem., 33 (1971) 3191.
7. S. SIEKIERSKI, I. FIDELIS, J. Inorg. Nucl. Chem., 34 (1972) 2225.
8. I. FIDELIS, Bull. Acad. Polon. Sci. Ser. Sci. Chim., 20 (1972) 605.
9. D. F. PEPPARD, G. W. MASON, J. L. MAIER, W. J. DRISCOLL, J. Inorg. Nucl. Chem., 4 (1957) 334.
10. E. CERRAI, G. GHERSINI, J. Chromatog., 24 (1966) 383.
11. E. CERRAI, G. GHERSINI, Advances in Chromatog., 9 (1970) 3.
12. A. I. MIKHAILICHENKO, R. M. PIMENOVA, Radiokhimiya, 11 (1969) 8.
13. T. HARADA, M. SMUTZ, R. G. BAUTISTA, Proc. Int. Solv. Extr. Conf., The Hague, Vol. 2, p. 950, 1971.
14. E. HESFORD, E. E. JACKSON, H. A. C. McKAY, J. Inorg. Nucl. Chem., 9 (1959) 279.
15. J. W. O'LAUGHLIN, Progress Nucl. Energ., Series IX, Anal. Chem., Vol. 6, p. 95, 1966.
16. G. R. CHOPPIN, S. L. BERTHA, J. Inorg. Nucl. Chem., 35 (1973) 1309.
17. Y. MARCUS, I. ABRAHAMER, J. Inorg. Nucl. Chem., 22 (1961) 141.
18. E. P. HORWITZ, C. A. A. BLOOMQUIST, L. S. SAURO, D. J. HENDERSON, J. Inorg. Nucl. Chem., 28 (1966) 2313.
19. F. L. MOORE, Anal. Chem., 36 (1964) 2158.
20. P. Th. GERONTOPULOS, L. RIGALI, P. G. BARBANO, Radiochim. Acta, 4 (1965) 75.

21. F. L. MOORE, Anal. Chem., 38 (1966) 510.
22. F. L. MOORE, Anal. Chem., 33 (1961) 748.
23. D. F. PEPPARD, G.W. MASON, I. HUCHER, J. Inorg. Nucl. Chem., 18 (1961) 245.
24. B. K. KETELLE, G. E. BOYD, J. Am. Chem. Soc., 69 (1947) 2800.
25. B. H. KETELLE, G. E. BOYD, J. Am. Chem. Soc., 73 (1951) 1862.
26. E. J. WHEELWRIGHT, F. H. SPEDDING, G. SCHWARZENBACH, J. Am. Chem. Soc., 75 (1953) 4196.
27. H. BOMMER, Z. Anorg. Chem., 241 (1939) 273.
28. S. G. THOMPSON, B.B.CUNNINGHAM, G.T. SEABORG, J. Am. Chem., Soc., 72 (1950) 2798.
29. G. H. HIGGINS, K. STREET, Jr., J. Am. Chem. Soc., 72 (1950) 5321.
30. D. F. PEPPARD, W. J. DRISCOLL, R. J. SIRONEN, S. McCARTY, J. Inorg. Nucl. Chem., 4 (1957) 326.
31. M. TAUBE, Nukleonika, 7 (1962) 389.
32. G. W. MASON, S. LEWEY, D. F. PEPPARD, J. Inorg. Nucl. Chem., 26 (1964) 2271.
33. D. F. PEPPARD, G. W. MASON, S. LEWEY, J. Inorg. Nucl. Chem., 27 (1965) 2065.
34. I. FIDELIS, S. SIEKIERSKI, J. Inorg. Nucl. Chem., 29 (1967) 2629.
35. S. SIEKIERSKI, J. Inorg. Nucl. Chem., 32 (1970) 519.
36. I. FIDELIS, J. Inorg. Nucl. Chem., 32 (1970) 997.
37. S. SIEKIERSKI, J. Inorg. Nucl. Chem., 33 (1971) 377.
38. S. HUBERT, M. HUSSONNOIS, R. GUILLAUMONT, J. Inorg. Nucl. Chem., 35 (1973) 2923.
39. I. FIDELIS, S. SIEKIERSKI, X. Rare Earth Research Conference, Carefree, Arizona, 1973.
40. I. FIDELIS, Proc. XV Int. Conf. Coord. Chem., Moscow, Vol, 2, p. 534, 1973.
41. T. MIODUSKI, S. SIEKIERSKI, Proc. XV Int. Conf. Coord. Chem., Moscow, Vol. 2, p. 514, 1973.
42. S. SIEKIERSKI, I. FIDELIS, XXIV IUPAC Congress, Hamburg, 1973.
43. D. L. G. ROWLANDS, J. Inorg. Nucl. Chem., 29 (1967) 809.
44. I. FIDELIS, Proc. XIII Int. Conf. Coord.Chem., Kraków-Zakopane, Poland, Vol. 2, p. 195, 1970.
45. D. F. PEPPARD, C. A. A. BLOOMQUIST, E. P. HORWITZ, L. LEWEY, C. W. MASON, J. Inorg. Nucl. Chem., 32 (1970) 339.
46. D. H. HARMON, J. R. PETERSON, W. J. McDOWELL, C. F. COLEMAN, J. Inorg. Nucl. Chem., 34 (1972) 1381.
47. I. FIDELIS, S. SIEKIERSKI, INR Report 1212/V/C, Warsaw, 1970.
48. C. K. JØRGENSEN, J. Inorg. Nucl. Chem., 32 (1970) 3127.
49. C. K. JØRGENSEN, Modern Aspects of Ligand Field Theory, North Holland Publ. Co., Amsterdam, 1971.
50. A. PALCZEWSKI, S. SIEKIERSKI, Proc. XIII Int. Conf. Coord. Chem., Kraków-Zakopane, Vol. 2, p. 197, 1970.
51. I. STROŃSKI, Chromatographia, 2 (1969) 285.
52. J. Van OOYEN, Proc. Int. Solv. Extr. Conf., Gotenburg, p. 485, 1966.
53. I. FIDELIS, INR Report. 1393/V/C, Warsaw, 1972.
54. I. FIDELIS, INR Report. 1394 (V)C, Warsaw, 1972.
55. J. W. O'LAUGHLIN, D. F. JENSEN, J. Chromatog., 32 (1968) 567.
56. E. A. HUFF, J. Chromatog., 27 (1967) 229.
57. E. CERRAI, C. TESTA, C. TRIULZI, Energia Nucleare, 9 (1962) 377.
58. H. F. ALY, M. RAIEH, Anal. Chim. Acta, 54 (1971) 171.
59. E. P. HORWITZ, C. A. A. BLOOMQUIST, D. J. HENDERSON, J. Inorg. Nucl. Chem., 31 (1969) 1149.
60. E. P. HORWITZ, C. A. A. BLOOMQUIST, D. J. HENDERSON, D. E. NELSON, J. Inorg. Nucl. Chem., 31 (1961) 3255.
61. E. P. HORWITZ, C. A. A. BLOOMQUIST, Inorg. Nucl. Chem. Letters, 5 (1969) 753.
62. E. P. HORWITZ, L. J. SAURO, C.A.A. BLOOMQUIST, J. Inorg. Nucl. Chem., 29 (1967) 2033.

63. J. M. WINCHESTER, USAEC Report, CF-58-12-43, 1958.
64. J. M. WINCHESTER, J. Chromatog., 10 (1963) 502.
65. J. M. WINCHESTER, J. Chinese Chem. Soc., Ser.II, 10 (1963) 80.
66. T. B. PIERCE, P. F. PECK, Nature, 194 (1962) 84.
67. T. B. PIERCE, P. F. PECK, Nature, 195 (1962) 597.
68. E. CERRAI, C. TESTA, C. TRIULZI, Energia Nucleare, 9 (1962) 193.
69. E. CERRAI, C. TESTA, J. Inorg. Nucl. Chem., 25 (1963) 1045.
70. R. SOCHACKA, S. SIEKIERSKI, J. Chromatog., 16 (1964) 376.
71. H. GROSSE-RUYKEN, J. BOSHOLM, J. Prakt. Chem., 25 (1964) 79.
72. J. BOSHOLM, H. GROSSE-RUYKEN, J. Prakt. Chem., 26 (1964) 83.
73. H. GROSSE-RUYKEN, J. BOSHOLM, J. Prakt. Chem., 30 (1965) 77.
74. H. GROSSE-RUYKEN, J. BOSHOLM, Kernenergie, 8 (1965) 224.
75. H. GROSSE-RUYKEN, J. BOSHOLM, G. REINHARDT, Isotopenpraxis, 1· (1965) 124.
76. K. WATANABE, J. Nucl. Sci. Techn., (Tokyo), 2 (1965) 45.
77. K. WATANABE, J. Nucl. Sci. Techn., (Tokyo), 2 (1965) 112.
78. L. M. MOSKVIN, B. K. PREOBRAZHENSKI, Radiokhim. Metody Opred, Mikroelementov, Akad. Nauk SSSR, Sb. Statei, p. 85, 1965.
79. I. STROŃSKI, Radiochim. Acta, 13 (1970) 25.
80. M. T. RICCATO, G. HERMANN, Radiochim. Acta, 14 (1970) 107.
81. R. F. HORNBECK, J. Chromatog., 30 (1967) 447.
82. E. HERMANN, H. GROSSE-RUYKEN, N. A. LEBEDEV, V. A. KHALKIN, Radiokhimiya, 6 (1964) 756.
83. G. I. SHMANENKOVA, M. G. ZEMSKOVA, Sh. G. MELAMED, G. P. PLESHAKOVA, G. V. SUHKOV, Zavodsk. Lab., 35 (1969) 897.
84. A. I. MIKHAILICHENKO, R. M. PIMENOVA, Zh. Prikl. Khim., 42 (1969) 1010.
85. ORNL Report-4145, 219 (1967).
86. I. FIDELIS, S. SIEKIERSKI, J. Chromatog., 17 (1965) 542.
87. L. R. HUNT, Nucl. Sci. Abstr., 21 (1967) 30346.
88. B. KOTLIŃSKA-FILIPEK, S. SIEKIERSKI, Nukleonika, 8 (1963) 607.
89. J. BRANDSTETR, T. S. WANG, K. A. GAVRILOV, E. GWOZDS, J. MALY, M. TAUBE, Radiokhimiya, 6 (1964) 26.
90. J. MIKULSKI, K. A. GAVRILOV, V. KNOBLOCH, Nukleonika, 9 (1964) 785.
91. P. G. BARBANO, L. RIGALI, J. Chromatog., 29 (1967) 309.
92. C. TESTA, Anal. Chem., 34 (1962) 1556.
93. J. R. PETERSON, B. B. CUNNINGHAM, Inorg. Nucl. Chem. Letters, 3 (1967) 327.
94. J. R. PETERSON, B. B. CUNNINGHAM, J. Inorg. Nucl. Chem., 30 (1968) 823.
95. J. KOOI, Radiochim. Acta, 5 (1966) 91.
96. K. A. GAVRILOV, E. GWOZDZ, J. STARÝ, WANG TUNG SENG, Talanta, 13 (1966) 471.
97. J. KOOI, R. BODEN, Radiochim. Acta, 3 (1964) 221.
98. F. L. MOORE, A. JURRIAANSE, Anal. Chem., 39 (1967) 733.
99. A. PLUYM, Thesis, Mol, Belgium, 1970.
100. S. F. MARSH, Anal. Chem., 39 (1967) 641.
101. S. A. ALY, S. H. EBERLE, Inorg. Nucl. Chem. Letters, 7 (1971) 153.
102. G. F. BEST, E. HESFORD, H. A. C. McKAY, J. Inorg. Nucl. Chem., 12 (1959) 136.
103. D. M. ZIV, B. I. SESTAKOV, I. A. SESTAKOVA, Radiokhimiya, 10 (1968) 738.
104. E. P. HORWITZ, C. A. A. BLOOMQUIST, K. A. ORLANDINI, D. J. H. HENDERSON, Radiochim. Acta, 8 (1967) 127.
105. M. G. ZEMSKOVA, N. A. LEBEDEV, Sh. G. MELAMED, D. F. SAUNKIN, V. A. KHALKIN, E. HERRMANN, G. I. SHMANENKOVA, Zavodsk. Lab., 33 (1967) 667.
106. O. B. MICHELSEN, M. SMUTZ, J. Inorg. Nucl. Chem., 33 (1971) 265.
107. O. B. MICHELSEN, M. SMUTZ, Proc. Int. Solv. Extr. Conf. The Hague, Vol. 2, p. 939, 1971.
108. L. M. MOSKVIN, S. B. TOMILOV, Radiokhim. Metody Opred. Mikroelementov, Akad. Nauk SSSR Sb. Statei, p. 93, 1965.

109. L. N. MOSKVIN, V. T. NOVIKOV, Radiokhim. Methody Opred. Mikroelementov, Akad. Nauk SSSR Sb. Statei, p. 95, 1965.
110. D. C. HOFFMAN, O. B. MICHELSEN, W. R. DANIELS, USAEC Report La-Dc-7869, 1965.
111. I. P. ALIMARIN, A. Z. MIKLISHANSKII, Yu. V. YAKOVLEV, J. Radioanal. Chem., 4 (1970) 75.
112. B. TOMAŽIČ, S. SIEKIERSKI, J. Chromatog., 21 (1966) 98.
113. B. TOMAŽIČ, Anal. Chim. Acta 49 (1970) 37.
114. B. F. RIDER, J. P. PETERSON, Jr., C. P. RUIZ, Trans. Am. Nucl. Soc., 7 (1964) 350.
115. M. R. MONSECOUR, A. C. DEMILDT, Anal. Chem., 41 (1969) 27.
116. W. SMUŁEK, Nukleonika, 14 (1969) 521.
117. T. ZELENAY, Radiochem. Radioanal. Letters, 2 (1969) 33.
118. R. DENIG, N. TRAUTMAN, G. HERRMANN, J. Radioanal. Chem., 6 (1970) 371.
119. K. JOON, R. den BOEF, R. de WIT, J. Radioanal. Chem., 8 (1971) 101.
120. E. HERRMANN, J. Inorg. Nucl. Chem., 30 (1968) 498.
121. E. P. HORWITZ, C. A. A. BLOOMQUIST, J. Inorg. Nucl. Chem., 34 (1972) 3851.
122. T. B. PIERCE, P. F. PECK, R. S. HOBBS, J. Chromatog., 12 (1963) 74.
123. S. SIEKIERSKI, R. SOCHACKA, J. Chromatog., 16 (1964) 385.

CHAPTER 9

EXTRACTION CHROMATOGRAPHY OF FISSION PRODUCTS

M. BONNEVIE-SVENDSEN, K. JOON

1. INTRODUCTION

Extraction chromatography has proved to be a useful means of solving various problems encountered in fission product analysis. Since practically all groups of the periodical system are represented in a typical fission product mixture, the extraction chromatographic behaviour of fission elements will also be treated in other chapters of this book. This chapter will concentrate upon typical problems in the main fields where analysis, separation, isolation, purification or characterization of fission products are required.

In reprocessing analysis, extraction chromatography is used both for quantitative radiochemical analysis and for control of the extractability and chemical state of fission products and actinides in process solutions.

For burn-up determinations, the method can be used to prepare pure fractions of typical burn-up monitors.

In connection with fundamental studies of nuclear reactions, extraction chromatography is utilized for rapid separations of short-lived isotopes, for the purification of fractions for β-decay measurements, for the enrichment of isotopes with low fission yields.

The method may also be used to isolate long-lived or stable fission products from waste solutions, environmental and biological samples.

Extraction chromatography has several characteristic merits which make it specially suited for this type of analysis:

1. the high separation factors achieved by multistage contacts facilitate the separation of minor amounts of fission products from macro concentrations of fissile materials;

2. the method is easily adapted to microscale analysis and carrier-free separations. The samples can be so small that even highly radioactive solutions may be analyzed without elaborate shielding and safety precautions. The volumes of radioactive analytical wastes can be limited to a minimum;

3. the analogy to multistage liquid-liquid extraction makes it possible to utilize extraction chromatography for microscale model tests on extraction systems, to study the influence of chemical process variables and possible anomalies in feed and process solutions.

2. SEPARATION SYSTEMS AND PROCEDURES

Fission product analyses for nuclear power programs (burn-up and reprocessing analyses) are mainly concerned with highly radioactive samples from spent reactor fuel containing percent concentrations of mixed long-lived and stable fission products. On the other hand, in fundamental studies about nuclear reactions, the main emphasis is on short-lived fission products with half-lives ranging from fractions of a second to a few minutes.

A fission product mixture may contain more than 30 fission elements with atomic numbers ranging from 30 to 66. Most of the primary fission products are radioactive, and more than 300 different isotopes have been observed in irradiated fuel. Their half-lives range from fractions of a second to 10^7 years (for ^{129}I). The main fission products with half-lives exceeding a few days are listed in Table 1.

Fission product samples may contain fission materials, uranium and plutonium as major components. Other actinides, process reagents, alloying elements, and constituents from canning materials may be present. The absolute and relative abundances of the different fission products will depend upon the irradiation history and cooling time of the samples.

Most of the analyses involve separations. The requirements on separation factors and recoveries vary with the applied measuring technique, the desired accuracy, and the purpose of the analyses. Some efforts have been made to establish general separation schemes for sequential fission product analyses; such schemes are useful as a general guide, but their applications are limited. As a consequence of the large differences in sample compositions, radioactivities, analytical requirements and purpose of analysis, a great variety of analytical approaches has been utilized. These are mostly modifications of systems applied for naturally occurring elements.

Tri-n-butyl phosphate—nitric acid($TBP—HNO_3$)has been the preferred system for the separation of fission products from actinides and for studies about the chemical state and the extraction behaviour of fission products and actinides. Di(2-ethylhexyl) phosphoric acid—nitric (or hydrochloric) acid ($HDEHP—HNO_3$ (HCl)) has

TABLE 1

(a) Major radioactive fission products with $t_{1/2} \geqslant 2$ d

Isotope	Half-life	Isotope	Half-life
^{89}Sr	51 d	^{136}Cs	13 d
^{90}Sr	28 y	^{137}Cs	30 y
^{91}Y	58 d	^{140}Ba	12.8 d
^{95}Zr	65 d	^{140}La	1.6 d
^{95}Nb	35 d	^{141}Ce	35.5 d
^{103}Ru	40 d	^{144}Ce	285 d
^{106}Ru	1 y	^{147}Pm	2.6 y
^{125}Sb	2 y	^{155}Eu	1.7 y
^{131}I	8 d		

(b) Concentration of long-lives and stable fission products in high burn-up nuclear reactor fuel[1]*

Element	g/ton	Element	g/ton	Element	g/ton
Se	68	Pd	1701	Ce	3518
Rb	462	Ag	68	Pr	1610
Sr	1201	Cd	103	Nd	5453
Y	624	In	2	Pm	127
Zr	5045	Sn	78	Sm	1023
Nb	9	Sb	23	Eu	273
Mo	4841	Te	704	Gd	127
Tc	1153	Cs	4003	Tb	2
Ru	3154	Ba	2134		
Rh	519	La	1729	Total	$\sim$ 40000

*Light water reactor (LWR), exposure 45000 MWd/ton at 30 MW/ton, 6 months of cooling

been used for most separations of mixed fission products from each other. Other stationary phases used for these separations are: di-n-octyl phosphoric acid, (HDOP), tri(2-ethylhexyl) phosphate (TEHP), trilaurylamine (TLA), tri-n-octylamine (TNOA), Aliquat-336, dipicrylamine (DPA), methyltripicrylammonium chloride (MTCA), α-benzoinoxime, 2-thenoyltrifluoroacetone (HTTA), diisobutylcarbinol (DIBC).

In addition to different concentrations of nitric and hydrochloric acid, and of their sodium and ammonium salts, several more specific elution mixtures (containing HF, H_2SO_4, oxidizing, reducing or complexing agents) have been applied as the mobile phase.

Polytrifluorochloroethylene (PFCE, Kel-F, Hostaflon, Voltalef, Plaskon) or kieselguhr (SiO_2^+; diatomaceous earth, Hyflo Super Cel, Celite, Chromosorb), in some cases silica gel (SiO_2-gel),are the main supporting materials in extraction column chromatography of fission products. Also glass beads, glass powder, swollen styrene-divinylbenzene copolymers and open capillary tubes have been used as supports.

The different procedures used for fission product separations are listed in Table 2. In the following chapters, the methods are presented according to the purpose of analysis.

TABLE 2

Column extraction chromatographic separation

of fission products and actinides

Separated elements	Support	Stat. phase	Mobile phase	Remarks	Ref.
U–FP	Kel-F	TBP	HNO_3		[1a]
U–FP	Kel-F	TBP	5.5 M HNO_3	Ce reduction to Ce(III)	[2]
U–Pu–FP	SiO_2^+	TBP	HNO_3	Micro-scale Purex process	[3]
Pu–FP	Voltalef	TBP	5 M, 3 M, 1 M HNO_3	Analyses of impurities in high-purity Pu	[4]
U,Pu–FP	Polyethylene tubes	TBP	HNO_3.–$NaNO_3$	Open wall-coated tubes	[5]
Np–U–FP	SiO_2^+	TBP	HNO_3, var. conc.	Investigation of difficult FP	[6]
Th, Pu, U–FP	Voltalef	TBP	5.5 M HNO_3	Hot-cell sep., oxid.with $K_2Cr_2O_7$	[7]
U–FP	Styrene-di-vinyl-benzene copolymers	TBP+var. org. solv.	HNO_3, var. conc.	Investigate the influence of swelling	[8]
U–FP		TBP	HNO_3	Investigate the influence of swelling	[9]
Np,U–Pu–FP	Kel-F	TLA	1 M HNO_3 – $Fe(SO_3NH_2)_2$ 0.5 M H_2SO_4 –0.18 M HNO_3	Selective separation of Np(IV) from U (U/Np $> 10^{10}$), Pu and FP	[10]
Pu–FP	Paper	TNOA	7.5 M HNO_3	Stationary phase eluted with liquid scintillator-soln.	[11]
U–FP	SiO_2^+	HDEHP	HNO_3, var.conc.	Increased temp. (77 $^\circ$C)	[12]
U–Zr,Pu,Fp	Plaskon	95%TEHP+ 5% HDEHP	FP etc. 8 M HNO_3 Pu: 0.4 M HNO_3^+ + 0.02 M HF U+Zr: 0.5 M HNO_3^+ 3 M HF	Analyses of impurities in Pu–U–Zr alloys	[13]
U–Np–FP	Hostaflon	TBP, HDEHP	8 M HNO_3 ($NaClO_4$) 8 M HNO_3(SO_2) 0.1–12 M HCl	40–60 $^\circ$C, multi-column system for sequential sep. of U, Np, FP	[14]

(Table 2 cont.)

Separated elements	Support	Stat.phase	Mobile phase	Remarks	Ref.
U—various FP	Hostaflon	TBP	HCl, var. conc.	Sep. scheme for FP	[15]
Ac—Th—Pa—U—Np—FP	Hostaflon	HDEHP, TBP,TNOA	HCl, var. conc.	Mother-daughter sep. e. g. Zr—Nb	[16]
Np,U,Zr rest. FP	PFCE	TBP	8 M HNO_3—	Sep. scheme—multi col-	[17]
Nb,Sb,I— rest FP	PFCE		$NaClO_3$	umn system; 50 $^\circ$C, 70 $^\circ$C	
RE, Mo— rest. FP	PFCE	HDEHP	0.1 M HCl		
Te, Tc— rest. FP	PFCE	TBP	6 M HCl—$NaClO_3$		
Am, Cm—FP	SiO_2^+	HDEHP—Solvesso 100	0.1 M HNO_3, 0.1 M $H_2C_2O_2$— 0.01 M HNO_3	Am+Cm: DTPA-lactic acid	[18]
FP from each other	SiO_2^+	HDEHP	0.15—6 M HCl	Used together with other methods for sep. schemes	[19]
FP from each other	SiO_2^+	HDEHP		Review of methods for sequential sep.	[20]
Cs—other FP	Kel-F	Dipicryl-amine	0.01 M EDTA, 0.5 M HNO_3	pH = 7	[21]
Ce—other FP	SiO_2^+	HDEHP	5 M HNO_3 + 0.1 M ascorbic acid		[22]
RuNO complexes	SiO_2^+	TBP	HNO_3 (var. conc.)	Sep. of individual nitrato compounds	[23]
RuNO complexes	SiO_2^+	TBP	HNO_3 (var. conc.)	Determination of extract-ability	[24]
RuNO complexes	SiO_2^+	TBP	HNO_3 (var. conc.)	Sep. of individual nitro compounds	[25]
RuNO complexes	SiO_2^+	TBP	HNO_3 (var. conc.)	Isolation of polymeric nitro and nitrato comp.	[26]
RuNO complexes	SiO_2^+	TBP	HNO_3 (var. conc.)	Complex formation of nitro and nitrato comp.	[27]
Mo, Np—U, FP	Glass powder	HTTA-xylene	0.5—1 M HCl	Burn-up, reactor lattice parameters	[28]
Mo—U,FP	Chromosorb	α-ben-zoinoxime	1 M HCl	Burn-up	[29]
Mo—U, Pu, FP	SiO_2^+	TBP	3 M HCl—3 M HNO_3	Burn-up fast reactor fuel	[30]
Cs—FP	Kel-F	TBP—amyl-acetate	0.01 M EDTA—1 M HCl, acetic acid	Burn-up, health-physics	[31]
Eu—U,FP	SiO_2^+	HDEHP	HNO_3 (var.conc.)	Burn-up	[32]

258

(Table 2 cont.)

Separated elements	Support	Stat.phase	Mobile phase	Remarks	Ref.
Ce−FP	SiO_2^+	HDEHP	5 M HNO_3 + ascorbic acid	Ce(IV)	[33]
Ce− U, FP	Chromosorb	HDEHP	2 M HNO_3− $NaBrO_3$	Burn-up, Ce(IV)	[34]
RE−U, Pu, FP	Chromosorb	HDEHP	Lactic acid+DTPA	Burn-up, hot-cell operation	[35]
RE−FP	SiO_2^+	HDEHP	HNO_3 (var. conc.)	Burn-up	[36]
Zr, Th, U,Pu, Nd−irr.fuel	Chromosorb	HDEHP	HNO_3−HCl+I⁻, HNO_3	Burn-up, hot-cell operation	[37]
Nd− U, Pu,FP	SiO_2^+	HDEHP	HCl (var.conc.)	Burn-up	[38]
Nd−FP	SiO_2^-	HDEHP	HCl (var. conc.)	Burn-up	[39]
Nd− U, Pu,FP	Chromosorb	HDEHP	0.3 M HCl	Burn-up	[40]
Nd−FP	Chromosorb	HDEHP	Dilute acid	Burn-up	[41]
Pr−Ce	SiO_2^+	HDOP	6 M HNO_3 + + $KBrO_3$	Nucl.prop.of ^{144m}Pm	[42]
RE−each other	SiO_2^+	HDEHP	Dil. HCl at 80 °C	Fission yields	[43]
Ce−Pr	Hostaflon	HDEHP	HNO_3	Nucl. prop. of ^{145,146,147}Ce	[44]
Mo−Tc	Hostaflon	HDEHP	HNO_3	Nucl. prop. of Ce and Mo isotopes	[45,46]
Dy− FP, Tb,Ho	Silica gel	HDEHP	HCl	Yield, new fission chain	[47]
Sb−FP	Hostaflon	HDEHP, DIBC	9 M HCl−H_2O	Nucl. prop. of ^{131,132,133}Sb	[48]
Fe−Sb	Hostaflon	HDEHP	9 M HCl	─── " ───	[48]
Zr−FP	Hostaflon	TBP	8 M HNO_3−H_2O	Nucl. prop. of ^{98}Zr	[49]
Nb− Zr	Hostaflon	TBP	8 M HNO_3	Nucl.prop. of ^{98}Nb	[49]
Zr−FP	Plastic grains	TBP	7.5 M HNO_3+ $KBrO_3$	Nucl. prop. of ^{99,101}Zr	[50]
Tc−FP (Pu)	Plastic grains	Tetraphenyl-arsonium chloride	0.1 M HNO_3− 2 M HNO_3	Nucl. prop. of ^{106,107,108}Tc	[50]
As−Sb	Plastic grains	HDEHP	9 M HCl + $KClO_3$	Delayed neutron emission from ^{85}As	[50]
Pd−FP	Plaskon	Aliquat-336	8 M HNO_3− 1.5 M NH_4OH	Commercial recovery of Pd	[51]

3. REPROCESSING ANALYSIS

3.1. SPECIFICATION OF PROBLEMS AND SAMPLES

The development and control of extraction procedures for reprocessing of irradiated nuclear fuel has introduced a series of rather complicated analytical problems. Reprocessing is carried out to remove neutron poisons, to purify and to recover the fissile materials. Most processes are based on the separation of uranium, plutonium and fission products by liquid-liquid extraction, as illustrated by the simplified Purex flow-diagram in Fig. 1. After dissolution in nitric acid, uranium and plutonium are simultaneously separated from fission products and other actinides by extraction with TBP in kerosene; U and Pu are subsequently separated from each other by selective backwash into the aqueous phase. Actual processes are far more complex and involve several purification and recovery cycles, head- and tail-end treatments. In addition to TBP, several other extractants, such as amines or ketones, are or have been used. The process samples may also contain other acids, process reagents, alloying elements from the fuel and dissolved canning materials.

Analyses are required to control decontamination factors, recoveries, releases and waste streams, for product specifications and safety supervisions. A certain control of the chemical state and extractability of fission products and actinides is also needed. Typical problems are connected with the high radioactivity of the analytical samples, the complicated chemistry of the elements involved, the determination of trace elements in concentrated technical solutions.

The fuel is mostly cooled for several months (e. g. 100 days) before it is processed. Thus, short-lived fission products ($t_{1/2} < 2-3$ days) are not present in samples from a

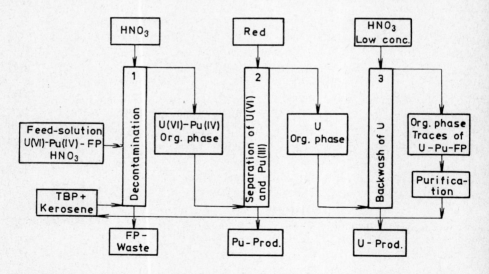

Fig. 1. Simplified Purex flow-diagram for separation of uranium, plutonium and fission products by liquid-liquid extraction with TBP

reprocessing plant, but the concentration of mixed long-lived and stable fission products in the fuel may amount to several percent, and the activity of the samples from feed and fission product streams may be so high that most analyses must be carried out by remote operations in shielded boxes.

Most processes are based on the separation of extractable U(VI) and Pu(IV) from essentially inextractable fission products. The presence of fission products with increased extractability, of certain nitro- and nitrato complexes of nitrosylruthenium, of polymerized or colloidal Zr and Nb species, may impair the decontamination from fission products. Losses of Pu to the fission product streams may occur, if Pu species with reduced extractability (Pu(III) or Pu(VI)) are present.

Even the chemical state of fission products and actinides should therefore be controlled. For fundamental studies, it is desirable to have quantitative analyses of all important fission product and actinide species, of a variety of ruthenium complexes, of several more or less undefined Zr and Nb species, of Pu(III), Pu(IV), Pu(VI). For practical purposes, simple and rapid model tests giving a fractionation according to extractabilities may be preferred.

Extraction chromatography can be used to solve many of the problems encountered in reprocessing analysis. A rather extensive development of analytical methods has been carried out parallel to the development of the actual process, and has given rise to many of the extraction chromatographic procedures which are now used for other purposes. The method has been used for elemental analysis, to separate fission products from actinides, from other process components, and from each other. Being analogous to multistage liquid-liquid extraction, the method is specially suited for fundamental studies on the complex chemistry and extraction behaviour of fission products and actinides, for model tests, small scale separations and fractionation of various species according to their extractability.

However, it should be emphasized that extraction chromatographic procedures worked out for synthetic fission product mixtures may completely fail when applied to the actual samples from a reprocessing plant. The above discussed species with "anomalous" extractabilities may also have an adversary effect on the analytical separations and may lead to tailing, bad reproducibility, or even completely erroneous results. A strict control of feed adjustments and column preparation, a thorough testing of possible sources of error and controlled measurement of a large number of process samples is required before a method can be adopted for quantitative routine analysis of plant samples.

3.2. THE SEPARATION OF FISSION PRODUCTS FROM ACTINIDES AND FROM EACH OTHER

Various modified adaptations of reprocessing flow-sheets, such as the Purex flow-sheet shown in Fig. 1, are used for the extraction chromatographic separation of fission products (see Table 2).

In most cases, 100% TBP supported on organic polymers (Kel-F, Hostaflon, Voltalef), siliconized silica gel or kieselguhr is used as the stationary phase. The separations are based on the elution of fission products with varying concentrations of nitric acid, while the actinides are retained on the column as TBP-extractable species (U(VI), Pu(IV), Np(IV)).

In 1960–61, Hamlin et al.[1a,2] reported the quantitative recovery of microgram to gram quantities of U from highly impure solutions in a form suitable for α-counting, volumetric or colorimetric analysis: U was retained by a column of Kel-F supporting TBP. Fission products together with other impurities were eluted with 5.5 M nitric acid, and U was eluted with water. Th(IV) and Ce(IV) interfere, but Ce interference was avoided by reduction to Ce(III) prior to the separation. The authors emphasized that the use of column extraction chromatography reduces the hazard of manipulating radioactive materials and makes it possible to limit the volume of analytical radioactive waste to a minimum.

Eschrich[3] developed a small scale chromatographic "Purex Process". The feed solution of irradiated (200–350 MWd/t) uranium, cooled for 250–300 days, was adjusted to 6 M in nitric acid and 0.1 M in sodium nitrate. After standing for 24 hours, the solution was filtered through a silica-gel-asbestos column to remove suspended particles and an essential part of Zr–Nb. The effluent was introduced into the extraction chromatographic column consisting of undiluted purified TBP supported on siliconized kieselguhr, preequilibrated with 6 M HNO_3. Most of the fission products were eluted by washing the column first with 6 M and then with 1 M HNO_3; plutonium was stripped with 1 M HNO_3 containing 0.05–1 M ferrous sulphamate, and uranium was eluted with water (Fig. 2). Residual ruthenium was removed from the column by washing with 0.3 M NaOH followed by water. Gross gamma decontamination factors of 10^3-10^4 and $5 \cdot 10^4$ were obtained for the plutonium and uranium products, respectively.

In another method, the selective separation of U from nuclear fuel reprocessing solutions was achieved by using open capillary columns, wall-coated with TBP[5]. It was shown that these columns have certain advantages compared to packed columns, especially in view of the remote analysis of highly radioactive solutions. Thus, a fully automated remotely controlled extraction chromatographic procedure for the separation and subsequent quantitative determination of U was developed.

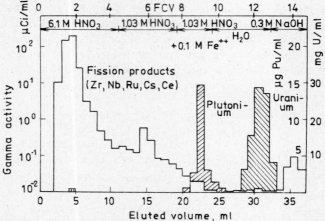

Fig. 2. Elution behaviour of uranium, plutonium and fission products in a chromatographic "Purex Process" TBP-kieselguhr column, free column volume (FCV): 2.5 ml; Flow rate: 2–4 ml·cm^{-2}·min^{-1} (Ref. [3])

Beranova and coworkers[8,9] made extensive investigations on the use of styrene-divinylbenzene copolymers swollen in various extractants for the separation of U from fission products. By using TBP diluted with $CHCl_3$ as the swelling agent, and by adjusting the aqueous phase to 5.0–5.3 M HNO_3, high loading factors and a quantitative separation of U from Ru, Zr–Nb, Sr–Y, Ce and Cs was achieved.

For the separation of trace amounts of fission products from macroamounts of U, HDEHP on kieselguhr can also be applied[12]. The presence of U on the column, however, impairs the retention of some of the fission products. Increasing the temperature to 77 °C and decreasing the flow rate to 0.2 $ml \cdot cm^{-2} \cdot min^{-1}$ improves the retention, and results in 95% recovery of the fission products.

For the specification analysis of high-purity plutonium materials and the analysis of process solutions occurring in the final purification cycle of nuclear fuel processing plants, Eschrich and Hundere[4] developed an extraction chromatographic procedure for the separation of total impurities from plutonium. In a 5 to 7 M nitric acid solution, plutonium is quantitatively converted to the tetravalent oxidation state by addition of hydrogen peroxide; then it is separated from its impurities by extraction chromatography, employing purified TBP supported on polytrifluoromonochloroethylene (Voltalef, 300 Micro Pl) particles as the stationary phase, and nitric acid of various molarities as the mobile phase. Plutonium(IV) is retained on the column while most of the common impurities appear in the effluent on washing the column successively with 5, 3 and 1 M nitric acid. The eluate containing the impurities is evaporated to a small volume (1 to 2 ml) and analyzed by suitable methods. More than 60 elements in the range of parts per billion to per cent of Pu can be separated from plutonium and recovered with a yield exceeding 90%.

Gross[11] separated Pu from fission products by sorption on the nitric acid salt of TNOA coated on an inert support. The amine salt containing Pu was then dissolved by a toluene-based scintillator solution, and Pu was determined by liquid scintillation counting. Only two hours are required to complete an analysis. The method chosen for Pu determination has the advantage that residual fission product activity causes only minor interferences[11,52,53].

A method for the determination of Np in reprocessing solutions was developed by Gourisse and Chesné[10]. The chromatographic separation was performed by passing 2 M HNO_3 + 0.1 M iron(II) solution, containing up to 100 g/l of uranium, through a column of TLA–HNO_3–Kel-F beads. After washing the column with nitric acid containing iron(II) sulphamate, neptunium is eluted with a sulphuric acid–nitric acid mixture. The method is very selective for neptunium(IV), and allows for a good separation from fission products, plutonium and uranium; it was applied to solutions in which the weight ratio U:Np was greater than 10^{10}, and to the preparation of radiometrically pure ^{239}Np.

Np can also be recovered quantitatively and radiochemically pure from irradiated U by means of TBP–kieselguhr columns and elution with different HNO_3 concentrations[6]. On the column, Np is selectively reduced to the pentavalent state. The behaviour of interfering fission products is discussed in the same paper.

A sequential scheme for the isolation of Np, U and various fission products has been developed by Denig et al.[14,17]. In the first extraction chromatographic column, Zr, U and Np are extracted with TBP from 8 M HNO_3–$NaClO_3$. In the second column, Nb, Sb and I are extracted by HDEHP from 9 M HCl–$NaClO_3$. Rare earths and Mo are extracted in the third column (HDEHP) from 0.1 M HCl, and finally, in the fourth column (TBP), Te and Tc are isolated from 6 M HCl. Further separations within each group are achieved by selective elution from the respective columns. The entire separation procedure, carried out at elevated temperature, requires less than 12 hours including the radiochemical determinations[14]. Some of the more important fission products are determined by X-ray fluorescence[17].

Another separation procedure for multielement analysis uses only one extraction chromatographic column[13]. The extractant is a mixed organic phase containing 95% tri-(2-ethylhexyl) phosphate (TEHP) and 5% HDEHP. After the sample is added to the column, fission products and metal impurities are eluted with 8 M HNO_3: Pu is recovered by elution with 0.4 M HNO_3 + 0.2 M HF, and U + Zr by 0.5 M HNO_3 + 3 M HF. The method was developed for tracer impurity analyses by the copper spark spectrographic method in Pu–U–Zr alloys, but can easily be modified for reprocessing analysis.

In waste streams from reprocessing plants, the determination of Am and Cm contents is often required. A suitable column extraction chromatographic method to separate these elements from bulk-fission products is the sorption on 0.3 M HDEHP supported on kieselguhr[18]. Elution with 0.1 M HNO_3 followed by 0.05 M oxalic acid + 0.01 M HNO_3 removes a number of fission products. Am and Cm are recovered using 0.05 M DTPA + 1 M lactic acid at pH = 3 as the eluent, leaving the lanthanides and some other fission products on the column.

A procedure for the separation of Th, Pu and U from fission products was developed for hot-cell operation in connection with studies on actinide composition in irradiated THTR fuel[7]. The method can easily be adapted to meet reprocessing requirements (process control); it involves extraction by TBP supported on Voltalef. After removal of Pa by sorption on Vycor glass, the filtered fuel solution is adjusted to 5.5 M HNO_3. Actinides are oxidized to Th^{4+}, UO_2^{2+}, NpO_2^{2+} and PuO_2^{2+}, respectively, by heating with $5 \cdot 10^{-4}$ $K_2Cr_2O_7$. The oxidized solution is poured onto the column, and the fission products are eluted with 8 column volumes of 5.5 M HNO_3. The actinides are then eluted together with 0.01 M HNO_3. The decontamination factor achieved was about 5000. The main residual fission products were Zr–Nb.

In process control of reprocessing plants, high efficiency separations are mainly needed for decontamination of actinides from radioactive fission products in order to reduce radiation hazards. This enables the final determination of the actinides outside the glove box or hot cell. For the determination of fission products the situation is different. The most important of them (^{137}Cs, [144]Ce, [95]Zr–Nb, [103-106]Ru–Rh) have characteristic gamma energies suitable for direct determination by gamma spectrometry; if a separation is necessary, simple liquid-liquid extraction is generally sufficient. In case of product specifications, however, high separation factors may also be needed for the fission products and other impurities. Separation processes based on column extraction chromatography and applicable for this purpose have been described[4,13,15,17]. In additio a review by Fidelis[20] of methods for sequential separations should be mentioned. Sequenti separation schemes of which only a part is based on column extraction chromatography have also been developed: in most of these cases, extraction chromatography is only used for the isolation and purification of the lanthanides[19]. Some of such mixed procedures and methods for isolation of specific fission products will be described in the sections on burn-up and nuclear chemistry.

3.3. THE CHEMICAL STATE AND EXTRACTABILITY OF FISSION PRODUCTS AND ACTINIDES

3.3.1. RUTHENIUM SPECIES

In reprocessing of spent nuclear fuels by TBP extraction, one of the limiting factors with respect to the decontamination of U and Pu is fission product ruthenium. The reason is that during dissolution of the irradiated material in boiling nitric acid, part of the Ru is converted to trivalent nitrosyl-Ru ($RuNO^{3+}$) that forms several series of coordination compounds of the general composition

$$\left\{Ru(NO)(NO_2)_x\,(NO_3)_y\,(H_2O)_{5-x-y}\right\}^{+(3-x-y)}$$

with x = 0, 1, 2, y = 0, 1, 2, 3, 4, and $0 < x + y \leqslant 4$. The relative abundances of the series having x = 0,1 or 2 depend upon dissolution conditions. In addition to this, within each series nitration–aquation equilibrium reactions occur. The result is a complex mixture of species with extraction coefficients depending upon the number of coordinated nitrate and nitrite groups. A further consequence is that the overall extraction and back-extraction of the fission product ruthenium is a function of time, temperature, and NO_3^- concentration.

Although valuable information has been obtained with other methods, extraction chromatography proved to be most versatile in the study of the Ru complexes. By means of paper impregnated with TBP and elution with nitric acid of different concentrations, Scargill[54] separated individual nitrato (x = 0) and dinitro (x = 2) complexes of RuNO, and was also able to partially resolve a mixture of these compounds. At about the same time, Eschrich[23] reported the separation and identification of five individual nitrato complexes, as well as the equilibrium compositions of their mixture in different nitric acid media. The separations were performed by column extraction chromatography using TBP supported on hydrophobized kieselguhr as the stationary phase, and 0.5, 3.5, 5 and 10 M nitric acid as successive eluents. Figure 3 is an example of such separations, where 2 M instead of 3.5 M HNO_3 was used as the second eluent. The respective fractions contain the mononitrato complex, which runs unretained through the column, followed by one and then by the other of the isomeric dinitrato, by the trinitrato, and finally by the tetranitrato RuNO-compounds. The non-nitrato and the pentanitrato species have not been identified. Extraction coefficients of the individual complexes decrease with increasing nitric acid concentration in the aqueous phase, but increase with the number of coordinated nitrate groups. Moreover, the composition of an equilibrium mixture of RuNO nitrato compounds shifts towards the better extractable species with increasing HNO_3 concentration.

The same technique was used by Joon[25] in a study of the RuNO-nitro complexes

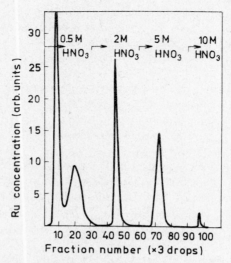

Fig. 3. Separation of RuNO-nitrato complexes aged in 3 M HNO₃.

TBP-kieselguhr column, 120 mm long and with 3.5 mm diameter. Temperature: 0 °C; Flow rate: 8-10 drops/min; Fraction volume: 90 μl (Ref. [27])

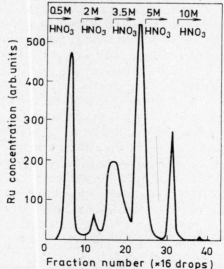

Fig. 4. Separation of RuNO-nitro complexes aged in 5 M HNO₃.

TBP-kieselguhr column, 120mm long and with 3.5 mm diameter. Temperature: 0 ;C; Flow rate: 10-12 drops/min; Fraction volume: 0.32 ml (Ref. [25])

(x = 1, 2 and y=0,1,2). A mixture of the compounds, prepared by treating nitrato complexes with nitric oxides, can be separated into five main fractions (Fig. 4). The mononitro and two dinitro isomers of RuNO are not retained on the column, and are collected together in the first fraction. The second fraction contains an isomer of mononitro-mononitrato

266

RuNO. The second mono-mono isomer is found in the third fraction together with a di-nitro-mononitrato complex. Another isomer of the latter constitutes the fourth fraction. The last main fraction, eluted with 5 M HNO_3, consists of a mononitro-dinitrato complex. The other mono-di isomer has not been found, nor were indications obtained on the existence of +3 or anionic compounds. By repeated column extraction chromatographic separations, these nitro and mixed nitro-nitrato species could be purified. This enabled the determination of their compositions, of their aquation and nitration rates and equilibrium concentrations in different nitric acid media, and of their extraction behaviour with respect to TBP.

In addition to the above mentioned compounds, that impair the extractive decontamination of the main fuel constituents, fission product ruthenium also forms very highly extractable, rather stable compounds, which are difficult to remove from the extractant, TBP. Apparently, only minor amounts of ruthenium are involved but since in practice the extractant is recycled, such species may accumulate, leading to enhanced radiolytic degradation. Column extraction chromatography has been applied also to investigate this problem: again TBP on hydrophobized kieselguhr was the stationary phase, and HNO_3 the eluent. Extraction—elution experiments with each of the main fractions of the nitro-, mixed nitro-nitrato, and nitrato compounds of RuNO showed that small amounts of Ru may be retained by TBP provided the main compound has been extracted[26,55]. This supports Fletcher and Scargill's theory about the formation of compounds in which TBP is directly bonded to the metal[56]. In industrial reprocessing, the total amount of RuNO extracted in this way will be less than 0.05%, and the compounds are decomposed within some hours when in contact with 0.5 M HNO_3[26].

In another series of experiments Joon and coworkers[26] aged pure RuNO nitro (x = 1,2) or nitrato (x = 0) complexes in the presence of different concentrations of HNO_3 and HNO_3+$LiNO_3$. Aliquots of the resulting equilibrium mixtures were sorbed on TBP columns. The main constituents were then eluted with 10 M HNO_3 after which any remaining Ru was washed out with fuming HNO_3. The amount of Ru in the latter fraction could be related to the concentrations of RuNO tetranitrate and NO_3^- in the stock solutions in the case of pure nitrato compounds, and to the concentrations of the mono-nitro-dinitrato complex and H^+, NO_3^- and H_2O in the case of nitro compounds. It was concluded that in the aqueous solutions very highly extractable dimeric species of the following compositions

$$[Ru(NO) (H_2O) (NO_2) (NO_3)_2]_2(OH)^-$$

$$[Ru(NO) (NO_2) (NO_3)_2(OH)]_2$$

$$[Ru(NO) (H_2O) (NO_3)_3]_2$$

might have been present. These are very difficult to remove or decompose. In industrial reprocessing, about 0.5% of the fission product ruthenium present as RuNO may be retained by the TBP in this form.

Because of the troubles caused by nitrosyl-Ru complexes in reprocessing, several techniques have been investigated to reduce their extractability. In one case, column extraction chromatography was applied to study the influence of organic masking reagents on the extraction of the different RuNO species into TBP[27]. The results indicated that remarkable reductions in the extraction could be obtained. The effect is stronger for nitro than for nitrato compounds, and is most pronounced for the cis-dinitro species and at lower nitric acid concentrations. In some cases the ruthenium compounds formed extractable complexes with the masking agents.

3.3.2. PROCESS SAMPLES

For a systematic investigation of process variables, as well as for process control in reprocessing plants, it is often desirable to know the composition of the mixture of ruthenium species, e. g. in the feed solution. Extraction chromatography offers an excellent analytical tool for classification of the different species, for instance as inextractable, moderately extractable and extractable fractions. Bonnevie-Svendsen and Martini[24] developed a method for pilot-plant analyses in which such a classification is obtained simultaneously with the determination of another important parameter, i. e. the plutonium valency states. The method uses TBP-columns to which the sample containing U, Pu, Ru and other fission products in 2 M HNO_3 is added. Elution with 3 M HNO_3, as shown in Figure 5, produces two fractions containing inextractable Ru and Pu(III) species, and moderately extractable Ru and Pu(VI) species, respectively. Thereafter, extractable Ru is eluted with 7 M HNO_3 and Pu(IV) with ferrous sulphamate in 2 M HNO_3; Ru is determined by gamma spectrometry, Pu by alpha counting.

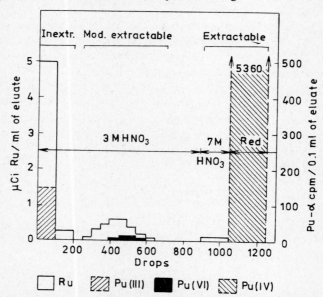

Fig. 5. Separation of ruthenium and plutonium species in pilot-plant feed solutions. TBP-kieselguhr column, 120 mm long and with 3.5 mm diameter; Flow rate: 0.15−0.2 ml·min^{-1} Drop volume 18 μl (Ref. [24])

268

The determination of plutonium valency states was based on a method developed by Siekierski and Gwozdz. Their procedure could not be directly applied to plant samples because a light-induced reduction of Pu on the column occurred in the presence of excess uranyl nitrate; there were indications that even the state of Ru and Nb was influenced by similar effects. Therefore, all separations were performed under exclusion of light.

With some additional precautions, the extractability of Zr and Nb may be controlled in the same operation. Pu(III), (VI) and (IV) can serve as markers to evaluate distribution coefficients of unknown fission product species.

4. BURN-UP ANALYSIS

4.1. BURN-UP MONITORS

Burn-up can be defined as the number of fissions per 100 heavy nuclide atoms initially present in the fuel. The total number of fissions is a measure of the thermal energy produced by the fuel. Burn-up may also be expressed in MWd/ton.

Burn-up analyses are needed to evaluate fuel performances and characteristics, to establish operating parameters for the irradiation behaviour of experimental fuel.

The burn-up can be determined by analysis of heavy nuclides or of selected fission products. The latter approach is considered to be the most accurate for thermal reactor fuels. Only a limited number of fission products are suited as burn-up monitors. The fission yield of a burn-up monitor must be accurately known and should be essentially independent of the neutron energy and nearly equal for the major fissionable nuclides in a fuel. Migration of the burn-up monitor and its precursors in the fuel must be negligible. Furthermore, it should be stable or have a long half-life, while all precursors should be short-lived. The neutron capture cross-section should be low and reliable methods for accurate analyses of the burn-up monitor in the fuel must be available.

4.2. NEODYMIUM

At present, [148]Nd is probably the most popular burn-up monitor. It can be determined by both activation analysis and isotope dilution mass spectrometry. These methods put somewhat different requirements to the decontamination of neodymium from irradiated samples.

Generally, lanthanide fission products, except cerium, are first separated as a group, from which Nd is then isolated. For the latter step, column extraction chromatography has two advantages compared to ion exchange:
- the separation time is shorter, because Nd is recovered at the beginning of the elution instead of at the end,
- the eluate is not contaminated with organic salts.

The group separation may be performed in different ways, but here also extraction chromatography (with HDEHP on kieselguhr) can be applied[38,57]. An aliquot of the dissolved fuel sample is brought on the column in 0.1 M HNO_3 or HCl and washed with the same acidity. Next, the lanthanides, Am, Cm and part of the Ru are eluted with 1 M acid; Ce(IV), Pu and Ru (partially) can then be removed from the eluate on an anion-exchange resin mixed with PbO_2[38], or Ce(IV) and Ru can be sorbed on a lead dioxide–lead iodate column.

The lanthanides and the bulk of the fuel and cladding materials and fission products can also be separated by a single anion-exchange step[40,41], or by solvent extraction with HDEHP–hexane from highly acidic bromate solution[36,39].

In all cases, the final purification and recovery of Nd is performed by column extraction chromatography with HDEHP on kieselguhr. The lanthanide mixture is brought on the column in 0.1 N acid. After washing with the same acid, the individual elements are eluted at different acidities in the range of 0.2–0.4 N. Gross alpha and gamma decontamination factors are generally in the order of $10^5 - 10^7$, and the separation of Nd from neighbouring rare earths is quite sufficient. The chemical yield is up to 100% dependent upon the part of the final neodymium fraction collected. However, some contamination with Am and Cm may occur.

4.3. MOLYBDENUM

The fission product molybdenum is one of the most reliable burn-up indicators. Some of the methods for its isolation incorporate a column extraction chromatographic step.

In one of these procedures[28], the irradiated sample is dissolved in nitric acid. A predecontamination, mainly to get rid of the bulk uranium, is performed by precipitation of Mo as $PbMoO_4$. After redissolution in concentrated sulphuric acid, the Mo is extracted from 0.1 M H_2SO_4 in tri-n-octylamine supported by hydrophobized glass-powder, and recovered by elution with 2 M HCl. A final purification by scavenging with $Fe(OH)_3$ may be necessary. A high purity product is obtained with a chemical yield of 60–70%. The method was developed for the measurement of microscopic reactor lattice parameters by means of the radiochemical determination of ^{99}Mo, but can be adapted to burn-up determination.

In another method, Mo is extracted together with the actinides from a 3 M HCl solution into TBP supported on hydrophobized kieselguhr[30]. Washing with 3 M HCl removes such fission products as Cs, Ce, Ru, Zr, Nb and Sr; molybdenum is eluted with 3 M HNO_3. The decontamination factor for alpha activity is reported to be in the order of 10^4, and the chemical yield for Mo to be near 100%. As the method is intended for burn-up determination of fast reactor fuels, decontamination factors for fissile materials of at least 10^7 are required. It is therefore proposed to repeat this cycle and perform a final purification by solvent extraction from 3 M H_2SO_4 into acetylacetone (50% in chloroform) and back-wash with ammonia. Mo is determined by activation analysis of ^{100}Mo.

According to Yoshida et al.[29], extraction chromatography can also be used for the extractive purification of fission product molybdenum with α-benzoinoxime from hydrochloric–oxalic acid medium. The method has been successfully applied for burn-up measurements by means of the isotopic dilution mass spectrometric determination of molybdenum.

270

4.4. CESIUM

Also the radiochemical determination of ^{137}Cs has been applied for burn-up measurements. The most suitable method for reducing the interference of other radioactive fission products is probably solvent extraction of cesium into tetraphenylboron. In order to increase the efficiency, the method has been modified into an extraction chromatographic technique[31]. The fuel sample is dissolved in nitric acid and the resulting solution adjusted to pH = 6.8 and made 0.01 M in EDTA; to complex Zr and Ru completely, this solution is heated prior to sorption of Cs on a column of Kel-F powder supporting tetraphenylboron dissolved in amylacetate. Other fission products are quantitatively washed out with 0.01 M EDTA, after which Cs is eluted with a 1:1 mixture of HCl and acetone. The cesium recovery is 100%, independent of the addition of carrier.

This method for burn-up determination has two obvious disadvantages:
- cesium is not always a correct burn-up indicator due to its migration within the fuel rod during irradiation;
- the extraction chromatographic column can be used only once because TPB is washed out together with Cs.

4.5. CERIUM

Compared to cesium, a better suited radiochemical burn-up monitor is [144]Ce. Krtil and Bezdek described its separation and purification from irradiated fuel, based on extraction chromatography with HDEHP–siliconized kieselguhr columns[33,34].

After dissolution of the fuel sample, cerium is quantitatively converted to Ce(IV) in 5 M HNO_3 + 0.5 M $NaBrO_3$. From this solution, Ce(IV) is retained on the column, which then is washed with the oxidation mixture. After washing with 5 M HNO_3, cerium is eluted with 5 M HNO_3 containing 0.1 M ascorbic acid. Under these conditions, fission products like Cs, Sr, some Zr–Nb, and the trivalent lanthanides are not retained on the column, while over 90% of the Zr–Nb is retained irreversibly. The most important contaminant is [103]Ru, the interference of which can be strongly reduced by boiling with $NaNO_2$ prior to the separation (which converts Ru to nitrosyl-nitro compounds), or by its complete removal through evaporation with perchloric acid.

The method gives 100% recovery of radiochemically pure Ce, provided the column loading with uranyl ions does not exceed 25%. The overall accuracy of the burn-up determination is reported to be 4%[34].

4.6. EUROPIUM

The [154]Eu to [155]Eu ratio was also proposed as a burn-up monitor. In connection with the radiochemical determination of these isotopes, a thorough study was made of the europium purification and recovery from irradiated fuel by means of extraction chromatography with columns of HDEHP supported on siliconized kieselguhr[32].

The three-column separation procedure consists of adding an aliquot of the dissolved fuel sample in 2 M HNO_3 to the first column, followed by elution with the same acid. Uranium and zirconium, the major part of niobium, and part of yttrium and ruthenium are retained. The eluate is evaporated to dryness and then added to the second column in 0.3 M HNO_3. Washing with seven free column volumes of 0.3 M HNO_3 removes Cs, Ba, Sr, most of the lighter rare earths and the remainder of Ru and Nb. Eu is then eluted with 2.5 M HNO_3. For final purification the last step is repeated with a fresh column.

Apart from the requirement of high radiochemical decontamination factors, the method involves tedious calculations and opportunities for systematic errors. This is due to the low fission yields, to the differences in fission yields for ^{235}U and ^{239}Pu, and to complex nuclear relationships that result from often very high neutron capture cross-sections of the nuclides in this region. In addition, the fresh fuel has to be absolutely free from europium. The accuracy of the Eu determination improves with increasing burn-up and longer cooling times of the fuel. Isotopic dilution mass spectrometry would probably result in more accurate data and simplification of the separation procedure[58].

4.7. THE LANTHANIDE GROUP

Most of the lighter fission product lanthanides are suitable burn-up monitors when determined by isotopic dilution mass spectrometry. Column extraction chromatography is an excellent tool for their separation and purification.

Marsh[35] developed a method in which a first column separates the lanthanide group from the trivalent actinides and all other fission products. The stationary phase is HDEHP in diisopropylbenzene, supported on chromatographic grade Chromosorb; the eluent is lactic acid + sodium diethylenetriaminepentaacetate. The second column is made with undiluted HDEHP on Chromosorb and is used to selectively retain Ce(IV) from strong HNO_3 + $KBrO_3$, as trivalent lanthanides run through. On the final column, made of cation-exchange resin, the individual lanthanides are separated by elution with α-hydroxyisobutyrate at controlled pH.

Joon and coworkers [36] separated the individual lanthanide fission products from each other by extraction chromatography after they had been isolated as a group by solvent extraction. The method was developed for analysis of irradiated pyrolytic carbon—SiC-coated PuO_2 fuel. From the dissolved fuel sample, Zr, Nb, and Sb were first removed by sorption on a silica gel column, and Pu by anion exchange. The other fission products were then separated by liquid-liquid extraction of the lanthanides into HDEHP in n-hexane. After back extraction with 10 M HNO_3, the lanthanides were separated from each other on an HDEHP—hydrophobized kieselguhr column by gradient elution with dilute nitric acid (Fig. 6). This effectively decontaminates them from small amounts of Ru which follow the lanthanides in the batch extraction. Consistent results were obtained in four different hot runs[58].

4.8. MANY COMPONENTS

Because most methods for burn-up determination suffer from some disadvantages, a number of them are often used in combination[36,37].

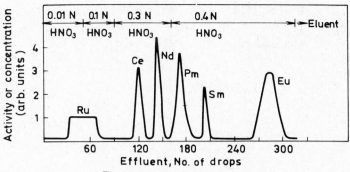

Fig. 6. Separation of rare earths.

HDEHP-kieselguhr column, 120 mm long and with 3 mm diameter; Temperature:75 °C; Flow rate: 2−3 drops/min; Drop volume: 30 μl (Ref. [61])

As an example, the single separation procedure developed by Marsh and coworkers [67] enables radiochemical determination of ^{137}Cs, the mass spectrometric determination of ^{148}Nd and the mass spectrometric determination of changes in the isotopic composition of the fuel constituents. U, Pu, Am-Cm, Cs, and Nd are sequentially separated on two extraction chromatographic and two inorganic ion exchange columns. Zr, Th, U and Pu are retained from nitric acid on the first extraction column, containing HDEHP on Chromosorb. Subsequently, U and Pu are sequentially eluted with HCl and HCl containing iodide. The second column (lead dioxide−lead iodate) removes unwanted Ce and Ru. The third column (potassium cobaltoferrocyanide) selectively retains Cs. The buffered effluent is loaded on the final HDEHP extraction column, Am-Cm are eluted with glycolic acid−sodium diethylenetriaminepentaacetate, and Nd, Pm, and Sm are eluted with dilute nitric acid. The procedure, which is applicable to Th, U, and Pu fuels, requires about 4 hours for a complete separation.

5. FISSION PRODUCTS IN NUCLEAR CHEMISTRY

The separation of fission products is an important factor also in nuclear chemistry, where it may be necessary in connection with:
— identification of (new) fission nuclides,
— determination of nuclear properties such as half-life and decay scheme,
— determination of fission yield or fission cross-section,
— production of target material for renewed irradiation, in order to obtain additional information on the nuclear properties of the actual fission nuclide and/or its reaction product(s).
The information obtained in this way finds application in studies of nuclear structure and, ultimately, in nuclear physics calculations for reactor design and operation.
In the following, examples will be given to demonstrate the applicability of column extraction chromatography for such purposes. Generally, extraction chromatography is only a part of a sequence of separation procedures. Often the techniques described have

been developed for one special purpose only, but they may sometimes with slight modifications also be applied for other purposes.

It should be noted here that fission product nuclides are not the only source of information in the fields mentioned above. However, non-fission product nuclides fall outside the scope of the present Chapter.

Radiochemical and mass spectrometric study of the independent fission yields of rare earth isotopes requires fast separation of a given rare earth element with maximum decontamination from neighbouring elements. For this purpose, Crouch and Fryer[43] applied gradient elution from HDEHP columns at 80 °C: in this way they separated the lanthanides from La to Eu in 45 min, and all rare earths in 3.5 hours.

The same technique was applied by Joon and coworkers [36] for determination of the burn-up in high-exposure PuO_2 fuel. By comparison of measured and published data for a relatively large number of nuclides, additional information could be obtained on the fission yields of some of these nuclides from [239]Pu and [241]Pu[58]. Prior to the column separation, Zr, Nb, Sb, Pu and Ce had been removed.

Münze et al. also removed the bulk of the irradiated material (uranium) prior to extraction chromatographic fractionation of the lanthanides by gradient elution[47]. The fraction containing Tb, Dy and Ho was thereafter separated into its components on a fresh HDEHP column by elution with 0.9 M HCl at 41 °C, and the Dy fraction was collected. In this way they were able to identify a new fission chain (mass 166) and determine its fission yield.

Studies of the decay of neutron-rich cerium isotopes are particularly interesting from the standpoint of nuclear structure because these isotopes lie in the transition region between the spherical nuclei around 82 neutrons and those with substantial deformation beginning around 90 neutrons. With suitable fast separation techniques, these nuclides offer the opportunity of studying the effect of adding a neutron at a time, all the way from 82 neutrons on (stable [140]Ce) up to 89 and 90 neutrons ([147]Ce and [148]Ce). For this purpose, Michelsen and Hoffmann[45] developed a 4-minute liquid-liquid extraction procedure for isolating Ce from irradiated uranyl nitrate. The Ce was then sorbed on an HDEHP column, and interfering short-lived Pr daughters were continuously removed by elution with $6 M HNO_3 + NaBrO_3$. Decay studies of [145]Ce, [146]Ce and [147]Ce could thus be made by gamma spectrometric measurements of the column activity[44,46]. The decay of [147]Ce was studied for the first time, while decay schemes for the other two isotopes mentioned could be proposed by combination with other radiochemical measurements.

The same milking technique was applied in a search for an isomeric state in [144]Pr, the only differences being that HDEHP was replaced by dioctyl phosphoric acid and that measurements were carried out on the eluate. An approximate half-life of 0.1 second for [144m]Pr could thus be deduced[42].

The technique can be applied with advantage also for investigation of short-lived molybdenum fission product isotopes. The irradiated uranyl nitrate (enriched to 93% [235]U) is dissolved in 0.03 M HNO_3, and Mo is extracted into 0.5 M HDEHP in heptane. After back extraction with 0.02 M HNO_3 containing 7.5% H_2O_2, the peroxide is destroyed with $NaBrO_3$ and Mo is loaded onto an HDEHP–Hostaflon column. Two Tc daughters are removed by continuous washing with nitric acid at concentrations ranging from 0.03 to 11 M[45].

274

A completely different application is the fission product molybdenum purification for determination of microscopic reactor lattice parameters through radiochemical assay of ^{99}Mo, described by Perricos and Thomassen[28]. In combination with the assay of ^{239}Np, this enables the determination of such parameters as the ratio of epi-Cd to sub-Cd neutron capture in ^{238}U and fission in ^{235}U, the fission rate ratio for ^{238}U and ^{235}U, and the initial conversion ratio. The main Mo purification is obtained by sorption on a tri-n-octylamine extraction chromatographic column from 0.1 M H_2SO_4. The complete procedure has been described in the section on Mo as a burn-up monitor (Paragraph 4.3.).

Fission product isotopes of antimony can be recovered from a fission product mixture within 1.5 min with sufficient purity to study their decay behaviour. For this purpose, the dissolved irradiated sample is pretreated to remove interfering iodine. Sb is then oxidized with chlorate, and sorbed on a diisobutylcarbinol column from 9 M HCl, after which interfering Np is removed with 6 M HCl. Sb is recovered by elution with water[48]. The same method can also be applied for the recovery of antimony from neutron irradiated tellurium. In addition, it may be modified into a milking technique, where the Te isotopes are eluted with 9 M HCl to give information on their mother Sb isotopes still fixed on the column[48,49]. Studies on the decay properties of Sb isotopes in the mass range of 126 to 133 have been performed in this way, resulting in the determination of the half-life of ^{131}Sb (23 min)[48] and the discovery of four gamma lines and of a half-life of 3.1–3.7 min originating from either ^{132}Sb or ^{133}Sb[49].

Also zirconium can be separated from fission product mixtures by column extraction chromatography. After pretreatment with AgCl to remove halogens, Zr is retained on a TBP column from 8 M HNO_3. It can be eluted with water. By applying this technique, Trautmann[49] was able to determine the half-life and maximum beta energy of ^{98}Zr. No gamma lines could be observed. Herrmann[50] used the same technique: by irradiating an uranium solution, automating the operations and analyzing the Zr still on the column, he was able to start measuring only 4 seconds after the end of the irradiation. From these experiments the half-lives of ^{99}Zr and ^{101}Zr could be estimated, and a gamma ray energy assigned to ^{99}Zr. If the zirconium fission product isotopes are left on the column, their niobium daughters can be milked by elution with 8 M HNO_3. This enabled Trautmann [49] to determine the half-life, three maximum beta energies, and eight gamma lines of the isomeric state of ^{98}Nb.

As in the case of the cerium isotopes, the decay properties of the fission product isotopes ^{106}Tc and ^{108}Tc are of interest because they fall into the transitional region between spherical and deformed nuclear shapes. For such a study, a fast method for the separation of Tc from irradiated Pu was developed[50]. Again halogens are removed by exchange with AgCl, after which Tc is retained as pertechnetate from 0.1 M HNO_3 by tetraphenylarsonium-chloride in $CHCl_3$ on plastic grains; it is accompanied by part of the noble gases. Technetium is then eluted with 2 M HNO_3 containing perrhenate, and coprecipitated with this carrier in aqueous tetraphenylarsoniumchloride. The precipitate is filtered off and counting may start 7.5 seconds after the end of the irradiation. By application of this technique, the half-life and gamma transitions of ^{106}Tc, ^{107}Tc and ^{108}Tc could be investigated.

For the fast separation of arsenic from irradiated ^{235}U, the former element is volatilized as the hydride from a strongly acid solution. The gases are passed through a KOH—ethanol solution dispersed on quartz wool to remove the co-volatilized hydrides of Se, Te, and Sb, and AsH_3 is decomposed in HCl containing $KClO_3$. The final decontamination from Sb is obtained by retaining this element on an extraction chromatographic HDEHP layer. This procedure requires only four seconds; the method enabled one of the first studies of neutron spectra from delayed-neutron emitters in the medium mass region, i. e. the spectrum of 2 seconds ^{85}As[50].

6. COMMERCIAL FISSION PRODUCT RECOVERY

Commercial recovery of fission products from spent nuclear fuels is rather limited, and applications of column extraction chromatography for this purpose only sporadic. Nevertheless, one interesting example should be mentioned here.

It is connected with the proposed commercial recovery of fission product palladium and rhodium. These are among the elements produced with highest yield in the fission process. After a suitable cooling time their radioactivity can be neglected because of the long half-life and low energy of the remaining active nuclides. Taking into account the high commercial value of these metals, their recovery from spent nuclear fuels opens possibilities for supplementing natural sources. A study[59] indicated that the annual rhodium production in U. S. power reactors by 1978 will equal the 1968 consumption of that country. A similar situation for palladium is expected to occur by about 1990.

Colvin[51] described a method based on column extraction chromatography for the recovery of Pd, which at the same time is a necessary step in a recovery procedure for Rh. The starting material was a high-salt alkaline (pH = 11—12) waste solution, generated by processes involving recovery of actinides from irradiated uranium. This solution, stored for years and containing 10 mg/l of both metals, is directly fed to a column of tricapryl monomethyl ammonium chloride on an inert support, where Pd is retained quantitatively. The eluate is a feed for the recovery of Rh (and Tc) by ion exchange[60]. A pure palladium product was obtained from the extraction chromatographic column by elutions with 8 M HNO_3 followed by 1.5 M NH_4OH. After three elution cycles, 97% of the Pd had been recovered. The whole process was carried out at about 25 °C. Loading and stability characteristics of the column were excellent.

REFERENCES TO CHAPTER 9

1. K. I. SCHNEIDER, Reactor Technology, 13 (1970—71) 391.
1a. A. G. HAMLIN, B. J. ROBERTS, Nature, 185 (1960) 527.
2. A. G. HAMLIN, B. J. ROBERTS, Anal. Chem., 33 (1961) 1547.
3. J. VAN OOYEN, R. BAC, M. BONNEVIE—SVENDSEN, H. ESCHRICH, Proc. Third Int. Conf. Geneva 1964, Vol. 10, p. 402—411, P/758, 1966.
4. H. ESCHRICH, I. HUNDERE, ETR—222, 1968.
5. H. ESCHRICH, ETR—239, 1969.
6. S. LIS, E. T. JOZEFOWICZ, S. SIEKIERSKI, J. Inorg. Nucl. Chem., 28 (1966) 199.
7. U. WENZEL, H. -I. RIEDEL, IAEA—Symp. on Analytical Methods in the Nuclear Fuel Cycle, Vienna, 1972.
8. H. BERANOVA, M. NOVAK, Coll. Czech. Chem. Commun., 30 (1965) 1073.

9.   H. BERANOVA, M. TEJNECKY, UJV 2557-Ch, 1971.
10.   D. GOURISSE, A. CHESNÉ, CEA−R 3245, 1967.
11.   D. J. GROSS, NDL−TR−103, 1968.
12.   B. TOMAŽIČ, S. SIEKIERSKI, J. Chromatog., 21 (1966) 98.
13.   E. A. HUFF, S. J. KULPA, Anal. Chem., 38 (1966) 939.
14.   R. DENIG, N. TRAUTMANN, G. HERRMANN, AED−Conf, 1966−306−6.
15.   R. DENIG, Thesis, Mainz Univ., 1967.
16.   R. DENIG, N. TRAUTMANN, AED−Conf 68-196-014.
17.   R. DENIG, N. TRAUTMANN, G. HERRMANN, J. Radioanal. Chem., 6 (1970) 331.
18.   M. Y. MIRZA, G. KOCH, Proc. 6th Int. Symp. für Mikrochemie, Graz, Vol. E, p. 169−74, 1970.
19.   W. J. de WET, E. A. C. CROUCH, J. Inorg. Nucl. Chem., 27 (1965) 1735.
20.   I. FIDELIS, W. SMUŁEK, Report "P", No. 1357/V/C.
21.   C. TESTA, C. CESARANO, J. Chromatog., 19 (1965) 594.
22.   J. KŘTIL, M. BEZDEK, J. MENCL, J. Radioanal. Chem., 1 (1968) 369.
23.   J. VAN OOYEN, R. BAC, M. BONNEVIE−SVENDSEN, H. ESCHRICH, Proc. Third Int. Conf. Geneva 1964, Vol. 10. p. 402−411, P/758 1966.
24.   M. BONNEVIE-SVENDSEN, V. MARTINI, KR−81, 1966.
25.   K. JOON, in Solvent Extraction Research, A. S. KERTES, Y. MARCUS, Eds., Wiley-Interscience London, 1969.
26.   K. JOON, R. den BOEF, G. A. NEEFJES KR−143, 1971.
27.   K. JOON, KR−119, 1967.
28.   D. C. PERRICOS, J. A. THOMASSEN, KR−83, 1964.
29.   H. YOSHIDA, C. YANEZAWA, K. GUNJI, Japan Analyst, 19 (1970) 813.
30.   W. D'OLIESLAGER, M. D'HONT, J. Nucl. Mat., 44 (1972) 185.
31.   C. CESARANO, G. PUGNETTI, C. TESTA, J. Chromatog., 19 (1965) 589.
32.   W. SMUŁEK, Nukleonika, 14 (1969) 521.
33.   J. KŘTIL, M. BEZDEK, J. Radioanal. Chem., 1 (1968) 369.
34.   J. KŘTIL, M. BEZDEK, Jad. Energiyà, 16 (1970) 181.
35.   S. F. MARSH, Anal. Chem., 39 (1967) 641.
36.   K. JOON, R. den BOEF, J. HAALAND, D. E. STIJFHOORN, R. de WIT, KR−140, 1970.
37.   S. F. MARSH, 155th ACS National Meeting, San Francisco, California, USA, April 1−5, 1968.
38.   M. R. MONSECOUR, A. C. DEMILDT, Anal. Chem., 41 (1969) 27.
39.   T. ZELENAY, Radiochem. Radioanal. Letters, 2 (1969) 33.
40.   B. F. RIDER, Trans. Am. Nucl. Soc., 7 (1964) 350.
41.   J. KŘTIL, V. BULOVIČ, J. MENCL, A. MORAVEC, Radiochem. Radioanal. Letters, 9 (1972) 335.
42.   P. C. HANSEN, H. L. NIELSEN, K. WILSKY, AERE−M−1078, p. 63, 1962.
43.   E. A. C. CROUCH, J. R. FRYER, AERE−M−1078, p. 91, 1962.
44.   D. C. HOFFMANN, O. B. MICHELSEN, W. R. DANIELS, LA−DC−7869, 1965.
45.   O. B. MICHELSEN, D. C. HOFFMANN, Radiochim. Acta, 6 (1966) 165.
46.   O. B. MICHELSEN, D. C. HOFFMANN, KR−76, 1965.
47.   R. MÜNZE, H. GROSSE−RUYKEN, G. WAGNER, Kernenergie, 12 (1969) 380.
48.   R. DENIG, N. TRAUTMANN, G. HERRMANN, Z. Anal. Chem., 216 (1966) 41.
49.   N. TRAUTMANN, Dissert. Mainz Univ. 1968.
50.   G. HERRMANN, AED−Conf.71-386-002, 1971.
51.   C. A. COLVIN, ASH−SA−28, 1969.
52.   K. JOON, P. A. DEURLOO, I. A. HUNDERE, KR−100, 1965.
53.   K. JOON, P. A. DEURLOO, J. Radioanal. Chem., 1 (1968) 73.
54.   D. SCARGILL, AERE−R−4583, 1964.
55.   K. JOON, Z. Phys. Chem., 243 (1970) 249.
56.   J. M. FLETCHER, D. SCARGILL, in Solvent Extraction Chemistry of Metals, H. A. C. McKAY Ed., Macmillan, London-New York, 1965.
57.   W. N. BISHOP, S. F. MARSH, R. C. WILLIAMS, G. E. WOLFE, BAW−3809−7, 1969.
58.   K. JOON, R. den BOEF, J. HAALAND, D. E. STIJFHOORN, R. de WITH, RCN−157, 1970.

277

59. G. A. ROHRMAN, Engelhard Industries Technical Bulletin, Vol. IX, 2. p. 59, 1968.
60. J. K. PANESKO, ASH—733, 1968.
61. K. JOON, R. den BOEF, R. de WIT, J. Radioanal. Chem., 8 (1971) 101.

CHAPTER 10

USE OF EXTRACTION CHROMATOGRAPHY
IN RADIOTOXICOLOGY

C. TESTA

1. GENERAL REMARKS

1.1. INTRODUCTORY NOTES ABOUT RADIOTOXICOLOGY AND THE METHODS USED FOR THE EVALUATION OF INTERNAL CONTAMINATIONS FROM RADIONUCLIDES

As it is supposed that this Chapter is addressed especially to people who are familiar with chemistry rather than with radiation protection, a few introductory notes about radiotoxicology are given at first.

Radiotoxicology is a new branch of science which is interested in the prevention and determination of internal contaminations from radioactive substances.

When the contaminating radionuclide is a gamma or an X-ray emitter, a technique can be used which provides the direct determination of body burden by means of gamma spectrometric analysis on the contaminated subject: the complete apparatus is named Whole Body Counter.

On the contrary, if the incorporated radionuclide is an alpha or a beta emitter, no direct method can be used, and the body burden can be calculated only by indirect methods taking into account a relation between the body burden and the excretion patterns of the contaminant. Generally, the biological sample which is analyzed for this purpose is urine.

As the maximum permissible body burdens (MPBB) for the various radionuclides have been calculated[1], and the urinary excretion patterns of the radionuclides with time are generally known[2], it is possible to indirectly obtain the body burden by determining the content of the radionuclide in the daily urine sample. A simple example will certainly clear this statement.

The MPBB for tritium is 1000 μCi; the per cent retention R(t) of tritium in the body as a function of time (t) is expressed by the relation[2]

$$R(t) = 100 \exp\left(-\frac{0.603}{t}\right);$$

as a 60% of tritium leaves the body by the urines, the urinary excretion with time, obtained by the derivative of R(t) multiplied by 0.6, is

$$U(t) = 3.5 \exp\left(-\frac{0.693}{t}\right).$$

From this relation it appears that a 3.5% of the tritium present in the body is excreted daily with the urines. Therefore, the urinary excretion corresponding to the MPBB is $0.035 \cdot 1000$ μCi = 35 μCi (action level) and the body burden can be calculated always by multiplying the tritium daily urinary excretion by $\frac{100}{3.5} = 28.6$.

Tritium is really a clear and simple case; the urinary excretion patterns for the various radionuclides are generally more complicated and they can be represented by a sum of two or more exponential functions, by one or more power functions, and finally, by a mixture of exponential and power functions[2,3]. In any case it is always possible to derive some action levels for the urines and to roughly calculate the body burden following the urinary excretion curve of the radionuclide.

Therefore, periodical radiotoxicological analyses have to be carried out on the technical personnel which is exposed to a risk of internal contamination from radionuclides. Except for natural uranium and natural thorium, the final determination is made by counting the alpha or beta emitters previously separated from the urine.

The many advantages of extraction chromatography as a separative tool in the field of inorganic chemistry were checked by the author in a five year experience during which several chromatographic separations were carried out with paper, cellulose powder and polytrifluorochloroethylene loaded with organic extractants such as tri-n-octylamine, tri-n-octylphosphine oxide and di(2-ethylhexyl) orthophosphoric acid[4].

Therefore the extrapolation of such an experience to the field of radiotoxicology was not difficult and quite natural.

1.2. FEATURES OF RADIOTOXICOLOGICAL TECHNIQUES FOR ROUTINE PURPOSES

The radiotoxicological analyses have to be precise, sensitive, selective and possibly also simple, cheap and rapid. Because of the several chemical steps which are often necessary, it is almost impossible to obtain a quantitative final yield, and a mean recovery must be previously calculated carrying out many analyses on urines spiked with the involved radioun ide [5,6].

Sometimes the final activity levels to be counted are so low that it is necessary to carry out many blanks on the urines of non-exposed persons and to use the most pure reagents.

Furthermore the final alpha counting has to be performed with sources having a minimum degree of self-absorption and preferably prepared by electroplating.

Because of the great number of the routine radiotoxicological analyses, it is highly desirable that many analyses be carried out contemporaneously during the day-work, performing the long counting procedures over-night or during the week-end days.

A wet or dry mineralization of the urine has often to be obtained in order to clear the sample and to be sure that the metabolised radionuclide is in the ionic form.

1.3. INERT SUPPORTS AND STATIONARY PHASES
EMPLOYED FOR COLUMN AND BATCH EXTRACTION
PROCEDURES IN RADIOTOXICOLOGY

The inert supports which have been used in our laboratory for radiotoxicological analyses are polytrifluorochloroethylene (Kel-F) and microporous polyethylene (Microthene−710).

Kel-F (100−170 mesh) showed very good properties[4] of supporting irreversibly a volume of the stationary phase up to a ratio of 1:1 (v/w). Unfortunately, Kel-F is quite expensive and not easily available; therefore, it was replaced in 1967 by Microthene-710 which is supplied at a very low cost: the 50−100 U.S. mesh powder can support about the same volume of stationary phase as the Kel-F, although a ratio of 4:5 (v/w) is preferred[7]. Several other supports (Table 1) have been tested in our laboratory[4], but no technical reason was found to replace Microthene-710 for raditoxicological purposes.

The stationary phases have been chosend in order to obtain a selective isolation of the radionuclides from urines wet-mineralized with nitric acid: to this purpose, the existing analytical and technical results of the extractants were taken into account.

Except for tri-n-butylphosphate (TBP), the other extractants, tri-n-octylamine,(TNOA); tri-n-octylphosphine oxide, (TOPO); neo-tridecano hydroxamic acid; di(2-ethylhexyl) phosphoric acid (HDEHP) were diluted with cyclohexane or xylene.

The concentration of the supported extractant in the diluent ranged from 0.1 M to 1.5 M and was selected in order to obtain a sufficiently good extraction of the radionuclide and a prompt and selective elution from the column.

TABLE 1

Materials tested as supports for columns

Trade name	Chemical denomination	Formula	Supplier
Kel-F	Polytrifluorochloroethylene	$[-CFCl-CF_2-]_n$	M. M. M. (U. S. A.)
Algoflon	Polytetrafluorochloroethylene	$[-CF_2-CF_2-]_n$	Montedel (Italy)
Moplen	Isotactic polypropylene	$[-CH-CH_2-]_n$ with CH_3	Montedel (Italy)
Corvic	Polyvinylchloride-polyvinylacetate-copolymer	$[-CH_2-CH-CH_2-CH-]_n$ with Cl and $O-C=CH_3$, O	I. C. I. (Great Britain)
Vipla	Polyvinylchloride	$[-CHCl-CH_2-]_n$	Montedel (Italy)
Microthene-710	Microporous polyethylene	$[-CH_2-CH_2-]_n$	Columbia Org. Chem.(U.S.A.)

282

1.4. COLUMN EXTRACTION CHROMATOGRAPHY AND BATCH SEPARATION PROCESSES

Both column extraction chromatography and batch separation processes have been used to isolate the radionuclides from biological samples. The choice of either method depends on the sample volume to be analyzed, on the wanted decontamination factors, on the possibility of separating and determining in the same sample more than one radionuclide, on the desired recovery and on the time to be devoted to the analysis. On the first ocassion only column extraction chromatography was used, but successively the batch extraction procedures were preferred both for routine purposes and for rapid analyses[6,8].

TABLE 2

Batch extraction of Y(III) from 500 ml urine (pH = 0.5) by
5 g of Microthene supporting 5.0 ml of the HDEHP solution
as a function of the HDEHP concentration and of the stirring time

Stirring time (minutes)	Yttrium extraction, %		
	HDEHP 10%	HDEHP 30%	HDEHP 50%
2	3.3	28.2	40.0
5	21.3	30.0	51.3
10	25.4	30.9	87.8
20	34.4	50.9	94.5
30	42.6	58.1	99.1

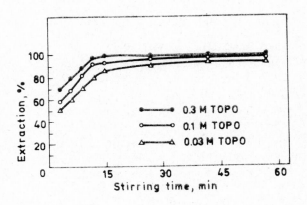

Fig. 1. Pu(IV) batch extraction by Microthene-TOPO as the function of the TOPO concentration and of the stirring time

283

Table 2 and Fig. 1 show that an almost quantitative extraction of a radionuclide from urines can be obtained in batch if a suitable concentration of the supported extractant and a sufficient stirring time are employed. This procedure takes a shorter time than column chromatography when a great sample volume is involved; on the contrary, batch extraction is not advisable with small liquid volumes and with poor extraction coefficients. In fact, after the batch extraction is carried out, the slurry is transferred into a chromatographic column for the successive washing and elution; the extracted element is homogeneously distributed along the column and a fraction could be lost from the lower part of the bed if the extraction coefficient is not so high. Therefore, it was sometimes necessary to place a small quantity of the original slurry in the bottom of the column so as not to loose a significant fraction of the radionuclide previously isolated by batch extraction from the urine.

The preparation of the slurry to be used for extraction chromatography is very simple. The stationary phase is added drop by drop, with gentle stirring, to the inert support in the wanted ratio; the slurry is then stirred for about one hour with a sufficient volume of the conditioning solution in order to obtain an homogeneous product and to discard the floating particles. Generally, 2–5 grams of the support loaded with 2–5 ml of stationary phase are sufficient both for batch extraction and for column procedure.

The diameter of the utilized beds ranges from 8 to 25 mm, and the height from 30 to 150 mm; the flow-rate ranges from 0.5 to 5 ml·min^{-1}.

1.5. PRELIMINARY CHEMICAL TREATMENT OF THE ORGANIC SAMPLES BEFORE EXTRACTION CHROMATOGRAPHY

Before the isolation of the radionuclide by extraction chromatography, the urine samples must be mineralized in order to obtain a clear solution and to be sure that the element is in the wanted ionic form.

Many radiotoxicological methods are based on a dry mineralization of the sample in order to obtain a white residue of inorganic salts. These methods, which generally imply a first wet mineralization with nitric acid and a final calcination at 400–500 °C in a muffle, are quite time-consuming and often furnish a non-completely soluble residue. Therefore a wet mineralization with nitric acid and eventually with H_2O_2 generally was preferred; in fact, nitric acid works very well as a wet mineralizing agent and gives rise to a clear solution; on the contrary, hydrochloric acid and sulphuric acid are not so suitable, whereas perchloric acid was not taken into consideration because it does not give rise to extractable complexes. All the determinations of the actinide elements in the urines by extraction chromatography with supported TOPO, TNOA and TBP are based on the formation of nitric complexes; therefore it is clear that nitric acid presents the double advantage to be the mineralizing and the complexing agent at the same time.

When the whole 24-hour urine sample (~1250 ml) has to be analyzed by column extraction chromatography, a coprecipitation of the radionuclides with calcium and magnesium phosphate can be made in order to reduce the volume to be passed through the column.

Sometimes the sample volume is reduced also by boiling the urine with nitric acid for a suitable time. For each procedure given in the following section, the reproducibilities are given of the results obtained, as the standard deviations σ of a single result, obtained from ten independent determinations on the average.

2. ANALYTICAL PROCEDURES

2.1. DETERMINATION OF THORIUM IN URINE BY A KEL-F-TOPO COLUMN FOLLOWED BY COLORIMETRIC DETERMINATION WITH THORON[9]

2.1.1. PRELIMINARY CONSIDERATIONS

The MPBB for natural thorium, related to bone, is 0.01 μCi (90 mg); the biological half-life is high ($5.7 \cdot 10^4$ days) and the urinary excretion very poor; therefore an action level of 10 μg per day could be derived[2].

Persons who work with thorium and its compounds are generally controlled by means of urine analysis; because of the low specific activity of natural thorium ($1.1 \cdot 10^{-7}$ Ci/g) the final determination is carried out by colorimetry.

In 1963 a method was prepared in our laboratory by using a small column of Kel-F supporting TOPO for the isolation of the radionuclide from the urine: thorium was then determined with Thoron[10] ($\epsilon = 13,000$).

A nitric acid medium has been chosen for the extraction taking into account the following favourable conditions: (a) the wet mineralization of the urine by this acid; (b) the large extraction of thorium by TOPO from nitric acid[11]; (c) the poor extraction of the majority of cations present in the urines. Thoron was selected for the colorimetric determination because at that time it seemed to be a sensible and selective reagent[10].

2.1.2. REAGENTS AND EQUIPMENT

Kel-F (polytrifluorochloroethylene), grade 300, low density (100—170 U. S. mesh) was supplied by Minnesota Mining Manufactury (USA).

Tri-n-octylphosphine oxide was supplied by Eastman Organic Chemicals (USA).

Thoron, [1-(o-arsono-phenyl-azo)-2-naphthol-3:6 disulphonic acid-Na] was supplied by BDH (U. K.).

The spectrophotometer was an Unicam Mod. SP 600 equipped with 400 mm cells.

The chromatographic column had an internal diameter of 10 mm and a fritted glass disc at the bottom.

2.1.3. ANALYTICAL PROCEDURE AND EXPERIMENTAL RESULTS

100 ml of conc. HNO_3 are added to 500 ml of urine and the solution boiled for 30'. After cooling, the solution is passed at 5 ml·min^{-1} through a column consisting of 2 g of Kel-F supporting 2 ml of 0.1 M TOPO in cyclohexane. After washing with 50 ml 4 M HNO_3 and 50 ml H_2O at the same flow-rate, thorium is eluted with 20 ml of 6 M HCl and 70 ml 0.5 M HCl at a flow-rate of 0.5 ml/min. After adding 1 g of magnesium as $MgCl_2$, the eluate is taken to dryness; 1 ml conc. $HClO_4$ is added to destroy completely the possible traces of organic substances; then 1 ml of water and 1 ml of 20% $HClO_4$ are added, and the solution transferred quantitatively into a 40 mm cell; finally, 0.5 ml of 0.1% Thoron in water and 2.0 ml of water are added to reach the final volume of 5.5 ml.

After 10' the colorimetric determination is performed at 545 mμ against a blank containing the reagents. The amount of thorium can be calculated by a standard curve.

On the basis of the results obtained with urines spiked with 0.5—5 μg of thorium, a 97.5 mean recovery ($\sigma = 6.9$) and a sensitivity limit of 1 μg thorium per litre were found.

One analysis can be carried out within eight hours.

2.2. DETERMINATION OF ENRICHED URANIUM IN URINE BY A KEL-F-TBP COLUMN AND ALPHA COUNTING[12]

2.2.1. PRELIMINARY CONSIDERATIONS

The analysis of natural uranium in the urines is carried out by fluorimetry; on the contrary, the persons who have a risk of internal contamination from enriched uranium must be controlled by the radiometric determination of the urinary alpha activity due to the presence of uranium isotopes (^{233}U, ^{234}U, ^{235}U) having a higher specific activity; the MPBB related to the bone is 0.05 μCi[2] and the urinary action level 50 dpm per day.

A method based on the retention of uranium by a Kel-F-TBP column was devised in 1964 in our laboratory[12] to improve the existing techniques.

This method was derived from the paper of Hamlin et al.[13] on the utilization of TBP supported on Kel-F for the isolation of uranium from aqueous solutions.

2.2.2. REAGENTS AND EQUIPMENT

Tri-n-butylphosphate (TBP) was supplied by BDH (U. K.).
The chromatographic column had an internal diameter of 8 mm.
The alpha counting was performed with a scaler connected to a 2" ZnS(Ag) detector.

50 ml conc. HNO_3 are added to 200 ml of urine; the solution is boiled for about 60'
to reach a volume of about 40 ml. After cooling, the concentration of HNO_3 is brought
to 7.5 M and the solution passed at 1 ml·min⁻¹ through a column consisting 3 g of
Kel-F supporting 3 ml of TBP. After washing with 20 ml of 7.5 M HNO_3, uranium is
eluted with 15 ml of water (Fig. 2). The solution is concentrated up to 1−2 ml and
quantitatively transferred to a platinum dish for the alpha counting by a ZnS(Ag) detector.

A final yield of 91.1% and a sensitivity limit of 5 dpm were obtained within 3
hours.

A greater sensitivity can be obtained if uranium is coprecipitated from one litre of
urine with calcium and magnesium phosphates by adding NH_4OH up to pH = 8.5. The
precipitate is centrifuged, washed twice with 3 M NH_4OH and dissolved in 20 ml of
7.5 M HNO_3 before the chromatographic isolation. In this case, a recovery of 95% and
a sensitivity limit of 1 dpm/litre were obtained within 4 hours.

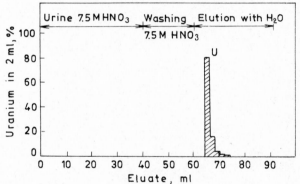

Fig. 2. Elution diagram of enriched uranium from a Kel-F-TBP column

2.3. DETERMINATION OF PLUTONIUM IN URINE BY A KEL-F-TNOA
COLUMN AND ALPHA COUNTING[14]

2.3.1. PRELIMINARY CONSIDERATIONS

Plutonium is a highly radiotoxic radionuclide[15]: the MPBB related to the bone is
only 0.04 μCi. Furthermore the urinary excretion as a function of time is so poor that
an action level of 2 pCi/day could be derived[3]; consequently the sensitivity limit of the
method has to be lower than 0.1 pCi.

As reported by Keder and coworkers[15] plutonium(IV) in 2 M HNO_3 is highly extracted by TNOA; on the contrary uranium(VI), thorium(IV), americium(III), curium(III),
radium (II), protactinium(V) are not extracted to a considerable degree. A column of
Kel-F supporting TNOA was therefore prepared in our laboratory in 1964 to selectively
isolate plutonium from urine[14].

The alpha determination was carried out by counting the electroplated plutonium with a solid state detector having a very low alpha background.

2.3.2. REAGENTS AND EQUIPMENT

Tri-n-octylamine (TNOA) was supplied by Fluka (Switzerland).

The electroplating procedure was carried out by an apparatus supplied by C. Erba (Milan); the 16×60 mm cylindrical cells were made of plexiglass and stainless steel.

The alpha counting was performed by means of a scaler provided with a 200 mm² solid state detector (Molechem Inc., USA).

2.3.3. ANALYTICAL PROCEDURE AND EXPERIMENTAL RESULTS

One litre of urine is acidified with 50 ml of conc. HNO_3 and 50 ml of conc. $HClO_4$; after boiling, the solution is cooled, and NH_4OH is added up to pH = 8.5. The precipitate is let to settle overnight and centrifuged; after washing twice with 3 M NH_4OH, the residue is dried, mineralized with conc. HNO_3, and dissolved in 50 ml of 2 M HNO_3. Plutonium is brought to Pu(IV) by adding NH_2OH and $NaNO_2$, and the solution is passed at 1 ml·min⁻¹ through a column of 3 g of Kel-F supporting 3 ml of TNOA 50% in xylene. After washing with 50 ml of 1 M HNO_3, plutonium is eluted by reduction to Pu(III) with 100 ml of a saturated H_2SO_3 solution (Fig. 3). The eluate is taken to dryness, mineralized with some conc. HNO_3 and taken to dryness again; 0.3 ml of conc. H_2SO_4 and water are added and the pH is adjusted to 4 with NH_4OH; the solution is transferred into the electrolytic cell for the electroplating procedure.

The cathode is a stainless steel disc having an electrodeposition surface of 40 mm², and the anode is a platinum wire; the current intensity is fixed at 400 mA for 4 hours.

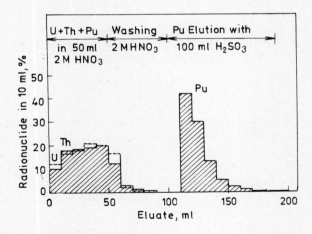

Fig. 3. Elution diagram of plutonium from a Kel-F-TNOA column

When the source is dry, the alpha counting is performed by using a scaler supplied with a solid state detector. The background should range from 0.001 to 0.01 cpm; the counting time ranges from 100 to 1000 min depending on the activity to be counted.

The final recovery resulted 90.5% with a sensitivity limit of 0.04 pCi/litre; the time to carry out one analysis is of about two working days.

2.4. DETERMINATION OF NATURAL THORIUM IN URINE BY BATCH EXTRACTION WITH MICROTHENE-TOPO FOLLOWED BY COLORIMETRIC DETERMINATION WITH ARSENAZO III[17]

2.4.1. PRELIMINARY CONSIDERATIONS

The radiotoxicology of thorium and some important aspects of these analyses have been already reported in Paragraph 2.1.1.

In 1968 a new technique was devised in order to save time and to obtain a greater sensitivity[17]: column chromatography was replaced by batch extraction chromatography and Thoron was replaced by Arsenazo III[18]. Furthermore Kel-F was replaced by Microthene-710 taking into account the considerations reported in Paragraph 1.3.

The complex thorium-Arsenazo III in 8 M HCl shows an $\epsilon = 130,000$, and the colorimetric reaction is very selective.

2.4.2. REAGENTS AND EQUIPMENT

For TOPO, the chromatographic column and the spectrophotometer see in Paragraph 2.1.2.

Microthene-710 (microporous polyethylene), 50—100 U. S. mesh, was supplied by Columbia Organic Chemicals (USA).

Arsenazo III ($C_{22}H_{16}A_{12}N_4O_{14}S_2$) was a BDH (Great Britain) product; the solution was prepared by dissolving 0.5 g of Arsenazo in 1 litre of 8 M HCl.

2.4.3. ANALYTICAL PROCEDURE AND EXPERIMENTAL RESULTS

To a 500 ml urine sample, 10 ml of H_2O_2 (120 V) and 100 ml of conc. HNO_3 are added; the volume is concentrated to 400 ml by boiling for 1 hour and 100 ml of conc. HNO_3 are added. After cooling, the solution is stirred by a magnetic stirrer with 2 g of Microthene supporting 2 ml of 0.5 M TOPO in cyclohexane. After 20 min the slurry is transferred into a chromatographic column and washed with 50 ml of 4 M HNO_3; thorium is finally eluted with 40 ml of 0.3 M H_2SO_4 at a flowrate of 1 ml·min^{-1}.

The eluate is evaporated to dryness and 4.5 ml of 8 M HCl and 1 ml of Arsenazo III solution are added. The colorimetric determination is performed in 40 mm cells at 665 mμ, against a blank.

20*

The final yield resulted 98.2% ($\sigma = 11.5$) with a sensitivity limit of about 0.2 μg of thorium per litre.

The time required for one analysis is about 5 hours.

2.5. DETERMINATION OF ENRICHED URANIUM IN URINE BY BATCH EXTRACTION WITH MICROTHENE-TOPO AND ALPHA COUNTING[17]

2.5.1. PRELIMINARY CONSIDERATIONS

Some considerations about the radiotoxicology of the enriched uranium have already been reported in Paragraph 2.2.1.

As TOPO is a good extractant for uranium in a nitrate medium[19], a batch extraction with a slurry of Microthene -TOPO has been devised[17].

The greatest difficulty was met in obtaining a good elution of uranium from the column, and a solution of HF was finally chosen for this purpose.

2.5.2. REAGENTS AND EQUIPMENT

For TOPO and Microthene see Paragraphs 2.1.2. and 2.4.2., respectively; for the alpha detector see Paragraph 2.2.2.

The chromatographic column (10 mm i. d.) was made of teflon.

2.5.3. ANALYTICAL PROCEDURE AND EXPERIMENTAL RESULTS

The wet mineralization of the urine and the batch extraction are the same as described for thorium in Paragraph 2.4.3. After washing the column with 50 ml of 4 M HNO_3, uranium is eluted with 40 ml of 1 M HF at 1 ml·min^{-1}; the solution is collected in a platinum vessel and evaporated to dryness. Some 8 M HNO_3 is added, and the resulting solution is transferred to a 5 cm watch glass; after drying, the alpha counting is performed for 60 min.

The final yield was 70.0% ($\sigma = 3.0$) with a sensitivity limit of 1 dpm/l of urine.

2.6. DETERMINATION OF GROSS BETA ACTIVITY, NATURAL THORIUM AND ENRICHED URANIUM IN A SINGLE URINE SAMPLE BY A MICROTHENE-TOPO COLUMN [17]

2.6.1. PRELIMINARY CONSIDERATIONS

The technical staff engaged with the reprocessing of irradiated nuclear fuel coming from a ^{232}Th breeder reactor may have a risk of internal contamination from natural thorium, enriched uranium (^{232}U, ^{233}U, ^{234}U, ^{235}U) and beta-gamma fission products.

290

The gamma emitters can be determined directly by a gamma spectrometric determination on the urine samples; the beta emitters (^{89}Sr, ^{90}Sr, ^{90}Y, [91]Y, [147]Pm, etc.) can be detected by a coprecipitation with calcium and magnesium phosphates and by a counting with a low background beta detector: the gross counting is all attributed to ^{90}Sr which is considered to be the most radioactive beta emitter. Natural thorium and enriched uranium can be determined in the urine by an extraction chromatography method with Microthene-TOPO. Therefore a method was devised in order to perform the three analyses using the same urine sample[17].

2.6.2. REAGENTS AND EQUIPMENT

For TOPO, Arsenazo III, Microthene, the alpha detector and the chromatographic column see Paragraphs 2.1.2., 2.2.2., 2.3.2., 2.4.2., and 2.5.2. respectively.

An anticoincidence low background beta detector with two 2″ plastic scintillators has been used for the beta counting.

2.6.3. ANALYTICAL PROCEDURE AND EXPERIMENTAL RESULTS

50 ml of conc. HNO_3 and 10 ml of H_2O_2 (120 V) are added to 500 ml of urine; after a 15′ boiling the solution is cooled and some conc. NH_4OH is added up to pH = 8.5. The precipitate is allowed to settle overnight and washed twice with 3M NH_4OH; the residue is then transferred to a dish and mineralized with conc. HNO_3 until it is completely white. The residue is crushed in a mortar and an aliquot (100−200 mg) is counted in a low background beta detector having a background of 1−2 cpm. The beta activity is calculated taking into account the correction factor due to self-absorption, and it is all attributed to [90]Sr.

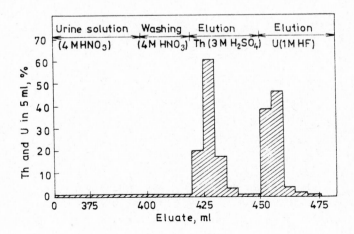

Fig.4. Simultaneous separation of thorium and enriched uranium by a column of Microthene-TOPO.

After the dissolution of the whole residue by heating with 50 ml of conc. HNO_3, 150 ml of water are added and the solution is passed at 5 ml·min^{-1} through a column of Microthene-TOPO (see Paragraph 2.4.3.). After washing with 20 ml of 2 M HNO_3 thorium is eluted with 40 ml of 3 M H_2SO_4 at 1 ml·min^{-1}; in these conditions uranium remains on the column and can be eluted afterwards with 40 ml of 1 M HF (Fig. 4).

Thorium is finally determined by colorimetry with Arsenazo III as described in Paragraph 2.4.3., and enriched uranium is alpha-counted as reported in Paragraph 2.5.3.

The final yield found with spiked urines was 98% for natural thorium and 77% for enriched uranium.

2.7. RAPID DETERMINATION OF PLUTONIUM IN URINE BY BATCH EXTRACTION WITH MICROTHENE-TOPO AND ALPHA COUNTING[20]

2.7.1. PRELIMINARY CONSIDERATIONS

Plutonium radiotoxicology has been described briefly in Paragraph 2.3.1. The retention of the radionuclide in the body can be reduced by some chelating agents, such as ethylenediaminotetraacetic acid (EDTA) and particularly diethylenetriaminopentaacetic acid (DTPA) which increase the urinary excretion of plutonium. However, these therapeutical agents have to be used only in case of ascertained internal contaminations because of their toxicity; furthermore, it has been demonstrated[21,22] that the increase of the urinary excretion is directly related to the rapidity of the treatment. Some simple methods with a sensitivity limit of 1–10 pCi per sample are available to detect plutonium in the urines within a few hours[23].

A very simple and selective method[7,20] was devised in our laboratory in 1970 by a batch extraction of plutonium from the urines with a slurry Microthene-TOPO. In fact, TOPO is a good extractant for plutonium(IV) in a HNO_3 medium[11,24]; radium, americium, curium are not extracted, while uranium, thorium, protactinium and neptunium remain on the column when plutonium is eluted after reduction to Pu(III).

2.7.2. REAGENTS AND EQUIPMENT

For TOPO, Microthene, the alpha detector and the chromatographic column see Paragraphs 2.1.1., 2.2.2., 2.4.2. and 2.5.2. The HI solution (Analar), containing less than 0.0002% of phosphorous compounds, was supplied by BDH (Great Britain).

2.7.3. ANALYTICAL PROCEDURE AND EXPERIMENTAL RESULTS

170 ml of conc. HNO_3 are added to 500 ml of urine and the solution is boiled for 5′; after cooling, 10 ml of 1 M $FeCl_2$ are added and the solution is stirred for 5′; then 10 ml of $NaNO_2$ 25% in water are added and the solution stirred again for 5′. At this

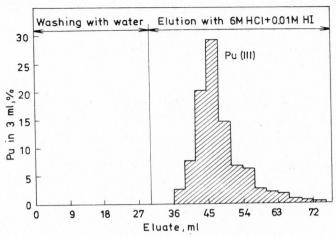

Fig. 5. Elution diagram of plutonium from a Microthene-TOPO column

point plutonium is surely present as Pu(IV) and the concentration of HNO_3 is about 4 M.

2.5 g of Microthene supporting 2.0 ml of 0.3 M TOPO in cyclohexane are added to the wet mineralized urine, and the slurry is stirred for 30' by a magnetic stirrer (see Fig.1). The solution is then eliminated by filtering on a G–1 buchner; after washing with 30 ml of 4 M HNO_3, the slurry is transferred into a chromatographic column, where 1 g of Microthene supporting 0.8 ml of TOPO solution had previously been placed. The column is washed with 30 ml of water at 2 ml·min⁻¹, and plutonium is eluted with 45 ml of 6 M HCl + 0.01 M HI at 1 ml· min⁻¹. (Fig. 5). After adding 5 ml of conc. HNO_3 the solution is boiled and concentrated up to about 2 ml, which are transferred onto a 2″ watch glass; after drying, a 15' alpha counting is carried out by a 2″ ZnS(Ag) detector having a background lower than 0.1 cpm.

The final yield resulted 76.5% ($\sigma = 5.7$) with a blank of about 0.1 pCi per sample.

The time to carry out one analysis is 3–4 hours.

A greater sensitivity limit can be achieved for routine purposes by analyzing 1 litre of urine and by performing the electrodeposition of plutonium on a 20 mm stainless steel disc before counting the planchet with a 400 mm² solid state detector; in case of high values an alpha spectrometric analysis can be carried out.

By using this alternative method[25] a final yield of 72.5% ($\sigma = 5.35$) was obtained with a blank of 0.07 pCi for a 24 hour sample.

2.8. DETERMINATION OF NEPTUNIUM–237 IN URINE BY BATCH EXTRACTION WITH MICROTHENE-TOPO AND ALPHA COUNTING[26]

2.8.1. PRELIMINARY CONSIDERATIONS

[237]Np has to be considered a radionuclide having a very high radiotoxicity. The maximum permissible body burden related to bone is 0.06 μCi; the biological half-life is high and the daily urinary excretion is poor[2]: therefore the action level was fixed at 6 pCi per day.

In 1972 a new method[26] was devised based on the extraction of Np(IV) with Microthene - TOPO from urine samples acidified with 6 M in HNO_3; the elution of neptunium was carried out by oxidizing the radionuclide to Np(V) with chlorine at 70 °C; in these conditions americium, curium and plutonium(III) are not extracted and high decontamination factors from these transuranium elements can be obtained.

2.8.2. REAGENTS AND EQUIPMENT

For the majority of the reagents and equipment see the preceeding paragraphs.
The solution of iron(II) sulphamate has to be prepared daily.

A superscaler with a single channel analyzer (Berthold, West Germany) equipped with an Ortec 400 mm^2 solid state detector was used for alpha counting.

For the electroplating procedure, a C. Erba apparatus equipped with plexiglass cells was used: a 20 mm stainless steel disc was used as the cathode and a platinum wire as the anode.

A 10 mm i. d. column with a thermostating jacket was used for the chromatographic separation.

2.8.3. ANALYTICAL PROCEDURE AND EXPERIMENTAL RESULTS

300 ml of conc. HNO_3 are added to 500 ml of urine and the solution is heated until it is clear. After cooling 10 ml of 1 M iron(II) sulphamate are added slowly, and the solution is stirred for 15 min; in these conditions neptunium is completely reduced to Np(IV) and plutonium to Pu(III). 2.5 g of Microthene (100–170 U. S. mesh) supporting 2.0 ml of 0.1 M TOPO in cyclohexane are then added, and the solution stirred for 45 min on a magnetic stirrer (Fig. 6). The aqueous solution is eliminated and the slurry transferred into the chromatographic column by means of 6 M HNO_3; after washing with 2 M HNO_3, the column is thermostated at 70 °C, and neptunium eluted by oxidation to Np(V) with 70 ml of 2 M HCl + chlorine water mixture (1:1) at 1 ml·min^{-1}[28] (Fig. 7). The eluate is evaporated to dryness, and 5 ml of 6 M HCl are added; the solution is then transferred into an electroplating cell containing 0.5 ml of 12 M KOH. After adding one drop of methyl red, the solution is neutralized with conc. NH_4OH and acidified again with 1 ml of 1 M HCl. Electroplating is performed at 600 mA for 2.5 hours. The disc is then removed and the alpha counting carried out for 200 min with a low background solid state detector.

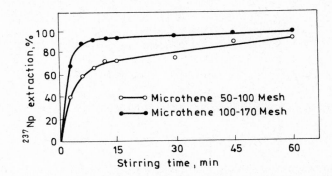

Fig. 6. Extraction of neptunium with Microthene-TOPO as a function of the Microthene granulometry and of the stirring time

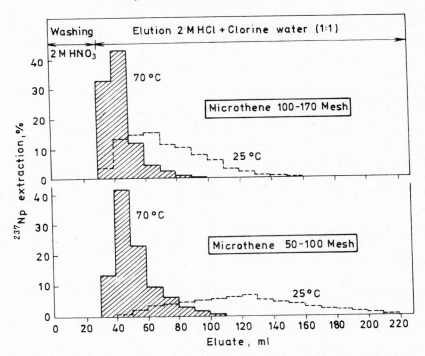

Fig. 7. Elution diagram of neptunium from Microthene-TOPO as a function of the Microthene granulometry and of the temperature

The final yield resulted to be 83.2% ($\sigma = 6.5$) with a blank value of 0.026 pCi. The decontamination factors were 10^4 for americium and curium; 10^2 for plutonium and 30 for uranium.

A work-day is sufficient to carry out the analysis, if the counting time is not taken into account.

2.9. DETERMINATION OF PLUTONIUM AND NEPTUNIUM IN URINE BY A COLUMN OF MICROTHENE SUPPORTING NEO-TRIDECANO-HYDROXAMIC ACID [29]

2.9.1. PRELIMINARY CONSIDERATIONS

It is well known that the hydroxamic acids form stable complexes with different cations[30].

A new hydroxamic acid, containing an aliphatic group derivative of neotridecanoic acid, has been recently synthesized at CSN-Casaccia[31].

The neo-tridecano-hydroxamic acid (HX70) was found to be a very useful and selective extractant for Pu(IV) and Np(IV) also at very low pH values[32].

As this compound is sufficiently insoluble in water, a column of Microthene-HX70 was devised to isolate neptunium and plutonium from urines[29]; the low solubility of the compound suggested using moderate volumes of aqueous solution and performing a preliminary coprecipitation of the two radionuclides with calcium and magnesium phosphates.

2.9.2. REAGENTS AND EQUIPMENT

The neo-tridecano-hydroxamic acid ($C_{12}H_{25}CONHOH$) was synthesized, purified and characterized in the Chemical Laboratories of CSN-Casaccia, CNEN[31].

The slurry Microthene-HX70 was prepared by mixing 5 g of Microthene (50–100 U. S. mesh) with 4 ml of 0.3 M HX70 in cyclohexane.

For the electroplating apparatus and for the alpha counting see Paragraph 2.8.2.

The chromatographic column had a 10 mm internal diameter.

2.9.3. ANALYTICAL PROCEDURE AND EXPERIMENTAL RESULTS

100 ml of conc. HNO_3 and 15 ml of H_2O_2 (12 Vol) are added to 1 litre of urine and the solution is boiled for two or three hours until it is clear. After cooling, the phosphates are precipitated by adding conc. NH_4OH up to pH = 8.5–9.0. After centrifugation, the precipitate is washed with 0.3 M NH_4OH and then dissolved with 7.5 M HNO_3; repeated mineralizations with 7.5 M HNO_3 in the presence of H_2O_2 are necessary to obtain a white solid residue which is dissolved in 20 ml of 2 M HNO_3.

After adding 100 mg of $NaNO_2$, the solution is boiled for 15 min; finally, urea is added to eliminate the excess of nitrous acid. The solution is then percolated at 1 ml·min^{-1} through the Microthene-HX70 column preequilibrated with 20 ml of 2 M HNO_3. After a washing with 20 ml of 2 M HNO_3, plutonium is eluted by reduction to Pu(III) with 50 ml of 2 M HNO_3 containing 0.15% hydroquinone; the column is washed again with 20 ml of 2 M HNO_3 and neptunium is eluted with 30 ml of 0.1 M oxalic acid, (Fig. 8). The plutonium solution is completely evaporated and mineralized with 7.5 M HNO_3 and H_2O_2 to obtain a white residue. Electroplating of plutonium is carried out from an $(NH_4)_2SO_4$ solution at 500 mA for 5 hours.

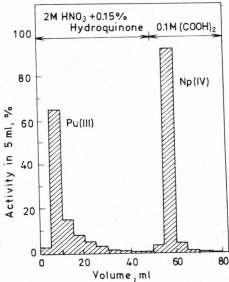

Fig. 8. Separation of plutonium from neptunium by a column of Microthene supporting the neo-tridecano-hydroxamic acid

Neptunium solution is partially evaporated and made 0.25 M in ammonium oxalate
(20 ml); electroplating of neptunium is carried out at 600 mA for 5 hours.

The final yield for plutonium was 73.5% ($\sigma = 3.0$); the medium value of five blank
analyses was 0.078 pCi per 24 hour urinary excretion. The final yield for neptunium was
82.3% ($\sigma = 3.4$) with a blank value of 0.044 pCi per 24 hour urinary excretion.

2.10. DETERMINATION OF AMERICIUM AND CURIUM IN URINE
BY BATCH EXTRACTION WITH MICROTHENE-HDEHP[33]

2.10.1. PRELIMINARY CONSIDERATIONS

Americium and curium also are highly radiotoxic, their MPBB being very similar to
plutonium (0.05 μCi related to bone). The urinary excretion is poor and therefore an
action level of 2 pCi per day was fixed[2,15].

Both elements show a three-valence state and are not extracted by TOPO from acid
solution; on the contrary, the liquid cation exchanger di (2-ethylhexyl) orthophosphoric
acid (HDEHP) can extract americium and curium from an aqueous solution having pH
ranging from 2 to 4[34].

A method was therefore worked out in 1972[33] which is based on a batch extraction of americium from the urine at pH = 3 by means of a Microthene-HDEHP slurry.

For Microthene and the alpha counting see Paragraph 2.8.2. HDEHP was supplied by K and K Lab. (Hollywood, Ca., USA).

2.10.3. ANALYTICAL PROCEDURE AND EXPERIMENTAL RESULTS

50 ml of conc. HNO_3 and 20 ml of H_2O_2 are added to 500 ml of urine and the solution boiled until it is clear. The pH is brought to 3 with NH_4OH (Fig. 9) and 3.5 g of Microthene supporting 2.5 ml of HDEHP 50% in toluene are then added. After stirring for 45 min, the solution is filtered through a G.1 buchner and the slurry transferred to a 1 cm chromatographic column by means of 0.001 M HNO_3. After wasing with 50 ml of 0.001 M HNO_3, americium is eluted with 45 ml of 3 M HNO_3.

The eluate is brought to dryness, and 5 ml of 6 M HCl are added; the solution is transferred into an electrolytic cell containing 0.5 ml of 12 M KOH, and NH_4OH is added to neutrality with methyl red.

The electroplating is carried out for 2.5 hours at 600 mA using a 20 mm i. d. stainless steel disc.

The alpha counting is carried out for 200' with a solid state detector.

The final yield resulted to be 85.9% ($\sigma = 7.6$); the blank value was 0.04 pCi/24 hours; one analysis is carried out within 8 hours.

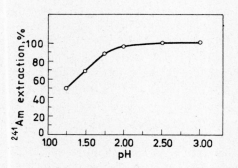

Fig. 9. Extraction of americium by Microthene-HDEHP as a function of pH

2.11. DETERMINATION OF STRONTIUM-90 IN URINE BY A
BATCH EXTRACTION OF YTTRIUM-90 WITH MICROTHENE-HDEHP
AND BETA COUNTING[35]

2.11.1. PRELIMINARY CONSIDERATIONS

Strontium-90 is a highly radiotoxic fission product having a physical half-life of 10^4 days and a biological half-life of $1.8 \cdot 10^4$ days; the MPBB related to the bone is 2.0 μCi and urinary excertion after the first days is small[2], with an action level of 2.0 nCi per day.

The determination of gross beta activity reported in Paragraph 2.6.3. can be used to detect strontium-90 in the urines, but it is not selective at all. Therefore, a method was devised[35] in 1969 to determine only strontium-90 by a batch extraction of its daughter yttrium-90 with Microthene supporting HDEHP. In fact, this extractant is very selective for yttrium at pH 0.5[36]; a 6 M HCl solution can easily be used to strip the element which is successively precipitated as yttrium oxalate and counted in a low background beta detector.

2.11.2. REAGENTS AND EQUIPMENT

For Microthene and the low background beta detector see Paragraph 2.6.2. The HDEHP was supplied by K and K Lab. (Hollywood, Ca. USA).

The chromatographic column had a 25 mm i. d.

2.11.3. ANALYTICAL PROCEDURE AND EXPERIMENTAL RESULTS

To 500 ml of urine 5 mg of stable strontium and yttrium, 25 ml of conc. HNO_3 and 5 ml of H_2O_2 (120 V) are added. The solution is heated gently until it is clear, and some diluted NH_4OH or HNO_3 is added up to pH = 0.5. Then a batch extraction (Table 2) is carried out for 20' with 5 g of Microthene supporting 5 ml of 50% HDEHP in toluen; this step is considered as the zero time for the yttrium-90 decay. The slurry is transferred into the chromatographic column and washed with 150 ml of 0.3 M HNO_3 at 10–15 ml·min^{-1}; finally, yttrium is eluted with 100 ml of 6 M HCl at 5 ml·min^{-1} (Fig. 10).

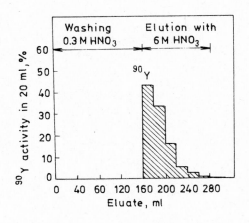

Fig. 10. Elution diagram of yttrium from a Microthene-HDEHP column

299

The solution is concentrated up to 10 ml and 20 ml of 8% oxalic acid are added; then pH is brought to 1.5 with NH_4OH in order to quantitatively precipitate yttrium oxalate.

The oxalate is filtered through a G—4 buchner and washed three times with methyl alcohol; the precipitate is fixed with 1% formwar and counted for 60' in a low background beta detector. The counting is repeated 4—5 times in a week to follow the decay of activity as a function of time (T_f of yttrium-90 = 65 hours). The activity at zero time is referred to yttrium-90 in equilibrium with strontium-90.

A final yield of 92% ($\sigma = 3.5$) and a sensitivity limit of 5 pCi per 24 hour sample were found; the decontamination factors were the following: $> 10^4$ from ^{59}Fe, ^{60}Co, ^{85}Sr, ^{90}Sr, ^{95}Zr, ^{106}Ru, ^{137}Cs; $> 10^3$ from natural uranium and ^{226}Ra; $> 10^2$ from natural thorium, ^{144}Ce, ^{152}Eu; > 30 from ^{147}Pm and ^{210}Pb.

Only 4 hours are necessary to carry out the analysis, if only the first beta counting is taken into account.

3. CONCLUSIONS

3.1. FINAL CONSIDERATIONS ON THE OBTAINED RESULTS

Table 3 summarizes all the radiotoxicological analyses based on extraction chromatograp and reports some technical results. Several techniques are used day by day to control the personnel exposed to risk of internal contamination from radionuclides. About 10,00 analyses have been carried out at CSN Casaccia from 1963, and at present about 100 analyses based on extraction chromatography are performed monthly. The recoveries, the standard deviations and the sensitivity limits are good; this was confirmed also by an international intercomparison on human urines organized by the International Atomic Energy Agency. The decontamination factors are sufficiently high: Table 4 reports the decontamination factors for some actinide elements obtained with the suggested methods The times to carry out the analyses are generally shorter than those implying other techniques.

3.2. ADVANTAGES OF EXTRACTION CHROMATOGRAPHY OVER
OTHER SEPARATION TECHNIQUES

The advantages of extraction chromatography as over liquid-liquid extraction are the following: (a) the possibility of using a greater volume ratio aqueous phase/organic phase; (b) the elimination of foams which often form when a urine sample is shaken with an organic solution (c) the possibility of transforming the liquid-liquid extraction into a column chromatographic procedure and of separating two or three radionuclides by using suitable eluting solutions.

The advantages of extraction chromatography over the ion-exchange resins are: (a) the possibility of having faster exchange rates; (b) the possibility of supporting many different organic extractants having suitable extraction features for a quantitative and selective

300

TABLE 3

Determinations of radionuclides in urine by means of extraction chromatography techniques

Radio-nuclide[a]	Pretreat-ment[b]	HNO$_3$ Molarity in urine	Inert support[c]	Stationary phase[d]	Chroma-tog. pro-cedure[e]	Eluting agent	Final recov-ery, %	Sensitivity limit	Detec-tion system[f]	Time, hrs.	Ref.
Th	W.M.	4.0 M	K.	0.1 M TOPO	C.C.	0.5 M HCl	97.5	1 μg/l	Col.	8	[9]
	W.M.	4.0 M	M.	0.5 M TOPO	B.E.	0.3 M H$_2$SO$_4$	98.2	0.2 μg/l	Col.	4	[17]
U	W.M.	7.5 M	K.	conc. TBP	C.C.	Water	91.1	2 dpm/l	ZnS	4	[12]
	W.M	4.0 M	M.	0.5 M TOPO	B.E.	1 M HF	70.0	1 dpm/l	ZnS	4	[17]
Pu	P.P	2.0 M	K.	1 M TNOA	C.C.	conc. H$_2$SO$_3$	90.5	0.04 pCi/l	SSD	16	[14]
	W.M	4.0 M	M.	0.3 M TOPO	B.E.	6 M HCl+0.01 M HI	70.5	0.07 pCi/l	SSD	8	[25]
	W.M.	2.0 M	M.	0.3 M TOPO	B.E.	6 M HCl+0.1 M HI	76.5	0.10 pCi/l	ZnS	4	[20]
	P.O.	2.0 M	M.	0.3 M HX70	C.C	2 M HNO$_3$+0.15% hydroquinone	73.5	0.08 pCi/l	SSD	16	[29]
Np	W.M.	6.0 M	M.	0.1 M TOPO	E.E.	6 M HCl+ Cl$_2$	83.2	0.05 pCi/l	SSD	8	[26]
	P.P.	2.0 M	M.	0.3 M HX70	C.C.	0.1 M oxalic acid	82.3	0.04 pCi/l	SSD	16	[29]
Am (Cm)	W.M.	0.001 M	M.	1.5 M HDEHP	B.E.	3 M HNO$_3$	85.9	0.05 pCi/l	SSD	8	[33]
Y	W.M.	0.3 M	M.	1.5 M HDEHP	B.E.	6 M HCl	92.0	5 pCi/l	β	4	[35]

[a]Th = Natural thorium; U = enriched uranium; Pu = ^{239}Pu, ^{240}Pu; Np = ^{237}Np; Am = ^{241}Am; Y = ^{90}Y, for the determination of ^{90}Sr

[b]W.M. = Wet mineralization; P.P. = Phosphates precipitation

[c]K. = Kel-F; M. = Microthene-710

[d]TOPO = Tri-n-octylphosphine oxide; TBP = Tri-n-butylphosphate; TNOA = Tri-n-octylamine; HX70 = Neo-tridecanohydroxamic acid; HDEHP = Di(2-ethylhexyl) phosphoric acid

[e]C.C. = Column chromatography; B. E. = Batch extraction

[f]Col. = Colorimetry; ZnS = Alpha counting with a ZnS(Ag) detector; SSD = Alpha counting with a solid state detector after electroplating; β = low background beta counting

301

TABLE 4
Decontamination factors of Pu, Np, Am and Cm from some actinide elements obtained
with the suggested methods

Radio-nuclide	Supported extractant	Th	Pa	U	Np	Pu	Am(Cm)	Ref.
Pu (rapid)	TOPO	30	50	200	15	*	1,000	[20]
Pu	TOPO	100	2,000	600	300	*	4,000	[20]
Pu	HX70	1,000	–	1,000	500	*	200	[29]
Np	TOPO	–	–	30	*	100	10,000	[26]
Np	HX70	1,000	–	1,000	*	20	200	[29]
Am(Cm)	HDEHP	70	60	80	100	100	*	[33]

isolation of an ion; (c) the possibility of changing the quantity and the concentration
of the organic extractant in order to obtain the best conditions of extraction and elution.

REFERENCES TO CHAPTER 10

1. ICRP Publ. n. 2., Pergamon Press, London, 1959.
2. ICRP Publ. n. 10, Pergamon Press, London, 1968.
3. S. JACKSON, G. W. DOLPHIN, Health Physics, 12 (1966) 481.
4. C. TESTA, CNEN Report RT/PROT (65) 33, 1965.
5. C. TESTA, Proc. Conf. on Assessment of Radioactive Contamination in Man, IAEA, Vienna, p. 405, 1972.
6. A. DELLE SITE, G. SANTORI, C. TESTA, Regional Conference on Radiation Protection, Jerusalem, March 1973, to be published.
7. C. TESTA, L. STACCIOLI, Analyst, 97 (1972) 527.
8. C. TESTA, Symp. Det. Radionuclides in Environmental and Biological Materials, London, April, 1973.
9. C. TESTA, Proc. Conf. on Radiological Health and Safety in Mining and Milling of Nuclear Materials, IAEA Vienna, Vol. II, p. 489, 1964.
10. W. Ho, Methods of Radiochemical Analysis, WHO/FAO Report No. 173, p. 95, 1959.
11. J. C. WHITE, W. J. ROSS, USAEC Report NAS–NS 3102, 1961.
12. C. TESTA, G. MASI, Minerva Nucleare, 9 (1965) 22.
13. A. G. HAMLIN, Anal. Chem., 33 (1961) 1547.
14. C. TESTA, Giornale di Fisica Sanitaria e Protezione contro le Radiazioni, 10 (1966) 202.
14. ENEA, Conf. on Radiation Protection Problems Relating to Transuranium Elements, Karlsruhe, September, 1970.
15. W. E. KEDER, J. C. SHEPPARD, A. S. WILSON, J. Inorg. Nucl. Chem., 12 (1960) 327.
16. C. TESTA, D. De ROSA, A. SALVATORI, CNEN Report RT/PROT(68)6, 1968.
17. S. B. SAVVIN, Talanta, 8 (1961) 673.
19. C. A. HORTON, J. C. WHITE, Anal. Chem., 30 (1958) 1779.
20. C. TESTA, G. SANTORI, Proc. XVI. Nat. Congress of Ital. Health Phys. Assoc., Firenze, September, 1972.
21. V. VOLF, see Ref. 14, p. 223.
22. J. C. NENOT, R. MASSE, J. LAFUMA, Health Phys., 2 (1967) 297.
23. W. B. SIELKER, Health Phys., 11 (1965) 965.
24. G. M. COLEMANN, USAEC Report NAS-NS-3058, 1965.
25. C. TESTA, G. SANTORI, Minerva Fisiconucleare, 16 (1972) 1.

26.   G. SANTORI, C. TESTA, J. Radioanal. Chem., 14 (1973) 37.
27.   B. WEAVER, O. E. HORNER, J. Chem. Eng. Data, 5 (1960) 260.
28.   L. B. MAGNUSSON, C. HINDMAN, J. J. La CHAPELLE, The Transuranium Elements, Div. IV, Vol. 14b, McGraw-Hill, N.Y., 1949.
29.   A. DELLE SITE, J. Radioanal. Chem., 14 (1973) 45.
30.   G. GROSSI, CNEN Report RT/CHI(70)15, 1970.
31.   G. GROSSI, CNEN Report RT/CHI(70)27, 1970.
32.   F. BARONCELLI, G. GROSSI, Ital. Patent No. 48741/71, March, 1971.
33.   G. SANTORI, A. DELLE SITE, C. TESTA, to be published.
34.   M. H. CAMPBELL, Anal. Chem., 36 (1964) 2065.
35.   C. TESTA, G. SANTORI, Energia Nucleare, 17 (1970) 320.
36.   D. F. PEPPARD, G. W. MASON, G. L. MAIER, W. DRISCOLL, J. Inorg. Nucl. Chem., 4 (1957) 334.

CHAPTER 11

CHELATING AGENTS AS STATIONARY PHASE IN EXTRACTION CHROMATOGRAPHY

F. ŠEBESTA

1. INTRODUCTION

The first application of chelating agents in extraction chromatography was reported in 1952 from Chemical Research Laboratory, Teddington[1]. Carrit[2], in 1953, described an extraction column with dithizone for the separation and concentration of traces of metals from natural waters. Later, in 1961, Pierce and Peck[3,4] separated indium on the column with the same agent. 2-Thenoyltrifluoroacetone was used in 1962 by Preobrazhenskii and Katykhin[5] for the separation of zirconium and niobium. Some examples of chromatography using undissolved solid chelating compounds loaded on the surface of some supports are available in the literature[6–11], exploiting some retention ability for the ions originally present in aqueous phases put in contact with them. However, the retention of metals studied has to be ascribed to solid-liquid interface phenomena (precipitation), so that they cannot be considered as extraction chromatographic systems, basically involving the partition of elements between two liquid phases.

In the further period mainly neutral and acidic alkylphosphoric extractants were used as a stationary organic phase. This was especially due to the fact that these extractants are liquids and can be used in the pure state without diluents, allowing for very simple preparation of the extraction column.

Chelating agents are mostly solid, and therefore they have to be used dissolved in appropriate organic solvents. In comparison with liquid extractants, the capacity of the stationary organic phase is smaller. After the evaporation of the solvent, precipitation of the solid agent as well as of the metal chelates can occur. Probably for these reasons Stronski[12] stated in his rewiev that chelating agents cannot be considered as suitable extractants for chromatographic separation.

The solvent extraction of metal chelates has some advantages in comparison with other types of extractants. The overall two-phase heterogeneous reaction can be described by relatively simple equations, that relate the distribution ratio D with pH, concentration of the extractant in organic phase, and concentration of masking agents in aqueous phase. Extraction constants of different metal chelates have been compiled[13] which help to develop simple extraction-chromatographic separations. Many metal chelates are coloured and therefore separation can be controlled visually. As a result, chelating agents have been largely used in extraction chromatography for separations related to activation analysis, for concentration of metals from dilute solutions, preparation of radiochemically pure or carrier-free radionuclides.

In the last four years the number of papers dealing with extraction chromatography using chelating agents is steadily increasing. The technique has been improved for the preparation of the extraction column. The reliability was also confirmed for the calculation of the optimum conditions for the separations from the published values of the extraction constants.

Classification of the acidic organophosphorus extractants is rather complicated. In Starýs [14] monograph they are listed between chelating agents. According to Marcus and Kertes[15] these extractants form an independent group, and on this basis they are treated in Chapter 4, in spite of their chelating properties. Alkylphosphoric acids were used most frequently for the extraction-chromatographic separations of lanthanides and actinides, which were described in detail in previous chapters.

2. THEORY OF EXTRACTION BY CHELATING AGENTS

2.1. INFLUENCE OF THE EXTRACTANT CONCENTRATION IN THE ORGANIC PHASE AND INFLUENCE OF THE AQUEOUS PHASE COMPOSITION

The quantitative treatment of the solvent extraction of metal chelates has been given in detail e. g. by Starý[14] and Marcus and Kertes[15]. The extraction of a metal ion M^{m+} with a chelating agent HX, dissolved in an inert organic solvent, can be generally described by the following equation

$$M^{m+}_{aq} + m\ HX_{org} \rightleftharpoons MX_{m,org} + m\ H^+_{aq}$$

The equilibrium constant of the above two-phase reaction is called the extraction constant K_{ex}

$$K_{ex} = \frac{[MX_m]_{org}[H^+]^m_{aq}}{[M^{m+}]_{aq}\ [HX]^m_{org}} \qquad (1)$$

and at a given temperature, the extraction constant depends only on the ionic strength of the aqueous phase. The order of extractability of various metal chelates is characterized by the value $\frac{1}{m} \log K_{ex}$ in Table 1 the extraction constants of metals are listed with typical chelating extractant dissolved in frequently used solvents.

If the metal is present predominantly as the metal ion M^{m+} in the aqueous phase, then equation (1) may be re-written in the form

$$K_{ex} = D \frac{[H^+]^m_{aq}}{[HX]^m_{org}} \qquad (2)$$

where D is the distribution ratio of the metal. Expressing equation (2) in the logarithmic form, we obtain

$$\log D = \log K_{ex} + m\ pH + m\ \log [HX]_{org} \qquad (3)$$

from which it follows that log D is directly proportional to m pH and m $\log[HX]_{org}$.

TABLE 1

Decadic logarithms of distribution constants $K_D(HX)$, dissociation constants K_a and K_H of chelating agents HX and logarithms of extraction constants K_{ex} of metal chelates with 2-thenoyltrifluoroacetone (HTTA), 8-hydroxyquinoline (HOx), N-nitrosophenylhydroxylamine (cupferron, HCup), N-benzoyl-N-phenylhydroxylamine (HBPHA), diphenylthiocarbazone (dithizone, H_2Dz) and diethyldithiocarbamic acid (HDDC) · (20−25°C, I = 0.1) (from Ref[13]).

$\log K_D(HX)$	1.62	2.66	2.18	2.33	4.34	2.39
$\log K_a$	6.23	9.66	4.16	8.15	4.46	3.82
$\log K_H$		5.00		?		
Metal chelate	HTTA C_6H_6	HOx $CHCl_3$	HCup $CHCl_3$	HBPHA $CHCl_3$	H_2Dz CCl_4	HDDC CCl_4
LiX	−10.16	N	N	N	N	N
NaX	−11.16	N	N	N	N	N
KX	−11.16	N	N	N	N	N
CsX	−10.2	N	N	N	N	N
BeX_2	− 3.2	− 9.62	−1.54	E	N	N
MgX_2	−10	−15.3			N	N
CaX_2	−12	−17.89*	N		N	N
SrX_2	−12	−19.7**	N	N	N	N
BaX_2	−14.4	−20.9**	N	N	N	N
AlX_3	− 5.25	− 5.22	−3.5	−7?	N	N
GaX_3	.	3.72	4.92	E	−1.3	P
InX_3	− 4.34	0.89	2.42	−1.74	7.2	12
TlX	− 5.1	− 9.4	P	−7.58	−3.8	−0.5
TlX_3	E	5	3		N	E
PbX_2	− 5.24	− 8.04	−1.53	−8.2	1.0	8.0
AsX_3		N	N		N	E
SbX_3			E		N	E
BiX_3	− 1.9	− 1.2	5.08	5.3	10.8	16.8
CuX_2	− 0.53	1.77	2.66	−0.66	10.4	14.0
AgX		4.5*	P	P	8.9	11.9
ZnX_2	− 8.0	{− 5.2* − 2.4**	P	−9.94	2.3	2.8
CdX_2		− 5.29**	P	−12.06	2.0	5.8
HgX_2		P	0.91	P	26.8	30
ScX_3	− 0.3	− 6.64	3.34		N	N
YX_3	− 6.8	−13.0	−4.74		N	N
LaX_3	−10.51	−15.66	−6.22	−13.59*	N	N
CeX_3	− 9.43			−13.12	N	N
PrX_3	− 9.0				N	N
NdX_3	− 8.76	−14.7			N	N

306

Metal chelate	HTTA C_6H_6	HOx $CHCl_3$	HCup $CHCl_3$	HBPHA $CHCl_3$	H_2Dz CCl_4	HDDC CCl_4
PmX_3	−7.82	−15			N	N
SmX_3	−7.68		−5.8		N	N
EuX_3	−7.66		.		N	N
GdX_3	−7.57			−12.95	N	N
TbX_3	−7.51				N	N
DyX_3	−7.03				N	N
HoX_3	−7.25	−14			N	N
TmX_3	−6.96				N	N
YbX_3	−6.72				N	N
LuX_3	−6.77				N	N
TiX_4		2			N	N
ZrX_4	9.2		>10	E	N	N
HfX_4	7.9		10	12.6	N	N
ThX_4	1.0	−7.12	4.4	−0.65	N	N
VO_2X		$\begin{cases} 1.67 \\ 4.4* \end{cases}$			N	
VOX_2				1.5	N	
MoO_2X_2		9.88		E	N	N
UX_4	5.3	E			N	N
UO_2X_2	−2.6	−1.6*		−3.14	N	N
MnX_2		−9.32			P	−4
FeX_3		4.11	9.85	6 —	N	N
CoX_2	−6.7	−2.16**	−3.5	−0.5*	1.6	2.3
NiX_2	−6.6	$\begin{cases} -2.18 \\ -0.1* \end{cases}$	P	−9	−0.6	E
PdX_2		15			>26	>26
NpX_4	5.6				N	N
PuX_4	6.85			2.95	N	N
AmX_3	−7.48				N	N

*Complexes of the type $MX_m HX$ are formed; **Complexes of the type $MX_m 2HX$ are formed; N—No extraction occurs; P—Only partial extraction occurs; E—Extraction is complete

When the metal chelate is coordinatively unsaturated, an additive complex with the extractant may be formed according to the equation

$$M_{aq}^{m+} + (m + n)\, HX_{org} \rightleftharpoons MX_m\, (HX)_{n,org} + m\, H_{aq}^{+}$$

and in this case the D-value is proportional to the $(m + n)$th power of the concentration of reagent HX in the organic phase.

The equilibrium concentration of the chelating agent HX in the organic phase, for equal volumes of the organic and water phases, can be calculated from the initial concentration c_{HX} of the chelating agent according to the equation[14]

$$[HX]_{org} = c_{HX} / \left\{ 1 + K_D(HX)^{-1}(1 + K_a[H^+]_{aq}^{-1} + K_H^{-1}[H^+]_{aq}) \right\}$$

where $K_D(HX)$, K_a and K_H are defined as follows:

$$K_D(HX) = \frac{[HX]_{org}}{[HX]_{aq}}$$

$$K_a = \frac{[H^+][X^-]}{[HX]}$$

$$K_H = \frac{[H^+][HX]}{[H_2X^+]}$$

Within the pH region $(\log K_D(HX) + pK_H + 2) > pH < (\log K_D(HX) + pK_a - 2)$, the equilibrium concentration of $[HX]_{org}$ is equal to c_{HX}, assuming that $K_D(HX) > 100$ and the total concentration of metal can be neglected in comparison with c_{HX}.

When the metal to be extracted is present in the aqueous phase as different complex species, e. g. hydroxocomplexes $M(OH)_p^{m-p}$, complexes with chelating anions MX_u^{m-u} and other ligands ML_i^{m-i}, then

$$D = \frac{[MX_m]_{org}}{[M^{m+}]_{aq} + \sum_1^p [M(OH)_p^{m-p}]_{aq} + \sum_1^i [ML_i^{m-i}]_{aq} + \sum_1^u [MX_u^{m-u}]_{aq}} =$$

$$= \frac{[MX_m]_{org}}{[M^{m+}]_{aq} \left(1 + \sum_1^p \beta_p [OH^-]_{aq}^p + \sum_1^i \beta_i [L^-]_{aq}^i + \sum_1^u \beta_u [X^-]_{aq}^u\right)}$$

(4)

where β_p, β_i and β_u denote stability constants of the complexes $M(OH)_p^{m-p}$, ML_i^{m-i} and MX_u^{m-u}. Combining equations (1) and (4) one obtains

$$D = \frac{K_{ex}[HX]_{org}^m}{[H^+]_{aq}^m \left(1 + \sum_1^p \beta_p[OH^-]_{aq}^p + \sum_1^i \beta_i [L^-]_{aq}^i + \sum_1^u \beta_u [X^-]_{aq}^u\right)}$$

and

$$\log D = \log K_{ex} + m\, pH + m \log [HX]_{org} - \log(1 + \sum_1^p \beta_p [OH^-]_{aq}^p + ..)$$

(5)

308

In extraction chromatography, the distribution ratio D, for various pH and extractant concentration, can be determined from the relationship

$$D = (V_m - F)/V_{org} \qquad (6)$$

where V_m is the retention volume of the metal (volume of eluent at the maximum of the elution peak of the metal); F is the volume of the mobile aqueous phase in the column (free column volume); V_{org} is the volume of the stationary organic phase on the column support. In this way the extraction-chromatographic behaviour can be compared with the theory of solvent extraction of metal chelates.

In Fig. 1 the dependence on pH is given of the logarithm of D as calculated from chromatographic results for the system zinc—dithizone—carbon tetrachloride—oxalic acid[16]. The values of the extraction constant and stability constants β_1, β_2 of the complexes $Zn(C_2O_4)$ and $Zn(C_2O_4)_2^{2-}$, as calculated from the above results, are in good agreement with the published data. A good agreement was also obtained in the study of the extraction behaviour of copper, indium and thorium on the columns loaded with N-benzoyl-N-phenylhydroxylamine in chloroform[17,18]. From these examples it can be seen that the distribution ratio D, as calculated from equations (3) and (5) can be used for the prediction of optimum conditions for the separation of various metals using extraction chromatography.

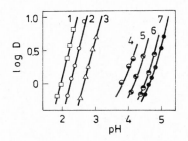

Fig. 1. The plots of log D vs. pH for different dithizone (H_2Dz) concentrations in CCl_4 and for different oxalic acid concentrations in the elution solutions[16].
Curve $1 - 1 \cdot 10^{-3}$ M H_2Dz; $2 - 5 \cdot 10^{-4}$ M H_2Dz; $3 - 2 \cdot 10^{-4}$ M H_2Dz; $4 - 5 \cdot 10^{-4}$ M H_2Dz, $3 \cdot 10^{-2}$ M $C_2O_4H_2$; $5 - 5 \cdot 10^{-4}$ M H_2Dz, $5 \cdot 10^{-2}$ M $C_2O_4H_2$; $6 - 5 \cdot 10^{-4}$ M H_2Dz, $8 \cdot 10^{-2}$ M $C_2O_4H_2$; $7 - 5 \cdot 10^{-4}$ M H_2Dz, $1 \cdot 10^{-1}$ M $C_2O_4H_2$

2.2. INFLUENCE OF KINETIC FACTORS

Being a dynamic process, extraction chromatography is affected by the kinetics of extraction and re-extraction of the metal investigated. However, only scarce information is available about the rate of extraction and particularly re-extraction processes. From the published data it appears that the rate of extraction of metal chelates depends on the rate of their formation in aqueous phase, rather then on the rate of their transfer into the organic phase, and that the rate determining step is the formation of the complex with the first molecule of ligand. It is generally confirmed that the rate of extraction

309

depends on the nature of metal chelate; it increases with increasing pH, concentration of extractant HX, and dissociation constant of extractant K_a. With increasing values of the distribution constant of extractant $K_D(HX)$, the rate of extraction decreases. Hydrolysis and formation of stable complexes of the metal also decrease the rate of extraction.

2.3. REPLACEMENT OF METAL CHELATES

In some cases, e. g. when the stability against decomposition of the chelating agent is low or when a group separation is desirable, a metal chelate MX_m can be advantageously substituted for the free chelating agent in the solution used as the stationary phase in extraction chromatography. In this case the value of $\frac{1}{m} \log K_{ex}$ for the metal chelate used has to be lower than those of the metal ions exchanged.

The pH range in which the metal is not transferred from its chelate in organic phase into the aqueous phase can be derived in analogy to what is done in the case of substoichiometry[19]. However, the resulting exact equation for the dynamic extraction-chromatographic process would be very complicated. For the sake of simplicity one can suppose that the column is eluted with a solution having volume V. In the assumption that more then 99% of metal chelate MX_m remains on the column, the following equations apply:

$$[M^{m+}]_{aq} \leqslant 0.01 \; c_{MX_m} \; (V_{org}/V)$$

$$[HX]_{org} \leqslant 0.01 \; c_{MX_m}$$

$$[MX_m]_{org} = c_{MX_m} - [M^{m+}]_{aq} \; (V/V_{org}) = c_{MX_m} - 0.01 \; c_{MX_m} \doteq c_{MX_m}$$

and combining the above equations with equation (1), the expression for the threshold pH can be obtained:

$$pH \geqslant \frac{1}{m} \log c_{MX_m} - \frac{1}{m} \log K_{ex} - \log 0.01 \; c_{MX_m} - \frac{1}{m} \log 0.01 \; c_{MX_m} \; (V_{org}/V)$$

$$pH \geqslant (2m + 2) \; / \; m - \frac{1}{m} \log K_{ex} - \log c_{MX_m} + \frac{1}{m} \log (V/V_{org})$$

This latter equation is exactly valid only when the dissociation of HX in the aqueous phase can be neglected, and when the metal ion is present in aqueous phase predominantely as ion M^{m+}. The first condition is fulfilled for the pH values[19]:

$$pH \leqslant pK_a + \log K_D(HX) + \log (V_{org}/V)$$

Replacement of the metal M from its chelate by the metal N can be generally described by the following exchange equation:

$$n \; MX_{m,org} + m \; N^{n+}_{aq} \rightleftharpoons m \; NX_{n,org} + n \; M^{m+}_{aq}$$

whose equilibrium constant is

310

$$K_{exch} = \frac{[NX_n]_{org}^m \, [M^{m+}]_{aq}^n}{[MX_m]_{org}^n \, [N^{n+}]_{aq}^m} = \frac{K_{ex}^m (NX_n)}{K_{ex}^n (MX_m)}$$

from which it follows that the replacement of metals depends on the ratio of their extraction constants.

For the quantitative replacement of the metal M from its chelate MX_m by substoichiometric amounts of metal N, the following equation has been derived for the batch extraction[20]:

$$m \log K_{ex} (NX_n) - n \log K_{ex} (MX_m) \geqslant 3 \, m \tag{7}$$

The advantage of replacement extraction chromatography lies in the fact that the quantitative replacement can be reached even when the ratio of the extraction constants is smaller than required by equation (7).

2.4. SYNERGISM IN THE EXTRACTION OF METAL CHELATES

The term synergism denotes the enhancement of the extraction of a metal from aqueous solution by a mixture of extractants in appropriate diluent in comparison with the extraction of the same metal by each of the two extractants separately. Marcus and Kertes[15] distinguish four types of synergistic systems, and two of them involve chelating extractants:

(a) chelating agent—neutral solvating extractant
(b) two chelating agents

The extraction of metal by a mixture of two chelating agents has been hardly studied and no marked synergistic effect has been observed. On their side, the former systems are simple and the experimental data can be easily interpreted: due to the larger synergistic enhancement of extraction, these systems were most frequently studied and were also applied in extraction chromatography. As chelating extractants β-diketones were used, mainly 2-thenoyltrifluoroacetone.

Extraction equilibria in these systems can be described by the following equation:

$$M_{aq}^{m+} + m \, HX_{org} + x \, S_{org} \rightleftharpoons MX_m S_{x,org} + m \, H_{aq}^+ \tag{8}$$

where HX is the chelating extractant, and S is the neutral donor ligand (e. g. organophosphorus compound, alcohol, ketone, heterocyclic base, sulphoxide, amine). The equilibrium constant of equation (8) is given by the extraction constant K_{ex}^x

$$K_{ex}^x = \frac{[MX_m S_x]_{org} \, [H^+]_{aq}^m}{[M^{m+}]_{aq} \, [HX]_{org}^m \, [S]_{org}^x} \tag{9}$$

311

For the metal and chelating agent studied, the value of the equilibrium constant K_{ex}^x depends on the basicity, and in some cases on the structure of the ligand S and on the nature of the diluent. From the K_{ex}^x values it appears that synergistic enhancement increases with the increasing basicity of the donor ligand[15], e. g. in the order phosphate $<$ phosphonate $<$ phosphinate $<$ phosphine oxide. The influence of basicity of the ligand may be depressed in the case of bulky neutral ligands, where for steric hindrance the stability of the synergistic adducts is low.

Extractability of the synergistic adducts is also influenced by the diluent employed. Healy[21] has shown that synergistic enhancement is higher in diluents with lower water solubility (or with lower polarity). The distribution ratio increases in the order chloroform $<$ benzene $<$ carbon tetrachloride $<$ hexane $<$ cyclohexane.

From equation (9) it follows that with increasing concentrations of donor ligand $[S]_{org}$, at constant concentration of chelating agent $[HX]_{org}$ and constant pH, the distribution ratio of metal increases. This is true only when the ligand S is not in great excess as compared to the chelating agent. The excess of neutral donor ligand can produce destruction of synergism. According to Healy and coworkers[21], the amount of water in the organic phase increases when the amount of the neutral ligand in the organic phase is increased. The greater content of water in organic phase is probably due to destruction of species produced by the synergistic effect. Mechanism of destruction of synergism is not yet finalized and it is very likely that it may vary from system to system.

3. EXPERIMENTAL TECHNIQUE

As it was mentioned in the introduction, special attention has to be given to the preparation of columns containing chelating extractants, to avoid the evaporation of the diluent and precipitation of the solid agent as well as of the metal chelates. This phenomenon often was not taken into account, especially in the very first works where granular cellulose or silica gel were used, treated with saturated solution of dithizone in carbon tetrachloride or chloroform[2—4,22]. Diluents were nearly evaporated and impregnated supports were packed into columns. Although the authors noted that great care had to be exercised not to over-dry the support, the columns prepared in that way still contained an excess of the solid reagent. Actually, the results obtained for the distribution of mercury and copper between silica gel treated with dithizone solutions indicate that the solid metal chelates were precipitated[23,24]. Also in the separation of indium from cadmium the metals have been eluted from the column in a sequence which is not in agreement with the corresponding extraction constants[4].

The above-mentioned circumstances account for the unpopularity of chelating agents in laminar chromatography, where the evaporation of volatile organic solvents cannot be avoided.

In this connection, it is also important to draw attention to those studies of synergism in extraction chromatography, where the supports were impregnated only with a mixture of chelating agent and liquid neutral donor ligand, without any diluent: in such cases quantitative comparison can hardly be made between chromatographic behaviour and solvent extraction, where inert diluents are generally used.

The best way for column preparation is the following: the dry support is poured into the column, light tapping on the tube wall being applied in order to achieve a tight and homogeneous filling. The solution of the chelating agent HX is then pipetted onto the top of the column, and the organic phase is immediately covered with the aqueous eluting solution. After opening the column stopcock the solutions are sucked through the column under moderate vacuum, until the organic phase is sorbed onto the whole support. When the diluent used is lighter than water, the solution of HX is first soaked into the column support, and then covered with washing solution. Instead of moderate vacuum, the overpressure can be used too, to reach a high flow-rate. The volume of the organic phase sorbed on the support depends on the flow-rate of the eluent and therefore it is advisable to use higher flow-rates in the preparation of the column to avoid the undesirable loss of the organic phase from the column upon the subsequent elution experiments.

For the preparation of columns with known volume of organic phase, it is necessary to pipette the suitable volume of the HX solution on the column, as determined from preliminary experiments. When the accurate value of V_{org} is not required, an excess of HX solution can be used, the surplus being washed out from the column.

The above mentioned procedure can be applied for supporting materials such as hydrophobized kieselguhr, glass powder or comercial plastics. Only in the case of swelling macro-porous styrene-divinylbenzene the described procedures were different[25-28]: the support was shaken with the solution of dithizone in carbon tetrachloride for 24 hours, the excess of solution was removed and the swelled gel was packed into the column.

Extraction columns can be used several times; their life predominantly depends on the stability of the reagent against decomposition and on the volume of the eluting solutions. After several cycles air bubbles are formed on the upper part of the column bed. To renew the column it is usually sufficient to wash out the organic phase with ethanol or acetone. After drying the supporting material with warm air, a new solution of organic reagent can be sorbed on the cleaned support.

Elution and washing solutions are pre-equilibrated before use with a reagent solution having the same composition of the stationary phase or with the pure solvent. In this way aqueous solutions are saturated of the organic solvent and reagent, and at the same time purified from all trace impurities. Pre-equilibration is especially important when unbuffered solutions are used, e. g. mixture of $NaClO_4$ and $HClO_4$, and when one wishes to know the equilibrium pH value for the calculation of the extraction constant.

To reach a good reproducibility of the experimental results, it is necessary to exclude the influence of kinetic factors. The elution volume and shape of elution curves must not be influenced by the flow-rate of the eluate. The asymmetric shape of the elution curves may often be ascribed to slow kinetics.

The behaviour of the elements studied is also influenced by the loading of the stationary phase with respect to metals. It is generally assumed that a loadings lower than 1% of total capacity of the column does not influence the chromatographic behaviour of the element. If the loading of the column is high, a local decrease may occur of the reagent concentration in the organic phase. As a consequence, the distribution ratio is diminished, and the metal is eluted from the column with smaller retention volume. For even higher loadings of the column, the solid metal chelate may precipitate when the solubility of the

313

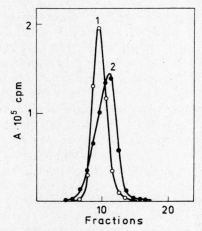

Fig. 2. Dependence of the shape of the elution curves on the loading of the column with copper[17]
1 – Loading 0.06% of theoretical capacity, V_m – F = 20 drops; 2 – Loading 10% of theoretical capa-
ity, V_m – F = 29 drops; Column of $5 \cdot 10^{-2}$ M HBPHA in chloroform supported by kieselguhr; Colu-
dimensions; 5 · 70 mm; Flow rate: 0.46 ml · cm^{-2} · min^{-1}; Volume of fractions: 5 drops

metal chelate is lower in comparison with the concentration of HX used. For example,
a study on the loading of the column with copper was carried-out, using N-benzoyl-N-
phenylhydroxylamine in chloroform[17,18]. In contrast to theoretical expectation, it
was found that the elution volume increased with increasing loading of the column (Fig.
From the shape of the elution curves, it was possible to deduce that these differences
were due to the kinetic factors during the re-extraction of copper. When a very small
flow-rate was used, the elution curve was symmetrical and the elution volume V_m was
smaller; this was explained by the precipitation of solid metal chelate in the organic pha-
at the higher loadings, the rate of re-extraction then also depends on the rate of dissolution of the metal chelate.

4. CHELATING AGENTS AND SYSTEMS USED

4.1. β-DIKETONES

The most important reagent of this group is 2-thenoyltrifluoroacetone (HTTA), which
has found the greatest use in the liquid-liquid separation of actinide elements. HTTA is
freely soluble in a variety of organic solvents. Because of its sensitivity to light, the reagent or its solutions should be stored in the dark. The use of HTTA in extraction chromatography is limited by the slow establishment of extraction equilibria, so that it has
been especially used so far in the studies of synergism and for the selective separation
of neptunium from uranium and fission products.

The use of 0.5 M HTTA in benzene was described by Preobrazhenskii and Katykhin
for the separation of Zr-Nb[5]. Polytetrafluoroethylene (Ftoroplast-4) was used as the

314

support. In the procedure described, zirconium is retained on the column from 6 M HCl, whereas niobium is washed out. Zirconium can be recovered with 2.5 M HF. It was noted that with 0.1 M HTTA in benzene the kinetics of the zirconium extraction is rather slow.

Kawamura et al.[29] studied the chromatographic behaviour of 14 inorganic ions on paper pre-treated with cellulose acetate and impregnated with 0.2 M HTTA in benzene, using acetate buffers with pH ranging from 3 to 6.5. Separations of Pb-Cu, Co-Cu, Zn-Cu and Fe-Co-Ni were performed in this system. Pre-treatments of filter papers were also examined with vaseline or silicone, and silanization or acetylation; these procedures were found to be unsuitable.

The technique of extraction chromatography was successively used for the separation of ^{239}Np and ^{99}Mo from uranium and fission products[30,31]. The procedure described was developed in order to assist in the measurement of some reactor lattice parameters. For the separation, siliconized glass powder supporting 0.5 M HTTA in xylene was used. Neptunium(IV) was selectively extracted on the column from $0.5-1$ M HCl containing hydroxylamine hydrochloride. Uranium(VI) and fission products (with the exception of zirconium) were washed out by 0.5 M HCl + 0.1 M $NH_2OH.HCl$; ^{239}Np was then recovered with 6 M HCl or ethanol. When the separation of ^{239}Np from irradiated enriched uranium was carried-out, the activity of the zirconium isotopes interfered in the radiochemical determination of neptunium and the procedure described has to be improved: zirconium was separated before the reduction of Np(VI) to Np(IV) on the same extraction column[31]. The above mentioned selective separation of ^{239}Np was also adapted for the determination of submicrogram quantities of uranium by neutron activation analysis[32,33].

The separation of alkaline earth metals (Ca, Ba, Sr) was described by Akaza[34], using a column of polytrifluoromonochloroethylene (Kel- F) on which 1.5 M HTTA in methyl isobutyl ketone (MIBK) was supported. Elution was performed with buffers consisting of acetic acid, ammonium acetate and aqueous ammonia. Barium from strontium and strontium from calcium were separated, at pH = 6.5 and 5.5, respectively.

Sweet et al.[35] used dipivaloylmethane as the stationary phase supported by polytrifluoromonochloroethylene (Kel-F) in the research on the separation of the rare earths. 0.5 M Acetic acid-sodium acetate buffer (pH = 6.75) was used as elution solution. Even at very moderate flow-rate of the eluate, the separation of europium from terbium was not quite successful.

Further papers reported the synergistic effect in extraction chromatography. Lee[36] separated lithium from other alkali metal ions on the column of 0.1 M dibenzoylmethane (HDBM) + 0.1 M trioctylphosphine oxide (TOPO) in dodecane or 0.18 HDBM in (1:1) tributylphosphate and dodecane, supported by polytetrafluoroethylene (Haloport F). Great synergistic effect was established especially for lithium. Using the former system, lithium was extracted on the column from 3.2 M NH_4OH and thus separated from heavier alkali metals which were not retained. Lithium was recovered with 0.6 M HCl. Cvjetičanin[37,38] studied the retention of Am, Ce, La, U and Th on paper treated with HTTA or tri-n-octylphosphate (TOP) and with a mixture of both extractants. In agreement with solvent extraction, synergistic effect was established and the composition of species formed in organic phase was determined. Conditions for the separation of ura-

nium from thorium and their mutual separation from some tervalent lanthanides were
also defined. Aly and coworkers [39,40] reported the synergistic effect for europium,
samarium, curium and californium in the system HTTA and dibutyldiethylcarbamoyl
phosphonate, supported by hydrophobized Celite. Separations of Eu-Sm and Cf-Cm were
carried out using nitric acid solution at pH = 1.5. Parameters affecting the separations
were also studied.

4.2. OXIMES

In order to form a chelate ring, the molecule of the oxime must contain another
electron donor group as it is in the case of dioximes or hydroxyoximes. Dioximes are prefer
red for the selective extraction and determination of nickel and palladium. Clingman and
Parrish[41] tested a number of various compounds as selective extractants, using extraction chromatography to eliminate the formation of emulsion. Between the reagent studied, heptadecane-2,3-dione dioxime was also used. They found that at pH $>$ 7 cobalt,
copper and nickel are completely extracted on the column with the polystyrene cross-
linked with 4% of divinylbenzene as the support, swelled with a solution of dioxime in
xylene or in a 1:1 mixture of cyclohexanone and carbon tetrachloride. Nickel and copper were eluted from the column with 2 M HCl or HNO_3. Cobalt could not be eluted
from the column probably because of its oxidation.

α-Benzoinoxime was used for the selective separation of molybdenum. Malvano and
coworkers[42] described the separation of molybdenum from different elements on a
column of powdered polytetrafluoroethylene (Algoflon-F) supporting α-benzoinoxime
in chloroform, for the purpose of activation analysis. The only serious interference was
found to be from tungsten. Yoshida et al.[43] also studied the separation of molybdenum from fission products by solvent extraction and extraction chromatography using
0.2% α-benzoinoxime in chloroform. The separation of $^{95}Zr+^{95}Nb$ from ^{99}Mo is described
^{99}Mo is extracted on the column from 1 M HCl containing oxalic acid. $^{95}Zr + ^{95}Nb$ is
eluted from the column by this solution and ^{99}Mo is recovered by concentrated nitric
acid.

Recently, selective extractants for the recovery of copper from industrial waters were
studied. One of these extractants sold under the trade name Lix-64 was also used in
extraction chromatography. It is believed, that this liquid compound is probably a mixture of 2-hydroxy-5-dodecylbenzophenone oxime and 5,8-diethyl-7-hydroxy-6-dodecano-
ne oxime. Cerrai and Ghersini[44-46] used Lix-64 as the stationary phase in extraction
chromatography on paper and cellulose powder columns. They reported the R_f-spectra
of iron(III), iron(II), copper, cobalt, nickel, manganese(II), chromium(III), vanadium(IV),
molybdenum and tungsten using ammonium sulphate or ammonium chloride solutions
of various pH. These results were utilized for the column separation of copper from som
metals (Fig. 3). The capacity of the columns with Lix-64 was rather high (6 mg of copper per cm^3 of bed). Copper is fully retained from solutions with pH $>$ 2. Even thick
filter paper discs treated with Lix-64 are able to retain copper from remarkably large
volumes (1—2 litres) of aqueous solution. The use of treated thick filter paper discs for
the isolation of copper has found a useful application in the colorimetric determination
of copper in dilute solutions (0.2 — 1 μg/l)[44,46]. Copper is eluted from the paper

316

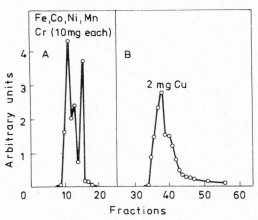

Fig. 3. Separation of copper from iron(II), cobalt, nickel, manganese and chromium on column of
cellulose powder treated with Lix-64[45]
A – 0.5 M $(NH_4)_2SO_4$ at pH = 3; B – 2 M H_2SO_4; Column dimensions: 12·270 mm; Flow rate:
1 ml·min^{-1}; Volume of fractions: 2 ml

discs by 4 M H_2SO_4 and colorimetrically determined using zinc dibenzyldithiocarba-
mate.

Fritz and coworkers [47] examined also two samples of Lix-64, which had different
copper(II) capacities and somewhat different physical properties. Authors decided to
synthethize pure α-hydroxyoximes, 10-hydroxyeicosan-9-one oxime (HEO) and 5,8-diethyl-
7-hydroxy-6-dodecanone oxime (DHDO). Prepared α-hydroxyoximes were used as the
stationary phase in extraction chromatography. They reported R_f-spectra of 32 metal
ions on the paper Whatman 1, impregnated with 10% toluene solution of DHDO. In the
column separations, copper was successfully separated at the pH = 5 from 18 other metal ions on the column of 20% w/w solution each of the oximes in toluene, supported by
inert styrene-divinylbenzene copolymer (Amberlyst XAD-2). Copper was recovered from
the column by 6 M HCl. The method was also used in the determination of copper in
standard samples. Separation of Mo(VI) from each of 10 different metal ions was studied,
using the same extraction column[48]. Molybdenum was selectively extracted on the column from 0.01 N H_2SO_4 and after washing of the column with the same solution, molybdenum was recovered with ammonium hydroxide (pH ~ 12). This method was successfully verified in the analysis of standard steel samples.

4.3. HYDROXAMIC ACID

The most important reagent of this group is N-benzoyl-N-phenylhydroxylamine
(HBPHA), which is a crystalline solid, stable towards heat, light and air. The reagent
is soluble with difficult in water, but freely soluble in chloroform (0.74 M). HBPHA
has found the use as the reagent for the colorimetric and gravimetric determination of
some metals. Recently, the systematic study of the extraction of metals with HBPHA
was performed.

317

N-Oleoylhydroxylamine was tested among other extractants by Clingman and Parrish[4] Using 10% w/v of the reagent in CCl_4 supported by polystyrene-divinylbenzene copolymer, iron(III), aluminium and copper(II) were retained on the column at pH = 4. Nickel and cobalt could be washed out of the column with a buffer at pH = 4, but 2 M nitric acid and 5 M hydrochloric acid were required to elute copper and iron, respectively.

Tserkovnitskaya and Luginin[49] used for the separation of vanadium(V) and (IV) saturated solution of benzhydroxamic acid in tributyl phosphate, supported by polytetrafluoroethylene (Ftoroplast-4). Vanadium(V) was retained on the column from mineral acid solution: 5–0.5 M HCl, $HClO_4$, H_2SO_4 and 10–1 M H_3PO_4. Vanadium(IV) is not extracted under these conditions. Vanadium(V) could be recovered with concentrated H_3PO_4. It is very probably that synergistic extraction of V(V) takes place with mixture of benzhydroxamic acid and tributyl phosphate.

A new derivative of hydroxamic acid, neo-tridecanohydroxamic acid (HX70) has recently been used as the stationary phase in extraction chromatography by Delle Site[50] for the Am-U-Th and Pu-Np separations. 0.3 M HX70 in cyclohexane supported by polyethylene (Microthene-710) was used. The HX70 was proved to be a selective extractant for Pu(IV) and Np(IV) at pH = 0 and therefore selective method fot the determination of ^{237}Np and ^{239}Pu in urine was worked out.

The first evaluation of HBPHA as a reagent for extraction chromatography has been reported by Fritz and Sherma[51]. They studied the chromatographic behaviour of some 35 different metal ions on paper impregnated with HBPHA in 2-octanone, using eluents of various acidity (from 3 M acid to pH = 3). A number of separations can be accomplish in this system.

A detailed study of the use of HBPHA in extraction column chromatography was performed by Láznička and Šebesta[17,18]. The extraction chromatographic behaviour was investigated of copper, indium and thorium on a column of hydrophobized kieselguhr supporting the solution of HBPHA in $CHCl_3$. From the results obtained, the extraction constants K_{ex} of metal chelates were calculated, which were in good agreement with K_{ex} values, determined in batch extraction. From known K_{ex} values, the conditions for the separation of different metals were foreseen. The separations of Zn-Cu, Co-Fe, Co-Cu-Fe, U-Th and Th-Pa were reported. The described procedures were used for the preparation of carrier-free ^{234}Th(UX$_1$) from uranyl nitrate and ^{233}Pa from irradiated thorium.

The separation of metals with strong tendency to hydrolysis is rather complicated. These metals are usually extracted from mineral acid solutions and K_{ex} values are unknown, so that the development of the chromatographic separations can be based only on the data concerning complete extraction and re-extraction of these metals. In this way Caletka[10] achieved the separation of ^{95}Zr - ^{95}Nb (Fig. 4). The possibility of the separation of Hf, Nb, Ta and Pa using HBPHA in $CHCl_3$ was also studied by Pošta and Šebesta[52,53]. It was proved that recommended solutions for re-extraction of niobium and tantalum were unsuitable for the extraction column, consisting of polytetrafluoroethylene powder supporting the solution of HBPHA. The best separations of Hf-Ta, Pa-Ta Pa-Nb, Nb-Ta and Pa-Nb-Ta were achieved using hydrochloric and hydrofluoric acids as eluting solutions. The separation of ^{233}Pa - ^{95}Nb-^{182}Ta is presented as an example in Fig. 5. The readiochemical purity of the separated fractions was checked by measuring of γ-ray

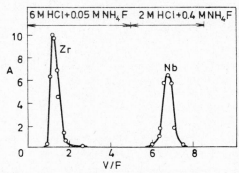

Fig. 4. Separation of ^{95}Zr - ^{95}Nb on column of 0.2% HBPHA in chloroform supported by polytetrafluoroethylene[40].

A – Radioactivity in relative units; V – Volume of effluent; F – Free volume of the column; Column
dimensions: 3·80 mm; Flow rate: 0.2–0.3 ml·cm^{-2}·min^{-1}

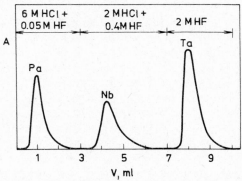

Fig. 5. Separation of ^{233}Pa - ^{95}Nb - ^{182}Ta on column of $1·10^{-2}$ M HBPHA in chloroform supported by
polytetrafluoroethylene[53].

A – Radioactivity; V – Volume of effluent; Column dimensions: 5·100 mm; Flow rate: 0.2 ml·cm^{-2}·min^{-1}

spectra and decontamination factors were better than 10^3 for ^{233}Pa and ^{182}Ta, and
about 10^2 for ^{95}Nb. The separation of pertechnetate and perrhenate using an extraction
column with HBPHA has also been descirbed by the same authors[54]. Technetium was
extracted on the column, whereas rhenium has been washed out with 5 M HClO$_4$. Elution of technetium can be performed with 0.05 M HClO$_4$. The separation of Re-Tc was
applied to the determination of traces of rhenium in molybdenite by activation analysis.

4.4. DIPHENYLTHIOCARBAZONE (DITHIZONE)

Dithizone (H$_2$Dz) is a versatile reagent, used in solvent extraction for the separation
and determination of a number of heavy metals. The reagent is oxidized under weak
oxidizing conditions to diphenylcarbodiazone and therefore it has to be purified before
use. H$_2$Dz has been used relatively frequently in extraction chromatography, in spite of

its disadvantages: low solubility of reagent and metal dithizonates in organic solvents, formation of two types of chelates.

Carrit[2] used a chromatographic column packed with H_2Dz in CCl_4 supported by granular cellulose acetate for the concentration and separation of lead, zinc, manganese(II), cadmium, cobalt and copper from natural waters. At pH = 7, the complete recovery of the above mentioned metals was obtained, even from 10 l samples. Metal dithizonates were eluted from the column with 1 M hydrochloric acid or concentrated ammonia, and concentration factors up to 10^3 were achieved.

Pierce and Peck[3], and Mapper and Fryer[22] used columns packed with cellulose acetate treated with H_2Dz in (1:1) CCl_4 and $CHCl_3$. After the irradiation and dissolution of samples, [116]In was retained on the column at pH = 5. Column was then washed with a 5% w/v solution of the disodium salt of ethylenediaminetetraacetic acid in water, and by 0.01 M $HClO_4$. [116]In was recovered with 1 M HCl. A column containing H_2Dz in CCl_4 or $CHCl_3$ on silica gel was used to separate [115]In from parent [115]Cd after their sorption on the column at pH = 5. [115]In was eluted by 0.002 M $HClO_4$, whereas [115]Cd remained on the column[4].

Spěvačková and Křivánek[25–28] studied the applicability of extraction chromatography using macroporous styrene-divinylbenzene copolymer swollen with H_2Dz in CCl_4. In preliminary experiments, the extraction constant K_{ex} of zinc dithizonate was determined from the elution volumes V_m accordint to equations (2) and (6). The best agreement with K_{ex} values determined in batch extraction was reached when the volume of swelled gel (support + diluent + dithizone) was taken as the volume of the stationary organic phase. Although the authors explain these results with the formation of "inner solution of dithizone in gel", other possible explanations of the results can be given.

The above described extraction column was used for the determination of copper, zinc and iron in samples of mussels by activation analysis[26,27]. Samples irradiated for 10 hrs were immediately dissolved and transferred into small volume of acetate buffer at pH = 4.9, which was poured on to the column. After washing of the column with the same buffer, [69m]Zn was eluted with 0.1 M HCl and ^{64}Cu was estimated measuring γ-ray spectra of the column bed. Samples irradiated for 80 hrs and cooled for one week were used for the determination of iron and zinc. After dissolution, they were transferred into a small volume of acetate buffer (pH = 4.9) and poured into the column. [59]Fe was eluted with the same buffer, and [65]Zn was again recovered with a solution of hydrochloric acid. The results obtained were in good agreement with data given by IAEA. The above mentioned authors[27,28] also described a method for the determination of low gold content in platinum. Since the reaction [197]Au(n, γ) [198]Au was employed, the production of [198]Au and [199]Au from platinum interfered: for this reason, gold(III) was first isolated from the sample by retention on the column from 0.1 M HCl; platinum(IV) is not extracted at this condition. The column, made from a polyethylene tube, was then sealed and irradiated. The same procedure was applied to standard samples and to a blanc, and by measuring the activity of [198]Au in the column beds, the content of gold was estimated. In further experiments, the group separation on the column was studied with solution of lead dithizonate[26,27]. A column with dithizone was also applied for the removal of ^{60}Co from the preparation of [55+59]Fe, which may contain ^{60}Co as radiochemical impurity[25]. For the conditions given, Co(II) forms a stable dithizonate, whereas Fe(II) can be quantitatively recovered from the column.

The chromatographic behaviour of zinc was also studied on a column made of dithizone in carbon tetrachloride supported by hydrophobized kieselguhr[16]. Figure 1 shows the dependence of log D on pH for different dithizone concentrations in organic phase (curve 1,2,3) and for different concentrations of oxalic acid in elution solutions (curve 4,5,6,7). In full agreement with equation (3), the slope of the curves log D = f(pH) in absence of oxalic acid were equal to two. The difference between $pH_{1/2}$ (D = 1) for 10^{-3} M and $2 \cdot 10^{-4}$ M $H_2 Dz$ solutions was 0.7 units as it was theoretically expected. The determined values K_{ex} and stability constants of zinc oxalate complexes were in agreement with the published data, demonstrating the possibility of predicting the extraction-chromatographic separations of various metals from known values of extraction and stability constants.

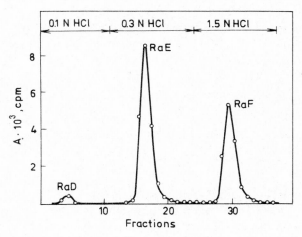

Fig. 6. Separation of RaD(^{210}Pb) - RaE(^{210}Bi) - RaF(^{210}Po) on column of $5 \cdot 10^{-4}$ M $H_2 Dz$ in carbon tetrachloride supported by kieselguhr[55].

Column dimensions: 5·70 mm; Flow rate: 0.5 ml · cm^{-2}·min^{-1}; Volume of fraction: 5 drops

Again, using $H_2 Dz$ in CCl_4 supported by kieselguhr, the separations of Zn-Cd, Ag-Hg, Cd-Ag and Pb-Bi-Po were reported[55]. For example (Fig. 6), the separation of RaD(^{210}Pb)-RaE(^{210}Bi)-RaF(^{210}Po) was performed using different hydrochloric acid solutions and $5 \cdot 10^{-4}$ M $H_2 Dz$ in CCl_4. The applicability of an extraction column for the radioanalytical determination of mercury was also studied[56]. The aim of the experiments was to employ the heterogeneous isotope exchange between a solution of ^{203}Hg(HDz)$_2$ in CCl_4 supported by kieselguhr and mercury ions Hg^{2+} in aqueous sulphuric acid solutions. The results were not quite satisfactory, probably because of uncontrolled sorption of mercury on the walls of the vessels and also on the supporting material.

The quick spectrophotometric determination of mercury in biological samples using extraction column with $H_2 Dz$ was described by Woidich and Pfannhauser[57]. In the method proposed, the mercury is extracted on the column with 3% $H_2 Dz$ in $CHCl_3$ from sulphuric acid solutions. The excess of dithizone is washed from the organic phase with a solution of sodium hydroxide. Mercury dithizonate is treated with 10% acetic acid and recovered from the column by chloroform. The absorbancy of $Hg(HDz)_2$ in $CHCl_3$ is measured at 483 nm and corrected for blanc.

Ueno et al.[58] described the separation of mercury from zinc, cadmium and lead on the column of styrene-divinylbenzene copolymer impregnated with 0.3% H_2Dz in chlorobenzene. All elements studied were extracted at pH ~ 5.2 on the column and Zn, Cd and Pb were recovered from the column with 0.1 M HCl.

4.5. DIETHYLDITHIOCARBAMIC ACID

Diethyldithiocarbamic acid (HDDC) is readily soluble and extractable by organic solvents. It is very unstable even in weakly acidic media and it is therefore of limited use in acid solution: actually the rate of its decomposition is directly proportional to the hydrogen ion concentration. However if the reagent is dissolved in an organic solvent, its stability is much higher. From the K_a and $K_D(HX)$ values it is apparent that, at a pH lower than 4, more than 99% of HDDC is present in the carbon tetrachloride phase; conversely, it is transferred into aqueous phase at pH higher than 8. HDDC can be stripped from the organic phase into solution of sodium hydroxide, as sodium diethyldithiocarbamate (NaDDC). NaDDC is a crystalline compound soluble in water and only slightly soluble in organic solvents.

Various metal diethyldithiocarbamates or ammonium salt (DDDC) can advantageously substitute HDDC. Solutions of DDDC in organic solvents are rather stable and they can be used for the extraction of metals even from acid medium.

Alimarin and coworkers[59–61] have worked-out the method for the substoichiometric displacement of various metals on a column prepared with zinc diethyldithiocarbamate $(Zn(DDC)_2)$ in chloroform, supported on porous polytetrafluoroethylene (Ftoroplast PF-4). The method was used for the radioactivation determination of zinc, copper, cadmium, silver, mercury, manganese, cobalt and iron traces in different matrices, namely yttrium, molybdenum, zirconium, niobium and beryllium. The sensitivity of the determination was $10^{-4}-10^{-8}\%$ for 0.1–1 g samples irradiated for 20 hrs with a neutron flux of $1.2 \cdot 10^{13} n \cdot cm^{-2} \cdot sec^{-1}$. Among the many chelating agents studied (dithizone, cupferron, 8-mercaptoquinoline and sodium diethyldithiocarbamate), NaDDC was found to be the most suitable. For the determination of zinc, the heterogeneous isotope exchange between $Zn(DDC)_2$ in chloroform and radioactive zinc isotopes in the aqueous phase (pH = 6–7) was studied, and the conditions for the isotope exchange were determined.

In the recommended procedure, metals to be determined were separated from the dissolved matrix, after the addition of carriers, by extraction as diethyldithiocarbamates into $CHCl_3$. The organic phase was evaporated and, after mineralization, the metals were transferred into 0.5 ml of water (pH = 6–7). This solution was pipetted on the top of the column with $Zn(DDC)_2$ in chloroform. Cadmium, copper, silver and mercury displace zinc in the upper part of the column and zinc is distributed oven the whole column due to the isotopic exchange. Manganese, cobalt and iron pass through the column and are determined in the effluent. Zinc is then eluted by a substoichiometric amount (50%) of cadmium.

After washing of the column with water, the excess of cadmium (25%) is passed through the column to remove all zinc; the column is again washed with water. In the identical way the displacement of cadmium by substoichiometric amount of copper, copper by substoichiometric amount of silver and silver by substoichiometric amount of mercury was carried-out. Mercury was determined after the quantitative elution from the column by 1 M KI solution at pH = 1.

Thanks are due to Professor Jiří Starý for his interest and valuable criticism of the manuscript.

REFERENCES TO CHAPTER 11

1. Chemical Research Laboratory, Teddington, Nature, 170 (1952) 650.
2. D. E. CARRIT, Anal. Chem., 25 (1953) 1927.
3. T. B. PIERCE, P. F. PECK, Analyst, 86 (1961) 580.
4. T. B. PIERCE, P. F. PECK, J. Chromatography, 6 (1961) 248.
5. B. K. PREOBRAZHENSKII, G. S. KATYKHIN, Radiokhimiya, 4 (1962) 536.
6. H. ERLENMAYER, H. DAHN, Helv. Chim. Acta, 22 (1939) 1369.
7. F. BURRIEL-MARTI, F. PINOPEREZ, Anal. Chim. Acta, 3 (1949) 468.
8. A. M. GURVICH, Zh. Anal. Khim., 11 (1956) 437.
9. P. F. ANDREEV, L. T. DANILOV, G. O. KOSHISHYAN, Zh. Prikl. Khim. 34 (1961) 2419.
10. R. CALETKA, Chemické listy, 62 (1968) 669.
11. E. A. KOGAN, D. Ya. EDVOKIMOV, Ukr. Khim. Zh., 34 (1968) 1089.
12. I. STRONSKI, Österr.Chem. Ztg., 68 (1967) 5.
13. J. STARÝ, H. FREISER, Equilibrium Constants of Liquid-Liquid Distribution Reaction, Part III, The Chemical Society London, in preparation.
14. J. STARÝ, Extraktsiya Khelatov, Mir, Moscow, 1966.
15. Y. MARCUS, A. S. KERTES, Ion Exchange and Solvent Extraction of Metal Complexes, Wiley-Interscience, London, 1969.
16. F. ŠEBESTA, J. Radioanal. Chem., 6 (1970) 41.
17. A. DAŇKOVÁ, Thesis, Faculty of Nuclear Science and Physical Engineering, Prague, 1971.
18. F. ŠEBESTA, A. LÁZNÍČKOVÁ, J. Radioanal. Chem., 11 (1972) 221.
19. J. RŮŽIČKA, J. STARÝ, Substoichiometry in Radiochemical Analysis, Pergamon Press, London, 1968.
20. J. STARÝ, J. RŮŽIČKA, Talanta, 18 (1971) 1.
21. T. V. HEALY, Solvent Extraction Research, p. 257, A. S. KERTES, Y. MARCUS, Eds. Wiley-Interscience, London, 1969.
22. D. MAPPER, J. R. FRYER, Analyst, 87 (1962) 297.
23. T. B. PIERCE, Anal. Chim. Acta, 24 (1961) 146.
24. T. B. PIERCE, P. F. PECK, Anal. Chim. Acta, 26 (1962) 557.
25. V. SPĚVÁČKOVÁ, M. KŘIVÁNEKí Radiochem. Radioanal. Letters, 3 (1970) 63.
26. V. SPĚVÁČKOVÁ, M. KŘIVÁNEK, Proceedings of the III. Anal. Chem. Conference, Budapest, p. 121, 1970.
27. V. SPĚVÁČKOVÁ, Thesis, Nuclear Research Institute, Řež near Prague, 1972.
28. V. SPĚVÁČKOVÁ, M. KŘIVÁNEK, J. Radioanal. Chem., in press.
29. S. KAWAMURA, T. FUJIMOTO, M. IZAWA, J. Chromatography, 34 (1968) 72.
30. J. A. THOMASSEN, H. H. WINDSOR, Kjeller Report, KR−44, 1963.
31. D. C. PERRICOS, J. A. THOMASSEN, Kjeller Report, KR−83,1964.
32. D. C. PERRICOS, E. P. BELKAS, Talanta, 16 (1969) 745.
33. H. WEHNER, Al−MURAB, M. STOEPPLER, Radiochem. Radioanal. Letters, 13 (1973) 1.
34. I. AKAZA, Bull. Chem. Soc. Japan, 39 (1966) 980.

35. T. R. SWEET, U. S. Air Force Report ARL 68–0210, 1968.
36. D. A. LEE, J. Chromatography, 26 (1967) 342.
37. N. CVJETIČANIN, J. Chromatography, 34 (1968) 520.
38. N. CVJETIČANIN, ve Symp. Chromatog. Electrophorese, Bruxelles, p. 186, 1969.
39. H . F. ALY, M. A. El–HAGGAN, Radiochem. Radioanal. Letters, 3 (1970) 249.
40. H. F. ALY, M. RAIEH, Anal. Chim. Acta, 54 (1971) 171.
41. A. L. CLINGMAN, J. R. PARRISH, J. Appl. Chem., 13 (1963) 193.
42. R. MALVANO, P. GROSSE, M. ZANARDI, Anal. Chim. Acta, 41 (1968) 251.
43. H. YOSHIDA, C.YONEZAWA, K. GUNJI, Japan Analyst, 19 (1970) 813.
44. E. CERRAI, G. GHERSINI, Second SAC Conference, Nottingham, 1968.
45. E. CERRAI, G. GHERSINI, Analyst, 94 (1969) 599.
46. E. CERRAI, G. GHERSINI, National Analytical Conference, Ljubljana, 1972.
47. J. S. FRITZ, D. R. BEUERMAN, J. J. RICHARD, Talanta, 18 (1971) 1095.
48. J. S. FRITZ, D. R. BEUERMAN, Anal. Chem., 44 (1972) 692.
49. I. A. TSERKOVNITSKAYA, V. A. LUGININ, Vestn. Leningradskogo Universiteta, Ser. Fis. Khim., (2) (1969) 162.
50. A. DELLE SITE, J. Radioanal. Chem., 14 (1973) 45.
51. J. S. FRITZ, J. SHERMA, J. Chromatography, 25 (1966) 153.
52. S. POŠTA, Thesis, Faculty of Nuclear Science and Physical Engineering, Prague, 1972.
53. F. ŠEBESTA, S. POŠTA, Radiochem. Radioanal. Letters, 14 (1973) 183.
54. F. ŠEBESTA, S. POŠTA, Z. Randa, Radiochem. Radioanal. Letters, 11 (1972) 359.
55. F. ŠEBESTA, J. Radioanal. Chem., 7 (1971) 41.
56. A ZEMAN, F. ŠEBESTA, unpublished results.
57. H. WOIDICH, W. PFANNHAUSER, F. Z. Anal. Chem., 261 (1972) 31.
58. K. UENO, T. YANO, T. KOJIMA, Anal. Letters 5 (1972) 439.
59. Yu. V. YAKOVLEV, O. V. STEPANETS, Zh. Analit. Khim., 25 (1970) 578.
60. O. V. STEPANETS, Yu. V. YAKOVLEV, I. P. ALIMARIN, Zh. Analit. Khim., 25 (1970) 1096.
61. I. P. ALIMARIN, Yu. V. YAKOVLEV, O. V. STEPANETS, J. Radioanal. Chem., 11 (1972) 209.

CHAPTER 12

USE OF EXTRACTION CHROMATOGRAPHY
FOR TRACE METAL PRECONCENTRATION AND SEPARATION

I. P. ALIMARIN, T. A. BOLSHOVA

1. INTRODUCTION

Analytical chemists often face the problem of determining extremely small amounts of impurities over high background levels of the main substance. When the sensitivity of direct methods for trace analysis in the presence of macrocomponents is insufficient, preconcentration of the trace impurities is necessary to increase the sensitivity of their determination.

In analytical chemistry, a number of techniques are available for the preconcentration of trace amounts of various elements, the most important being coprecipitation, ion-exchange and liquid-liquid extraction. One of the most widely used methods is extraction, which may yield a fairly high enrichment. However, liquid-liquid extraction techniques do not usually provide a sufficiently high degree of absolute preconcentration. The preconcentration of small amounts of impurities by ion-exchange has been little used for separating microtraces from large amounts of bulk material. In many cases, this type of analytical problem can be solved by the method of extraction chromatography.

Extraction chromatography is suitable for the separation of elements with closely similar properties, and is finding expanding application in various fields of radiochemistry, analytical chemistry and nuclear technology. It is doubtless superior to single-step liquid-

liquid extraction, as this repetitive technique permits preconcentration of trace elements with a very high enrichment coefficient as referred to the ratio of the concentration of the trace element of interest and that of its matrix. This is of particular importance in the analysis of the high-purity materials applied in the electronics industry. There are numerous publications dealig with the study of this method, and Refs. [1–6] are some examples. It has often been pointed out that an efficient separation of components with very similar distribution coefficients can be achieved by repeated separation on a chromatographic column, even if this objective is not obtainable by liquid-liquid extraction under static conditions. For the same reason, the separation of complex compounds with similar stability constants is also possible.

Extraction chromatography has mainly been applied to the quantitative separation of components present at comparable levels, but its multi-step character also allows the separation of components present in micro to macro ratios, with a simultaneous absolute preconcentration of the microcomponent.

Combined with various physical and physico-chemical methods, it ensures a considerable enhancement of the sensitivity of trace element analysis, and in the authors'opinion, preconcentration of trace elements is one of the most interesting possibilities offered by this technique. However, extraction chromatography has only rarely been applied to such problems, in spite of the very promising results reported by some authors. For example, the effective preconcentration of trace amounts of platinum in nickel and of manganese(II) in permanganate have been reported[7,8] on chromatographic columns consisting of Ftoroplast-4, with an enrichment factor of 10^4. The present authors succeeded[2] in achieving even higher enrichment factors, of the order of 10^7-10^8.

During the past several years, a number of papers have appeared on the development and improvement of the general technique of extraction chromatography. For example, the possibility of separating multicomponent mixtures by liquid-liquid chromatography has been surveyed[9]. Many authors consider this method indispensable when it is necessary to achieve maximum chemical purity of the products together with a high degree of enrichment.

A novel direction is the application of extraction displacement chromatography based on the different stabilities of metal chelate complexes. The authors have used the exchange reaction of diethyldithiocarbamate dissolved in chloroform and loaded on Ftoroplast-4 [10]: this allowed the preconcentration of trace metal components in the activation analysis of pure yttrium and molybdenum. Numerous other metal exchange reactions of chelate complexes dissolved in organic solvents may also be employed[11,12]. Extraction chromatography utilizing chelate compounds has been dealt with in several papers [13,14], and is treated in detail in Chapter 11.

2. EXPERIMENTAL TECHNIQUES

The theoretical basis of extraction chromatography is discussed in the preceding chapters of this book, and also in Ref. [15]. As known from theoretical considerations, the partition coefficients of dissolved substances depend on the natures of the interactions between the individual components and the mobile and stationary phases. The success of the choice of the optimum conditions for chromatographic preconcentration of the impurities is therefore decisively influenced by the properties of the solvents to be used. Full exploitation of the properties of the stationary and mobile phases, however, is seriously limited by the number of high-quality supports available.

2.1. SUPPORTS

The supports applicable in extraction chromatography should meet the following requirements:

1. They should be inert, with practically no tendency to adsorb the substances to be separated or the mobile phase, and no chemical reaction should occur on their surface.

2. The supports should bind the organic phase strongly to their surface, without being soluble in the solvents used.

Let us briefly consider the most common supports applied in column chromatography for the individual or collective preconcentration of trace impurities.

Beginning from 1959, a number of papers have appeared, suggesting the use of hydrophobic supports. Such materials can retain the organic solvent, and this results in a stationary phase present on the surface of the support in the free state, and in permanent contact with the mobile aqueous phase.

The separation is thus purely of the extraction type with no interference from other processes.

The requirement of a purely liquid-liquid extraction mechanism in the column is due to the fact that in such cases the procedure for chromatographic separation and preconcentration can be based on available literature data referring to the extraction of elements. This possibility is limited because of the lack of high-quality supports. The most widely employed supports for organic phases are hydrophobic silica gel, alumina, polytrifluorochloroethylene (Kel-F) and polytetrafluoroethylene (Ftoroplast-4, Teflon). Synthetic ion-exchangers are also to be regarded as potentially useful supports in partition chromatography. However, the styrene-divinylbenzene copolymer, which swells in organic solvents, has been found to change the properties of the absorbed solvent, thereby influencing the extraction process.

In the following, a brief characterization of some supports will be given.

2.1.1. SILICONIZED SILICA GEL

This support is often applied by the present authors for individual or collective preconcentration. To prepare the support, commercial silica gel is milled, the grain fraction with a diameter of 0.08 mm is washed free of impurities, then treated with dimethyldichlorosilane and dried at 100 °C. The hydrophobic silica gel obtained by this treatment is carefully mixed with the organic solvent required for the chormatographic process and introduced into the column. Such columns have been used for the reversed-phase separation of impurities from rare-earths, niobium, gallium, indium, etc. The capacity of hydrophobic silica gel for TBP and TNOA is 0.6—0.7 g/g.

2.1.2. FTOROPLAST-4 (POLYTETRAFLUOROETHYLENE)

The application of this support in extraction chromatography has proved to be very efficient. The rigid structure and high porosity allow the uniform filling of the column, which ensures a high number of theoretical plates in the chromatographic process [16,17]. The chemical stability of this polymer is superior to that of all other known materials: it is resistant to aggressive media except molten alkalies and molecular fluorine, and swells only slightly in organic solvents. The organic phase covers the powdered Ftoroplast-4 as a film, which ensures rapid distribution of the given substance between the stationary phase and the mobile aqueous solution. Polytetrafluoroethylene is considered to be the best organic phase support in partition chromatography, although its capacity is somewhat lower than that of hydrophobized silica gel, not exceeding 0.3–0.4 g/g for TBP and TNOA.

Ftoroplast has been successfully used to separate trace amounts of some elements (Se, Fe, Ga, In, etc.) from relatively large quantities of other elements with closely similar chemical properties.

2.2. STATIONARY AND MOBILE PHASES

Together with that of a suitable support, the selection of the proper liquid phases is of paramount importance in chromatographic analysis. They should be selected in such a manner that the distribution coefficeints of the impurities differ by a factor of 10 to 100 from those of the bulk constituent. The extraction system should conform to a number of requirements:

1. The substances to be separated should be readily extracted into the stationary phase, but their distribution coefficients should be different; the distribution coefficient of the microimpurity to be preconcentrated should be higher than that of the bulk constituent.

2. Both phases should be mutually saturated, so as not to change the composition or amount during the chromatographic process.

3. There should be no interaction between the liquid phases and the support.

Various extractants can be applied as the stationary phase for the separation and preconcentration of trace elements; tributyl phosphate (TBP)[18-21], di (2-ethylhexyl) phosphoric acid[7, 22, 23] and other organophosphorus compounds[24-26], diethylsulfoxide[27], high-molecular weight amines[28-30] and methyl isobutyl ketone[31]. It is worth pointing out that preconcentration by column chromatography can also be performed using solvents that form stable emulsions[24].

The experimental set-up for the process is not more complex than that applied in liquid-liquid extraction.

The columns employed by the present authors for the separation and preconcentration of microimpurities are of the usual type, with a length of 10–15 cm and an internal diameter of 0.4–0.8 cm. Before the column is packed with the solid support, its bottom part is filled with a thin layer of glass wool.

Microcolumns are sometimes used, with lengths of 5–10 cm and internal diameters of 0.2—0.3 cm. It should be noted that the separation processes performed on micro-columns at atmospheric pressure require a longer time, owing to the low flow rates through the column. The flow rate of the mobile phase can be increased by operating the column at elevated (2—3 atm) or reduced pressure[15], which leads to a 2- or 3-fold decrease in time requirement.

One of the most delicate operations involved in the chromatographic separation of microamounts is the filling of the column.

This can be done by several techniques. The column may be packed with the solid support, specially prepared for chromatographic purposes, and subsequently loaded by elution with the extractant selected as the stationary phase. Alternatively, the support may previously be mixed with the extractant, the mixture being transferred into the column. In all cases, however, it is imperative that a homogeneous column be formed without air bubbles or channels. For this purpose, the support in the column should be compacted by carefully tapping or by percolating an aqueous phase under slight pressure. After this, the filled column is washed with the mobile phase saturated with the extractant.

The packing of the support in the column is of decisive importance for determining the shape of the elution curves. If the support layer is not sufficiently compact, the elution curves tail off. On the other hand, too compact packing reduces the flow rate through the column, or even makes flow impossible. In some cases the flow through the column is assisted by slight pressure or vacuum.

After suitable preparation of the column, the solution containing the elements to be separated is brought onto the upper layer by means of a pipette with a drawn-out end. There is another technique of introducing the solution into the column: the sample solution is mixed with a small amount of the dry support, and the mixture is quantitatively transferred onto the column bed. The column is then eluted with the mobile phase.

The solution leaving the column is collected in portions of a given volume, in which the concentration of the element coming off the column is determined by a suitable physico-chemical method. Narrow peaks are usually obtained.

The flow rate of the mobile solvent through the column is of great significance, as it determines the time of contact between the stationary and mobile phases. At higher flow rates the partition equilibrium is sometimes not reached and no sharp separation is obtained. If the flow rate is too low, the time requirement of the analysis increases. It is sometimes necessary to perform preliminary experiments, therefore, to study the kinetics of distribution of the various elements in the column and establish the optimum flow rate. This usually ranges between 0.5 and 1.5 ml·min^{-1}, so that separations are obtained in a relatively short period of time (2—3 hr).

3. INDIVIDUAL AND COLLECTIVE PRECONCENTRATION AND SEPARATION OF TRACE IMPURITIES

Theoretical considerations as to the mechanism of the distribution of the elements and their different abilities to interact with the stationary phase in the column allow one to find the optimum conditions for preconcentration and separation from the bulk component.

As already mentioned, the preconcentration of trace impurities is one of the most interesting applications of extraction chromatography, in particular when combined with physico-chemical methods of analysis, such as mass-spectroscopy, activation analysis, atomic absorption spectroscopy, emission spectroscopy, spectrophotometry, luminescence and polarography.

From this viewpoint extraction chromatography has special advantages, since the separated elements are usually obtained in small volumes of solution that can be directly analyzed, so that the technique often appears the simplest and fastest way to solve the problem of both individual and collective preconcentration of microimpurities.

In this section, a number of examples are given of such applications.

Kuzmin and Meshchankina[35] carried out the group preconcentration of the elements occurring as impurities in water and $CuCl_2$ solutions. The stationary phase was a 0.1 M solution of 8-hydroxyquinoline in a mixture of carbon tetrachloride and isopentanol supported on Ftoroplast-4.

In order to increase the sensitivity of the determination of microimpurities in gallium arsenide, the above authors[36] performed collective preconcentration of the traces on a chromatographic column containing decanol as the stationary phase. The sensitivities for Ga, Fe(III), Sn(IV) and In were $10^{-4} - 10^{-6}\%$, referred to the individual elements in 1 mg samples.

In the activation analysis of lutetium in ytterbium, and of holmium in dysprosium[37], the impurities were preconcentrated on a column of hydrophobized silica gel loaded with di(2-ethylhexyl) phosphoric acid.

The neutron activation analysis of the impurities in antimony[38] and in niobium and its compounds[39] was performed by means of preconcentration using extraction chromatography. Enrichment factors obtained were 10^9 for antimony, and $7 \cdot 10^5 - 1.10^6$ for niobium and its compounds. The results showed that the method ensures practically complete separation of the microcomponents in a single chromatographic step.

The system TBP—HCl has been successfully applied to the separation and preconcentration of 19 elements from large amounts of iron[40].

Cerrai and Ghersini[41] suggested an original method for the determination of extremely small amounts of copper in water, based on chromatographic preconcentration using a selective extractant for copper, and subsequent spectrophotometric analysis with zinc dibenzyl dithiocarbamate. The extractant used, Lix—64, is described in Ref.[42]. The authors suggest the use of a thick filter paper loaded with the extractant as support. The proposed technique allows the determination of $0.2-8 \mu g/1$ copper without interference from various ions present in water.

Other authors have applied extraction chromatography to separate[43] traces of cobalt from nickel in 1 g samples containing these elements in the ratio of 1 to $2 \cdot 10^7$, of traces of iron from manganese[44] at a ratio of 1 to 10^4, and of plutonium from uranium[46] at a ratio of 1 to 10^6. The preconcentration of platinum traces in a nickel matrix and manganese(II) traces in $KMnO_4$ was carried out in chromatographic columns filled with Ftoroplast-4 tablets[7,8]; the enrichment factor achieved was $1 \cdot 10^4$.

The use of extraction chromatography has been proposed for the collective preconcentration of Cu, Zn, Cd, Ba, Ga, In, Tl, Ge, Sn, Pb, Cr, Mn, W, Mo, Fe, Co and Ni in the analysis of high-purity tantalum[31]. Ftoroplast-4 was used as the support, 100% TBP as the stationary, and 1 N HF + 1 N HNO_3 as the mobile phase.

Gibalo and Mymrik[30] have developed a reliable method for the separation and pre-concentration of niobium from Mo, W and Zr, using Ftoroplast-4 as the support and 0.5 M tri-n-octylamine (TNOA) in benzene as the stationary phase.

The application of extraction chromatography in the neutron activation analysis of arsenic, antimony, gallium arsenide, niobium, niobium pentoxide, niobium pentachloride and lithium niobate has led to efficient and simple procedures of high sensitivity for the determination of impurities, e. g. $10^{-4}-10^{-6}\%$ of Mn, Ni, Cu and Zn in GaAs[38,39].

The present authors have employed extraction chromatography fot the collective and individual preconcentration of traces of gallium, iron(III), cobalt and copper(II) in the anylysis of pure aluminium and its salts, and for the separation of indium and cadmium from large amounts of zinc.

Indium and zinc, present in the ratio 2 to 10^7 were separated with 0.8—1.0 M HBr. If the mobile phase is 1 M HBr, traces of indium can be preconcentrated from zinc sul-phate and chloride without prior conversion into bromides. For example, 2.0—5.0 μg of indium was separated from 10—15 g of $ZnSO_4$, and 10 μg of gallium from 500 g of $ZnCl_2$. For the rapid chromatographic separation of microamounts of indium from rela-tively large quantities of gallium, the system 2 M HBr -tri (2-ethylhexyl) phosphate was used: 2 μg of indium was successfully separated and concentrated from 2 g of pure gallium, with an enrichment factor of 10^4-10^6.

The collective preconcentration of a group of elements has been performed using 1 M ammonium thiocyanate as the mobile, and 100% TBP as the stationary phase. The chromatographic behaviour of some elements was utilized for the collective preconcen-tration of trace impurities in the analysis of metallic aluminium and of its salts from 100—200 g samples (Table 1), and also for the extensive purification of ammonium thiocyanate from traces of iron[33].

Using literature data and the present authors' own experimental results, a survey was made of the chromatographic behaviour of the majority of the elements in the periodic system which, when present at the microtrace level, have been preconcentrated and sep-arated from macrocomponents by means of extraction chromatography. The results are shown in Fig. 1. Separation and preconcentration was attained by the application of selective extractants with high partition coefficients for the microelement, with practically no extraction of the macrocomponent. The individual or collective preconcentration of the majority of elements is possible only by employing different extraction systems. For example, the present authors have shown that traces of Ga and Fe(III) can be precon-centrated and separated from the bulk matrix (alkali and alkaline earth metals, Al, Mn(II), Co, Zn, Ni, Fe(II), Cd, etc.) in the systems HCl—TBP, MCl_n—TBP, HCl—MCl_n—TBP, etc., where M stands for elements of Groups I and II of the periodic system. The collective preconcentration of traces of Ga, Fe(III), Co and Cu(II) can be performed in the NH_4SCN—TBP system.

The information given in this section was meant only to illustrate briefly the possible applications of extraction column chromatography to the individual and collective pre-concentration and separation of microimpurities from a macrocomponent. The examples presented and the literature data show that extraction chromatography will doubtless find widespread application in analytical and preparative chemistry as an interesting and effi-cient method of solving such problems.

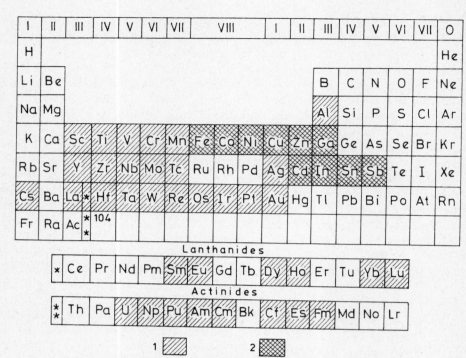

Fig. 1. Preconcentration of trace elements by extraction chromatography.
1 — Methods described in the literature; 2 — Methods developed by the authors

TABLE 1

Preconcentration of trace elements by extraction chromatography*

Substance	Sample weight, g	4 M HCl–TBP found, %		1 M NH₄SCN–TBP found %			
		Ga(III)	Fe(III)	Ga(III)	Fe(III)	Co(II)	Cu(II)
Al	10	$(6\pm1.2)\cdot10^{-5}$	$3\cdot10^{-5}$	$(5\pm0.8)\cdot10^{-5}$	$2\cdot10^{-3}$	—	$5\cdot10^{-3}$
Zn	10	$(2\pm0.2)\cdot10^{4}$	—	—	—	—	—
Cd	10	$(5\pm0.5)\cdot10^{-4}$	—	—	—	—	—
AlCl₃	230	$(3\pm0.5)\cdot10^{-6}$	$6\cdot10^{-6}$	$(4\pm0.2)\cdot10^{-6}$	$5\cdot10^{-6}$	$4\cdot10^{-6}$	$3\cdot10^{-6}$
ZnCl₂	500	$2\cdot10^{-6}$	$3\cdot10^{-6}$	$3\cdot10^{-6}$	$4\cdot10^{-6}$	—	—

*The confidence limits of the gallium results were calculated on a statistical basis (95% confidence level) from the averages of several determinations.

4. EFFECTS OF CERTAIN FACTORS ON THE PRECONCENTRATION AND SEPARATION OF ELEMENTS BY COLUMN EXTRACTION CHROMATOGRAPHY

In connection with the development of methods for the separation and preconcentration of trace impurities, considerable efforts have been expended to elucidate the effects of various factors on the efficiency of the column. As the mechanism of the process in the column is of the liquid-liquid extraction type, the distribution of elements between the two phases depends on the factors that govern the liquid-liquid extraction process under static conditions, i. e. the nature of the extractant, the composition of the aqueous phase, the concentrations and oxidation states of the components, and the temperature of the column. In the selection of the conditions for the separation of impurities from macrocomponents, it is necessary, in addition to the optimum composition of the aqueous phase, to take into account the capacity of the stationary phase for the macrocomponent, the number of theoretical plates and the thickness of the stationary phase layer on the support. Many systematic studies of the factors determining the behaviours of various elements in the chromatographic column are available, but few of them have been carried out with the aim of optimizing preconcentrations of elements.

The following is an attempt to summarize the available material from this latter viewpoint.

4.1. THE EFFECT OF THE CONCENTRATION AND NATURE OF THE ACID IN THE MOBILE PHASE ON THE DISTRIBUTION OF MICROCOMPONENTS IN THE COLUMN

This effect is one of the most important factors determining the optimum conditions for the preconcentration of impurities and their separation from the bulk material.

In the theory of liquid-liquid extraction, detailed treatments are available concerning, for instance, the dependence of the partition coefficients of microtraces on the pH of the aqueous solution, in both the absence and the presence of macrocomponents[47]. When acidic extractants are used, the theory foresees, and practice confirms, that in the absence of a macrocomponent the extraction behaviour of the microelement exhibits a steep decrease of the distribution coefficient with increasing initial acid concentration in the aqueous phase. However, the actual shape of the extraction curve will depend on the ability of the metal to form complexes with the anion of the acid, causing differences with respect to the extractable forms of the various elements.

Information on the extraction behaviours of the different elements in the extraction systems applied to chromatographic separations is to be found in Chapter 4. Here, some specific examples are given to illustrate the correlations between the liquid-liqud extraction behaviours of elements and the possibility of their effective separation and preconcentration by extraction chromatography. They refer to work carried out by the present authors.

Literature data referring to liquid-liquid distribution coefficients of zinc, cadmium

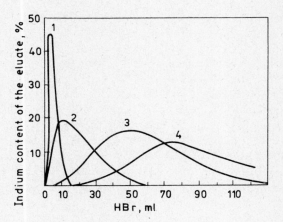

Fig. 2. Elution curves of indium in the system TBP–HBr. C_{In} – $8.6 \cdot 10^{-4}$ M; 1 – 0.1 M HBr; 2 – 0.2 M HBr; 3 – 0.3 M HBr; 4 – 0.5 M HBr

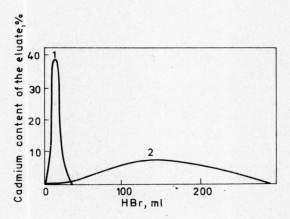

Fig. 3. Elution curves of cadmium in the system TBP–HBr. C_{Cd} – $8.8 \cdot 10^{-4}$ M; 1 – 0.1 M HBr; 2 – 0.3 M HBr

and indium show that the stabilities of halide complexes determine the differences in the behaviour of microamounts of these elements in the column if the stationary phase is a neutral organophosphorus compound, such as TBP.

In Figs 2–4, the peaks of microamounts of In, Zn and Cd are shown, obtained on elution from a TBP column with HBr at different concentrations.

As expected, the retention volumes of all three elements increase with increasing HBr concentration in the mobile phase. It appears that at low bromide concentration (0.3 M), the retention volumes of the elements decrease in the order Cd > In > Zn, in accordance with the decreasing stability of their bromide complexes: in fact, a linear log-log de-

334

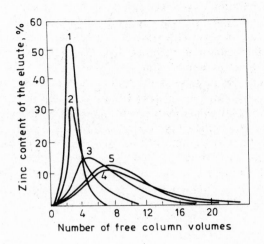

Fig. 4. Elution curves of zinc in the system TBP–HBr. C_{Zn} – $1.5 \cdot 10^{-3}$ M; 1 – 0.1 M HBr; 2 – 0.3 M HBr; 3 – 0.6 M HBr; 4 – 0.8 M HBr; 5 – 1.0 M HBr

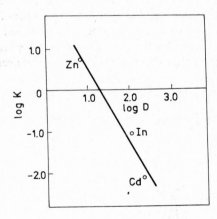

Fig. 5. Dependence of the degree of extraction on the stabilities of the bromide complexes (K = I/β)

pendence was found between stability constants and distribution coefficients (Fig. 5). With 0.3 M HBr, the separation factors β for pairs of these elements have the following values: $\beta_{In/Zn}$ = 31.1; $\beta_{Cd/Zn}$ = 56.4; and $\beta_{Cd/In}$ = 1.8, and allow their chromatographic separation (Fig. 6).

From the preconcentration viewpoint, Figs 3–5 show that the increase of the HBr concentration results not only in a larger retention volume for each metal, but also in a broadening of the elution peak, in accordance with plate height theory. Furthermore, larger peaks tend to form "tails" in their rear front. In practice, therefore, the higher the acid concentration, the poorer the elution of the element from the column, requiring also larger acid volumes. On increase of the acidity of the solution (0.8 M HBr), indium and cadmium can be quantitatively retained by the organic phase of the column, and this can be utilized for their separation and preconcentration from non-extractable elements. However, when they have to be recovered from the column, the lowest concentration is advisable, to keep to a minimum the volume of solution necessary for quantitative elution. The behaviours in the column of the elements Ga, Fe(III), Cd, Zn, Al and Cu(II) in the HCl–TBP system have also been studied in detail. The literature data on the liquid-liquid extraction of these elements with TBP indicate that, with the exception of Al, they are transferred into the organic phase with large distribution coefficients, so that their separation, and especially the preconcentration of their traces, is not feasible by liquid-liquid extraction methods.

Study of the behaviour of these elements in a column containing TBP, as a function of the hydrochloric acid concentration, established the order Ga > Fe(III) > Cd > Zn > > Cu(II) $\approx$ Al for the interaction of chloride complexes with the stationary phase, in accordance with liquid-liquid extraction data. At 3 M and higher HCl concentrations, gallium is selectively retained in the column with a very high distribution ratio. Iron(III) is quantitatively retained only from 5–7 M HCl, whereas Zn, Cu(II) and Al are not retained at all. On this basis, several variants have been devised for the preconcentration of Ga traces, and for their separation from Fe(III), Cd, Zn, Cu(II) and Al. The separation factor is 10^7.

Traces (2 μg) of indium have been preconcentrated and separated from large amounts of zinc ($\sim$ 10 g) in the system TBP–0.8 M HBr. The system is applicable for the preconcentration of indium from zinc salts: 160 μg of indium was preconcentrated from 10 g of

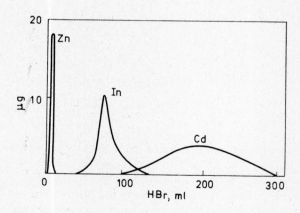

Fig. 6. Elution curves of zinc, cadmium and indium in the system 0.3 M HBr – TBP

zinc sulphate without prior conversion of the salt into the bromide. The confidence limit of the determination of In was ±13.1%. The same system was also applied to the preconcentration of indium from drosses of the copper industry and from natural iron-based materials. The method developed for the preconcentration of indium has a number of advantages over earlier liquid-liquid extraction techniques, in which, prior to a double extraction with TBP, iron and other elements have to be precipitated as the hydroxides.

In some cases it is not necessary to determine the elution curves to evaluate the behaviour of an element in a given chromatographic column, because recourse can be made to known values of D (determined under single-step liquid-liquid extraction conditions as a function of the various parameters) to calculate the retention volume V_R from the well-known relationship $D = (V_R - V_m)/V_s$, where V_m and V_s are the volumes of the mobile and stationary phases, respectively, in the column (see Chapter 2). Once V_R is known, the partition of the microelement and its possible preconcentration and separation from a macrocomponent can be foreseen.

For example, V_R was calculated by the present authors for certain elements, on the basis of D values published by Ishimori and Watanabe[48] for the TBP–HCl system, for a column having V_m = 1 ml and V_s = 1.2 ml. For such a column the retention volumes are related to D by the formula V_R = 1 + 1.2 D.

For copper and aluminium, whose distribution coefficients are around 1 and 10^{-2}, respectively, V_R is small, not exceeding 2 ml for 3–7 M HCl.

For cadmium and zinc ($D_{Zn} \approx 2.5$ and $D_{Cd} \approx 50$), V_R ranges between 3 and 60 ml for 4–7 M HCl.

The value of D is very large for gallium and iron(III) in 4–6 M HCl solutions, and V_R approaches infinity, indicating selective retention by the stationary phase of the column and good separation from large amounts of Cu, Al and Zn. In connection with the preconcentration of gallium and iron traces, such approximate calculations, based on literature values of D, have been advantageously applied to determine the optimum conditions of analysis.

4.2. THE EFFECTS OF MACROCOMPONENTS ON THE EXTRACTION CHROMATOGRAPHIC DISTRIBUTION OF MICROCOMPONENTS

These effects are of extreme importance when this method is used for the preconcentration of trace impurities.

The mutual influence of micro- and macrocomponents in single-step liquid-liquid extraction has been extensively studied, and is systematically discussed by Zolotov and Kuzmin [47]. In a detailed study concerned with the effects of non-extractable and extractable macrocomponents on the extraction of microelements, the above authors have shown that the effects can be fairly different, and have developed a theory correlating the distribution coefficient (D) of the microcomponent with various parameters of the system, e. g. pH of the medium, concentrations of the reagent and the macrocomponent, etc. The effects of strong electrolytes in large amounts on the process of extraction are known to be very diverse. The anion of the salt may form a non-extractable complex with the metal to be extracted, thereby impeding extraction. On the other hand, the salt often acts as a dehydrating agent, in this way promoting extraction. The addition of salts to the solution changes the dielectric constant and ionic strength of the aqueous phase, causing changes in all the characteristic constants of the equilibrium. Obviously, the effects on distribution, resulting from the variation of the salt concentration in the aqueous phase, can be completely different, and even opposite, depending on the nature of the organic phase involved. In the view of the authors, these problems are of equal importance in the study of extraction chromatography. There are numerous studies in which trace impurities are separated from large amounts of a macrocomponent but, unfortunately, the above factors are often neglected.

Extraction chromatography has been applied[49] for the separation of microamounts of lanthanides from macroamounts of neighbouring rare earths, and for the separation of the matrix interfering with the analysis of Dy and Yb, with subsequent determination of these microelements. For the spectrophotometric determination of small amounts of Y, the latter was separated from relatively large quantities of dysprosium and terbium[50]

$LiNO_3$ or LiCl solutions are often suggested as mobile phases[51]. With variation of the concentrations of these salts, closely-resembling lanthanides were more effectively separated on long-chain amine columns than in the corresponding free acids.

The present authors have studied the effects of extractable and non-extractable macro salts on the chromatographic separation of microelements in the systems involving TBP and metal halides, namely the chlorides or bromides of Li, Na, K, Mg, Ca, Zn and Cd. The selected microelements were Ga, In, Cd and Zn. The retention of gallium appeared to depend on the salt concentration; absorption is greater at higher salt concentrations, as in the case of single-step liquid-liquid extraction. However, under dynamic conditions (i.e. high flow-rate) the percentage retention is much smaller, especially from $MgCl_2$ and $CaCl_2$ solutions, whereas practically no retention occurs from solutions of $ZnCl_2$ and $CdCl_2$ (known to be extractable by TBP).

The retention of gallium from dilute salt solutions decreases in the order $HCl > LiCl >$ $> NaCl > MgCl_2 > CaCl_2 > ZnCl_2 > CdCl_2$. With the exception of $ZnCl_2$ and $CdCl_2$, salt concentrations higher than 5 N bring about quantitative retention of the microelement in the MCl_n–TBP and HCl–MCl_n–TBP systems. The addition of 3 M or more hydrochloric acid increases the retention of gallium. For example, this element is quantitatively retained in the column if HCl is added to the solution of $ZnCl_2$ and $CdCl_2$. This has been utilized for the preconcentration of gallium traces.

The same phenomenon is observed in the extraction chromatography of microamounts of indium, cadmium and zinc ($8 \cdot 10^{-4}$ M) from solutions of bromide salts not extractable with TBP[52] (Fig. 7) and in the retention of indium from chloride solutions (Fig. 8). The distribution coefficient of indium increases in the order $LiBr > NaBr > KBr$. Extractable macrocomponents suppress indium retention. Among the tested metals the following sequence was found: $CdBr_2 > GaBr_3 > FeBr_3 > ZnBr_2$. Thus, the easier the extraction of the macrocomponent, the greater its effect on the retention of the microelement in the column. The addition of hydrogen bromide to the salt solutions increases the amount of indium retained.

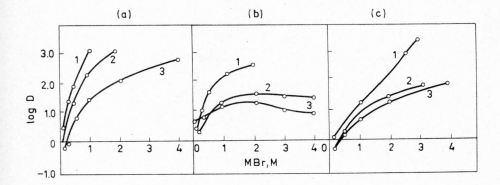

Fig. 7. Dependence of log D on the MBr concentration for (a) indium, (b) cadmium and (c) zinc

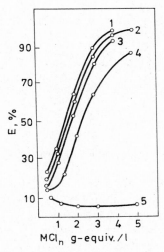

Fig. 8. Effect of metal chlorides on the extraction of indium.
1—Li;Cl; 2—MgCl$_2$; 3—AlCl$_3$; 4—NaCl; 5—ZnCl$_2$

4.3. EFFECT OF TEMPERATURE

It is known that the choice of suitable temperatures in some cases improves the separation of elements at various stages of the single-step liquid-liquid extraction process. In ion-exchange chromatography, the temperature effect is sometimes manifested in narrower peaks of the elution curves and in a shorter time of separation. Analogous effects are generally found in extraction chromatography too.

In the experiments described below the temperature of the column was varied between 29 and 90°C with an accuracy of ±0.5 °C. The mobile phases used were 4 M hydrochloric acid, 1 M HBr and ammonium thiocyanate.

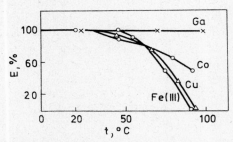

Fig. 9. Chromatographic distribution of elements as a function of the temperature

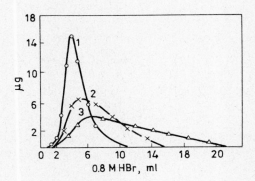

Fig. 10. Elution curves of zinc at different temperatures. 1–88 $^\circ$C; 2–39 $^\circ$C; 3–20 $^\circ$C

Figure 9 shows the distribution of 50 μg of different elements eluted with HCl on a TBP column at various temperatures. When the temperature is increased, only the percentage retention of gallium does not change appreciably, the amounts of the other elements in the stationary phase decreasing with increasing temperature. For example, while at 20 $^\circ$C iron(III) is quantitatively retained in the stationary phase, at 70 $^\circ$C only 60% is retained, and at 90 $^\circ$C it is completely washed out from the column. The release of zinc, cadmium and aluminium becomes faster with the rise of temperature, their elution peaks become narrower and, as a result, the volume of eluate required for complete recovery from the column decreases several-fold (Fig. 10).

The study of the behaviours of various elements in the column at higher temperatures led to the separation of small amounts of gallium from relatively large quantities of iron(III) without the addition of ascorbic acid or other reducing agents. Complete recovery of iron requires 5 times less eluate than at room temperature and the separation is much faster.

Gallium was separated from iron present in a ratio of 1 to 10^5, using the 4 M HCl–TB system in a column thermostated at 80 $^\circ$C. Through this column filled with Ftoroplast-4 20 ml of 4 M HCl solution containing 1 g of FeCl$_3$ · 6H$_2$O and 10 μg of gallium was percolated at a flow rate of 0.6 ml·min^{-1}. The column was then washed with 4 M HCl until complete elution of iron (about 140 ml), detected by the change in the colour of the column and through the negative test for iron with sulphosalicylic acid in the eluate. Gallium was recovered with 5 ml of 0.1 M HCl. The relative error of the determination was less than ±10%. Gallium can also be separated from iron with 1 M HCl solutions at 70–80 °C. In a thermostated column, gallium can be separated from large amounts of cadmium using hydrochloric acid solutions, whereas this is not possible at room temperature. Good results were obtained in the separation of Cd and In[53]. On the other hand, the experimental results indicate that the transfer of the majority of substances into the stationary phase becomes less favourable with increase of the temperature, which may be due not only to the exothermic nature of the extraction but also to the effect of the temperature on the activity coefficient of the complex in the system, as well as complex formation, dissociation, etc. Further work is needed to elucidate what factors contribute to the temperature-dependence of the distribution of complex acids extracted by a hydrate-solvate mechanism.

4.4. CHANGES IN THE OXIDATION STATE OF THE METAL

Another variant of preconcentration by extraction chromatography has recently been described, based on different oxidation states[35].

In the activation analysis of traces of yttrium, dysprosium, holmium, samarium and lanthanum in europium oxide (sensitivity $10^{-5} - 10^{-7}\%$), the main bulk of europium was separated by reduction to Eu(II) with metallic zinc in a Jones reductor, which was connected to the top of a chromatographic column filled with di(2-ethylhexyl) phosphoric acid-loaded Ftoroplast-4.

In this case, Eu(II) passes through the column, whereas all the other rare-earth traces in the 3+ oxidation state are retained in the column.

The sensitivity of the method and the analytical results on high-purity europium oxide are shown in Table 2.

Another example is the extraction chromatographic separation of gallium traces from large amounts of iron(III) after its reduction to iron(II). The enrichment factor reaches $10^6 - 10^8$.

TABLE 2

Analysis of samples

Element	Amount, %	Sensitivity, %
Yb	$(1.0\pm0.2)\cdot10^{-5}$	$1\cdot10^{-6}$
Dy	$(8.0\pm0.8)\cdot10^{-6}$	$1\cdot10^{-7}$
Ho	$(4.2\pm1.9)\cdot10^{-5}$	$2\cdot10^{-5}$
Sm	$(6.4\pm1.1)\cdot10^{-5}$	$2\cdot10^{-7}$
La	$(5.7\pm2.4)\cdot10^{-5}$	$1\cdot10^{-5}$

5. CONCLUSIONS

Extraction chromatography using columns filled with inert supports allows individual or collective preconcentration of trace elements with enrichment factors greater than 10^5, as well as rapid separation of similar metals in the form of complex compounds.

Since adsorption is negligible and only small amounts of extractants are used, the amounts of impurities introduced during the procedure are extremely small. In any case, the reagents and water applied may easily be subjected to preliminary purification from impurities, using the same method.

In order to extend the applicability of the method for the individual and collective preconcentration of impurities, it is necessary to study metal chelates with reference to their substitution by other metals in the organic phase of the column. In some cases it is apparently possible to separate similar elements utilizing the differences in the kinetics of complex formation[55].

Extraction chromatography has been found to be an excellent technique for the rapid separation of short-lived radioisotopes, with the special advantage that all the operations involved can be automated.

REFERENCES TO CHAPTER 12

1. G. S. KATYKHIN, Zh. Anal. Khim., 20 (1965) 615.
2. I. P. ALIMARIN, T. A. BOLSHOVA, N. I. ERSHOVA, M. G. POLINSKAYA, Zh. Anal. Khim., 24, (1969) 26.
3. I. P. ALIMARIN, T. A. BOLSHOVA, Z. Chem., 8 (1968) 411.
4. G. S. KATYKHIN, Zh. Anal. Khim., 27 (1972) 849.
5. J. STRONSKI, Rept. Inst. Fiz. Jadr. Krakowie, 803, 415, 1972.
6. I. P. ALIMARIN, T. A. BOLSHOVA, Pure Appl. Chem., 31, (1972) 493.
7. Yu. K. ZHUKOVSKII, G. S. KATYKHIN, A. L. MARINOV, M. K. NIKITIN, Radiokhimiya, 10 (1968) 252.
8. S. K. KALININ, G. A. YAKOVLEVA, M. K. NIKITIN, G. S. KATYKHIN, Trudi Instituta Gipronikel, 33. (1967) 192.
9. L. N. MOSKVIN, L. G. TSARITSINA, Radiokhimiya, 10 (1968) 740.
10. O. V. STEPANETS, Yu. V. YAKOVLEV, I. P. ALIMARIN, Zh. Anal. Khim., 25 (1970) 1906.
11. I. P. ALIMARIN, Yu. A. ZOLOTOV, Chem. Anal., (Warsaw), 13 (1968) 941.
12. Yu. A. ZOLOTOV, Ektraktsia vnutrikomplekshikh soedinenii, Izd. Nauka, Moscow, 1968.
13. T. B. PIERCE, P. F. PECK, J. Chromatogr., 6 (1961) 248.
14. T. ŠEBESTA, J. Radioanal. Chem., 7, (1971) 41.
15. P. MARKL, Extraktion und Extraktions-Chromatographie in der anorganischen Analytik, Frankfurt am Main, 1972.
16. G. S. KATYKHIN, M. K. NIKITIN, Radiokhimiya, 10 (1968) 474.
17. B. K. PREOBRAZHENSKII, A. V. KALYAMIN, O. M. LILOVA, L. N. MOSKVIN, B. S. USIKOV, Radiokhimiya, 10 (1968) 375; 10 (1968) 373.
18. I. P. ALIMARIN, T. A. BOLSHOVA, N. I. ERSHOVA, O. D. KISELEVA, Vestnik Moskovskovo Universiteta, Ser. II, Khimiya, 4 (1967) 55.
19. I. P. ALIMARIN, T. A. BOLSHOVA, Zh. Anal. Khim., 21, (1966) 411.
20. L. S. BARK, G. DUNCAN, R.Y.T. GRAHAM, Analyst, 92 (1967) 347.
21. L. N. MOSKVIN, Radiokhimiya, 6 (1964) 110.
22. E. HERRMAN, J. Chromatog., 38 (1968) 498.
23. E. L. MOORE, A. JURRIAANSE, Anal. Chem., 39, (1967) 733.

24. I. P. ALIMARIN, T. A. BOLSHOVA, T. G. DOLZHENKO, G. M. KANDYBINA, Zh. Anal. Khim., 26, (1971) 741.
25. I. P. ALIMARIN, T. A. BOLSHOVA, G. A. BAKHAREVA, Zh. Anal. Khim., 28 (1973), in press.
26. S. K. KALININ, G. S. KATYKHIN, M. K. NIKITIN, G. A. YAKOVLEV, Zh. Anal. Khim., 23 (1968) 1481.
27. J. S. FRITZ, D. C. KENNEDY, Talanta, 17 (1970) 873.
28. R. Y. T. GRAHAM, L. S. BARK, D. A. TISLEY, J. Chromatog., 35 (1968) 416.
29. I. M. GIBALO, N. A. MYMRIK, Zh. Anal. Khim., 25, (1970) 1747.
30. I. M. GIBALO, N. A. MYMRIK, Zh. Anal. Khim., 26 (1971) 918.
31. J. S. FRITZ, T. FRAZEE, G. L. GATWESEN, Talanta, 17 (1970) 757.
32. S. PRZESZLAKOWSKI, Chem. Anal., (Warsaw), 12, (1967) 1071.
33. I. P. ALIMARIN, N. I. ERSHOVA, T. A. BOLSHOVA, Vestnik Moskovskovo Universiteta, Ser. II. Khimiya, 6, (1969) 79.
34. N. I. ERSHOVA, T. A. BOLSHOVA, I. P. ALIMARIN, Zh. Anal. Khim., 26, (1971) 2113.
35. N. M. KUZMIN, S. V. MESHCHANKINA, Sbornik, Promishlennosty khimicheskikh reaktivov i osobo chistykh veshchestv, IREA, No. 20, p. 5, Izd. Khimiya, 1970.
36. N. M. KUZMIN, A. V. YEMELYANOV, S. V. MESHCHANKINA, V. L. SABATOBSKAYA, I. A. KUZOVLEV, T. I. ZAKHAROVA, Zh. Anal. Khim., 25, (1971) 371.
37. Yu. BOSHOLM, G. GROSSE-RUYKEN, N. A. LEBEDEV, E. HERMANN, V. A. KHOLKIN, Proc. XXth International Congress on Pure and Applied Chemistry, Section E, Izd. Nauka p. 91, 1965.
38. M. N. SHULEPNIKOV, G. I. SHMANENKOVA, B. P. SHCHELKOVA, Yu. V. YAKOVLEV, Zh. Anal. Khim., 28, (1973) 608.
39. G. I. SHMANENKOVA, V. I. FIRSOVA, V. P. SHCHELKOVA, M. N. SHULEPNIKOV, Zh. Anal. Khim., 28 (1973) 323.
40. I. AKAZA, T. KIBA, T. KIBA, Bull. Chem. Soc. Japan, 43 (1970) 2063.
41. E. CERRAI, G. GHERSINI, 2nd International Conference of the Society for Analytical Chemistry, Ljubljana, 1972.
42. E. CERRAI, G. GHERSINI, Analyst, 94 (1969) 599.
43. W. SMUŁEK, K. ZELENAY, Report JBJ. 677/V XMC, 1965.
44. W. SMUŁEK, Nukleonika, 20 (1966) 635.
45. J. TOMAZIĆ, S. SIEKIERSKI, J. Chromatog., 21, (1966) 98.
46. G. P. GORYANSKAYA, B. Ya. KAPLIN, I. V. KOVALIK, Yu. I. MERISOV, M. G. NAZA-ROVA, Zh. Anal. Khim., 27, (1972) 1498.
47. Yu. A. ZOLOTOV, N. M. KUZMIN, Ekstraktsionnoe kontsentrirovanie, Izd. Khimia, Moscow, 1971.
48. T. ISHIMORI, K. WATANABE, Bull. Chem. Soc. Japan, 33, (1960) 636.
49. G.I. SHMANENKOVA, M. G. ZEMSKOVA, Sh. G. MELAMED, G. P. PLESHCHAKOVA, G. V. SUKHOV, Zavodsk. Lab., 35 (1969) 897.
50. H. GROSSE-RUYKEN, H., J. BOSHOLM, J. prakt. Chem., 30 (1965) 77.
51. I. STRONSKI, Radiochem. Radioanal. Letters, 1, (1969) 191.
52. T. A. BOLSHOVA, T. G. DOLZHENKO, K. V. GRYSHINA, I. P. ALIMARIN, Zhurn. Anal. Khim., 28 (1973) 259.
53. T. G. DOLZHENKO, T. A. BOLSHOVA, I. P. ALIMARIN, Zhurn. Anal. Khim., 28 (1973) 263.
54. I. P. ALIMARIN, A. Z. MIKLISHANSKI, Yu. V. YAKOVLEV, J. Radioanal. Chem., 4 (1970) 75.
55. I. P. ALIMARIN, Pure Appl. Chem., 34, (1973) 1.

CHAPTER 13

USE OF CELLULAR PLASTICS IN EXTRACTION CHROMATOGRAPHY

T. BRAUN, A. B. FARAG

1. INTRODUCTION

The introduction, for use in column chromatography, of polytrifluorochloroethylene (Kel-F, Voltalef), polytetrafluoroethylene (Teflon, Haloport-F, Fluoroplast), the copolymer of polyvinyl acetate and polyvinyl chloride (Corvic), the copolymer of styrene and divinylbenzene, polyethylene (Mipor), and polyfluorocarbon represents marked progress[1]. These supports are applied in granular form, and the adsorption and elution of the components to be separated take place in a relatively thin layer of the stationary phase coated on the polymer surface. The considerable advantages of the application of such plastic supports are lessened by the necessity of using them as finely granulated particles, with large specific surface, in order to increase the capacity; this adversely affects the flow-rate through the column, and pressurized columns therefore have to be employed. Hornbeck[2] recognized this shortcoming ".... a pressurized column does little to simplify the operation or reduce the time and attention required on the part of the operator. Pressurized columns require close operator attention.... Loss of liquid at the top of a column bed results in air blowing through the column, which destroys the column operation and requires abandoning the sample or dismantling the column to recover it."

During the last two years systematic investigations have been carried out with cellular plastics as supports in extraction chromatography[3—7].

Cellular plastics (foams) can be defined as plastic materials in which a proportion of the solid phase is replaced by gas in the form of numerous small cells. The gas may be in a continuous phase to give an open-cell material, or it may be discontinuous, i. e. in the form of discrete, non-communicating cells. The cell structure is a function of the process by which the cellular material is made, and both rigid and flexible foams may be obtained with open and closed cells[8].

2. CELLULAR PLASTICS AS EXTRACTION CHROMATOGRAPHIC SUPPORTS

The basic idea of the application of cellular plastics as supports is that chromatographic adsorption, exchange and partition processes can be favourably influenced by giving the adsorbent a hollow spherical (cellular) form, and effecting the adsorption on the internal surface of the cells. This can be done by using solid, rigid or flexible open-cell-type foamed synthetic polymers as column fillings. As will be seen later, chromatographic columns packed with foamed polymers show excellent hydrodynamic properties and favourable kinetics of the partition processes which take place on the thin membranes separating the foam cells. One of the principal attractions of these foamed columns for routine work is that relatively high flow-rates can easily be attained simply by gravity flow.

It is of interest to mention that the application of a natural cellular material (in the form of sponge) as support in chromatography was probably tested more than four centuries ago. Bittel[9] in 1957 reported that Brunschwig[9a] in 1512 purified ethyl alcohol by distillation through an olive oil-soaked sponge. In fact this procedure can be classified as gas-liquid partition chromtography in which the sponge material was the support and the olive oil the stationary phase. Ethyl alcohol vapour can be regarded as the mobile phase. In 1962 Bayer[10] checked this ancient method and reported that it operates effectively.

Recently, however, several papers have been published[11—13] describing the possibility of using cellular plastics as supports in gas-solid chromatography. The application of foamed plastics for the adsorption and recovery of inorganic and organic materials was also reported recently. Bowen[14] examined the distribution ratios and adsorption capacities of polyurethane foams for a few selected elements. Later, [15] he recommended the application of polyurethane foam for the recovery of gold(III) from liquid mineral wastes. Schiller and Cook [16], following a recommendation by Bowen, applied polyurethane foam for the preconcentration of gold(III) from natural sweet waters.

The further application of polyurethane foam for the recovery of polychlorinated biphenyl (PCB) and organochlorine pesticides from aqueous wastes has been reported[17,18]. monitoring of organic matter in water supplied by polyurethane foam has been suggested too[19]. In a more recent publication[20], polyurethane foams with incorporated SH groups were examined for their ability to adsorb mercuric chloride and methyl mercuric chloride from aqueous solutions.

2.1. CHEMICAL AND PHYSICAL PROPERTIES OF CELLULAR PLASTICS

It has been reported[8] that every plastic material can be made into cellular form by either chemical, physical or mechanical methods. Table 1 shows the forms in which the most commonly used cellular plastics are available. It was mentioned[8] that as a general principle flexible materials tend to have an open-cell, and rigid materials a closed-cell structure. Some materials which can be made by more than one method can exist in both open and closed forms. Methods are also available whereby closed-cell structures can be converted into the open-cell form.

TABLE 1

Cellular forms in which plastics are normally encountered

Flexible		Rigid	
Closed-cell	Open-cell	Closed-cell	Open-cell
Polyvinyl chloride	Polyvinyl chloride	Polyvinyl chloride	Polythene sponge
	Polyurethane	Polystyrene	
	Polyvinyl formal	Polyurethane	
Silicone rubber	Rubber latex	Cellulose acetate	Polytetrafluoroethylene
		Polymethylmethacrylate	
		Phenol-formaldehyde	Urea-formaldehyde
		Urea-formaldehyde	
		Silicone	
		Ebonite	
		Polythene	
		Epoxides	

Preliminary experiments[3] on the different cellular plastics commercially available (polyvinyl chloride, viscose, rubber and polyurethane) showed that polyurethane foam was at that time the most suitable for application as support in extraction chromatography.

Polyurethane foams are in general copolymers of toluene diisocyanate and polyethers or polyesters cross-linked by amido-linkages[14]. The resulting material is a solid foam whose bulk density generally lies between 15 and 35 kg·m^{-3}.

Bowen[14] examined the chemical resistance of certain batches of polyurethane foams of polyether type and claimed that they are fairly stable and inert. He reported that the foam batches tested degrade when heated between 180 and 220 °C, and slowly turn brown in U. V. light. They are dissolved by concentrated sulphuric acid, destroyed by concentrated nitric acid and reduce alkaline potassium permanganate. They are mostly

346

unaltered, apart from reversible swelling, by water, hydrochloric acid up to 6 M, sulphuric acid up to 4 M, nitric acid up to 2 M glacial acetic acid, 2 M ammonia, 2 M sodium hydroxide and the following solvents: light petroleum, benzene, carbon tetrachloride, chloroform, diethyl ether, di-isopropyl ether, acetone, isobutyl methyl ketone, ethyl acetate, isopentyl acetate and alcohols. The change in weight after shaking with these reagents for 5 min and subsequent drying was less than 0.7%. It was also noted[14] that polyurethane foams can be dissolved in hot arsenic(III) chloride solution.

A detailed study[21] was made on the loading properties of different types of polyurethane foam to the organic extractant (TBP). It was found[8] that polyurethane foams generally absorb TBP much more efficiently than, for example, granular Voltalef which is considered one of the best granular supports. Polyurethane foams of polyether type retain TBP more efficiently than those of polyester type. The uptake of TBP by polyether foams is slightly increased by decreasing the cell dimensions (Table 2).

TABLE 2

The TBP loading capacities of various types of polyurethane foams* and Voltalef powder

Sample No.	Type of support	Weight of dry support, g	Weight of loaded support, g	Weight of TBP absorbed per 5g support	TBP in loaded support, %
1	Polyurethane foam, polyether, more than 95% open-cell (Hungarian)	5	14.9516	9.9516	66.6
2	Polyurethane foam, polyether, more than 95% open-cell (Hungarian)	5	15.9490	10.9490	68.7
3	Polyurethane foam, polyether, ca. 100% open-cell (Kunststoffbüro W 8100)	5	14.4372	9.4372	65.4
4	Polyurethane foam, polyether, ca. 100% open-cell, (Kunststoffbüro, W 8300)	5	14.5991	9.5991	65.8
5	Polyurethane foam, polyether, ca. 100% open-cell (Kunststoffbüro, W 8600)	5	15.8234	10.8234	68.4
6	Polyurethane foam, polyester, ca 100% open-cell (Eurofoam PPI 80nFR)	5	10.3270	5.3270	50.7
7	Voltalef powder grade 300 LDCHR 0.25 mm particle size (Plastimer, 5 Rue du Général Foy, Paris)	5	9.7540	4.7540	48.7

*The cell dimensions of sample 1 > sample 2 and sample 3 > sample 4 > sample 5.

2.2. HYDRODYNAMIC PROPERTIES OF FOAM-FILLED COLUMNS

In all types of chromatographic techniques column preparation is of prime importance. Cerrai and Ghersini[22] state: "Column preparation is a very critical step; it is the only one where 'reversed-phase art' plays an important role". Obviously, in order to use foam material as a support in reversed-phase partition chromatography, a suitable method of column packing had to be developed. The method devised must be very carefully selected to enable homogeneous preparation of the foam column by anyone and at the same time to produce columns with good performance and suitable flow characteristics. For this purpose a vacuum method was developed[4], whereby the dried, loaded foam material (in the form of small cubes or cylinders) was packed in the column, applying gentle pressure with a glass rod and avoiding formation of air bubbles during packing by using the following technique: Tap 1 (Fig. 1) was connected to a suction pump while tap 2 was

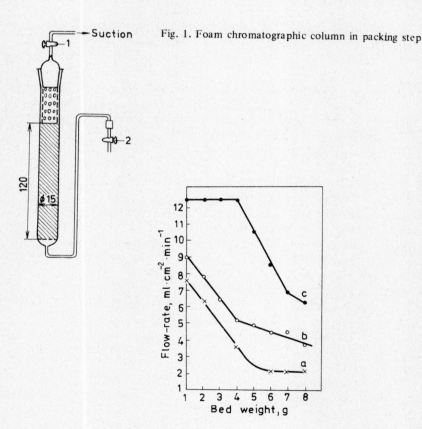

Fig. 1. Foam chromatographic column in packing step

Fig. 2. Flow-rate vs. weight of loaded foam and Voltalef powder
(a) Voltalef packed by the normal method; (b) Voltalef packed by the vacuum method; (c) Foam packed by the vacuum method

348

closed. The flat-bottomed connection of tap 1 prevented the foam material from upward movement during suction. After about 10 min of evacuation, distilled water was allowed to fill the column gradually through tap 2. The stopper of tap 1 was then replaced with a separatory funnel as a reservoir. Glass columns of various lengths and diameters (from 0.5 to 3 cm i.d.) can readily be filled by this method.

It is worth mentioning that application of this vacuum-packing technique for column fillings of TBP-treated granular Voltalef gave a good homogeneous bed with improved flow characteristics[23]. This technique is rapid, reproducible and simple and requires little technical skill.

Figure 2 shows plots of the flow rate (attained by gravity flow) as a function of the bed weight for TBP-loaded polyurethane foam and TBP-loaded granular Voltalef, using glass columns 10 and 15 mm in diameter and applying the above-mentioned vacuum-packing technique. Generally, the flow rate is decreased by the increase of the bed weight, but it is obvious from Fig. 2 that this is more pronounced in the case of Voltalef columns packed by the normal technique[23]. Foam columns generally allow higher flow rates.

The interstitial volumes of foam and Voltalef columns were also determined[24], and found to be 70–75% and 40–45%, respectively, of the original bed volume.

3. THE APPLICATION OF FOAM-FILLED COLUMNS IN SOME INORGANIC AND RADIOCHEMICAL SEPARATIONS

3.1. TRI-N-BUTYL PHOSPHATE (TBP) AS STATIONARY PHASE

3.1.1. THIOUREA - PERCHLORIC ACID MOBILE PHASE

The TBP – thiourea – perchloric acid system was used for the separation of nickel, bismuth and palladium[4] and also for chemical enrichment and separation of gold[25].

The rate of extraction of the palladium-thiourea complex by TBP-loaded polyurethane foam was investigated[4] by a batch technique. Figure 3 shows that the uptake of the complex by the loaded foam is fast and the half-life time of equilibrium adsorption, as calculated from the curves of Fig. 3, was 0.5 min. The half-life time of equilibrium adsorption of the palladium-thiourea complex on TBP-loaded Voltalef powder, under otherwise the same experimental conditions, was 1.6 min[24] (Fig. 4).

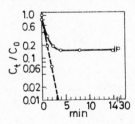

Fig. 3. Rate of adsorption of the palladium-thiourea complex on TBP-loaded polyurethane foam at room temperature.

Pd^{2+} – $1.44 \cdot 10^{-4}$ M; $NaClO_4$ – 1%; Thiourea – 3%; $HClO_4$ – 0.1 M

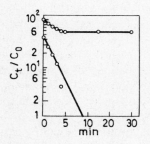

Fig. 4. Rate of adsorption of the palladium-thiourea complex on TBP-loaded Voltalef (0.16−0.25 mm particle size) at room temperature.

Pd^{2+} − $1.44 \cdot 10^{-4}$ M; $NaClO_4$ − 1%; Thiourea − 3%; $HClO_4$ − 0.1 M

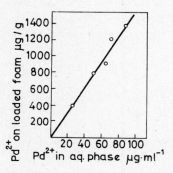

Fig. 5. Isotherm for the adsorption of the palladium-thiourea complex on TBP-loaded foam.

The uptake of the palladium-thiourea complex by the loaded foam[4] depends on the concentration of the complex in the aqueous phase. The isotherm (Fig. 5) shows that a good linear dependence was maintained between the concentration of palladium on the foam and its concentration in aqueous solution over a relatively wide range of palladium concentrations. In all cases equilibrium was approached only from one direction, that of the palladium-rich aqueous phase.

3.1.1.1. CHROMATOGRAPHIC BEHAVIOUR OF THE Pd - THIOUREA COMPLEX ON TBP-LOADED POLYURETHANE FOAM

Quantitative retention and elution of the palladium-thiourea complex was examined[4]. The results[4] indicated that all the palladium could be eluted with distilled water, in exactly the same manner as in the case of Voltalef columns[26], but with a relatively sharper peak[24]. The height equivalent to a theoretical plate (HETP), as calculated from the Glueckauf[27] equation, proved to be 1.7 and 3.0 mm for foam and Voltalef columns, respectively.

3.1.1.1.1. EFFECT OF TEMPERATURE AND FLOW-RATE ON THE HETP AND V_R OF ELUTION OF THE PALLADIUM–THIOUREA COMPLEX

The effect of flow-rates ranging between 0.9 and 6.4 ml·cm^{-2}·min^{-1} on the HETP was investigated[4] at 25, 35 and 45 °C. The results are shown in Fig. 6. The increase in plate height (decrease in column performance) with increasing flow-rate was pronounced only at 25 °C. On the other hand, only a slight increase in plate height with increased flow-rate occurred at higher temperatures, so that a relatively high flow-rate could be applied without any appreciable loss in column performance.

The V_R value was not affected by the flow-rate (broadening of the peak). In contrast, this value decreased with increasing temperature (Fig. 7) and at the same time the peak width decreased, resulting in sharper peaks. In general, increase of the temperature within the above-mentioned limits seemed to improve the column performance for palladium elution[4].

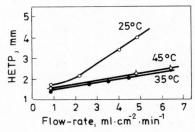

Fig. 6. Effects of temperature and flow-rate on the HETP of the palladium-thiourea complex using foam columns

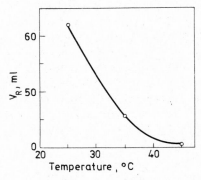

Fig. 7. Effect of temperature on V_R of the palladium-thiourea complex

3.1.1.1.2. CAPACITY

The practical usefulness of a column packed with TBP-loaded foam was tested by measuring its break-through capacity. Obviously, a column with a high break-through capacity will be used in many applications. The break-through capacity was defined as

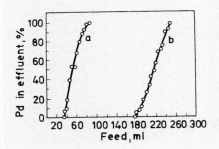

Fig. 8. Break-through curve of the palladium-thiourea complex on TBP-loaded foam (b) and TBP-loaded Voltalef (a). Flow-rate 1 ml·min^{-1}, Pd^{2+} – 0.309 mg·ml^{-1}. Foam column: weight of loaded foam: 3 g Voltalef column: Pd^{2+} –0.267 mg·ml^{-1}. Weight of loaded powder: 1 g

the amount of the palladium-thiourea complex that could be retained on the column when the complex was allowed to pass through at a rate of 1 ml·min^{-1} until first detected in the effluent solution.

After attainment of the break-through volume, elution was continued until the effluent solution concentration reached that of the feed solution. The curves of Fig. 8 represent both the break-through volume and the volume needed to reach bed saturation, for two columns packed with TBP-loaded polyurethane foam and granular Voltalef[4,23]. The break-through capacities as determined from the curves of Fig. 8 were 16.7 mg Pd^{2+}· g^{-1} loaded foam and 8.0 mg Pd^{2+}· g^{-1} loaded Voltalef. It is surprising, however, that whereas the percent of TBP present on loaded polyurethane foam and Voltalef powder was 68.6% and 48.7%, respectively, the break-through and overall capacities of the foam material were more than twice those of the Voltalef. This may be explained in light of the fact that foam materials have a relatively high surface area and most of these TBP-loaded surfaces are suitable for the adsorption of metal ion from the feed solution. On the other hand, a considerable amount of the TBP adsorbed on Voltalef powder does not have a complete chance to react with the metal ion in the external solution, perhaps because of the slowness of the diffusion processes which usually play an important role in this case.

3.1.1.1.3. SEPARATION OF NICKEL(II), PALLADIUM(II) AND BISMUTH(III)

As a consequence of the above characterization of the column performance for the palladium-thiourea complex, the separation of palladium(II) was tried[4] first from nickel(II), which does not form a complex with thiourea under the experimental conditions used, and then from bismuth(III), which readily forms a thiourea complex extracted on a TBP-loaded foam column. Separation of the three ions was also tried.

352

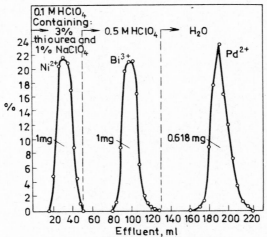

Fig.9. Separation of Ni(II), Bi(III) and Pd(II) on TBP—loaded foam at room temperature.
Bed height: 100 mm; Flow-rate: 2.5 ml·min^{-1}

Figure 9 shows the elution curves obtained[4] for the separation of a palladium — nick-el — bismuth mixture. Nickel(II) was found to move with the solvent front (0.1 M per-chloric acid — 3% thiourea — 1% sodium perchlorate), while the palladium(II) and bismuth(III) complexes were retained completely. The bismuth(III) complex was eluted with 0.5 M perchloric acid and then the palladium complex with water, at a flow-rate of 2.5 ml min^{-1} at room temperature.

3.1.1.2. CHEMICAL ENRICHMENT AND SEPARATION OF Au(III)

In a recent publication[25] it was proved that the gold-thiourea complex could be extracted with TBP under suitable experimental conditions. The formula of the extracted species was found to be probably $[Au(H_2N \cdot CS \cdot NH_2)]ClO_4 \cdot 4TBP$. The extraction of the gold-thiourea complex on TBP-loaded polyurethane foam was also tried[25]. The rate of extrac-tion of the gold complex by loaded foam was investigated by the batch technique. A series of these kinetic experiments was performed to determine the effects of various interfering elements on the rate of adsorption of the gold-thiourea complex on the TBP - loaded foam.

The elements whose interference with the separation of gold was tested were divided into three categories:

(a) Elements which do not interact with thiourea under the experimental conditions used, e.g. Zn^{2+}, Co^{2+} and Ni^{2+};

(b) Elements which form thiourea complexes, but are not or only partially extracted on TBP-loaded foam, e. g. Cu^{2+}, Fe^{3+} and Sb^{3+};

(c) Elements which form stable thiourea complexes which are strongly extracted onto the loaded foam material, e. g. Pd^{2+}, Bi^{3+} and Ag^+.

Zinc(II), iron(III) and bismuth(III) were selected to represent the three kinds of inter-fering elements. The results[25] show that the uptake of the gold complex by the loaded foam is fast and is not affected considerably by the presence of these elements. The half-life

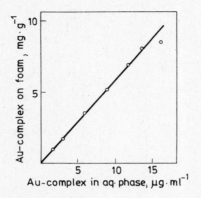

Fig. 10. Adsorption isotherm of the gold-thiourea complex on TBP-loaded foam

times of the equilibrium adsorption of gold alone and in the presence of Zn^{2+}, Fe^{3+} or Bi^{3+} were calculated to be 0.4, 0.6, 1.0 and 0.9 min, respectively.

As in the case of the palladium-thiourea complex, the uptake of the gold-thiourea complex by TBP-loaded foam depended on the concentration of the complex in the aqueous phase. Figure 10 shows a plot of the concentration of the gold-thiourea complex in the aqueous phase versus its concentration on the foam material. The isotherm indicates a good linear relationship over a relatively wide range of gold concentrations.

3.1.1.2.1. EFFECT OF THE FLOW-RATE

An important property of the foam column is the possibility of using high flow-rates, which can be attained simply by gravity flow, i. e. without applying any external pressure. A detailed study was therefore made of the effect of the flow-rate on the efficiency of retention of the gold complex; various concentrations of TBP in toluene were used to load the dry polyurethane foam (2 g) in a short column (5 cm bed height). Flow-rates from 10 to 150 ml·min⁻¹ were applied at room temperature. The results are shown in Fig. 11. Generally, the efficiency of extraction of 10 μg of gold from 100 ml of 0.1 M perchloric acid solution containing 3% thiourea and 1% sodium perchlorate was found to decrease on the increase of the flow-rate. It is clear from the curves of Fig. 11 that when undiluted TBP solution was used to load the foam material, relatively high flow-rates (up to 100 ml·min⁻¹) could be applied without appreciable loss in column performance.

The useful capacity of a column packed with 5 g TBP-loaded foam was determined at a flow-rate of 50—60 ml · min⁻¹ at room temperature. A total of 10 mg of gold(III) could be retained satisfactorily by the column from 1 l of aqueous solution.

354

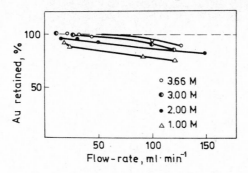

Fig. 11. Effect of flow-rate on the extraction efficiency of columns packed with TBP-loaded foam

3.1.1.2.2. SEPARATION OF GOLD FROM VARIOUS ELEMENTS

The practical utility of foam-filled columns was tested by studying their ability to separate gold quantitatively from various elements[25]: Zn(II), Co(II), Ni(II), Fe(III), Sb(III), Cu(II), Bi(III), Pd(II) and Ag(I).

It was possible to separate small amounts of Au(III) (10 μg) from high amounts of Zn(II), Co(II) or Ni(II) (1 g), because these elements moved with the solvent front, while the gold-thiourea complex was retained completely.

The gold-thiourea complex was separated from iron(III), antimony(III) or copper(II) by elution of these elements with water. It was also possible to separate the gold complex from the thiourea complexes of palladium(II) or bismuth(III) by elution of these latter complexes with water and 0.5 M perchloric acid, respectively. However, the silver-thiourea complex could not be completely removed from the column without affecting the gold complex.

A short foam column was found to be useful in chemical enrichment of small amounts of gold from dilute solution. On passage of 1 l of a solution containing 100 μg of gold(III) in 0.1 M perchloric acid, 3% thiourea and 1% sodium perchlorate, through a column filled with 5 g of TBP-loaded foam at a flow-rate of 50—60 ml·min^{-1}, 95—97% of the gold was retained by the column.

3.1.2. HYDROCHLORIC ACID MOBILE PHASE

3.1.2.1. SEPARATION OF Fe(III) FROM Cu(II), Co(II) AND Ni(II)

Separation of iron(III) from cobalt(II), copper(II) or nickel(II) has been investigated[7], by reversed-phase foam chromatography in a TBP-HCl system.

Before column chromatographic separation, liquid-liquid extraction experiments were conducted in order to determine the distribution ratios of HCl and the different metal ions between TBP and aqueous hydrochloric acid solutions. The distribution ratios of iron(III), cobalt(II) and copper(II) were further measured as a function of the hydro-

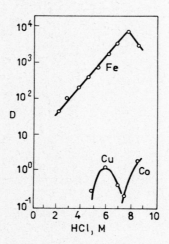

Fig. 12. Distribution ratios of cobalt, copper and iron in the TBP (polyurethane foam) — HCl system vs. hydrochloric acid concentration in the aqueous phase

chloric acid concentration, using TBP supported on polyurethane foam as the stationary phase. The results are shown in Fig. 12. The optimum hydrochloric acid concentrations obtained in liquid-liquid extraction were not significantly different in the case of TBP-loaded foam.

The chromatographic behaviours of iron(III), cobalt(II), nickel(II), and copper(II) were studied using columns filled with cylindrical pieces of polyurethane foam previously equilibrated with TBP. The retention of iron(III) is practically infinite from 4 M hydrochloric acid, whereas at this acid concentration nickel, copper and cobalt are completely eluted. Iron bound on the column could be eluted with 0.01 M hydrochloric acid. On the basis of these results, iron(III) was successfully separated from cobalt(II), nickel(II), or copper(II).

Figure 13, for example, shows the separation of iron(III) from nickel(II). Iron(III) could similarly be separated from copper(II) or cobalt(II)[7]. Separation of iron(III) from copper(II) or cobalt(II) using foam columns was also carried out at different metal ratios.

The separation of iron and cobalt was studied in greater detail because of its practical importance. Separation of ^{59}Fe and ^{58}Co was examined too. The purities of the separated isotopes were checked by means of their gamma-spectra taken with a Ge(Li) detector. The spectra of the initial solution and the solutions containing the separated isotopes are presented in Fig. 14. A quantitative yield of cobalt was obtained in the separation, while an iron recovery close to 100% was possible only in the case of larger amounts of iron. It was therefore recommended that the iron isotope must either be diluted with a carrier, or — if carrier-free isotope is required — one must be content with a yield of about 80%.

356

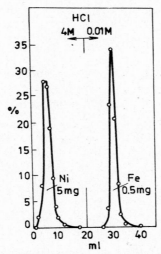

Fig. 13. Separation of iron from nickel.

Support: polyurethane foam; Stationary phase: TBP; Flow-rate: 1 ml·min⁻¹; Eluent for nickel: 4 M hydrochloric acid; Eluent for iron: 0.01 M hydrochloric acid ·

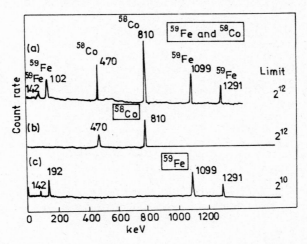

Fig. 14. Gamma-spectra of a mixture of ^{59}Fe and ^{58}Co and of the fractions obtained by separation on a TBP (polyurethane foam) chromatographic column; (a) before separation, (b) and (c) fractions obtained after separation

3.2. TRI-N-OCTYLAMINE (TNOA) AS STATIONARY PHASE

3.2.1. HYDROCHLORIC ACID MOBILE PHASE

The applicability of the foam chromatographic method to the tri-n-octylamine (TNOA – hydrochloric acid system has been investigated[5]. The separation of cobalt from nickel was studied in detail over a wide range of concentrations, as a representative separation model of practical importance.

The uptake of TNOA by polyurethane foam was found[5] to depend on the time of contact between the foam material and the TNOA solution. The amount of the amine loaded on the foam was determined gravimetrically and plotted as a function of the

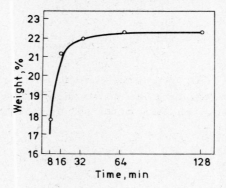

Fig. 15. Time-dependence of the tri-n-octylamine hydrochloride uptake by polyurethane foam

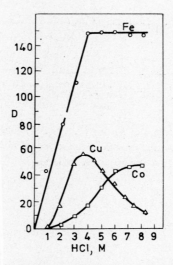

Fig. 16. Cobalt, iron and copper distribution ratios between a 67% TNOA loaded foam and hydrochloric acid

358

shaking time (Fig. 15). The curve shows that, using 0.1 M TNOA in benzene, the maximum possible load (22.4 wt% saturation value) was taken up by the foam, in about 30 min. On the other hand, TNOA uptake by the foam also depends on the concentration of the TNOA solution. The maximum uptake of TNOA (67 wt%) was obtained from a solution of 1 M TNOA in benzene previously equilibrated with the stoichiometric amount of 1 M hydrochloric acid.

3.2.1.1. SEPARATION OF Co (II) AND Ni(II)

The distribution ratios of cobalt(II), iron(III) and copper(II) were determined[5] as a function of the hydrochloric acid concentration using TNOA-loaded foam (67 wt%) as the stationary phase (Fig. 16). The distribution ratio of nickel was negligible at any acid concentration. It is clear from the distribution curves that higher acid concentrations favour the separation of cobalt from nickel.

Elution tests for cobalt(II), nickel(II), iron(III) and copper(II) in hydrochloric acid solution were carried out on columns filled with unloaded polyurethane foam, to see whether the support was completely inert or not, i.e. to check the zero retention in hydrochloric acid medium for each of the four metals. The results of these elution tests showed nearly zero retention of the four metal ions; thus, the support can be regarded as essentially inert towards them.

The chromatographic separation of cobalt from nickel on a TNOA — HCl — polyurethane column was first tried using foam material loaded with 67 wt% TNOA. Reasonably symmetrical peaks were obtained under these conditions, but only 85—88% of the metal could be eluted at a flow-rate of 1 ml·min^{-1}. This was explained by the probable retardation of the kinetic processes in the concentrated tri-n-octylamine hydrochloride on the foam. Attention was therefore directed towards the application of foam columns with lower percentage loads (3.8, 11.4 and 17.7%). Separation on the column loaded with 3.8% TNOA was incomplete, whereas the column with a load of 11.4% gave a good and complete sepa-

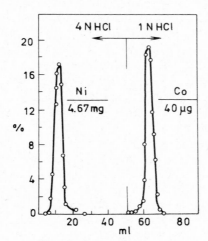

Fig. 17. Cobalt-nickel separation.

Stationary phase: 11.4% TNOA·HCl. Support: polyether-type polyurethane foam; Flow-rate: 1 ml·min^{-1}.
Recoveries: 94.5% for Ni and 99.5% for Co.

ration of cobalt and nickel (Fig. 17). Good results were also obtained[5] with the 17.7% load,
Fig. 18 shows the separation of carrier-free ^{58}Co from nickel.

The separation of cobalt and nickel was carried out over a very wide range of concentration

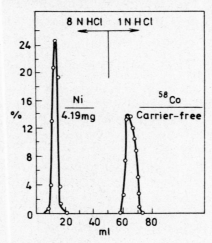

Fig. 18. Separation of carrier-free ^{58}Co from nickel.
Stationary phase: 17.7% TNOA·HCl. Support and flow-rate as for Fig.17. Recoveries: 99.6% for Ni,
and 101.6% for Co

3.3. TETRACHLOROHYDROQUINONE AS STATIONARY PHASE

3.3.1. REDOX REACTIONS ON NORMAL COLUMNS

Redox reactions have been examined on open-cell polyurethane foam columns supporting tetrachlorohydroquinone[6]. Polyurethane foam retains chloranil (tetrachloro-p-ben-
zoquinone) well and has been proposed for use in column operation in place of granular
Kel-F [28,29], with all the advantages of foam-filled columns.

The practical usefulness of a column packed with chloranil-loaded foam material was
tested by measuring its break-through capacity. Two columns were examined in which
the amount of dry foam was the same, but the chloranil loading solutions were of different concentrations. The columns were packed with 3 g of polyurethane foam previously treated with 1 g or 3 g of chloranil in 50 ml chlorobenzene. The break-through
curve of the former column loaded with chloranil solution of lower concentration was sharper
than that of the latter (Fig. 19). This was explained by the probable formation of larger
redox reagent crystals on the foam material treated with the more concentrated solution
(3 g chloranil/50 ml). The larger crystals of the redox reagent would be expected to react
more slowly than the very fine crystals observed to form in the case of foam material treated
with the more dilute chloranil solution (1 g/50 ml).

The effects of temperature and flow-rate on the redox column efficiency were
tested[6] by the reduction of cerium(IV), vanadium(V) and iron(III) for a wide range of

360

flow-rates at 35 °C and room temperature. Figure 20 presents the results obtained for the reduction of cerium(IV) at flow-rates ranging between 2 and 11 ml·min^{-1} at room temperature. It is clear that 2 ml of 0.0851 M cerium(IV) sulphate was completely re-

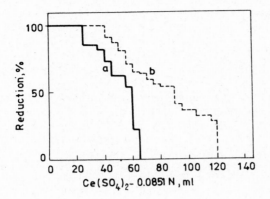

Fig. 19. Break-through capacity curve of cerium(IV) reduction on columns packed with open-cell polyurethane foam loaded with various amounts of chloranil.

a − 1 g chloranil/3 g dry foam; b − 3 g chloranil/3 g dry foam

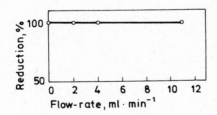

Fig. 20. Effect of flow-rate on the reduction of cerium(IV).

Ce: 0.1702 meq; Eluent: 1 M H_2SO_4; room temperature

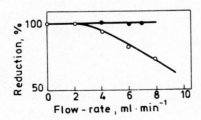

Fig. 21. Effects of flow-rate and temperature on the reduction of vanadium(V).

V − 0.1706 meq ○ − Room temperature; ● − 35 °C. Eluent: 1 M H_2SO_4

duced at all the flow-rates applied. In the case of vanadium(V), on the other hand, the reduction of 2 ml of 0.0853 N ammonium vanadate was quantitative only in the range 2–4 ml·min^{-1} (cf. Fig. 21). However, when the temperature of the foam-filled column was increased to 35 °C, quantitative reduction of vanadium(V) was obtained even if flow-rates as high as 7 ml·min^{-1} were applied (Fig. 21).

Application of the developed foam column for the reduction of cerium(IV), vanadium(V) and iron(III) gave quantitative results. It was possible to make a further measurement of the amount of the ion reduced: the amount of tetrachlorohydroquinone oxidized on the column was determined by passing a known volume of standard ascorbic acid solution through it and titrating the effluent with standard iodine solution.

3.3.2. REDOX REACTIONS ON PULSED COLUMNS

One of the useful developments of foam chromatography is the recent application of chloranil-loaded polyurethane foam in a pulsed column[30]. The pulsed column operation is mainly due to the flexible character of open-cell plastic foams, which is not found with any other kind of support, and to the rapid reaction between the metal ion in the solution and the redox reagent on the foam-sample. The proposed pulsed column method is simply realized after packing the chloranil-loaded foam material into a medical syringe. Pressing the glass plug of the pulsed column (syringe) will generally produce a compression of the foam material. If the tip of the column (syringe) is kept in a given solution and the plug is gradually released, the solution will penetrate into the column and the foam material will return to its original volume. Repetition of this process allows the external solution to come into contact with the redox reagent several times. This will result in reaction between the metal ion in solution and the redox reagent on the foam.

The effect of the number of pulses on the reduction efficiency of the pulsed column has been investigated[30] using two different columns. The first column was packed with foam material loaded by the previously described method[6] (i.e. 1 g chloranil/3 g foam). In contrast, the second column was filled with swollen foam prepared by mixing 3 g of the dried foam material with 50 ml chlorobenzene solution containing 1 g chloranil. The solvent was gradually removed by mixing the foam material, kept at ca. 80 °C in a drying oven, with a glass rod until a swollen foam remained with only a small volume of chloranil-chlorobenzene solution. The swollen foam was dried between two sheets of filter paper to remove the excess solution and then packed in the syringe in the usual manner. Using chlorobenzene and swollen or dry-loaded foam, a study was made of the effect of the number of pulses on the efficiency of reduction of Ce(IV)[30]. The efficiency of the pulsed column packed with the swollen foam material was found to be better than that of the column packed with the dry foam (Fig. 22).

This was explained by the more favourable kinetics expected for reaction between the metal ion in solution and the chloranil in the swollen foam.

362

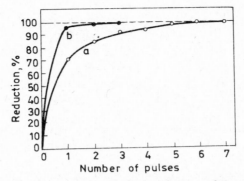

Fig. 22. Effect of the number of pulses on the reduction efficiency of the pulsed column.
a − With normally loaded foam; b − With swollen foam. Cerium(IV) sulphate concentration: 0.005 M;
Volume of solution; 5 ml; Room temperature

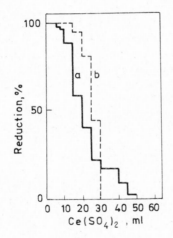

Fig. 23. Break-through capacity curves for cerium(IV).
a − At room temperature; b − After preheating the solution to ca. 80 °C. Weight of swollen foam:
2 g; Cerium(IV) sulphate concentration: 0.1 M

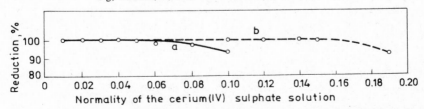

Fig. 24. Effect of the concentration of cerium(IV) sulphate on the efficiency of the pulsed redox column.
a − At room temperature; b − After preheating the solution to ca. 80 °C. Number of pulses: 5; Volume
of solution: 10 ml

The break-through capacity of the pulsed column packed with swollen foam material was measured[30] at room temperature and after preheating the solution before reduction to ca. 80 °C, using e.g. cerium(IV) solution. The results are presented in Fig. 23. It is clear that the break-through capacity curve obtained with preheating the solution before reduction is sharper than that obtained at room temperature. Moreover, the capacity obtained up to the break-through point depends on the temperature of the external solution. At room temperature it was 242.6 mg of cerium(IV), while with preheating the solution it was 606.5 mg.

The percentage reduction of the metal ion by the pulsed column (for 5 pulses) was found to vary considerably with the temperature and with the original concentration of the external solution. Figure 24, present the results obtained[30] for cerium(IV), reduced at different metal ion concentrations at room temperature and at ca. 80 °C. It is clear that preheating the metal ion solution before reduction generally increases the efficiency of the pulsed column.

Application of the pulsed foam column for the determination of cerium(IV), vanadium(V) and iron(III) gave quantitative results.

It is worth mentioning that homogeneous ion-exchange foams have been prepared[31] by introducing ion-exchange groups on previously prepared phenol-formaldehyde, polyurethane and polyethylene foams. Heterogeneous ion-exchange foams too were prepared by foaming a fine powder of a commercially available cation exchanger with the precursors of open-cell polyether-type polyurethane foam. A cation-exchange process was recently investigated[32] on the heterogeneous foam and seems to be very promising.

REFERENCES TO CHAPTER 13

1.   See Chapter 4.
2.   R. F. HORNBECK, J. Chromatog., 30 (1967) 438.
3.   T. BRAUN, A. B. FARAG, Talanta, 19 (1972) 828.
4.   T. BRAUN, A. B. FARAG, Anal. Chim. Acta, 61, (1972) 265.
5.   T. BRAUN, É. HUSZÁR, L. BAKOS, Anal.Chim. Acta, 64 (1973) 77.
6.   T. BRAUN, A. B. FARAG, A. KLIMES-SZMIK, Anal. Chim. Acta, 64 (1973) 71.
7.   T. BRAUN, L. BAKOS, Zs. SZABÓ, Anal. Chim. Acta, 66 (1973) 57.
8.   C. R. THOMAS, British Plastics, (1965) 552.
9.   A. BITTEL, Dissertation, Tübingen, 1957.
9a.   H. BRUNSCHWIG, Liber de Arte Distillandi, 1512.
10.   E. BAYER, Gaschromatographie, 2 Aufl. Springer-Verlag, Heidelberg, p. 4., 1962.
11.   I. O. SALYER, R. T. DAYTON, R. T. JEFFERSON, W. CAROLLTON, W. D. ROSS, US Pat. No. 3.580.843, May 25, 1971, USA.
12.   W. D. ROSS, R. T. JEFFERSON, J. Chromatogr. Sci., 8 (1970) 386.
13.   H. SCHNECKO, O. BIEBER, Chromatographia, 4 (1971) 109.
14.   H. J. M. BOWEN, J. Chem. Soc. A, (1970) 1082.
15.   H. J. M. BOWEN, Radiochem. Radioanal. Lett., 7 (1971) 71.
16.   P. SCHILLER, G. B. COOK, Anal. Chim. Acta, 54 (1971) 364.
17.   H. D. GESSER, A. CHOW, F. C. DAVIS, Anal. Lett., 4 (1971) 883.
18.   J. F. UTHE, J. REINKE, H. GESSER, Environmental Letters, 3 (1972) 117.

19. H. D. GESSER, A. B. SPARLING, A. CHOW, C. W. TURNER, J. Am. Water Works Assoc., 65 (1973).
20. A. CHOW, H. D. GESSER, M. MAZURSKI, Anal. Chim. Acta, 65(1973) 99.
21. T. BRAUN, A. B. FARAG, Anal. Chim. Acta, 66 (1973) 419.
22. E. CERRAI, G. GHERSINI, in J. C. GIDDINGS and R. A. KELLER Eds., Advances in Chromatography, Vol. 9, M. Dekker, New York, p. 16, 1970,
23. T. BRAUN, A. B. FARAG, Anal. Chim. Acta, 62 (1972) 476.
24. T. BRAUN, A. B. FARAG, unpublished results.
25. T. BRAUN, A. B. FARAG, Anal. Chim. Acta, 65 (1973) 115.
26. M. YUSAF, Z. ŠULCEK, J. DOLEŽAL, Anal. Lett., 4 (1971) 119.
27. E. GLUECKAUF, Trans. Faraday Soc., 51 (1955) 34.
28. E. CERRAI, C. TESTA, Anal. Chim. Acta, 28 (1963) 205.
29. R. BELCHER, J. R. MAJER, G. A. H. ROBERTS, Talanta, 14 (1967) 1245.
30. T. BRAUN, A. B. FARAG, Anal. Chim. Acta, 65 (1973) 139.
31. T. BRAUN, O. BÉKEF FY, I. HAKLITS, K. KÁDÁR, G. MAJOROS, Anal. Chim. Acta, 64 (1973) 45.
32. T. BRAUN, A. B. FARAG, Anal. Chim. Acta, 68 (1974) 119.

CHAPTER 14

LAMINAR TECHNIQUES AS AN AID IN PLANNING COLUMN EXTRACTION CHROMATOGRAPHIC SEPARATIONS

G. GHERSINI, E. CERRAI

1. INTRODUCTION

Planning column separations in extraction chromatography involves a search for the best conditions to resolve a known mixture of elements. Besides the possible problems related to the optimization of the column resolution power, that are beyond the scope of the present chapter, the search consists in the choice of the most suitable extraction system (i. e. nature of both organic and aqueous phases), and in the subsequent optimization of the concentration parameters involved, most often limited to the assessment of the best concentration values of the components of the aqueous phase.

These choices can be based on the general knowledge of the liquid-liquid extraction behaviour of the elements of interest, that certainly provides sufficient information for a wise choice of the extraction system, but often is insufficient to exactly predict the optimal concentration parameters involved.

The optimization of such parameters by means of column experiments would require cumbersome and long operations. Quick and simple methods, capable of giving almost all the information required, are paper or thin-layer extraction chromatography, which on the one hand retain all the fundamental features of extraction chromatography and, on the other hand, couple relatively rapid preparation and operation with little dimensions and economy of materials.

Speed and simplicity are the main advantages of such laminar extraction chromatographic systems. They present the highest grade of flexibility, but since they work on small spots, they are severely limited when the separations of massive amounts of elements, or the recovery of the separated substances are required.

Conversely, they result in a very attractive means to collect a great number of data in a short time and they actually have been thoroughly used for checking the applicability of well-known extractants as stationary phases in chromatography, for gaining information on the behaviour of particular elements in more or less known extraction systems, for investigating on the usefulness as extractants of new organic compounds, or specifically for planning column extraction chromatographic separations.

As a result, an enormous mass of data is available for elements eluted with different aqueous phases on paper or thin-layers impregnated with a variety of stationary phases. They cover not only almost all extraction systems that found so far actual application in columns, but also several attractive systems not yet exploited for column separations.

Most data are presented as R_f spectra, that is as R_f values of an element as a function of the concentration of the relevant component in the eluting solution. Since R_f values can be more or less quantitatively correlated to the retention volumes obtainable with a column having the same stationary and mobile phases, R_f spectra represent the most suitable indirect guide to the correct choice of the best conditions for a column separation.

In this chapter, a brief outline of the experimental techniques of paper and thin-layer extraction chromatography is followed by a discussion on the correlations of laminar R_f values to the distribution coefficients of the relevant liquid-liquid extraction system, and to the retention volumes of the corresponding chromatographic column. The principal features involved in planning a column separation from laminar R_f spectra are presented, and most R_f data are reviewed, available in the literature for the different extractants used as stationary phases.

No particular attention is given to the potential usefulness of the laminar techniques for identifying the components of unknown mixtures of elements to be separated, since very little examples of such application are found in the literature. Actually, most column extraction chromatographic separations involve mixtures of elements whose composition is already known either from their origin or from preliminary qualitative analysis made with more traditional techniques, such as spectrographic analysis or γ-spectroscopy of radioactive solutions.

2. EXPERIMENTAL TECHNIQUES IN LAMINAR
EXTRACTION CHROMATOGRAPHY

Experimental techniques for paper and thin-layer extraction chromatography are quite simple and all based on the fundamental operations of normal chromatography but for the preliminary treatment which is meant to ensure retention of the extractant (stationary phase) on the support itself.

Paper used in extraction chromatography is of normal chromatographic grade. In general, it is used as supplied, and only for particular purposes, e.g. when Ca or alkali metals are to be eluted, a preliminary washing is necessary. Treatment at 40—60 °C in an oven is often recommended to eliminate humidity before impregnation, that is accomplished by dipping the paper into a solution of the given extractant. Care has to be taken to obtain a uniform treatment. Extractant solutions have their maximum concentration limited by the wettability of the treated paper: usually, 0.01 to 0.5 M solutions in an organic diluent are used.

Dipping time is often carefully controlled. After impregnation, the treated paper is dried, usually with warm air.

To avoid tailing and irregular spots, generally the extractant solution for paper treatment is previously equilibrated with an aqueous phase having a composition equal to that of the eluent to be used. In practice, a strict coincidence is not imperative: often papers treated with an organic phase equilibrated with the relevant aqueous solution were eluted with eluents with the same components but at different concentrations over a reasonably wide range, without any great influence upon the results.

Although a given extractant undergoes different degrees of polymerization in different diluents, no particular effect of the diluent used is encountered.

Because of the low capacity of paper for the extractant, metals to be eluted are spotted in very low amounts, not to overload the stationary phase. Small drops are applied with a pipette or a glass rod. Metal solutions must be chosen which do not affect the composition of the eluent when passing over the spot: nevertheless, hydrolysis phenomena sometimes make the use of strongly acidic solutions unavoidable, and scattered results obtained with easily hydrolyzable ions can often be imputed to this feature. After application, spots are dried with a current of warm air.

Elution can be carried out with any of the well-known ascending, descending and radial chromatographic techniques. Most results, however, refer to the ascending technique.

After elution, the paper is dried in the usual way. Caution must be taken when $HClO_4$ is used as the eluent, to avoid dangerous reactions with cellulose.

The eluted elements are visualized with normal colorimetric reagents, of which comprehensive lists can be found in Refs [1—5]. It is necessary to point out that sometimes reagents, which are satisfactory in other chromatographic techniques, prove to be useless in extraction chromatographic work, since the complexes between the given metal and the stationary phase can be stronger than that of the chosen chromogenic reaction. Reagents, which base their detection ability on pH variations in the medium where a given metal is found, are normally useless because of the strong ionic character of the eluting solutions.

The great advantage of paper as the support for extraction chromatography lies in the great simplicity of the technique, that can easily give a noticeable number of data in a relatively short time. Actually, most R_f spectra were obtained by paper experiments. Provided that tailing or streaking phenomena do not occur, reproducibility is rather good.

On the other hand, paper suffers from a number of disadvantages. It has a low capacity toward the extractant and very small amounts of elements must be applied, which makes it difficult the subsequent detection of the eluted spot position. Elution rate varies with extractant loading or eluent ionic strength; it also varies as the eluent front moves along the strip. This rate is important because of the kinetic effects often involved in the retention mechanism. The phase volume ratio may vary along the chromatogram, and the concentration of a given component in the eluting solution may also vary.

In spite of all this, in most cases results obtained with paper compare favourably with those obtained with techniques which certainly are not affected by such phenomena, not only for the $R_M (= \log(1/R - 1))$ dependencies on the various parameters, but also for the separation factor obtained for two given elements.

Another disadvantage of paper is the lack of stability of cellulose toward some chemical reagents, e. g. concentrated $HClO_4$ or H_2SO_4. Glass-fibre paper can be used when such reagents are the eluents, impregnation being the same as described for cellulose paper. Cellulose can be used up to $3M$ $HClO_4$ or $4M$ H_2SO_4, but great care is necessary in the drying step.

Thin-layers have been obtained with supports analogous to those used for columns, mainly silica base compounds or cellulose. Support treatment is also very similar to that employed for columns. The thin-layers are prepared in the very same way as in traditional chromatography with either treated or untreated material. Binders are obviously to be avoided, if possible, when investigating the behaviour of the elements in a given system.

Elutions have been carried out with ascending, descending and horizontal techniques. Spotting and visualization of the elements are very similar to the methods already described for paper chromatography, and the same comments are true for the chromogenic reagents used to detect the eluted spot. The drying steps must be done in an oven, however, and not with warm air as can be done with the stiffer paper.

Many features of thin-layer extraction chromatography and comparisons with paper chromatography can be found in Refs [1,2]. Results obtained with thin-layers can be very reproducible (± 0.02 R_f units) [1], provided that tailing or streaking do not occur.

Thin-layers have the same advantages (ease and speed) as paper, but to a lesser degree because of the more complicated preparation of the stationary phase. However, they have a number of advantages over paper, because of their higher capacity. The phase volume ratio is probably more constant along the chromatogram than with paper, although no information on this feature is available; also a concentration gradient along the chromatogram is likewise absent. Furthermore, the variety of suitable supporting materials allows the use of eluents for which cellulose is unsuitable.

3. CORRELATION BETWEEN R_f VALUES
AND DISTRIBUTION COEFFICIENTS

As in column systems, laminar techniques in extraction chromatography can be discussed in terms of liquid-liquid extraction phenomena, so that R_f values may be related to distribution coefficients D under analogous conditions.

On the basic assumption that the process is fundamentally the same in both cases and that the spatial boundaries of the two phases are so well defined as to make meaningful the ratio of their cross-sectional areas, the relationship between D and R_f is the classical Martin and Synge equation

$$D = \frac{A_m}{A_s} (\frac{1}{R_f} - 1) \tag{1}$$

where A_m and A_s are the cross-sectional areas of the mobile (aqueous) and stationary (organic) phases, respectively, assumed to be constant along the whole chromatogram.

Equation (1) can be combined with the equations that are believed to relate the distribution coefficients to the various parameters involved in the extraction process considered, so as to obtain a relation between the thermodynamic parameters of the chromatographic system and the resulting R_f values. Actually, this was often done in paper and thin-layer studies, both to check the similarities between the chromatographic and liquid-liquid extraction systems, and to investigate on the retention mechanism of new compounds used as the stationary phases.

However, the applicability of the relations between the R_f values and the chromatographic parameters can be affected by a number of uncertainties deriving from the experimental set-up of the chromatographic system, in particular by the possible non-constancy of the $A_m : A_s$ ratio, and by the possible variation of the eluent composition along the chromatogram.

Although suitable impregnation techniques normally ensure a homogeneous loading of the whole support, the $A_m : A_s$ ratio may not remain constant along the whole chromatogram because of possible variations of the cross-sectional area of the mobile phase, especially in ascending chromatography. This may occur both in paper and thin-layers, but in the former case is likely to affect the results to a higher extent. Actually, spots in ascending paper chromatography seldom reach R_f values higher than $0.93-0.95$, even when the eluted element is known to have extremely low distribution coefficients in a large range of eluent compositions around the actual one involved. This suggests that the front of the eluent, as experimentally detected, may sometimes include a small portion of chromatogram where the support and stationary phase are only slightly wetted by the eluting solution. It is obvious how this could give rise to R_f values that scarcely fit in equation (1).

The few available experimental data of the actual volume of the mobile phase along a chromatogram refer to HDEHP-treated paper[6] and TBP-treated cellulose thin-layers[7] eluted with hydrochloric acid solutions by the ascending technique. They show that the above mentioned possible variations are rather small, anyway undetectable by the rather unsophisticated methods that were adopted for measurements: further more accurate experiments should be done, and extended to other systems.

An interesting experimental result, probably related to the ratio of the two phases, is the dependency of R_f values on the thickness of thin-layers prepared with the very same support and stationary phase, reported for HDEHP-treated kieselguhr[8]. It suggests a different role played in the chromatographic process by the portion of mobile phase that runs within the bulk of treated supporting material, and the portion that wets the outside parts of the chromatogram. In this case, however, the phenomenon is likely to result in coefficients in equation (1) slightly different from the actual $A_m : A_s$ ratio, but nevertheless constant for a given layer thickness. It should not affect, therefore, the usual deductions based on the slopes of R_M vs. log (concentration) plots.

The constancy of the eluent composition along the chromatogram should be ensured by a good pre-equilibration of the extractant used as the stationary phase, and of a suitable conditioning of the chromatographic chamber. In practice, however, the extractant is often pre-equilibrated with aqueous phases that are not strictly equal to those subsequently used as mobile phases, being averaged within the concentration range planned for elutions. Furthermore, even extractants equilibrated with an aqueous phase having the very same composition of the eluent may sometimes result in incorrectly pre-equilibrated stationary phases, since the impregnation step may involve selective processes.

Variations of the eluent composition due to extraction of the bulk components by the stationary phase are likely to be important in most chromatographic systems, and to affect in particular the highest R_f values and those systems where the R_f values decrease as the eluent concentration is increased. Also in this case, however, very few experimental data are available; they demonstrate a noticeable lowering of the hydrochloric acid concentration in the neighbours of the eluent front of ascending chromatograms on HDEHP-treated paper, involving the $R_f > 0.8$ portion of the final chromatogram[6].

Besides the possible limitations of equation (1) arising from the above mentioned experimental features, its overall applicability to extraction chromatographic systems has recently been under discussion. Actually, such systems differ from most of the traditional liquid-liquid chromatographic ones in the nature of their stationary phases, which have polar groupings that give them an electrical nature quite similar to that of the aqueous solutions used as the mobile phases. Furthermore, the inorganic ion-containing species that undergo partition between the two phases are often known to result from the simultaneous coordination of the ions by molecules of both the stationary and the mobile phase.

Both these peculiar properties led Bark and coworkers to have "some doubt concerning the absolute chromatographic and physical boundaries of the two phases"[7]. In the case a relevant interaction would occur between the two phases, then the systems could markedly depart from chromatographic ideality, and be unsuitable to be correctly represented by the various classical chromatographic equations, such as equation (1 all based on the ideal behaviour of two independent and discrete phases.

From experiments carried out with cellulose thin-layers loaded with different extractants[7,9-12] and eluted with hydrochloric acid, the above mentioned authors propose a more general linear correlation between R_f and D, of the type

$$D = A \left(\frac{1}{R_f} - 1 \right) - B \tag{2}$$

differing from equation (1) not only in the existence of an additional term B, but also in the possible remarkable disagreement between the A values and the corresponding ratio of the cross-sectional areas of the mobile and stationary phases, determined experimentally.

The same authors interpreted equation (2) in the light of both the known partial mutual solubility of the two phases and of the possible inactivity in the chromatographi process of part of the active groups of the stationary phase, consumed for binding to the cellulose support. They suggest that "A is a volume ratio term, not necessarily the ratio of the reversed-phase material to the amount of solvent covering a particular area of the plate and being concerned with the intermixing of the layers and the use of part of the stationary phase in binding to the cellulose layer; B is a term related to the interaction of the mobile and stationary phases"[7]. Only in systems where no interaction occurs between the two phases, and the stationary phase is all chromatographically active, equation (2) would therefore reduce to the ideal equation of Martin and Synge.

Obviously, the suggested departure from ideality is not limited to laminar systems, but applies also to inorganic reversed-phase extraction columns, and probably is strictly related to the general problem of the variations of phase volumes — and even third phase formation — of many liquid-liquid extraction systems. It perhaps could explain a number of those anomalous chromatographic results and of unexpected relations found in many extraction chromatographic works. However, in spite of the interesting implications of equation (2) both for laminar and column chromatography, no further investigation was carried out, to check Bark's interpretation or to better define the parameters possibly involved in the non-ideality of the systems. At present, therefore, equation (2) suggests what the investigated systems may not be, without explaining what they actually are.

In conclusion, R_f values can be affected by uncertainties deriving from the experimental set-up, mainly because of the possible variations of the eluent composition along the chromatogram. This must be kept in mind when using R_f values to calculate distribution coefficients or to investigate the retention mechanism of a given chromatographic system, and a reasonable allowance must be given for the accuracy of results obtained.

Within these limits, the validity of equation (1) to most extraction chromatographic systems is supported by the satisfactory agreement often found when the results obtained through it are compared to the corresponding liquid-liquid extraction data. Several examples of this are reported by Akaza in Chapter 2.

Bark's suggestion of a possible remarkable departure from ideality of many extraction chromatographic systems is undoubtedly very interesting, but certainly needs extensive further investigation before it could be applied to quantitatively correlate R_f data to distribution coefficients.

The separation factor between two given elements is defined in liquid-liquid extraction as the ratio of the distribution coefficients of the two elements in the very same extraction conditions. Analogously, in laminar extraction chromatography the separation factor is calculated by the ratio of the $(\frac{1}{R_f} - 1)$ values of the two elements eluted in the very same experimental conditions, assuming that equation (1) is valid in the chromatographic system considered.

Obviously, all comments already reported on the applicability of equation (1) to actual extraction chromatographic systems apply also to the possible correlations among separation factors obtained by liquid-liquid extraction and laminar extraction chromatography. In most cases, the resulting values are in good agreement among themselves; in some instances, however, relevant differences were found. A detailed discussion of the different cases can be found in Ref.[6].

4. CORRELATION BETWEEN R_f VALUES AND RETENTION VOLUMES

The R_f values for an element in a given laminar chromatographic system can be quantitatively correlated with the retention volume V_R obtainable for the same element in the analogous column system, through the relations of R_f and V_R to the distribution coefficient D of the element between the two chromatographic phases.

By assuming equation (1) to be valid for the laminar system considered, and recalling the analogous relation $D = (V_R - V_m)/V_s$ for columns (where V_m and V_s are the volumes of the mobile and stationary phases in the column, respectively), the following equation can be easily derived

$$\frac{V_R}{V_m} - 1 = \frac{A_m}{A_s} \frac{V_s}{V_m} \left(\frac{1}{R_f} - 1\right) = K \left(\frac{1}{R_f} - 1\right) \tag{3}$$

that points out an inverse relation between V_R and R_f values. The main consequence of this inverse relation is that the difference between V_R values, corresponding to a given difference between R_f results, will be greater the lower the R_f values involved. This is clearly illustrated in Fig. 1, showing the relation between the column peaks position (expressed in column void volume units $V_R : V_m$) and R_f values when $K = 1.5$.

The importance of the value of K is apparent in equation (3). In the experimental conditions normally used in extraction chromatography, both $A_m : A_s$ and $V_m : V_s$ ratios can vary of factors even greater than ten, mainly because of analogous variations of the volumes of the stationary phase, on their turn depending on method and conditions adopted for the impregnation of the supporting material. Although the value 1.5 for K, chosen for Fig. 1, could actually apply to the correlation of R_f values obtained with 0.1 M HDEHP-treated paper ($A_m : A_s = 15$)[6] with certain column results ($V_m : V_s = 10$)[13], in normal practice K values three- to five-fold greater or even ten-fold lower are likely to

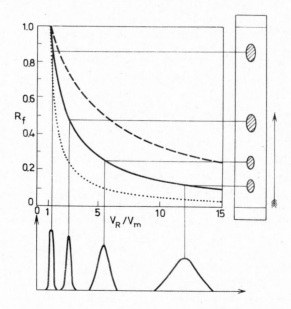

Fig. 1. Relation between laminar R_f values and column retention volumes for corresponding ideal chromatographic systems, when $K(=A_m V_s / A_s V_m) = 1.5$. The dashed and dotted lines refer to $K = 4.5$ and $K = 0.5$, respectively (see the text)

be encountered. In effect, the reported $A_m:A_s$ value probably approaches the minimum figure obtainable for laminar techniques (where the upper limit of A_s is dictated by an acceptable wettability of the paper or thin-layer by the aqueous mobile phase), while chromatographic columns can imply $V_m:V_s$ ratios lower or remarkably greater than 10, depending on whether they are prepared with the aim of greater capacity or maximum resolution power, respectively.

The relation between $V_R:V_m$ and R_f, for K equal to 4.5 and 0.5, are respectively shown in Fig. 1 as dashed and dotted lines.

An accurate analysis of Fig. 1 reveals the limitations of quentitative correlations between R_f values and retention volumes:

− for relatively low K values, such as K ⩽ 1.5, spots having relatively high R_f values, even when satisfactorily separated in the laminar chromatogram, correspond to peaks of the column having very small retention volumes, whose resolution is difficult;

− for relatively high K values, such as K ⩾ 4.5, spots having relatively low R_f values correspond to peaks having noticeably large retention volumes, often unacceptable in normal practice.

Possible uncertainties deriving from the experimental set-up of laminar chromatography may affect the correlation between column and laminar chromatographic results to the same extent as they affect the calculation of distribution coefficients from R_f data; again a reasonable allowance should be given for the accuracy of the resulting V_R values.

The information available at present does not foresee the consequences of a possible marked deviation from ideality of the chromatographic system, that should put under discussion the correlations between the distribution coefficient and the chromatographic parameter of both laminar and column systems.

To the knowledge of the authors, equation (3) has never been checked with purposely collected column and laminar data. Furthermore, the attempt to find the necessary information in the literature reveals that a very small number of data are suitable for comparison, because of the very little information on the phase volume ratios of both chromatograph systems, or because of the relevant differences in the operating procedure, e. g. temperature.

A comparison of actual column results with retention volumes calculated from laminar data is only possible with the R_f values obtained by Cerrai and coworkers[14] on paper treated with 0.1 M HDEHP solutions in cyclohexane and eluted with hydrochloric acid, and the column results reported by Siekierski and Sochacka[15,16] for rare earths eluted on HDEHP-columns with hydrochloric or nitric acid, at room temperature. Even in this case, the actual value of K is affected by the uncertainty deriving from the lack of direct information on the column phase volume ratio.

For the laminar system considered, an $A_m:A_s$ = 15 has been determined in specific experiments with 0.1 to 6 M HCl[6], essentially consisting in the determination of the amounts of HDEHP and hydrochloric acid present in the different portions of the developed chromatogram.

For the column (0.3x10 cm), V_s = 0.04 ml was calculated from the information that the column contained about 0.40 g of kieselguhr, and that the HDEHP to support ratio was 1:10[16]. The value V_m = 0.5 ml was obtained by comparing the position (in drops) of the thallium peak eluted with 5.12 M HNO_3 (Fig. 4 in Ref.[16]) with the $(V_R-V_m)/V_m$ datum for the same element eluted with 5.22 M HNO_3 (Table 5 in Ref.[15]), and calculating the drop volume from flow-rate figures (0.75 ml·cm^{-2}·min^{-1} ≅ 2 drops/min)[16]and column diameter: the sum of the volumes of support, extractant and mobile phase ($\frac{0.4}{2.26}$ + 0.04 + 0.5 = = 0.72 ml) agrees well with the total column volume (0.71 ml).

Table 1 reports the comparison of the $(V_R - V_m)/V_m$ values, obtained from the peak positions of the separations illustrated in Ref.[16] or directly reported in Ref.[15] (asterisks in Table 1) with those calculated with equation (3) (K = 1.2) from the corresponding R_f data. The experimental data include also results obtained with dilute nitric acid as the eluent, reasonably comparable to those obtainable with hydrochloric acid at the same concentration.

TABLE 1

Tentative comparison of experimental retention volumes
with those calculated from independent paper chromatographic experiments.
(See the text for details)

Eluent	Element	$(V_R - V_m)/V_m$	
		Found	Calculated
0.21 M HNO$_3$	La	1.1	1.7
	Ce	3.5	3.2
	Nd	5.6	5.9
0.39 M HNO$_3$	Sm	3.2	4.8
	Eu	8.2	7.5
	Gd	14.9	14.8
0.14 M HCl	Ce	6.0	8.7
	Pr	8.7	13.8
0.85 M HCl*	Eu	0.91	1.0
	Tb	6.8	2.9
2.3 M HCl*	Er	1.84	4.7
2.95 M HCl	Tm	1.8	3.6
	Lu	10	16
5.05 M HCl*	Tm	0.94	1.1

A satisfactory agreement is apparent in the table for results obtained with dilute acids, the only exception being the values for terbium (0.85 M HCl), whose anomalous behaviour on paper was often pointed out[6].

The agreement is no more found when more concentrated hydrochloric acid is involved, and figures calculated from R_f results are consistently greater than the experimental ones. This could derive from the inadequacy of the method adopted to determine the $A_m:A_s$ of paper in such conditions. A detailed discussion of this, however, is beyond the scope of the present chapter.

In spite of the above mentioned disagreement, in the opinion of the authors Table 1 demonstrates the validity of laminar results in foreseeing the retention volumes of a given column. In order to make full profit of this, however, it should be necessary that all new investigations on the behaviour of laminar extraction chromatographic systems provide information on the phase volume ratio of the system, in contrast with the practice up to now generally followed.

5. PLANNING COLUMN SEPARATIONS FROM LAMINAR RESULTS

The problems related to the separation of each component of a complex mixture of elements obviously consist of the combination of the problems related to two-by-two separations of all the elements involved. For the sake of simplicity, only the separation of two elements is discussed here below; the resulting comments and suggestions can easily be extended to more complex mixtures, as in the example reported at the end of the paragraph.

The criteria for the choice of the best conditions for the separation of two given elements by means of an extraction chromatographic column depend on the extent of the separation factor (S.F.) between the elements in the available extraction systems.

When the S.F. is sufficiently high — so as to imply a first elution through the column, on which one of the elements is strongly retained by the stationary phase while the other one runs with the eluent; and a second elution to recover the retained element from the column — then the problem simply consists in choosing a first eluting solution that ensures the minimum contamination of the eluate by the more retained element, and a second eluting solution that allows for a complete recovery of the more retained element, generally in the minimum volume of eluate.

When the S. F. is as low as to forbid a two-step operation as that described above, than the problem consists in the choice of an eluting solution that gives the highest possible S. F., and for an efficient exploitation of the resolution power (number of plates) of the column.

Both kinds of problems include a number of choices and laminar chromatographic results can be extremely useful in the basic one about the composition of the stationary and mobile phases suitable for the wanted separation.

Other choices and optimizations, such as those related to the capacity, stability or resolution power of the column, are beyond the scope of the present chapter.

The knowledge of the general behaviour of the elements in the different liquid-liquid extraction systems normally is the first guide to the choice of the nature of the stationary and mobile phases to be used. Laminar chromatographic results can be useful even in this very first choice, since they include systems for which little or any liquid-liquid extraction information is available.

When an extraction system is found that allows for sufficiently high separation factors (such as 10 or more)[17] then the elements are generally separated by means of the two-step operation, and the problem mainly consists in the choice of the suitable concentrations for both eluting solutions. Strictly speaking, a wise choice of the $V_m:V_s$ ratio in the column and possibly of the extractant to diluent ratio on the support is advised; however, since the optimization of these parameters is not imperative for obtaining good results with this kind of separation, little attention is generally given to them.

If paper or thin-layer R_f spectra are available for the two elements in the chosen chromatographic systems, their comparison makes the choice of the concentration of the first eluting solution very easy.

In the simplest cases, the more retained element has $R_f = 0$ within a certain concentration range, while the second element correspondingly has R_f values greater than zero. In this case, any concentration within the above mentioned range, and reasonably distant from its limits, usually ensures the quantitative retention of the first element on the column and the complete elution of the second one, irrespective of the difference in the phase volume ratios of the two chromatographic systems: obviously, the greater R_f value of the second element results in a smaller volume of the eluent necessary for its quantitative elution.

When both elements display R_f values greater than zero in the whole concentration range where their retention is different, the best results are usually obtained with the eluent concentration corresponding to the lowest R_f value of the more retained element. This may not be the case when the separation factor appreciably varies with the eluent concentration. In this case, a suitable compromise should be made between a sufficient separation factor and a low R_f value for the more retained element: although such compromise should take into account the phase volume ratios of the two systems, for average high-capacity columns and for most R_f spectra the dashed line in Fig. 1 may be sufficient for a good choice. In any case, it must be kept in mind that the higher is the R_f value of the more retained element, the more critical are the limitations on the eluent volume to avoid cross contamination; R_f values both greater that 0.5–0.6 generally do not give effective separations, unless columns are used having particularly low $V_m:V_s$ ratio.

Once the less retained element is completely eluted away, the best eluent concentration for the recovery of the more retained element from the column is easily chosen in the range of the highest R_f values of its spectrum. When for some reasons the eluent composition must be changed for the purpose (from hydrochloric to nitric acid, for instance) unexpectedly higher retention or bad tailing may sometimes occur, since R_f spectra are generally obtained in the absence of components extraneous to the investigated system, while the column can easily keep traces of the former eluent or retain the element because of the formation of particularly stable adducts containing the former component. In practice, however, this was very seldom observed.

When the two elements to be separated do not allow for S.F. to be sufficiently high to apply the two-step operation, the usefulness of laminar results to plan the column procedure is very limited.

In this case, the best conditions for the separation result from a compromise of three main requirements:
— maximum S. F. between the two elements;
— maximum resolution power of the column;
— suitable elution volumes so as to avoid possible contamination of too close peaks, and at the same time to keep elution volumes within reasonable values.

Elements that display only low S. F.'s generally are strictly similar in their chemical behaviour, so that their S. F. is seldom affected by variations of the concentration of the components in the two phases. In the few cases in which variations occur, laminar techniques can be useful to provide easy and fast information on the behaviour of the two elements. This can particularly be advantageous when multicomponent systems are involved, because of the great number of data necessary to obtain a complete information on the system: a thorough investigation of a synergistic systems as the stationary phase, for instance, could imply a tremendous amount of work when done with columns. Assuming equation (1) to be valid for the system under investigation, separation factors $(D_1 : D_2)$ are obtained from the ratios of the $(1/R_f - 1)$ values of the two elements eluted in the same given conditions. If the R_f spectra are available, a glance to the trends of the R_M vs. log (concentration) curves can often give a guide to the most advantageous concentration range.

It should be kept in mind, however, that the experimental uncertainties of the laminar techniques make it hard to evaluate actual small variations of S. F. sometimes very important for an effective separation. Another drawback of laminar techniques is the difficulty in obtaining data at temperatures higher than room temperature, while it is well known that many one-step separations are advantageously performed at such conditions, because of the positive effect on plate heights.

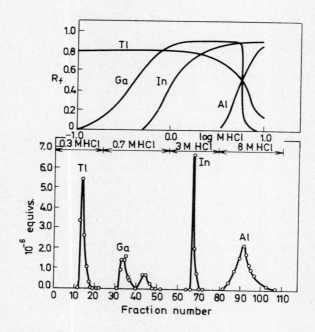

Fig. 2. Example of usefulness of the laminar chromatographic data in planning column separations. R_f spectra were obtained by ascending elution of Whatman CRL/1 paper treated with 0.1 M HDEHP in cyclohexane. The column (13X250 mm) was prepared with Whatman Standard Grade cellulose powder treated with 0.1 M HDEHP in cyclohexane.
Fraction volume = 2 ml; Flow rate: 0.5 ml·min^{-1} (from Refs [18,19])

Laminar results obviously do not help at all in defining the conditions for minimum plate height of the column. It is worth mentioning here that no direct relation exists between the resolution of two spots and that of the corresponding column peaks, not only because resolution is obviously strictly related to the plate heights, that can be very different in the two systems, but also because the spreading of the eluted element follows opposite trends in the two systems: the width of peaks coming out from a column is greater, the greater the retention volume; conversely, the spreading of the spots in the corresponding laminar chromatogram increases as R_f increases, that is as retention decreases.

As for the prevision of the elution volumes involved in the separation, the laminar results give little help, not only because of the scarce knowledge of the phase volumes ratios involved, but also as the optimization of the resolution performances of the column usually implies a number of experiments that automatically provide direct information on the column behaviour in this respect.

In conclusion, when one-step separations must be performed, the usefulness of the laminar chromatographic techniques is limited to a possible rapid investigation on the variation of S. F. with the composition of the two phases, that can provide reliable results only when relatively great variations occur.

Because of the just-reported severe limitations, no examples are found in the literature of one-step column separations based on laminar results. On the contrary, both paper or thin-layer data were often the basis of successful column separations performed with two (or more) elution steps.

To better illustrate the criteria followed in choosing the column elution conditions in this case and to point out the usefulness of laminar information in both planning and interpreting column results, the separation of aluminium, gallium, indium, and thallium(III) in the HDEHP–HCl system is taken here as an example [18,19]. It must be underlined, however, that this separation implies a number of peculiar features that make it rather more complicated than the usual cases.

The upper part of Fig. 2 reports the R_f spectra of the four elements of interest as obtained by ascending chromatography on paper treated with HDEHP. Additional information from paper results indicates remarkable irregularity and tailing in spots of aluminium at R_f values other than zero, and double spots occurring for gallium at hydrochloric acid concentrations within 5.5 and 6.5 M, that is in the neighbours of the steep decrease of its R_f values, due to strong solvation of the tetrachlorogallium complex by the P → O group of the extractant; in the above mentioned range, one spot is gradually more retained as the hydrochloric acid concentration is raised, while another spot stays at the vicinity of the eluent front.

The peculiar pattern of the R_f spectra of the four elements, resulting from the unusual combination of two completely different mechanisms of extraction, makes the HDEHP–HCl system suitable for a great number of different column separation schemes, involving two, three or all four elements in different sequences. The lower part of Fig. 2 illustrates the separation Tl–Ga–In–Al, that was performed on the basis of laminar data only, without any preliminary column experiment.

The column was prepared with HDEHP supported on cellulose powder: unfortunately, no data are available on the column characteristics, since the separation was carried out only to demonstrate the potentiality of the system. The feed solution was 0.01 M in hydrochloric acid, and contained $5 \cdot 10^{-6}$ moles of each of the elements; the amounts of the different elements in the single fractions were determined by colorimetry during the elutions.

To elute thallium first, any hydrochloric acid concentration sufficiently low to keep gallium on the column was suitable, and 0.3 M HCl was chosen to remain sufficiently apart from the lowest concentration investigated. At these conditions, gallium displays R_f values appreciably greater than zero (0.33), but this was not so important since it was the next element to be eluted; furthermore, a partial advancement of gallium through the column could have been advantageous in its successive separation from indium.

To recover gallium, 0.7 M acid was used: although it does not allow for the lowest retention volume for gallium, it corresponds to a sufficiently low R_f value for indium. Maybe $R_f \cong 0.3$ could also be acceptable, but it is always advisable to restrict the allowance for relatively high R_f values as the number increases of eluents that already passed through the column. Unexpectedly, gallium appeared as a double peak: we shall return to this later. It should be noticed that the first peak appeared early after the change of eluent concentration (as compared to that of thallium, having the same R_f value)

Indium was eluted with 3 M HCl, chosen at the upper limit of the acid concentration range where aluminium has $R_f = 0$. In its turn, aluminium was finally recovered with 8 M HCl: higher acid concentrations would have been more suitable, but could have implied a possible damage of the cellulose support. The aluminium peak was remarkably broad, as expected from the irregular spots obtained in paper chromatography.

As already pointed out, the R_f spectra reported in Fig. 2 show that the HDEHP–HCl system allows for a great number of different separation schemes of the elements, in addition to that illustrated in the figure. All pairs of elements, for instance, can be separated in either of the two elution sequences, the elution of gallium before indium being the only impossible one. Here below are given possible acid concentrations suitable for the different cases, as deducible from Fig. 2: In–Al: 3 M–8 M; Tl–In: 0.1 M–8 M; In–Tl: $\geqslant$ 8 M–0.1 M; Tl–Al: 0.1 M–8 M; Al–Tl : $>$ 9 M–0.1 M (complications could arise because of broad Al peak); Ga–In: 0.7 M–8 M; In–Ga: 8 M–2 M; Ga–Al: 2 M–8 M; Al–Ga: 8 M–2 M; Tl–Ga: 0.1 M–2 M; Ga–Tl could be resolved only with columns having a sufficiently low $V_m : V_s$ ratio, eluting Ga with $\sim$ 5 M HCl, and recovering Tl with more dilute acid.

Analogously, several combinations of three elements could also be separated. In some cases, however, the column should be carefully prepared so as to avoid excessive mixing of the first two consecutive eluents, that could result in unwanted elutions of the third elements, scar retained at acid concentrations intermediate between the two; in other cases, columns having $V_m : V_s$ ratio should be strictly necessary. In addition to the separation illustrated in Fig. 2, the sequence In–Al–Ga–Tl could also be tried, with a column of low $V_m : V_s$ ratio.

The double peak for gallium obtained in the above reported Tl–Ga–In–Al separation remained a question-mark for a long time, until a possible explanation was derived from paper chromatographic investigation, that is presented here to illustrate the help that laminar techniques can give not only in planning the column separation, but also in trying to interpret possible unexpected column results.

The double peak was obviously related to the double spots observed on paper eluted with concentrated hydrochloric acid, still unexplained at the time of the column separation. Paper chromatography finally threw light on the reason for the double spots. Experiments carried out with HDEHP-treated paper, spotted with gallium solutions of different hydrochloric acid concentrations and eluted with 5.5–6.5 M HCl before the spots went dry, showed that the developed spot running with the front of the eluent was not present when gallium was originally contained in solutions having acid concentration greater than 6.5 M. Furthermore, gallium contained in 6 M HCl solutions, prepared mixing slightly acidic gallium chloride solutions and 6.5 M HCl, gave a front spot gradually decreasing in intensity as the time increased between the preparation of the solution and the development of the chromatogram, disappearing however only when solutions were prepared several hours before elution. All this indicated that a slow kinetics of formation of the tetrachlorogallium complex in hydrochloric acid solutions lower than 6.5 M was most probably responsible for the double spot.

A slow kinetics of chloro complex formation appears therefore as a plausible explanation also for the double peak obtained with the column. R_M vs. log (H^+ activity) for gallium have slopes -3 up to about 0.1 M HCl, and -2 above that value[19]: this indicates that the exchange reaction most probably involves Ga^{3+}, or $GaCl^{2+}$, depending on the chloride concentration. When it was fed into the column in 0.01 M HCl, gallium was retained by HDEHP as Ga^{3+}, but it was subsequently recovered as $GaCl^{2+}$; if a slow kinetics is assumed for the formation of $GaCl^{2+}$ in 0.3 M HCl, then the first peak could correspond to the front of the $GaCl^{2+}$ band already eluted in 0.3 M HCl, and "pushed" by the more concentrated acid; the second peak, in its turn, could be the rest of gallium, complexed and eluted only by the more concentrated acid.

The suggested explanation finds support in the relative positions of the peak of thallium and of the first peak of gallium, that are consistent with those expected from a single-elution separation of two elements having 0.8 and 0.33 as R_f values (for K = 0.82; and V_m = 5.8 ml, in good agreement with the 6 ml value reported for a similar column)[20].

6. LITERATURE SURVEY OF AVAILABLE R_f SPECTRA

In this paragraph, information is given on what is available in the literature concerning the extraction chromatographic behaviour of elements in the different extraction systems, as obtained by paper or thin-layer extraction chromatography. The information is by no means complete, only those works being cited where a sufficient number of elements have been considered and reasonably large eluent concentration ranges have been investigated.

Some figures are included, where the R_f spectra are reported of representative systems, for some of which the distribution coefficients, obtained by liquid-liquid extraction, were already reported in Chapter 4. The reader is invited to compare the liquid-liquid extraction and chromatographic results, that clearly show their strict analogy. Other figures are devoted to extraction systems whose behaviour was only sketched in Chapter 4.

The great majority of both paper and thin-layer results refer to ascending chromatography. The diluents, possibly used with the extractant to treat the support, are not considered, since in most cases they are volatilized from the support before use, so as to exclude any effect on the retention features of the stationary phase. Noticeable ex-

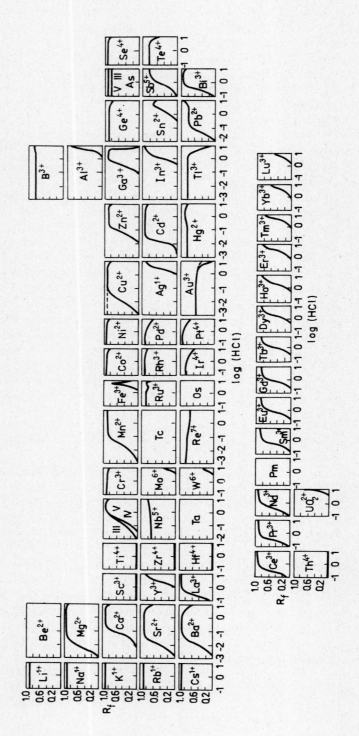

Fig. 3. R_f spectra of elements eluted with HCl on paper treated with HDEHP (reprinted from Ref.[3], p. 386–7, by courtesy of Elsevier Publ. Co.)

382

ceptions occur with solid extractants, such as phosphine oxides and fatty acids, which must be kept on the support together with some diluent as to result in a liquid stationary phase.

As already pointed out, practically no information is available on the $A_m:A_s$ ratios of the different laminar extraction systems, so that quantitative correlations can hardly be done to retention volumes of analogous columns. It should be kept in mind, however, that the $A_m:A_s$ ratios for paper should be generally greater than those for thin-layers, this assumption being confirmed in several instances, e. g. the reported generally lower retention power of amine-treated paper as compared to that of thin-layers loaded with the same amount of extractant[21].

6.1. ACIDIC EXTRACTANTS

Laminar chromatographic results with acidic organo-phosphorous compounds as stationary phases almost exclusively concern di(2-ethylhexyl) phosphoric acid (HDEHP).

The R_f spectra are available for 67 cations eluted with HCl ($1 \cdot 10^{-3}$ to 10 M) on paper treated with HDEHP[3], and are reported in Fig. 3. To the knowledge of the authors, these are the only data for which the $A_m:A_s$ ratio is known, at least approximately ($A_m:A_s \cong 15$ for 0.1 to 6 M HCl; see however the comments to Table 1 in Paragraph 5). The R_f patterns of 35 elements in the same system (0.1 to 8 M HCl) are reported in a paper issued in China[22]: from abstracted information, a general behaviour is derivable, analogous to that reported in Figure 3, although a number of sharp discrepancies are also found. Paper R_f spectra were also reported for alkali metals and alkaline earths eluted with acetic acid ($1 \cdot 10^{-2}$ to 2 M)[23], and for several representative elements eluted with LiCl ($1 \cdot 10^{-2}$ to 5 M) or NH_4SCN ($5 \cdot 10^{-3}$ to 7 M)[24].

The R_f spectra on HDEHP-treated silica gel thin-layers are also available, for 24 metals eluted with HCl ($1 \cdot 10^{-4}$ to 10 M) or HNO_3 ($1 \cdot 10^{-2}$ to 10 M), and for representative metals (Ag, Pb, Hg(II), Cd, Zn, Bi, Al, Cr(III), Mn(II), La, Ce, Ca) eluted with (0.01 to 5 M) KSCN + 0.1 M HNO_3[25].

The behaviour of rare earths was reported for a variety of chromatographic systems: HDEHP and HCl (paper) [3,14], HNO_3 (kieselguhr thin-layers)[8] and $Na_2SO_4 + CH_3COOH$ (paper)[26]; di-n-butylphosphoric acid and HCl (paper)[27]; 2-ethylhexyl-butylphosphoric acids and HCl (paper)[28]. As expected, the general patterns of results is nearly the same in all cases. Particular reference must be given to the rather unusual behaviour of rare earths when eluted with oxalic acid (0.1 to 1 M) on HDEHP-treated paper[29] (see Ref. [6]).

Figure 4 reports the R_f spectra of a number of elements eluted with HCl or HNO_3 on paper treated with dinonylnaphthalene sulphonic acid (HDNNS)[30]. Additional data are available for alkaline earths eluted with 0.1 to 0.5 M HCl or HNO_3[31]. The behaviour of rare earths was reported for HDNNS-treated paper eluted with HCl[32,33], lactic acid and hydroxyisobutyric acid solutions[34]: an increasing average separation factor between adjacent rare earths was obtained in the three systems, in very good agreement with results obtained with sulphonic cation-exchange resins.

The applicability of fatty (capronic, caprylic, pelargonic, and oleic) acids to extraction

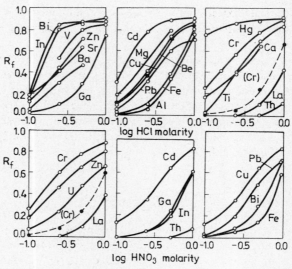

Fig. 4. R_f spectra of elements eluted with HCl or HNO_3 on paper treated with HDNNS (derived from the data of Ref.[30])

chromatography was investigated with paper eluted with chloride solutions (pH ranging from 1 to 9), and R_f spectra are available for about 15 elements[35]. Spectra for some representative elements (Fe(III), Co, Ni, Zn) were also reported for paper treated with stearic, elaidic or oleic acids eluted with acetate buffers (4 to 5.5 pH)[36]. In spite of their interesting selectivity for bismuth, iron(III) and thorium, fatty acids were never applied to column separations.

6.2. NEUTRAL ORGANOPHOSPHOROUS COMPOUNDS

R_f spectra for tri-n-butylphosphate (TBP) stationary phases eluted with HCl (0.5 to 12 M) are available for different supports, namely paper [22,37-40], cellulose[1,41,42] and silica gel[43] thin-layers. Results were also published for a few metals eluted on layers made of a copolymer of vinylchloride and vinylacetate (Corvic)[44]. A fair agreement is found among the various results, if the differences between the techniques are taken into account[6]: cellulose thin-layers, for instance, retain more TBP than silica gel ones, therefore displaying generally higher retention power. R_f spectra obtained with cellulose or silica gel thin-layers are collected in Fig. 5.

A systematic survey was made on the behaviour of a great number of elements eluted on TBP-treated paper[38–40] or silica gel thin-layers[45] with 0.5 to 12 M or 0.2 to 15 M HNO_3, respectively. The latter results are presented in Fig. 6 (curves 4) together with spectra referring to other stationary phases, dealt with later on in this paragraph. The R_f spectra are available of 24 metals eluted on TBP-treated silica gel thin-layers with 1 to 10 M LiCl, 1 to 5 M NaBr, or 0.5 to 6.6 M NaI, all eluting solutions being 0.1 M in $HClO_4$[45a].

384

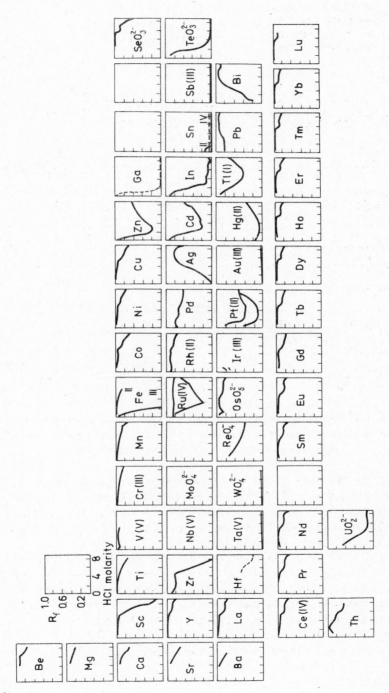

Fig. 5. R_f spectra of elements eluted with HCl on TBP-treated cellulose thin-layers[1]; dashed lines refer to elements eluted on TBP-treated silica gel thin-layers[43], (reprinted from Ref.[6], p. 130–1, by courtesy of Marcel Dekker, Inc.)

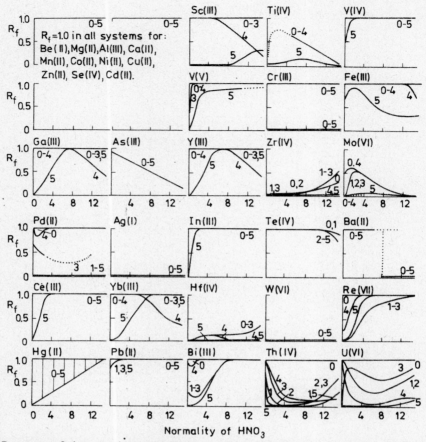

Fig. 6. R_f spectra of elements eluted with HNO_3 on silica gel thin-layers without any extractant (0), or impregnated with Alamine 336–S (1), Alamine oxide (2), tri-n-octylarsine oxide (3), TBP (4), or TOPO (5) (from Ref. [45], by courtesy of the authors)

Some R_f data obtained with TBP-treated paper were also reported for several representative elements eluted with $HClO_4$ (0.5 and 3 M)[39,40]; R_f spectra are available for noble metals eluted with NaBr (0.5 to 4 M), or (1 to 4 M) HCl + thiourea solutions[46]. Finally, U, Th, rare earths, Re, V, Mo, and W were eluted on TBP-treated paper with (0.1 to 0.5 M) HNO_3 + 0.2 M NH_4SCN solutions[47,48].

The behaviour of a number of elements eluted with HCl, HNO_3, $HClO_4$ and H_2SO_4 (0.1 to 10 N) on paper treated with tri-n-octylphosphate was reported[49], and the expected agreement was found with the corresponding results obtained with TBP. Some representative metals have been eluted with NH_4SCN ($1 \cdot 10^{-2}$ to 4 M) or LiCl ($1 \cdot 10^{-2}$ to 5 M) on paper treated with tri-i-octylphosphate[24]. The behaviour of the anionic chloride complexes of Au, Pt, and Pd is available, eluted with HCl, HNO_3 (0.5 to 10 M) or $HClO_4$ (0.1 to 2 M) on paper treated with triallylphosphate[50]; with the same sta-

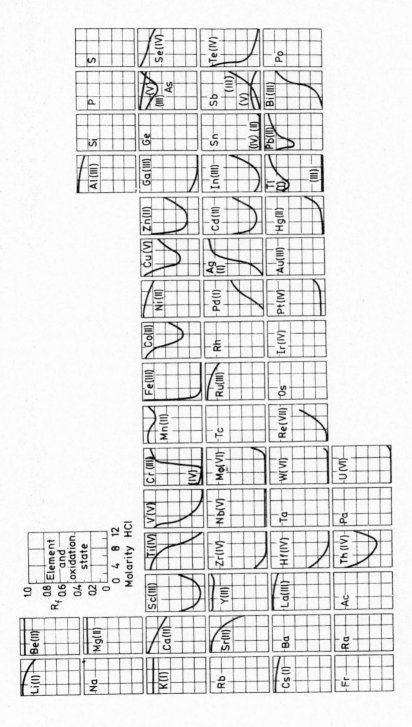

Fig.7. R_f spectra of elements eluted with HCl on paper treated with TOPO (reprinted from Ref.[4], p. 115, by courtesy of Elsevier Publ. Co.)

tionary phase, some R_f data were also reported for a few other metals eluted with HCl, HNO_3 (0.5 to 10 M) or H_2SO_4 (0.1 to 4 N)[51].

The behaviour of a large number of elements on paper treated with tri-n-octylphosphine oxide (TOPO) was independently investigated by different authors, using HCl (0.1 to 12 M) [4,39,40,52], and HNO_3 (0.5 to 14 M)[4,39,40] as the eluents; R_f spectra were also reported for TOPO-treated silica gel thin-layers eluted with HCl (0.1 to 10 M)[43] or HNO_3 (0.2 to 15 M)[45]. A satisfactory agreement is found again among the different results [6,43,45].

Results obtained in the TOPO–HCl system are reported in Fig. 7, while the spectra of Fig. 6 include the system TOPO–HNO_3 (curves 5).

Thorough information is available for TOPO-treated paper eluted with H_2SO_4 (0.1 to 4 M)[4], and for silica gel thin-layers eluted with HBr (0.2 to 9 M), and (2 to 10 M) LiBr + 0.5 M HBr[53]. Some data were also reported for $HClO_4$ (0.5 and 3 M) as the eluent[39,40].

Bifunctional neutral organophosphorous compounds have been extensively investigated as stationary phases in laminar extraction chromatography and a detailed review has been compiled[54]. Information is available on the R_f spectra of a number of elements eluted by HCl (0.5 to 12 M), HNO_3 (0.5 to 12 M) and $HClO_4$ ($5 \cdot 10^{-2}$ to 9 M) on paper treated with methylene bis(di-n-hexylphosphine oxide) or methylene bis(di(2-ethylhexyl) phosphine oxide)[39,40]. Particular attention was given to the behaviour of rare earths, and their detailed R_f spectra are available for both stationary phases[39,40,55,56].

Tetrabutylhypophosphate was also used as the stationary phase, and R_f spectra are available for several actinides eluted with 1 to 16 M HNO_3 on either paper or thin-layers[57].

6.3. AMINES AND QUATERNARY AMMONIUM COMPOUNDS

Long-chain amines and quaternary ammonium compounds are the most extensively investigated stationary phases in laminar extraction chromatography, and an exhaustive review on the numerous results obtained has recently been published[58]. Much effort was spent in the investigation on correlations between the class and structure of the extractants and their retention behaviour, so that the available results cover not only a large number of elements but also a variety of stationary and mobile phases. Particular mention must be given to the works of Brinkman's and Przeszlakowski's groups. Most attention has been devoted to the chloride system. With HCl as the eluent R_f spectra are available for almost all relevant elements on supports treated with Primene JM-T (paper[21], silica gel[2,58,59] and cellulose[60] thin-layers), hexadecylamine (paper)[61], dilaurylamine (paper)[61], Amberlite LA-1 (paper[21,62]and silica gel thin-layers[2,21,59,63], Amberlite LA-2 (paper[21,64,65], and silica gel thin-layers[21]), tri-n-octylamine (TNOA) (paper[5,21,63–68], silica gel[21], and cellulose[69] thin-layers), tri-i-octylamine (TIOA) (paper and silica gel thin-layers)[21], Alamine-336 (paper and silica gel thin-layers)[2,21,59], trilaurylamine (paper)[61,67], Aliquat-336 (paper[20,61,66] and silica gel thin-layers[59,70]).

Several data are also available for a great number of other amines, that for several

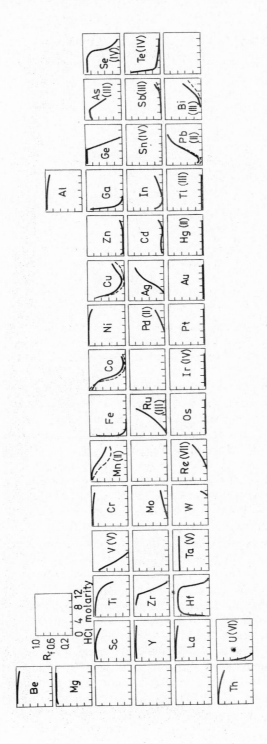

Fig. 8. R_f spectra of elements eluted with HCl on paper[5,64] (full lines) or silica gel thin-layers[21] treated with TNOA (reprinted from Ref.[6], p. 24-5, by courtesy of Marcel Dekker, Inc.).

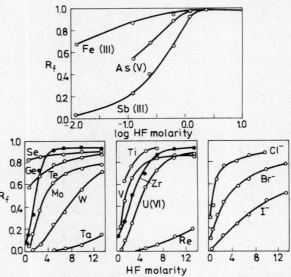

Fig. 9. R_f spectra of selected elements eluted with HF on paper treated with TNOA[73] (reprinted from Ref. [6], p. 39, by courtesy of Marcel Dekker, Inc.).

reasons proved to be scarcely attractive as stationary phases and did not find further applications[21,67]. The behaviour of several elements was also reported on paper treated with a mixture of Amberlite LA-2 and tributylamine (TBuA), eluted with HCl (2 to 8 M)[71].

A satisfactory agreement is generally found, not only among the R_f patterns involving the same stationary phase, obtained by different authors or with different techniques, but also among those involving amines of different classes, as expected for such extraction systems (see Chapter 4). The few discrepancies are discussed in Refs [6] and [58]. Figure 8 collects results obtained with paper and silica gel thin-layers treated with TNOA.

Several R_f spectra are available also for eluting solutions containing chloride salts instead of the acid. The behaviour of 18 elements was reported on silica gel thin-layers treated with TNOA or Amberlite LA-1, and eluted with LiCl (1 to 10 M, and 1 to 8 M, respectively) + 0.1 M $HClO_4$[63]. Few representative elements were investigated with slightly acidified LiCl as the eluent, the stationary phases being Amberlite LA-1, Alamine-336, or Adogen-464 (Cu(II), Co, Mn(II), Zn, Fe(III); paper and silica gel thin-layers; 0.5 to 11 M LiCl) [59,70], or Amberlite LA-2, TIOA or Aliquat-336 (Co, Zn; cellulose thin-layers; 0.1 to 8 M LiCl)[72]; the behaviour of 33 elements is available for TNOA-treated paper eluted with NaCl (0.5 to 4 M) solutions[5].

R_f spectra for some 40 elements were reported for paper treated with TNOA or Amberlite LA-2 and eluted with HF ($1.2 \cdot 10^{-2}$ to 13.4 M)[73], and Figure 9 reports the relevant spectra obtained with TNOA: Cd, Zn, As(III), Co, Cr(III), Cu(II), In, Mn(II), Ni, Tl(I), Th, La, Y and Sc were found at the front of the eluent ($R_f > 0.9$) in the whole range investigated, while Pt(IV), Au(III), and Ta(V) did not move from the start line. Closely analogous results were obtained with silica gel thin-layers treated with Primene JM-T, Amberlite LA-2, Alamine-336 and Aliquat-336[74].

390

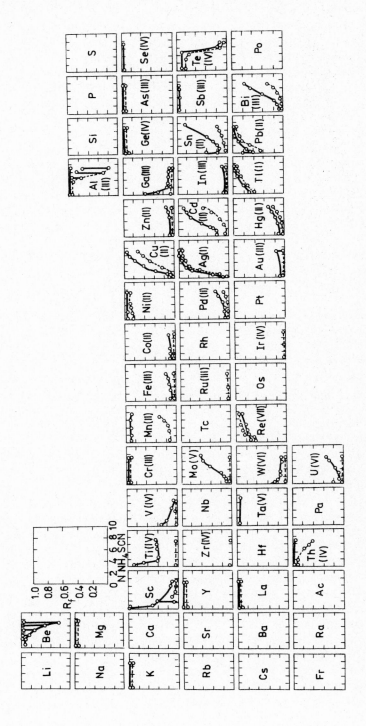

Fig. 10. R_f spectra of elements eluted with NH_4SCN (pH = 2 with H_2SO_4) on paper treated with Amberlite LA-2: (full lines) or TNOA (dashed lines) (from Ref. [64], by courtesy of the author).

A systematic survey was made on some 40 elements eluted with HBr (0.5 to 5 M), NaB (0.6 to 6 M; pH = 2) or NaI (0.5 to 7 M; pH = 2), on paper treated with TNOA or Amberlite LA-2[64,75] and with HBr (0.2 to 9 M) or (2 to 10 M) LiBr + 0.5 M HBr on silica gel thin-layers impregnated with Alamine -336, Amberlite LA-1 and Primene JM-T[53]. Eighteen metals were eluted on silica gel thin-layers treated with TNOA or Amberlite LA-1, and eluted with (1 to 6 M) NaBr + 0.1 M $HClO_4$, or (1 to 7 M) NaI + 0.1 M $HClO_4$[63]. Results are also available for representative elements (Ni, Cu(II), Sb, Fe(III), Co, Pb, Mn(II), Zn, Bi) eluted with HBr (0.5 to 8 M) or HI (0.5 to 7.5 M) on silica gel thin-layers impregnated with TIOA, Adogen-368, Aliquat-336 or Adogen-464[76].

Systems having thiocyanate solutions as the eluents were investigated by several authors. The behaviour is available of a great number of elements eluted with pure NH_4SCN (0.1 to 7 M) solutions on cellulose (incidentally also silica gel) thin-layers impregnated with Amberlite LA-2[77], and on paper treated with the same amine and eluted with ($1 \cdot 10^{-2}$ to 1 M) NH_4SCN + 0.5 M HCl or 0.1 M NH_4SCN + (1 to 6 M) HCl[78]. About 20 metals were eluted with KSCN (1 to 8 M) + 0.1 M HNO_3 on silica gel thin-layers treated with TNOA or Amberlite LA-1[63]. Some representative elements (see above for HI solutions) were eluted with 0.1 M HSCN+KSCN (0.1 to 7 M total) on paper or cellulose thin-layers treated with Amberlite LA-1, TIOA, Aliquat-336, Adogen-368 and Adogen-464[76].

Systematic investigations were also carried out with paper treated with TBuA, TIOA [79,80], Amberlite LA-2[64] and TNOA[64,79,80], eluted with NH_4SCN (0.5 to 7 M) at pH = 2 with H_2SO_4. Figure 10 reports the R_f spectra obtained with Amberlite LA-2 and TNOA in such conditions[64].

As Brinkman points out[58], in spite of the remarkable differences in acidity of the eluent, that are known to affect R_f values, an excellent agreement is found between the R_f patterns obtained in the different conditions. R_f spectra are available also for representative elements eluted on TBuA-treated paper, with 6 M NH_4SCN + ($1 \cdot 10^{-3}$ to 1 M) HCl[79,80] and with (1 to 7 M) NH_4SCN + 1 M HCl[79].

TNOA and Amberlite LA-1, supported on silica gel thin-layers, were investigated also with KCN (0.1 to 4 M, and 0.1 to 7 M, respectively) as the eluent, and the R_f spectra are available for some 20 elements[63].

When nitric acid solutions are used as the eluents, high R_f values are found for the majority of elements. A systematic survey is available of the behaviour of a great number of ions eluted with 1 to 10 M HNO_3 on paper treated with hexadecylamine, dilaurylamine, trilaurylamine, and Aliquat-336[61]. Very recently, the R_f spectra were reported of approximately 40 elements eluted with HNO_3 (0.2 to 14 M) on silica gel thin-layers treated with Alamine-336S[45], and they are included in Figure 6 (curves 1). R_f spectra are also available for several representative elements eluted with (0.5 to 10 M) HNO_3 on paper and silica gel thin-layers treated with Primene JM-T, Amberlite LA-1, TIOA, TNOA, Aliquat-336, Adogen-368, and Adogen-464[76]; on silica gel thin-layers treated with TNOA or Amberlite LA-1[63]; and for uranium, thorium[81,82], and other actinides[81] on TNOA-treated paper (0.5 to 10 M HNO_3).

The substitution of nitric acid by its salts generally increases the retention: this is well demonstrated by the R_f spectra reported for uranium, thorium and lanthanum eluted on TNOA-treated paper with slightly acidified ($5 \cdot 10^{-3}$ M HNO_3) solutions of

LiNO$_3$ (1 to 6 M), NaNO$_3$ (1 to 8 M), NH$_4$NO$_3$ (1 to 10 M), Ca(NO$_3$)$_2$ (0.5 to 3 M), and Al(NO$_3$)$_3$ (0.3 to 2 M)[82]. The behaviours on TNOA-treated paper are also available for rare earths and few other elements (Sc, Y, Th) eluted with (1 to 10 M) LiNO$_3$ + + 2·10^{-3} M HNO$_3$[83], (2 to 10 M) NH$_4$NO$_3$ (pH$\cong$ 2.4 with HNO$_3$) and 4 M NH$_4$NO$_3$ + + (2·10^{-3} to 1 M) HNO$_3$[84]; and for actinides, Np and Pu valency states, and some other elements (La, Ce(III), Y, Zr, Mo) eluted with (0.5 to 8 M) LiNO$_3$ + 0.5 M HNO$_3$ or (0.5 to 2 M) Al(NO$_3$)$_3$ + 0.5 M HNO$_3$[85,86].

The behaviour of 25 ions was investigated on silica gel thin-layers impregnated with Primene JM-T, Amberlite LA-1, Alamine-336 S and Aliquat-336, and eluted with H$_2$SO$_4$ (~0.025 to 5 M)[87]. R$_f$ spectra were also reported for U, Th, and Zr eluted on silica gel thin-layers treated with Amberlite LA-2, and eluted with 0.01 to 2 M H$_2$SO$_4$, or with (1·10^{-2} to 2 M) (NH$_4$)$_2$SO$_4$ + 2.5·10^{-2} M H$_2$SO$_4$[88]; in the same work, information is given on the behaviour of rare earths in these systems. Some representative elements were also eluted with 0.14 to 4 M H$_2$SO$_4$ on silica gel thin-layers treated with Amberlite LA-1 or TNOA[63].

The behaviour of few ions is available on paper or silica gel thin-layers treated with Amberlite LA-1 or Adogen-464, and eluted with HClO$_4$ (0.5 to 10 M; only to 8 M for paper)[76].

A systematic investigation was carried out on the behaviour of elements eluted with formic, acetic or monochloroacetic acids (5·10^{-2} to 6M) on paper impregnated with TNOA or Amberlite LA-2 [89]. A thorough study is also available on the behaviour of 18 elements on silica gel thin-layers treated with Amberlite LA-1 or TNOA, and eluted with aqueous solutions of a number of organic complexing agents (lactic, tartaric, citric, iminodiacetic, nitrilotriacetic, hydroxyethyliminodiacetic, ethylenediaminotetraacetic acids and glycine); R$_f$ spectra are available as a function of either the complexing agent concentration or pH[90].

6.4. CHELATING COMPOUNDS

Chelating agents were seldom used as stationary phases in laminar extraction chromatography.

The R$_f$ spectra are available for about 20 elements eluted with acetate buffers (0.1 to 3 M acetate; pH = 2.5 to 5.3) on cellulose acetate pre-treated paper, impregnated with 2-thenoyltrifluoroacetone (HTTA)[91].

The synergistic systems HTTA—tri-n-octylphosphate (Am, U(VI), and Th)[92] and HTTA—TOPO (Am, La, Ce(III)[93], and Ce(III), Sm, Er[94]) were also investigated by paper chromatography, and R$_f$ spectra are reported as functions of the composition of the stationary phase and of the concentration of the eluting HCl.

Especially synthesized 5,8-diethyl-7-hydroxydodecan-6-oxime was screened for its extraction behaviour by paper extraction chromatography[95,96] and the R$_f$ spectra are reported for some 30 metals eluted with acetate or tartrate buffers (pH $\cong$ 1.5 to ~6), that demonstrate the attractive selectivity for copper of such systems. R$_f$ spectra are also available for copper(II) and the main constituents of steels, eluted with 0.5 M (NH$_4$)$_2$SO$_4$ or NH$_4$Cl solutions (pH = 1 to 6) on paper treated with Lix-64, a commercial product believed to be a mixture of 2-hydroxy-5-dodecenylbenzophenone oxime and 5,8-diethyl-7-hydroxydodecan-6-oxime[97]: the R$_f$ spectra for Cr(III) and Mo eluted with (NH$_4$)$_2$SO$_4$ apparently are inverted.

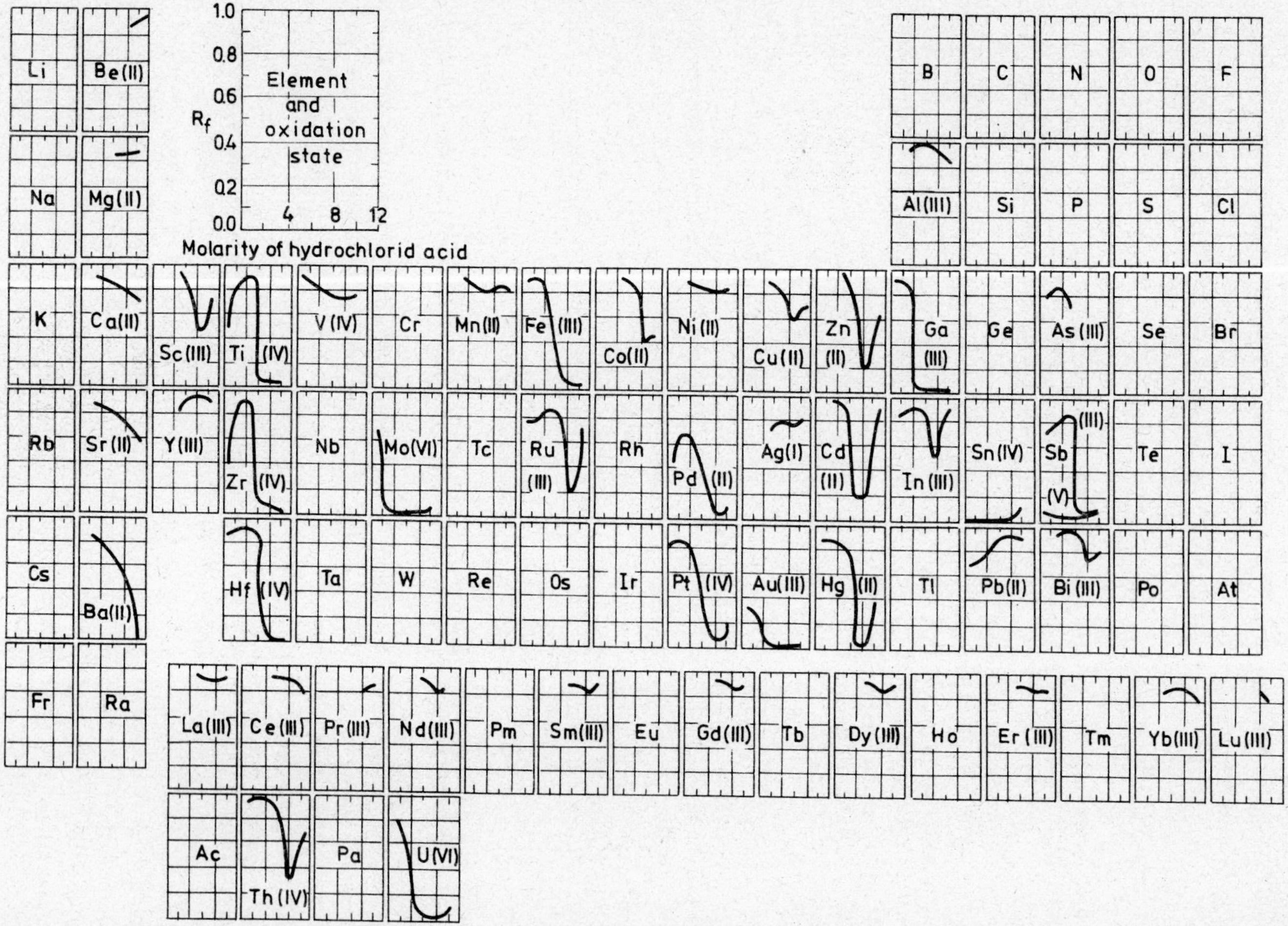

Fig. 11. R_f spectra of elements eluted with HCl on paper treated with di-n-octylsulphoxide (from Ref. [102], by courtesy of the authors).

The chromatographic behaviour of some 35 metals has been studied using paper impregnated with phenylbenzohydroxamic acid and 2-octanone, and extensive information is available for HCl and $HClO_4$ as the eluents[98]

6.5. MISCELLANEOUS EXTRACTANTS

Several data were collected for organic thiophosphorous compounds as stationary phases, that all display remarkable selectivity for silver and mercury(II).

The behaviour of 25 metals is available, eluted with HCl or HNO_3 (1 to 7 M) on paper treated with tri-n-octylphosphine sulphide[99], and of 29 metals eluted with HCl on paper treated with diheptylphenylphosphorodithioate or dibutylphosphorothioic acid[100]. Some representative elements were also studied eluting with HNO_3 ($1 \cdot 10^{-4}$ to 14.7 M) paper treated with methyl- or ethyl-parathion (0.0-dialkyl-0-p-nitro-phenylthiophosphate)[101].

Sulphoxides have been evaluated as extractants by means of paper extraction chromatography[102], and R_f spectra are available for purposely synthesized di-n-octyl sulphoxide (DOSO), bis(n-octylsulphinyl) methane (BOSM), bis(n-octyl-sulphinyl) ethane, and for p-tolyl sulphide, the eluent being HCl (1 to 10 M) or HNO_3 (1 to 11 M); the systematic survey was extended to DOSO and BOSM eluted with ($\sim 2 \cdot 10^{-2}$ to 2 M) $NH_4 SCN$ + + 1 M $HClO_4$, and to DOSO eluted with $HClO_4$ ($1 \cdot 10^{-2}$ to 4 M). Figure 11 reports the results obtained with the DOSO—HCl system.

The behaviour of silica gel thin-layers impregnated with tri-n-alkylamine or -arsine oxides was often compared to that of analogous systems involving amines or neutral organophosphorous compounds. Extensive R_f spectra are available for Alamine-366 S oxide and tri-n-octylarsine oxide, the eluents being HCl (0 to 12 M)[103], HBr (0.2 to 9 M), (2 to 10 M) LiBr + 0.5 M HBr[53], or HNO_3 (0.2 to 14 M)[45]. The latter data for both extractants are included in Fig. 6 (curves 2 and 3).

Table 2 collects the systems for which laminar R_f spectra are available.

TABLE 2

Summary of extraction systems for which R_f spectra are available.
Underlined references concern systematic surveys involving many elements

Extractant	Eluent	References
Di(2-ethylhexyl) phosphoric acid (HDEHP)	HCl	[3,14,22,25]
	HNO_3	[8,25]
	LiCl, NH_4SCN	[24]
	KSCN	[25]
	CH_3COOH	[23]
	$Na_2SO_4 + CH_3COOH$	[26]
	Oxalic acid	[29]
Di-n-butylphosphoric acid	HCl	[27]
2-Ethylhexyl butylphosphoric acid	HCl	[28]

(Table 2 continued)

Extractant	Eluent	References
Dinonylnaphtalene sulphonic acid (HDNNS)	HCl, HNO$_3$	[30, 31–33]
	Lactic, hydroxyiso-butyric acids	[34]
Fatty acids	Chloride solns.	[35]
	Acetate buffers	[36]
Tri-n-butylphosphate (TBP)	HCl	[1,22,37,38,39–41,42–44]
	HNO$_3$	[38,39,40,45]
	LiCl, NaI	[45a]
	NaBr	[45a,46]
	HCl + thiourea	[46]
	NH$_4$SCN + HNO$_3$	[47,48]
Tri-n-octylphosphate	HCl, HNO$_3$,HClO$_4$,H$_2$SO$_4$	[49]
Tri-i-octylphosphate	LiCl, NH$_4$SCN	[24]
Triallylphosphate	HCl, HNO$_3$	[50,51]
	HClO$_4$	[50]
	H$_2$SO$_4$	[51]
Tri-n-octylphosphine oxide (TOPO)	HCl	[4,39,40,43,52]
	HNO$_3$	[4,39,40,45]
	H$_2$SO$_4$	[4]
	HBr, LiBr+HBr	[53]
Methylene bis(di-n-hexyl-phosphine oxide)	HCl, HNO$_3$, HClO$_4$	[39,40,55,56]
Methylene bis(di(2-ethylhexyl) phosphine oxide)	HCl, HNO$_3$, HClO$_4$	[39,40,55,56]
Tetrabutylhypophosphate	HNO$_3$	[57]
Hexadecylamine	HCl, HNO$_3$	[61]
Primene JM-T	HCl	[2,21,58,59,60]
	HNO$_3$	[76]
	H$_2$SO$_4$	[87]
	HF	[74]
	HBr; NaBr+HBr	[53]
Dilaurylamine	HCl, HNO$_3$	[61]
Amberlite LA-1	HCl	[2,21,59,62,63]
	LiCl	[59,63,70]
	HBr, NaBr+HBr	[53]
	NaBr, NaI, KSCN, KCN	[63]
	HNO$_3$	[63,76]
	HClO$_4$, HSCN+KSCN	[76]
	H$_2$SO$_4$	[63,87]
	Various organic complexing agents	[90]

(Table 2 continued)

Extractant	Eluent	References
Amberlite LA-2	HCl	[21,64,65]
	LiCl	[72]
	HBr, NaBr, NaI	[64,75]
	HF	[73,74]
	H_2SO_4,$(NH_4)_2SO_4$+H_2SO_4	[88]
	NH_4SCN	[64,77]
	HCOOH,CH_3COOH,$CH_2ClCOOH$	[89]
Tri-n-butylamine (TBuA)	NH_4SCN	[79,80]
	NH_4SCN+HCl	[79,80]
Tri-n-octylamine (TNOA)	HCl	[2,21,63,64-69]
	NaCl	[5]
	LiCl, KSCN, KCN	[63]
	HBr	[64,75]
	NaBr, NaI(acidified)	[63,64,75]
	HF	[73]
	HNO_3	[63,76,81,82]
	$LiNO_3$,NH_4NO_3	[82,84]
	$NaNO_3$,$Ca(NO_3)_2$,$Al(NO_3)_3$	[82]
	NH_4NO_3 + HNO_3	[84-86]
	$Al(NO_3)_3$ + HNO_3	[85,86]
	H_2SO_4	[63]
	HCOOH, CH_3COOH,$CH_2ClCOOH$	[89]
	Various organic complexing agents	[90]
Tri-i-octylamine (TIOA)	HCl	[21]
	LiCl	[72]
	HBr, HI, HNO_3,HSCN+KSCN	[76]
	NH_4SCN	[79,80]
Trilaurylamine	HCl	[61,67]
Alamine -336	HCl	[2,21,59]
	LiCl	[59,70]
	HBr, LiBr+HBr	[53]
	HF	[74]
Alamine -336S	HNO_3	[45]
	H_2SO_4	[87]
Adogen -368	HBr,HI,HNO_3,HSCN+KSCN	[76]
Aliquat -336	HCl	[21,59, 61, 66,70]
	LiCl	[72]
	HBr, HI, HNO_3,HSCN+KSCN	[76]
	HF	[74]
	H_2SO_4	[87]

(Table 2 continued)

Extractant	Eluent	References
Adogen-464	LiCl	[59,70]
	HBr, HI, HNO_3, $HClO_4$, HSCN+KSCN	[76]
Amberlite LA-2 + TBuA	HCl	[65]
Other amines	HCl	[21,67]
2-Thenoyltrifluoroacetone (HTTA)	Acetate buffer	[91]
5,8-Diethyl-7-hydroxydodecan--6-oxime	Acetate, tartrate buffers	[95,96]
Lix-64	$(NH_4)_2SO_4.NH_4Cl$	[97]
Di-n-octylsulphoxide (DOSO)	$HCl,HNO_3,HClO_4.NH_4SCN$ + +$HClO_4$	[102]
Bis(n-octylsulphinyl) methane (BOSM)	HCl,HNO_3,NH_4SCN + + $HClO_4$	[102]
Bis(n-octylsulphinyl) ethane	HCl, HNO_3	[102]
p- Tolylsulphide	HCl, HNO_3	[102]
Alamine-336S oxide	HCl	[103]
	HBr,LiBr+HBr	[45]
Tri-n-octylarsine oxide	HCl	[103]
	HBr, LiBr+HBr	[45]
Tri-n-octylphosphine sulphide	HCl, HNO_3	[99]
Diheptylphenylphosphoro-dithioate	HCl	[100]
Dibutyl phosphorothioic acid	HCl	[100]
Methylparathion	HNO_3	[101]
Ethylparathion	HNO_3	[101]

7. CONCLUSIONS

Paper and thin-layer extraction chromatography are very simple and rapid techniques by which a large amount of information can be collected on the behaviour of the different elements in almost all extraction chromatographic systems.

In spite of a number of uncertainties related to the experimental set-up of the laminar techniques, and of possible deviations from ideality suggested for certain chromatographic systems, suitable correlations of laminar results with chromatographic parameters (such as concentrations in the eluent) most often provide reliable information on the chemical processes underlying the distribution of the element between the two phases.

However, the actual calculation of distribution coefficients from R_f data is severely limited by the very little available information on the phase volume ratios involved in the process. Also the exact prediction from R_f results of the retention volumes of a column is limited by the above mentioned lack of data.

The impossibility to calculate retention volumes exactly from R_f results does not appreciably affect the usefulness of R_f spectra in planning column separations involving elements with sufficiently high separation factors, in which case the laminar chromato- graphic data are often much more advantageous for assessing the best conditions for sepa- ration, as compared to liquid-liquid extraction ones. Conversely, when elements having low separation factors have to be separated, laminar results are of little help, besides few particular cases.

The enormous amount of R_f data collected in the literature ensures pertinent informa- tion for nearly all column separation problems. They can be particularly useful also in the very first choice of the most suitable extraction system for a given separation, since they include systems for which little, if any, liquid-liquid extraction information is available.

Laminar extraction chromatography could be advantageously applied in qualitative analysis, the elements being identified by their R_f values in proper chromatographic systems. The very few examples of this application support its promising features: thin-layer chro- matography in the Amberlite LA-1–HCl system, for instance, was successfully applied to the analysis of a large number of sulphidic and non-sulphidic ores[104], and an approach is in progress on the systematic qualitative analysis of the different groups of cations by paper extraction chromatography (see, e.g., Ref.[105]). Qualitative analysis could be an additional aid to laminar techniques in planning chromatographic separations of mixtures of unknown composition.

REFERENCES TO CHAPTER 14

1. L. S. BARK, G. DUNCAN, R. J. T. GRAHAM, Analyst, 92 (1967) 347.
2. U. A. Th. BRINKMAN, G. DE VRIES, E. VAN DALEN, J. Chromatog., 25 (1966) 447.
3. E. CERRAI, G. GHERSINI, J. Chromatog., 24 (1966) 383.
4. E. CERRAI, C. TESTA, J. Chromatog., 7 (1962) 112.
5. A. WAKSMUNDSKI, S. PRZESZLAKOWSKI, Chem. Anal., 11 (1966) 159.
6. E. CERRAI, G. GHERSINI, in J. C. GIDDINGS, R. A. KELLER, Eds., Advances in Chromatog- raphy, Marcel Dekker, New York, Vol. 9, p. 3, 1970.
7. L. S. BARK, G. DUNCAN, J. Chromatog., 49 (1970) 278.
8. H. HOLZAPFEL, L.– V. LAN, G. WERNER, J. Chromatog., 24 (1966) 153.
9. G. DUNCAN, M. Sc.Thesis, Salford, 1966.
10. L. S. BARK, F. B. BASKETTER, R. J. T. GRAHAM, Proc. 6th Int. Symposium on Chromatog- raphy and Electrophoresis, 1970, p. 375, Bruxelles.
11. L. S. BARK, G. DUNCAN, R. J. T. GRAHAM, N. HALSTEAD, D. McCORMICK, Proc. 6th Int. Symposium on Chromatography and Electrophoresis, 1970, p. 382, Bruxelles.
12. A. D. LEVAGGI, M. Sc. Thesis, Salford, 1969.
13. E. P. HORWITZ, C. A. A. BLOOMQUIST, D. J. HENDERSON, J. Inorg. Nucl. Chem., 31(1969) 1149.
14. E. CERRAI, C. TESTA, J. Chromatog.,8 (1962) 232.
15. S. SIEKIERSKI, R. J. SOCHACKA, J. Chromatog., 16 (1964)385.
16. R. J. SOCHACKA, S. SIEKIERSKI, J. Chromatog., 16 (1964) 376.
17. G. S. KATYKHIN, Zh. Anal. Khim., 27 (1972) 849.
18. E. CERRAI, G. GHERSINI, J. Chromatog., 16 (1964) 258.
19. E. CERRAI, G. GHERSINI, J. Chromatog., 18 (1965) 124.
20. E. CERRAI, C. TESTA, C. TRIULZI, Energia Nucleare, 9 (1962) 193.

21. U. A. Th. BRINKMAN, G.DE VRIES,E. VAN DALEN, J. Chromatog., 22 (1966) 407.
22. –, Yan Tsu Nang, 9 (1964) 854; Ref. Zh. Khim., 17G30 (1965)
23. E. CERRAI, G. GHERSINI, J. Chromatog., 13 (1964) 211.
24. E. SOCZEWINSKI, M. ROJOWSKA, S. PRZESZLAKOWSKI, Ann. Univ. Curie, 21 (1966) 21.
25. A. MUCHOVA, V. JOKL, Chem. Zvesti, 26 (1972) 303.
26. S. – W. PANG, S.–C. LIANG, Hua Hsueh Hsueh Pao, 29 (1963) 319; NSA, 19 (1965) 7298.
27. N. M. CVJETIČANIN, Bull. Boris Kidric Inst. Nucl. Sci., 15 (1964) 201.
28. S. FISEL, C. UNGUREASU, Gh. GROSU, Anal. Stiint. Univ. Al. I. Cuza, Sect. I, 12 (1966) 153.
29. S.–W. PANG, S. –C. LIANG, Hua Hsueh Hsueh Pao, 30 (1964) 237; CA 61 (1964) 10016a,
30. M. N. SASTRI, A.P. RAO, Z. Anal. Chem., 196 (1963) 166.
31. G. WERNER, R. HANNING, W. DEDEK, H. HOLZAPFEL, Proc. 3rd Anal. Chem. Conf., Budapest, Vol. 1, p. 75, 1970.
32. G. WERNER, Angew. Chem., 77 (1965) 920.
33. G. WERNER, Z. Chem., 5 (1965) 147.
34. G. WERNER, J. Chromatog., 22 (1966) 400.
35. E. SOCZEWINSKI, S. PRZESZLAKOWSKI, A. FLIEGER, Ann. Univ. Curie, 21 (1966) 7.
36. S. FISEL, Gh. GROSU, Anal. Stiint. Univ. Al. I. Cuza, Sect. I,12 (1966) 23.
37. C.–T. HU, H.–C. SHI, Ko Hsueh Tung Pao, (8) (1963) 54.
38. C.–T. HU, H.–C. SHI, Hua Hsueh Hsueh Pao, 30 (1964) 352.
39. J. W. O'LAUGHLIN, C. V. BANKS, Anal. Chem., 36 (1964) 1222.
40. J. W. O'LAUGHLIN, C. V. BANKS, U.S. At. Energy Comm. Rept. IS–737, 1963.
41. L. S. BARK, G. DUNCAN, R. J. T.GRAHAM, Analyst, 92 (1967) 31.
42. G. DUNCAN, L. S. BARK, R. J. T. GRAHAM, Proc. Soc. Anal. Chem. (London) 3 (1966) 145.
43. U. A. Th. BRINKMAN, H. VELTKAMP, J. Chromatog., 24 (1966) 489.
44. T. B. PIERCE, R. F. FLINT, J. Chromatog., 24 (1966) 141.
45. H. R. LEENE, G.DE VRIES,U.A. Th. BRINKMAN, J. Chromatog., 80 (1973) 221.
45a. A. MUCHOVA, V. JOKL, Chem. Zvesti, 27 (1973) 348.
46. C.T. HU, Ko Hsueh Tung Pao, 17 (1966) 166.
47. C.–C. CHANG, H.–H. YANG, Hua Hsueh Hsueh Pao, 31 (1965) 182; CA, 63 (1965) 3601g.
48. C.–C. CHANG, H.— H. YANG, Hua Hsueh Tung Pao, (1) (1965) 45; CA, 63 (1965) 7634d.
49. N. M. CVJETIČANIN, J. D. CVORIC, I. D. OBRENOVIC–PALIGORIC, Bull. Boris Kidric Inst. Nucl. Sci., 14 (1963) 83.
50. C. LITEANU, A. UNGUREANU, Rev. Roum. Chim., 16 (1971) 1769.
51. C. LITEANU, A. UNGUREANU, Rev. Roum.Chim., 13 (1968) 899.
52. C–C.CHANG, Hua Hsueh Tung Pao, (1963) 624.
53. U. A. Th. BRINKMAN, G. DE VRIES, H.R. LEENE, J. Chromatog., 69 (1972) 181.
54. J. W. O'LAUGHLIN, in D. C. STEWARD, H. A. ELION, Eds., Progress in Nuclear Energy, Series IX, Analytical Chemistry, Pergamon Press, Oxford, Vol. 6, p. 97, 1966.
55. J. W. O'LAUGHLIN, J. W. FERGUSON, J. J. RICHARD, C. V. BANKS, J. Chromatog., 24 (1966) 376.
56. J. W. O'LAUGHLIN, G. J. KAMIN, C. V. BANKS, J. Chromatog., 21 (1966) 460.
57. Yu. S. KOROTKIN, Radiokhimiya, 13 (1971) 137.
58. U. A. Th. BRINKMAN, in E. S. PERRY,C.J. VAN OSS, Eds., Progress in Separation and Purification, Wiley Interscience, Vol. 4. p. 241, 1971.
59. U. A. Th. BRINKMAN, Dissertation, Free University of Amsterdam, 1968.
60. R. J. T. GRAHAM, L. S. BARK, D. A. TINSLEY, J. Chromatog., 39 (1969) 200.
61. M.N. SASTRI, A. P. RAO, A. R. K. SARMA, J. Chromatog., 19 (1965) 630.
62. A. FLIEGER, S. PRZESZLAKOWSKI, Chem. Anal., 16 (1971) 1341.
63. A. MUCHOVA, V. JOKL, Chem. Zvesti, 25 (1971) 280.
64. S. PRZESZLAKOWSKI, Chem. Anal., 12 (1967) 1071.
65. S. PRZESZLAKOWSKI, Roczniki Chemii, 41 (1967) 1681.
66. E. CERRAI, G. GHERSINI, Energia Nucleare, 11 (1964) 441.
67. M. ISHIBASHI, H. KOMAKI, M. DEMIZU, Mitsubitshi Denki Giho, 39 (1965) 907.
68. C. TESTA, J. Chromatog., 5 (1961) 236.

400

69. D. McCORMICK, R.J.T. GRAHAM, L. S. BARK, Proc. 4th Int. Symposium on Chromatography and Electrophoresis, Bruxelles, 1966, Presses Academiques Européennes, Brussels, p. 199, 1968.
70. U. A. Th. BRINKMAN, G. DE VRIES, E. VAN DALEN, J. Chromatog., 31 (1967) 182.
71. F. V. CORNETT, T. W. GILBERT, Anal. Letters, 4 (1971) 69.
72. R. W. MURRAY, R. J. PASSARELLI, Anal. Chem., 39 (1967) 282.
73. S. PRZESZLAKOWSKI, Chem. Anal., 12 (1967) 321.
74. U. A. Th. BRINKMAN, P. J. J. STERRENBURG, G. DE VRIES, J. Chromatog., 54 (1971) 764.
75. S. PRZESZLAKOWSKI, E. SOCZEWINSKI, Chem. Anal., 11 (1966) 895.
76. U. A. Th. BRINKMAN, G.DE VRIES, E. VAN DALEN, J. Chromatog., 23 (1966) 287.
77. R. J. T. GRAHAM, A. CARR, J. Chromatog., 46 (1970) 301.
78. R. KURODA, K. OGUMA, K. NAKAMURA, Nippon Kagaku Zasshi, 91 (1970) 1155.
79. S. PRZESZLAKOWSKI, Chem. Anal., 12 (1967) 57.
80. A. WAKSMUNDZKI, S. PRZESZLAKOWSKI, in K. MACEK, I. M. HAIS, Eds., Stationary Phase in Paper and Thin Layer Chromatography, Elsevier, Amsterdam, p. 199, 1965.
81. W. KNOCH, B. MUJU, H. LAHR, J. Chromatog., 20 (1965) 122.
82. E. CERRAI, C. TESTA, J. Chromatog., 5 (1961) 442.
83. C. TESTA, Anal. Chem., 34 (1962) 1556.
84. S. – W. PANG, S. – C. LIANG, Acta Chim. Sinica, 30 (1964) 401.
85. W. KNOCH, H. LAHR, Radiochim. Acta, 4 (1965) 114.
86. W. KNOCH, B. MUJU, H. LAHR, J. Chromatog., 20 (1965) 122.
87. U. A. Th. BRINKMAN, G. DE VRIES, J. Chromatog., 56 (1971) 103.
88. T. SHIMIZU, R. ISHIKURA, J. Chromatog., 56 (1971) 95.
89. S. PRZESZLAKOWSKI, E. SOCZEWINSKI, A. FLIEGER, Chem. Anal., 13 (1968) 841.
90. A. MUCHOVA, V. JOKL, Chem. Zvesti, 26 (1972) 289.
91. S. KAWAMURA, T. FUJIMOTO, M. IZAWA, J. Chromatog., 34 (1968) 72.
92. N. CVJETIĆANIN, Summaries of 6th Int. Symposium on Chromatography and Electrophoresis, Bruxelles, p. 38, 1970.
93. N. CVJETIĆANIN, J. Chromatog., 34 (1968) 520.
94. N. CVJETIĆANIN, J. Chromatog., 74 (1972) 99.
95. D. R. BEUERMAN, Iowa State Univ. Thesis, IS–T–492, 1971.
96. J. S. FRITZ, D. R. BEUERMAN, J. J. RICHARD, Talanta, 18 (1971) 1095.
97. E. CERRAI, G. GHERSINI, Analyst, 94 (1969) 599.
98. J. S. FRITZ, J. SHERMA, J. Chromatog., 25 (1966) 153.
99. D. E. ELLIOTT, C. V. BANKS, Anal. Chim. Acta, 33 (1965) 237.
100. M. A. S. SCHMITT, Iowa State University Thesis, IS–T–144, 1967.
101. E. CERRAI, G. GHERSINI, J. Chromatog., 22 (1966) 425.
102. D. KENNEDY, J. S. FRITZ, Talanta, 17 (1970) 823.
103. H. R. LEENE, G.DE VRIES, U.A. Th. BRINKMAN, J. Chromatog., 57 (1971) 173; correction in J. Chromatog., 59 (1971) 516.
104. U. A. Th. BRINKMAN, G.DE VRIES, J. Chem. Educ., 49 (1972) 244.
105. S. PRZESZLAKOWSKI, A. FLIEGER, J. Chromatog., 81 (1973) 129.

CHAPTER 15

BIBLIOGRAPHY OF EXTRACTION CHROMATOGRAPHY

H. ESCHRICH, W. DRENT

1. INTRODUCTION

In the present chapter, we have tried to supply the reader with a comprehensive bibliography covering the relevant literat.re from the first application of extraction chromatography to inorganic analyis until the end of the year 1972. As far as available to us the papers published in 1973 were included.

Though this monograph is almost exclusively devoted to column techniques, the literature on laminar techniques — i. e. paper and thin-layer extraction chromatography — has been added for the sake of completeness.

All titles of publications are given in English. If the publication has not been written in English, the original language is indicated in brackets at the end of the reference. Due to space limitations, abstracts of the papers have been omitted.

The papers have been listed firstly according to the alphabetical order of the authors' names or the first name (senior author), if more than one author is given and secondly according to chronological order if an author or senior author has published more than one paper.

The translation into Latin script of names and titles originally written in Cyrillic, Japanese or Chinese script has most certainly introduced some faults and inconsistencies for which we beg to be excused by the authors concerned and the readers.

The bibliographic part is followed by a survey-table of experimentally important data on column extraction chromatographic investigations. Generally these data cannot be derived from the title or from the abstract of the publication. Respecting the scope of the present monograph, a corresponding table for paper or thin-layer extraction chromatographic separations is not contained in this chapter.

An index of the chromatographed species has been omitted as an "element" standing by itself without relation to another one or to the separation system appears to have little sense as regards practical usefulness. The above mentioned table gives therefore a more profitable survey of the separations or investigations already performed.

Publications dealing with column as well as laminar techniques are only listed under "column extraction chromatography"; however, the use of paper or thin-layers is indicated in the survey-table (Paragraph 7), in the row "support".

Furthermore two tables are attached summarizing information on extractants (Paragraph 8) and support materials (Paragraph 9) used in column extraction chromatography. The tables serve simultaneously as a cross-reference to the bibliographical part and the survey table (Paragraph 7) and moreover for statistical purposes. In the author index all

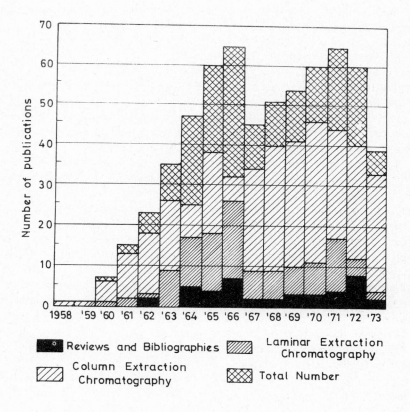

Fig. 1.   Number of publications dealing with extraction chromatography in the years 1958-1973.
(for 1973   incomplete)

authors are listed independent of whether their publications deal with column, or paper, or thin-layer chromatographic investigations, or other studies. The nature of their work appears from the attached reference number. The corporate author index lists, in alphabetical order, the organization or institution within the various countries responsible for the issue of the paper as indicated by its reference number.

Finally, in Fig. 1, a survey is given on the number of publications per year for the period 1958—1973 covered almost completely in this bibliography.

2. SELECTION OF PAPERS

When preparing this chapter we had to establish some rules for selecting the papers for their inclusion in the present bibliography.
According to these rules we included:
1. Bibliographies exclusively devoted to extraction chromatography (or reversed-phase partition chromatography).
2. Reviews and books which are not necessarily devoted to extraction chromatography alone, but contain an essential part dealing with extraction chromatographic investigations in a reviewing way.
3. All papers dealing with original extraction chromatographic investigations and separations of inorganic species.
4. Some papers, in which modifications or adaptations of known extraction chromatographic separations are described or in which known separation procedures are applied to solve a problem in a field other than originally described.
5. Papers of already known or very similar content, if not all the authors were the same as in the original ones or when these papers were published later in a journal more readily available to the public.
6. Conference papers and patents concerning extraction chromatography.
Excluded from the present bibliography were:
1. All papers in which merely the application of already published extraction chromatographic methods are described or only mentioned. (Following this rule approximately thirty papers were omitted.)
2. Contributions on extraction chromatographic investigations in "Quarterly Reports", "Semi-Annual Reports" or "Yearly Reports" which later have been published in form of a summarized report or in a journal.
3. Papers in which the extraction chromatographic part is of so little importance or value that their inclusion could hardly be justified.
Relevant publications not contained in this bibliography must not necessarily fall under the "exclusion rules" cited above, as we may either have overlooked them or had not sufficient information available to decide on their inclusion.

3. BIBLIOGRAPHIES

1. ESCHRICH, H. and DRENT, W.
    Eurochemic, Mol, Belgium
    BIBLIOGRAPHY ON APPLICATIONS OF REVERSED-PHASE PARTITION CHROMATO-
    GRAPHY TO INORGANIC CHEMISTRY AND ANALYSIS.
    Eurochemic Technical Report ETR–211 (1967), 99 p.

2. ESCHRICH, H. and DRENT, W.
    Eurochemic, Mol, Belgium
    EXTRACTION CHROMATOGRAPHY. – BIBLIOGRAPHY –
    Eurochemic Technical Report ETR–271 (1971), 130 p.

3. HEDRICK, C. E. and FRITZ, J. S.
    Inst. for Atomic Research and Dept. of Chemistry, Iowa State Univ., Ames, Iowa, USA
    BIBLIOGRAPHY OF REVERSED–PHASE PARTITION CHROMATOGRAPHY.
    IS–950. June 1, 1964, 20 p.

4. REVIEWS AND BOOKS

4. ALIMARIN, I. P. and BOLSHOVA, T. A.
    Chem. Dept., M. V. Lomonosov Moscow State Univ. Moscow, USSR
    SEPARATION AND CONCENTRATION OF ELEMENTS BY REVERSED PHASE PARTITION
    CHROMATOGRAPHY.
    Pure Appl. Chem., 31, 493–501 (1972)

5. BERANOVA, H. and TEJNECKY, M.
    Nuclear Research Inst., Řež, ČSSR
    PARTITION CHROMATOGRAPHY OF URANIUM AND FISSION PRODUCTS ON
    POROUS STYRENE–DIVINYLBENZENE COPOLYMER SWOLLEN IN THE SOLUTION OF
    TRIBUTYL PHOSPHATE. PART III. PROPERTIES OF COPOLYMER AS CARRIER OF
    STATIONARY PHASE AND PREPARATION OF SWOLLEN GELS. LITERATURE SURVEY.
    UJV–2557 Ch.1971, 38 p.

6. BRINKMAN, U. A. Th.
    Chemical Laboratory, Free University, Amsterdam, The Netherlands
    CHROMATOGRAPHY WITH LIQUID ION EXCHANGERS.
    Chem. Techn. Rev., 21, 529–33 (1966) (in Dutch)
    CA, 66, 40929 (1967)

7. BRINKMAN, U. A. Th
    Chemical Laboratory, Free University, Amsterdam, The Netherlands
    LIQUID ANION-EXCHANGERS IN ANALYTICAL CHEMISTRY. REVERSED-PHASE
    CHROMATOGRAPHY AND MECHANISM OF THE SORPTION PROCESS.
    Thesis, 1968. 113 p.

8. BRINKMAN, U. A. Th.
    Dept. Anal. Chem.,Free University, Amsterdam, The Netherlands
    USE OF LIQUID ANION EXCHANGERS IN REVERSED-PHASE EXTRACTION CHRO-
    MATOGRAPHY.
    Prog. Separ. Purif., 4, 241–96 (1971)
    CA, 75, 155182 (1971)

9. BRINKMAN, U. A. Th. and DE VRIES, G.
Free University, Amsterdam, The Netherlands
LIQUID ANION–EXCHANGERS IN REVERSED-PHASE EXTRACTION CHROMATOGRAPH
J. Chem. Education, 49, 244–8 (1972)

10. CERRAI, E.
CISE, Milan, Italy
LIQUID ION EXCHANGERS: SEPARATIONS ON INERT SUPPORTS IMPREGNATED
WITH LIQUID ION EXCHANGERS.
M. Lederer, ed. Chromatographic Reviews. 6. Elsevier Publishing Co., Amsterdam, London,
New York 1964. pp. 129–53.

11. CERRAI, E.
CISE, Milan, Italy
THE USE OF LIQUID ION EXCHANGERS IN PAPER CHROMATOGRAPHY.
K. Maček and I. M. Hais, eds. Stationary Phase in Paper and Thin-layer Chromatography.
Proceedings of the 2nd Symposium held at Liblice, June, 10–12, 1964. Elsevier Publishing
Company, Amsterdam, London, New York, 1965. pp. 180–98.

12. CERRAI, E.
CISE, Milan, Italy
REVERSED-PHASE PARTITION CHROMATOGRAPHY IN INORGANIC CHEMISTRY.
(A SURVEY ON RESULTS OBTAINED AT "CISE").
CISE-107. 1966. 46. p.

13. CERRAI, E. and GHERSINI, G.
CISE, Milan, Italy
REVERSED-PHASE EXTRACTION CHROMATOGRAPHY IN INORGANIC CHEMISTRY.
J. Calvin Giddings, Roy A. Keller, eds., Advances in Chromatography, Vol. 9, Marcel Dekker,
Inc., New York. 1970. pp. 3–189.

14. ESCHRICH, H.
Kjeller Research Establishment, Kjeller, Norway
REVERSED-PHASE PARTITION CHROMATOGRAPHY APPLIED TO INORGANIC
ANALYSIS.
KR-22. Quarterly Progress Report Jan. – March 1962, Annex I. 4 p.

15. FIDELIS, I. and SMUŁEK, W.
Dept. of Radiochemistry, Inst. of Nucl. Research, Warsaw-Zeran, Poland.
POSSIBILITIES OF APPLYING EXTRACTION CHROMATOGRAPHY FOR THE SEPA-
RATION OF SOME FISSION PRODUCTS.
INR-1357/V/C. 1971. 7 p.

16. GORET, J.
Univ. of Paris, Paris, France
CHROMATOGRAPHIC EXTRACTION BY MEANS OF
SYNTHETIC POLYMERS.
Thesis, 1966. 50 p. (in French)

17. GREEN, H.
British Cast Iron Research Association, Bordesley Hall, Alvechurch Birmingham, Great Britain
RECENT USES OF LIQUID ION EXCHANGERS IN INORGANIC ANALYSIS.
Talanta, 11, 1561–80 (1964)

18. GREEN, H.
Auckland Ind. Dev. Div. D. S. I. R., Auckland, New Zealand
USES OF LIQUID ION-EXCHANGERS IN INORGANIC ANALYSIS.
Talanta, 20, 139–61 (1973)

19. HULET, E. K. and BODE, D. D.
Lawrence Radiation Lab., Livermore, Calif., USA
SEPARATION CHEMISTRY OF THE LANTHANIDES AND TRANSPLUTONIUM ACTINIDES.
Bagnall, K. W. ed. Lanthanides and Actinides. pp.1–45,Baltimore University Park Press (1972)
CA, 76, 90714 (1972)
NSA, 27, 5011 (1973)

20. KATYKHIN, G. S.
A. A. Zhdanov State Univ., Leningrad, USSR
COLUMN PARTITION CHROMATOGRAPHY FOR INORGANIC SEPARATIONS.
Zh. Analit. Khim., 20 (5), 615–24 (1965) (in Russian)
CA, 63, 6296b (1965)

21. KATYKHIN, G. S.
A. A. Zhdanov State Univ. Leningrad, USSR
USE OF COLUMN PARTITION CHROMATOGRAPHY FOR SEPARATION OF IN-
ORGANIC SPECIES.
Zh. Anal. Khim., 27, 849–70 (1972) (in Russian)

22. KIBA, T. and AKAZA,I.
Univ. Kanazawa, Japan
INORGANIC REVERSED-PHASE CHROMATOGRAPHY.
Kagaku, 19, 984–90 (1964) (in Japanese)
CA, 64, 2721e (1966)

23. KIBA, T.
Department of Chemistry, Faculty of Sciences, Kanazawa, University, Japan
SEPARATION OF INORGANIC IONS BY REVERSED PHASE PARTITION CHRO-
MATOGRAPHY.
Bunseikii Kagaku, 15; 991–7 (1966) (in Japanese) ORNL-tr-1503,1967, 16 p. NSA, 20,45582(196(

24. KORKISCH, J.
Anal. Inst., University of Vienna, Austria
MODERN METHODS FOR THE SEPARATION OF RARER METAL IONS.
Pergamon Press, 1969, 620 p.

25. LEDERER, M. and MAJANI, C.
Lab. di Chromatografia del C. N. R., Rome, Italy
PAPER CHROMATOGRAPHIC DATA FOR INORGANIC SUSBTANCES. Chapter V.
CHROMATOGRAPHY OF METAL IONS ON PAPER IMPREGNATED WITH LIQUID ION
EXCHANGERS.
Chromatog. Rev., 12, 375–401 (1970)

26. LIU, C. L.
PARTITION CHROMATOGRAPHY WITH REVERSED PHASES.
Hua Hsueh Tung Pao, 7, 401–5 (1962)
KFK-tr-186 (in German, translation from Chinese) 1966, 17 p.

27. MARKL, P.
Anal. Institute of the University of Vienna, Austria
EXTRACTION AND EXTRACTION CHROMATOGRAPHY IN INORGANIC ANALYSIS.
"Methoden der Analyse in der Chemie" Vol. 13 . Akademische Verlagsgesellschaft, Frankfurt/Main
(1972)

28. MASSART, D. L.
INORGANIC ANALYTICAL CHROMATOGRAPHY. PART II. SEPARATION TECHNIQUES.
Mededelingen van de Vlaamse Chemische Vereniging, 1969, No. 1, 1–27. (in Dutch)

29. O'LAUGHLIN, J. W.
Iowa State University, Ames, Iowa, USA
SOME NEUTRAL BIFUNCTIONAL ORGANO-PHOSPHORUS COMPOUNDS AS SOLVENT
EXTRACTANTS.
D. C. Stewart, and H. A. Elion eds. Progress in Nuclear Energy. Series IX. Analytical Chemistry,
Vol. 6. Pergamon Press, Oxford, 1966. pp. 97–146.

30. PEPPARD, D. F.
Chem. Div. Argonne Nat. Lab., Argonne, Ill. USA
PARTITION CHROMATOGRAPHY.
Advances in Inorganic and Radiation Chemistry, H. J. Emeléus, and A. G. Sharpe eds. Vol. 9.
Academic Press, New York, 1966. pp. 57–74

31. POITRENAUD, C.
Inst. Nat. des Sciences et Techniques Nucléaires, Saclay, France
RECENT DEVELOPMENTS IN CHROMATOGRAPHIC EXTRACTION BY MEANS OF SYN-
THETIC POLYMERS AND THEIR APPLICATION TO INORGANIC ANALYIS.
Chim. Anal., 52, 39–47 (1970) (in French)

32. PREOBRAZHENSKII, B. K., MOSKVIN, L. N., and KATYKHIN, G. S.
DEVELOPMENT OF PARTITION CHROMATOGRAPHY OF INORGARNIC SUBSTANCES.
Paper presented at the XXth Congress on Pure and Appl. Chem. Moscow, USSR, 12–18 July
1965.   CONF 650723, E100, 178 pp. (1965)

33. SMALES, A. A.
AERE Harwell, Berks., Great Britain
ANALYTICAL CHEMISTRY – SCIENCE OR TECHNOLOGY?
Opening address at the SAC Conference, Nottingham, July 19th 1965
Proceedings of the SAC Conference Nottingham 1965 W. Heffer and Sons Ltd., Cambridge
1965, pp. 1–23.

34. SMUŁEK, W. and ZELENAY, K.
Inst. of Nuclear Research, Warsaw-Zeran, Poland
CURRENT WORK ON THE DEVELOPMENT OF RADIOCHEMICAL METHODS FOR THE
DETERMINATION OF NUCLEAR FUEL BURN-UP IN POLAND.
Analytical Chemistry of Nuclear Fuels. IAEA, Vienna, 1971. pp. 119–29.

35. SPEVÁČKOVÁ, V.
Ceskoslov. Akad. Ved, Řež/Prague, ČSSR
REVERSED-PHASE CHROMATOGRAPHY FOR THE SEPARATION OF SOME CATIONS.
Chem. Listy, 62, 1194-205 (1968) (in Czech)
C. A., 70, 70, 63726 (1969)

36. STARÝ, J.
    Department of Nuclear Chemistry, Faculty of Technical and Nuclear Physics, Prague 1,
    Czechoslovakia
    SEPARATION OF TRANSPLUTONIUM ELEMENTS
    Talanta, 13, 421–37 (1966)

37. STRONSKI, I.
    Inst. Kernphys. Krakow, Poland
    EXTRACTION CHROMATOGRAPHY OF METAL SALTS.
    Österr. Chem. Ztg., 68, 5–19 (1967) (in German)
    CA, 66, 72031 (1967)

38. STRONSKI, I.
    Inst. of Nuclear Physics, Krakow, Poland
    CHROMATOGRAPHIC COLUMN EXTRACTION OF INORGANIC COMPOUNDS.
    INP–803 C, 44 p., 1972 (in German)
    NSA, 27, 87, (1973)

39. STRONSKI, I.
    Inst. of Nuclear Physics, Krakow, Poland
    LONG-CHAIN ALKYL AMMONIUM SALTS AS EXTRACTING COMPOUNDS IN THE
    DYNAMIC METHODS OF SEPARATIONS OF METAL SALTS.
    Wiad. Chem., 26, 63–74 (1972) (in Polish)

40. TESTA, C.
    CNEN, Casaccia, Italy
    REVERSED-PHASE PARTITION CHROMATOGRAPHY AND ITS APPLICATION.
    RT/PROT (65) 33, 1965, 28 p.

41. TESTA, C.
    CNEN, Casaccia, Italy
    THE APPLICATION OF EXTRACTION CHROMATOGRAPHY TO THE DETERMINATION
    OF RADIONUCLIDES IN BIOASSAY.
    Paper presented at the Symp. on Determination of Radionuclides in Environmental and Biolo-
    gical Materials, London, April 2–3, 1973
    AED-Conf-73-085-016, 10 p., 5 figs., 2 tables, 13 refs., 1973

42. TSERKOVNITSKAYA, I. A. and LUGININ, V. A.
    Leningrad State Univ., USSR
    METHODS OF SEPARATING ELEMENTS OF VARIOUS OXIDATION STATES.
    Zh. Anal. Khim., 24, 54–68 (1969) (in Russian)
    NSA, 23, 13645 (1969)

5. ORIGINAL PUBLICATIONS ON LAMINAR
EXTRACTION CHROMATOGRAPHY

43. BABAYAN, Kh. S. and VARSHAL. G. M.
    Inst. of Geochem. and Anal. Chem., Moscow, USSR
    SELECTION OF CHROMATOGRAPHIC SYSTEMS FOR SEPARATING RARE-EARTH
    MIXTURES BY PARTITION CHROMATOGRAPHY.
    Arm. Khim. Zh., 25, 844–50 (1972) (in Russian)

44. BARK, L. S., DUNCAN, G. and GRAHAM, R. J. T.
    Dept. of Pure and Appl. Chem., Univ. of Salford, Salford, Lancs., Great Britain

INORGANIC THIN-LAYER CHROMATOGRAPHY. IV. REVERSED-PHASE THIN-LAYER CHROMATOGRAPHY OF SOME TOXICOLOGICALLY IMPORTANT METAL IONS.
Int. Symp. Chromatogr. Electrophor. IV, Brussels, Belgium, 1966, pp. 207–12,
Presses Académiques Européennes, Bruxelles, 1968

45. BARK, L. S., DUNCAN, G., and GRAHAM, R. J. T.
Dept. of Chemistry and Appl. Chem., The Royal College of Adv. Techn. Salford 5, Lancs.,
Great Britain
INORGANIC THIN-LAYER CHROMATOGRAPHY, PART II. CHROMATOGRAPHY OF SOME FIRST ROW TRANSITION METALS ON THIN LAYERS OF SUBSTRATES IMPREGNATED WITH TRIBUTYL PHOSPHATE.
Analyst, 92, 31–5 (1967)

46. BARK, L. S., DUNCAN, G., and GRAHAM, R. J. T.
Univ. of Salford, Salford, Lancs., Great Britain
THE REVERSED-PHASE THIN-LAYER CHROMATOGRAPHY OF METAL IONS WITH TRIBUTYL PHOSPHATE.
Analyst, 92, 347–57, (1967)

47. BARK, L. S.* and DUNCAN, G.**
* Dept. of Chemistry and Appl. Chemistry, Univ. of Salford, Salford, Lancs., Great Britain
** Dept. of Chemistry and Biology, Manchester Polytechnic, Manchester, Great Britain
THE CHROMATOGRAPHIC PROCESS IN A REVERSED-PHASE SYSTEM FOR THE SEPARATION OF SOME METAL IONS, BY ION-ASSOCIATION DISTRIBUTION.
J. Chromatog., 49, 278–89(1970)

48. BARK, L. S., BASKETTER, F. B., and GRAHAM, R. J. T.
Dept. of Chem. and Appl. Chem., Univ. of Salford, Salford, Lancs., Great Britain
A RAPID METHOD FOR THE SEPARATION OF AND IDENTIFICATION OF SOME METALS OF TOXICOLOGICAL AND POLLUTANT INTEREST.
Int. Symp. Chromatogr. Electrophor. VI, Brussels, Belgium, 14–16 Sept. 1970, pp. 295–301,
Presses Académiques Eruropéennes, Brussels, 1971

49. BARK, L. S., BASKETTER, F. B., and GRAHAM, R. J. T.
Dept. of Chem. and Appl. Chem., University of Salford, Salford, Great Britain
SOME CAUSES OF VARIATION IN THE DISTRIBUTION OF SOLUTES IN POLAR REVERSED-PHASE SYSTEMS.
Int. Symp. Chromatogr. Electrophor. VI, Brussels, Belgium, 14–16 Sept. 1970, pp. 375–81,
Presses Académiques Européennes, Brussels 1971.

50. BARK, L. S., DUNCAN, G., GRAHAM, R. J. T., HALSTEAD, N., and McCORMICK, D.
Dept. of Pure and Appl. Chem., Univ. of Salford, Salford, Great Britain
THE CHROMATOGRAPHIC PROCESS IN SOME REVERSED-PHASE SYSTEMS USED FOR THE SEPARATION OF METAL IONS.
Int. Symp. Chromatogr. Electrophor. VI, Brussels, Belgium, 14–16 Sept. 1970, pp. 382–7,
Presses Académiques Européennes, Brussels, 1971

51. BASKETTER, F. B.
Dept. of Chem. and Appl. Chem., University of Salford, Salford, Lancs., Great Britain
THIN LAYER CHROMATOGRAPHIC BEHAVIOUR OF METAL IONS.
Thesis, 1968. 141 p.

52. BEUERMAN, D. R.
Inst. for Atomic Research and Dept. of Chemistry, Iowa State Univ., Ames, Iowa, USA

USE OF ALIPHATIC α–HYDROXIMES FOR THE SEPARATION OF METAL IONS
IS–T–492. 1971. 112 p.

53.   BRINKMAN, U. A. Th. and DE VRIES, G.
      Chemical Laboratory, Free Univ., Amsterdam, The Netherlands
      CHROMATOGRAPHIC TECHNIQUES USING LIQUID ION–EXCHANGERS.
      J. Chromatog., 18, 142–4 (1965)

54.   BRINKMAN, U. A. Th., De VRIES, G., and VAN DALEN, E.
      Chemical Laboratory, Free Univ., Amsterdam, The Netherlands
      CHROMATOGRAPHIC TECHNIQUES USING LIQUID ANION EXCHANGERS. I. HCl
      SYSTEMS.
      J. Chromatog., 22. 407–24 (1966)

55.   BRINKMAN, U. A. Th., DE VRIES, G., and VAN DALEN, E.
      Chemical Laboratory, Free Univ., Amsterdam, The Netherlands
      CHROMATOGRAPHIC TECHNIQUES USING LIQUID ANION EXCHANGERS. II. STRONG
      MONOBASIC ACID SYSTEMS.
      J. Chromatog., 23, 287–97 (1966)

56.   BRINKMAN, U. A. Th. and VELTKAMP, H.
      Chemical Laboratory, Free Univ., Amsterdam, The Netherlands
      THIN-LAYER CHROMATOGRAPHY ON SUPPORTS IMPREGNATED WITH NEUTRAL
      ORGANOPHOSPHORUS COMPOUNDS.
      J. Chromatog., 24, 489–91 (1966)

57.   BRINKMAN, U. A. Th. and VELTKAMP, H.
      Chemical Laboratory, Free Univ., Amsterdam, The Netherlands
      THIN-LAYER CHROMATOGRAPHY ON SUPPORTS IMPREGNATED WITH NEUTRAL
      ORGANOPHOSPHORUS COMPOUNDS.
      J. Chromatog., 24, 489–91 (1966)

58.   BRINKMAN, U. A. Th., DE VRIES, G., and VAN DALEN, E.
      Chemical Laboratory, Free Univ., Amsterdam, The Netherlands
      CHROMATOGRAPHIC TECHNIQUES USING LIQUID ANION EXCHANGERS. III. SYS-
      TEMATIC THIN-LAYER CHROMATOGRAPHY OF THE ELEMENTS IN HCl SYSTEMS.
      J. Chromatog., 25, 447–63 (1966)

59.   BRINKMAN, U. A. Th., DE VRIES, G., and VAN DALEN, E.
      Chemical Laboratory, Free Univ., Amsterdam, The Netherlands
      THE NATURE OF SOME METAL-CHLORO ANIONS PRESENT IN THE ORGANIC PHASE
      IN REVERSED-PHASE CHROMATOGRAPHY INVOLVING LIQUID ANION-EXCHANGERS.
      J. Chromatog., 31, 182–95 (1967)

60.   BRINKMAN. U. A. Th. and KINGSMA. G.
      Chemistry Laboratory, Free Univ., Amsterdam, The Netherlands
      SMALL-SCALE THIN-LAYER CHROMATOGRAPHY OF INORGANIC CATIONS ON
      CELLULOSE LAYERS.
      Z. Anal. Chem., 249, 16–18 (1970)

61.   BRINKMAN, U. A. Th., STERRENBURG, P. J. J., and DE VRIES, G.
      Chemistry Laboratory, Free Univ., Amsterdam, The Netherlands
      CHROMATOGRAPHIC TECHNIQUES USING LIQUID ANION EXCHANGERS. IV.
      HF SYSTEMS.
      J. Chromatog., 54, 449–53 (1971)

62. BRINKMAN, U. A. Th. and DE VRIES, G.
Chemistry Laboratory, Free Univ.,Amsterdam, The Netherlands
CHROMATOGRAPHIC TECHNIQUES USING LIQUID ANION EXCHANGERS. V.
H_2SO_4 SYSTEMS.
J. Chromatog., 56, 103–9 (1971)

63. BRINKMAN, U. A. Th., DE VRIES, G., and LEENE, H. R.
Chemistry Laboratory, Free Univ., Amsterdam, The Netherlands
REVERSED-PHASE EXTRACTION CHROMATOGRAPHY USING Br-CONTAINING
ELUANTS.
J. Chromatogr., 69, 181–93 (1972)

64. BRINKMAN, U. A. Th., TAPHOORN, J. H., and De VRIES, G.
Chemistry Laboratory, Free Univ., Amsterdam, The Netherlands
REVERSED-PHASE EXTRACTION CHROMATOGRAPHY USING A TETRA-SUBSTITUTED
PYRAZOLE AS STATIONARY PHASE.
J. Chromatogr., 84, 407–11 (1973)

65. CALETKA, R.
Inst. of Nuclear Research, Prague, ČSSR
SEPARATION OF ^{95}Zr AND ^{95}Nb BY MEANS OF EXTRACTION CHROMATOGRAPHY.
Chem. Listy, 62, 669–73 (1968) (in Czech)
NSA, 22, 40029 (1968)
CA, 69, 40419 (1968)

66. CERRAI, E. and TESTA, C.
CISE, Milan, Italy
THE CHROMATOGRAPHIC SEPARATION OF URANIUM,THORIUM AND RARE EARTHS BY
MEANS OF PAPER TREATED WITH A LIQUID ANION EXCHANGER.
J. Chromatog., 5, 442–51 (1961)

67. CERRAI, E. and TESTA, C.
CISE, Milan, Italy
THE USE OF PAPER TREATED WITH TRI-n-OCTYLPHOSPHINE OXIDE FOR THE
CHROMATOGRAPHIC SEPARATION OF METAL IONS.
J. Chromatog., 7,112–9 (1962)

68. CERRAI, E. and TESTA, C.
CISE, Milan, Italy
CHROMATOGRAPHIC SEPARATION OF RARE EARTHS BY MEANS OF PAPER TREATED
WITH THE LIQUID CATION EXCHANGER DI-(2-ETHYLHEXYL) ORTHOPHOSPHORIC
ACID.
J. Chromatog., 8, 232–44 (1962)

69. CERRAI, E. and GHERSINI, G.
CISE, Milan, Italy
REVERSED-PHASE CHROMATOGRAPHY OF ALKALI METALS AND ALKALINE EARTHS
ON PAPER TREATED WITH BIS(2-ETHYLHEXYL) ORTHOPHOSPHORIC ACID IN
ACETATE MEDIUM.
J. Chromatog., 13, 211–23 (1964)
CA, 60, 9880e (1964)

70. CERRAI, E. and GHERSINI, G.
CISE, Milan, Italy

REVERSED-PHASE CHROMATOGRAPHY OF ALKALINE EARTHS ON PAPER TREATED WITH DI(2-ETHYLHEXYL) ORTHOPHOSPHORIC ACID IN CHLORIDE MEDIUM.
J. Chromatog., 15, 236–45 (1964)

71. CERRAI, E. and. GHERSINI, G.
CISE, Milan, Italy
SEPARATION OF ALUMINIUM, GALLIUM, INDIUM, AND THALLIUM BY REVERSED-PHASE CHROMATOGRAPHY ON PAPERS TREATED WITH DI(2-ETHYLHEXYL) ORTHOPHOSPHORIC ACID.
J. Chromatog., 16, 258–61 (1964)

72. CERRAI, E. and GHERSINI, G.
CISE, Milan, Italy
THE USE OF TRI-CAPRYLYL-MONOMETHYL AMMONIUM CHLORIDE IN REVERSED-PHASE CHROMATOGRAPHY OF URANIUM, ZIRCONIUM, HAFNIUM, AND TRANSITION METALS OF THE IRON GROUP.
Energia Nucl. (Milan), 11, 441–3 (1964)

73. CERRAI, E. and GHERSINI, G.
CISE, Milan, Italy
REVERSED-PHASE CHROMATOGRAPHY OF ALUMINIUM, GALLIUM, INDIUM, THAL-LIUM, AND THE TRANSITION METALS OF THE IRON GROUP ON PAPER TREATED WITH DI(2-ETHYLHEXYL) ORTHOPHOSPHORIC ACID IN CHLORIDE MEDIUM.
J. Chromatog., 18, 124–33 1965)

74. CERRAI, E. and GHERSINI, G.
CISE, Milan, Italy
REVERSED-PHASE PAPER CHROMATOGRAPHY OF SOME CATIONS EITH TWO NITROPHENYLTHIOPHOSPHATE DERIVATIVES AS STATIONARY PHASES.
J. Chromatog., 22, 425–30 (1966)

75. CERRAI, E. and GHERSINI, G.
CISE, Milan, Italy
REVERSED-PHASE PARTITION CHROMATOGRAPHY ON PAPER TREATED WITH DI(2-ETHYLHEXYL) ORTHPHOSPHORIC ACID: A SYSTEMATIC STUDY OF 67 CATIONS IN HCl.
J. Chromatog., 24, 383–401 (1966)

76. CERRAI, E., GHERSINI, G., LEDERER, M., and MAZZEI, M.
CISE, Milan and Laboratorio di Cromatografia del CNR, Rome (Italy)
ION EXCHANGE PAPER CHROMATOGRAPHY OF INORGANIC IONS. XXV. SOME CON-SIDERATIONS ON THE DETERMINATION OF THE "CHARGE" OF A METAL ION BY ION-EXCHANGE EQUILIBRIA.
J. Chromatog. 44, 161–6 (1969)

77. CHANG, C. C.
Northwestern Univ., China
ION-EXCHANGE PAPER CHROMATOGRAPHY.
Hua Hsueh Tung Pao 1963 (10) 624–31 (1963) (in Chinese)
CA, 60, 8609e (1964)

78. CHANG, C. C. and YANG, H. H.
Northwestern Univ., China

SEPARATION OF RHENIUM, VANADIUM, MOLYBDENUM, AND TUNGSTEN BY PAPER CHROMATOGRAPHY.
Hua Hsueh Tung Pao, No.1. 45−6 (1965) (in Chinese)
CA, 63, 7634d (1965)

79.   CHANG, C. C. and YANG, H. H.
Northwestern Univ., China
SEPARATION OF RARE EARTH METALS BY REVERSED-PHASE PAPER CHROMATO-GRAPHY. THE TRIBUTYL PHOSPHATE − NITRIC ACID − AMMONIUM THIOCYANATE SYSTEM.
Hua Hsueh Hsueh Pao, 31, 182−5 (1965) (in Chinese)

80.   CHANG, T−H. and CHOW, T−I.
Chem. Res. Center, Nat. Taiwan Univ., Taipei, Taiwan
STUDIES ON REVERSED-PHASE PARTITION THIN-LAYER CHROMATOGRAPHY OF RARE EARTH ELEMENTS.
Hua Hsueh, No 1−2, 18−26 (Jun. 1969) (in Chinese)
CA, 72, 96269 (1970)
NSA, 24, 16631 (1970)

81.   CORNETT, F. V. and GILBERT, Th. W., jr.
Dept. Chem., Univ. Cincinnati, Cincinnati, Ohio, USA
SEPARATION AND IDENTIFICATION OF METAL IONS BY REVERSED PHASE THIN-LAYER CHROMATOGRAPHY.
Anal. Lett., 4, 69−76 (1971)
CA, 74, 150694 (1971)

82.   CVJETICANIN, N., CVORIČ, J. D., and OBRENOVIC-PALIGORIC, I. D.
CHROMATOGRAPHIC SEPARATION OF METAL IONS BY MEANS OF PAPER TREATED WITH TRIOCTYL PHOSPHATE.
Bull. Boris Kidrich Inst. Nucl. Sci., 14, 83−93 (April 1963)

83.   CVJETICANIN, N.
CHROMATOGRAPHIC SEPARATION OF RARE EARTH ON PAPER TREATED WITH DI-n-BUTYL PHOSPHATE
Bull. Boris Kidrich Inst. Nucl. Sci., 15, n° 3, 201−3 (1964)
CA, 62, 15400 g (1965)

84.   CVJETICANIN, N.
Hot Laboratory Dept., Boris Kidric Inst. of Nuclear Sciences, Beograd. Yugoslavia
REVERSED-PHASE CHROMATOGRAPHY OF ALKALINE EARTHS ON PAPER TREATED WITH DI-n-BUTYL PHOSPHATE.
J. Chromatog., 32, 384−93 (1968)

85.   CVJETICANIN, N.
Hot Laboratory Dept., Boris Kidric Inst. of Nuclear Sciences, Beograd, Yugoslavia
SYNERGISM IN THE REVERSED-PHASE PARTITION CHROMATOGRAPHY OF AMERI-CIUM, CERIUM, AND LANTHANUM.
J. Chromatog., 34, 520−5 (1968)

86.   CVJETICANIN, N.
Hot Laboratory Dept., Boris Kidric Inst. of Nuclear Sciences, Beograd, Yugoslavia
SYNERGISM IN THE REVERSED-PHASE PARTITION CHROMATOGRAPHY OF URANIUM AND THORIUM.
Symp. V., Chromatog., Electrophorèse, Bruxelles, p. 186−96 (1969)

87. CVJETICANIN, N.
     Boris Kidric Inst. of Nucl. Sciences, Vinča, Yugoslavia
     REVERSED-PHASE PARTITION CHROMATOGRAPHY OF AMERICIUM, URANIUM AND
     THORIUM WITH A MIXTURE OF ORGANIC SOLVENTS.
     Paper presented at Intern. Symposium VI on Chromatography and Electrophoresis, Brussels,
     14–16 Sept. 1970.

88. CVJETICANIN, N.
     Boris Kidric Inst. of Nuclear Sciences, Beograd-Vinča, Yugoslavia
     REVERSED-PHASE PARTITION CHROMATOGRAPHY OF RARE EARTHS WITH A
     MIXTURE OF ORGANIC EXTRACTANTS.
     J. Chromatog., 74, 99–105 (1972)

89. DE, A. K. and BHATTACHARYYA, C. R.
     Dept. Chem. Visva-Bharati, Santiniketan, India
     REVERSED-PHASE PARTITION CHROMATOGRAPHY ON PAPER IMPREGNATED
     WITH 2-THENOYLTRIFLUOROACETONE.
     Separ. Sci., 6, 621–25 (1971)
     CA, 75, 14564 (1971)

90. DE VRIES, G. and SIJPERDA, W. S.
     Free Univ., Amsterdam, The Netherlands
     MINERAL ANALYSIS BY MEANS OF THIN-LAYER CHROMATOGRAPHY USING LIQUID
     ION EXCHANGERS. PART I: INTRODUCTION AND TECHNIQUE.
     Geol. Mijnbouw, 45, 275–80 (1966)

91. DE VRIES, G. and BRINKMAN, U. A. Th.
     Chemistry Lab., Free Univ., Amsterdam, The Netherlands
     APPARATUS FOR REVERSED-PHASE CHROMATOGRAPHY ON SMALL SCALE THIN-
     LAYER PLATES.
     J. Chromatog., 64, 374–7 (1972)

92. DUNCAN, G., BARK, L. S., and GRAHAM, R. J. T.
     Royal College of Adv. Techn., Salford, Lancs., Great Britain
     CHROMATOGRAPHY OF SOME FIRST AND SECOND ROW TRANSITION METALS ON
     THIN-LAYERS OF SUBSTRATES IMPREGNATED WITH TRIBUTYL PHOSPHATE.
     Proc. Soc. Anal. Chem. (London), 3, 145–6 (1966)

93. DUNCAN, G.
     Dept. of Chem. and Appl. Chem., Univ. of Salford, Salford, Lancs., Great Britain
     THIN-LAYER CHROMATOGRAPHY OF INORGANIC IONS.
     Thesis., Sept. 1966. 125 p.

94. ELLIOT, D. E.
     Iowa State Univ., Ames, Iowa, USA.
     PREPARATION AND SOME ANALYTICAL APPLICATIONS OF TRI-n-OCTYLPHOSPHINE
     SULFIDE. (thesis)
     IS-T-3. Feb. 1965. 55 p.
     NSA, 19, 19753 (1965)

95. ELLIOT, D. E. and BANKS, C. V.
     Iowa State Univ., Ames, Iowa. USA.
     TRI-n-OCTYLPHOSPHINE SULFIDE: A SELECTIVE ORGANIC EXTRACTANT.
     Anal. Chim. Acta, 33, n° 3, 237–44 (1965)
     CA, 63, 9038b (1965)

96. FISEL, S. and GROSU, Gh.
Univ. "Al. I. Cuza", Iasi, Roumania
REVERSED-PHASE CHROMATOGRAPHY OF INORGANIC SUBSTANCES. SYSTEMS
INVOLVING HIGHER FATTY ACIDS.
Anal. Stiint. Univ. Al. I. Cuza, 12, 23–6 (1966) (in French)

97. FISEL, S., UNGURENASU, T., and GROSU, Gh.
Univ. "Al. I. Cuza",Iasi, Roumania
REVERSED PHASE CHROMATOGRAPHY OF SOME RARE EARTHS ON PAPER TREATED
WITH DI-ALKYL ORTHOPHOSPHORIC ACIDS.
An. Stiint. Univ. "Al. I. Cuza," Iasi, Sect. IC, Chim., 12, 153–8(1966) (in French)
CA, 68, 26662 (1968)

98. FLIEGER, A. and PRZESZLAKOWSKI, S.
Katedra Chem. Nieorg., Akad. Med., Lublin, Poland
CHROMATOGRAPHIC ANALYSIS OF CATIONS OF THE SECOND ANALYTICAL GROUP
IN THE SYSTEM: LIQUID ANION EXCHANGER–HYDROCHLORIC ACID. I. SUBGROUP
IIA.
Chem. Anal. (Warsaw), 16, 1341–6 (1971) (in Polish)
CA, 79, 94125 (1972)

99. FLIEGER, A. and PRZESZLAKOWSKI, S.
Dept. of Inorg. and Anal. Chemistry, Academy of Medicine, Lublin, Poland
CHROMATOGRAPHIC ANALYSIS OF CATIONS OF THE SECOND ANALYTICAL
GROUP IN THE SYSTEM: LIQUID ANION EXCHANGER – HYDROCHLORIC ACID. II.
SUBGROUP IIB.
Chem. Anal. (Warsaw), 17, 333–7 (1972) (in Polish)

100. FRITZ, J. S. and SHERMA, J.
Inst. for Atomic Research and Dept. of Chemistry, Iowa State Univ., Ames, Iowa, USA.
REVERSED-PHASE PAPER CHROMATOGRAPHY OF METAL IONS WITH PHENYL-
BENZOHYDROXAMIC ACID.
J. Chromatog., 25, 153–60 (1966)

101. GRAHAM, R. J. T. BARK, L. S., and TINSLEY, D. A.
Univ. of Salford, Salford 5, Lancs., Great Britain
SOME ASPECTS OF THE USE OF LIQUID ANION EXCHANGER IN INORGANIC THIN
LAYER CHROMATOGRAPHY.
J. Chromatog., 35 416–22 (1968)

102. GRAHAM, R. J. T., BARK, L. S., and TINSLEY, D. A.
Univ. of Salford, Salford 5, Lancs., Great Britain
THE THIN-LAYER CHROMATOGRAPHY OF METAL IONS ON CELLULOSE IMPREG-
NATED WITH PRIMENE JM–T HYDROCHLORIDE,
J. Chromatog., 39, 200–10 (1969)

103. GRAHAM, R. J. T., BARK, L. S., and TINSLEY, D. A.
Univ. of Salford, Salford 5, Lancs., Great Britain
QUANTITATIVE THIN–LAYER CHROMATOGRAPHY ON LIQUID ANION EXCHANGERS.
PART I. AN INVESTIGATION INTO SOME OF THE PARAMETERS INVOLVED IN THE
DIRECT DENSITOMETRIC DETERMINATION OF ZINC.
J. Chromatog., 39, 211–7 (1969)

104. GRAHAM, R. J. T., BARK, L. S., and TINSLEY, D. A.
Univ. of Salford, Salford 5, Lancs., Great Britain
QUANTITATIVE THIN-LAYER CHROMATOGRAPHY ON LIQUID ANION EXCHANGERS.
PART II. A COMPARISON OF A SPOT REMOVAL METHOD WITH AN IN SITU DIRECT
DENSITOMETRIC METHOD.
J. Chromatog., 39, 218–27 (1969)

105. GRAHAM, R. J. T. and CARR, A.
Dept. of Chemistry and Applied Chemistry, Univ. of Salford, Salford 5. Lancs., Great Britain
THIN-LAYER CHROMATOGRAPHY OF METAL IONS ON CELLULOSE IMPREGNATED
WITH THIOCYANATE SALTS OF LIQUID ANION EXCHANGERS. I. EXPERIMENTAL
PARAMETERS INVOLVED IN THE SYSTEM AMBERLITE LA-2-THIOCYANATE-
AQUEOUS AMMONIUM THIOCYANATE.
J. Chromatog., 46, 293–300 (1970)

106. GRAHAM, R. J. T., and CARR. A.
Dept. of Chemistry and Applied Chemistry, Univ. of Salford, Salford 5. Lancs., Great Britain
THIN-LAYER CHROMATOGRAPHY OF METAL IONS ON CELLULOSE IMPREGNATED
WITH THIOCYANATE SALTS OF LIQUID ANION EXCHANGERS. II. SYSTEMATIC IN-
VESTIGATION OF METAL IONS IN THE SYSTEM AMBERLITE LA-2-THIOCYANATE-
AQUEOUS AMMONIUM THIOCYANATE.
J. Chromatog., 46, 301–11 (1970)

107. HIGASHI, K. and MIYAKE, Y.
Gov. Ind. Res. Inst., Osaka, Japan
SEPARATION AND ANALYSIS OF RARE EARTH ELEMENTS BY REVERSED-PHASE
PARTITION CHROMATOGRAPHY. I. MUTUAL SEPARATION WITH THIN LAYER
CHROMATOGRAPHY.
Osaka Kogyo Gijutsu Shikensho Kiho, 22, 6–10 (1971) (in Japanese)
CA, 75, 70949 (1971)

108. HIGASHI, K. and MIYAKE, Y.
Gov. Ind. Res. Inst., Osaka, Japan
SEPARATION AND ANALYSIS OF RARE EARTH ELEMENTS BY REVERSED-PHASE
PARTITION CHROMATOGRAPHY. II. COMPARISON OF EXTRACTING REAGENTS.
Osaka Kogyo Gijutsu Shikensho Kiho, 22, 282–7 (1971) (in Japanese)
CA, 76, 135296 (1972)

109. HOLZAPFEL, H., LE VIET LAN, and WERNER, G.
Inst. für Anorganische Chemie der Karl-Marx-Univ.,Leipzig. DDR (GDR)
SEPARATION OF RARE EARTHS BY LIQUID ION EXCHANGE. III. THIN-LAYER
CHROMATOGRAPHIC SEPARATION OF RARE EARTHS.
J. Chramatog., 20, 580–4 (1965)

110. HOLZAPFEL, H., LE VIET LAN, and WERNER, G.
Inst. für Anorganische Chemie der Karl Marx-Univ., Leipzig, DDR (GDR)
SEPARATION OF THE RARE EARTHS BY LIQUID ION EXCHANGE. V. THIN-LAYER
CHROMATOGRAPHIC SEPARATION OF MULTICOMPONENT MIXTURES OF RARE
EARTHS.
J. Chromatog., 24, 153–60 (1966)

111. HU. C. T. and HSIEH, H. C.
Univ. Lanchow, Taiwan
PAPER CHROMATOGRAPHIC DETERMINATION OF TITANIUM, ZIRCONIUM, AND
THORIUM.
Hua Hsueh Tung Pao, N° 3, 175–6 (1963)
CA, 59, 5767f (1963)

112. HU, C. T. and HSIEH, H. C.
Univ. Lanchow, Taiwan
SEPARATION OF VANADIUM, MOLYBDENIUM, AND TUNGSTEN BY (REVERSED-PHASE)
PAPER CHROMATOGRAPHY.
Hua Hsueh Tung Pao, N° 5, 312–3, 320 (1963)
CA, 60, 6d (1964)

417

113. HU, C. T. and HSIEH, H. C.
Univ. Lanchow, Taiwan
PAPER CHROMATOGRAPHIC SEPARATION OF SELENIUM, TELLURIUM, AND NOBLE
METALS.
Ko Hsueh Tung Pao, 30, N° 8,54 (1963)
CA, 60, 7432c (1964)

114. HU, C. T.
Univ. Lanchow, Taiwan
SEPARATION OF SELENIUM AND TELLURIUM BY REVERSED-PHASE PAPER CHRO-
MATOGRAPHY.
Hua Hsueh Hsueh Pao, 30, N° 4, 426–8 (1964)
CA, 61, 15324g (1964)

115. HU, C. T. AND MSIEH, H. C.
Univ. Lanchow, Taiwan
SEPARATION OF NOBLE METALS Se AND Te BY REVERSED-PHASE CHROMATOGRAPH
Hua Hsueh Hsueh Pao, 30, 352–6 (1964)
CA, 62, 3382e (1965)

116. HU, C. T. and LIU, C. L.
Univ. Lanchow, Taiwan
APPLICATION OF THIN-LAYER CHROMATOGRAPHY IN INORGANIC ANALYSIS.
II. SEPARATION AND DETERMINATION OF Se AND Te.
Ko Hsueh Tung Pao, 1964, 1103–4 (1964) (in Chinese)
CA, 63, 7f, (1965)
see also: Scientia Sinica, 14, 1235–7 (1965)

117. HU. C. T. and LIU, C. L.
Univ. Lanchow, Taiwan
APPLICATION OF THIN-LAYER CHROMATOGRAPHY IN INORGANIC ANALYSIS. III.
SEPARATION OF NIOBIUM AND TANTALUM, AND NOBLE METALS.
Hua Hsueh Hsueh Pao, 31, 267–9 (1965) (in Chinese)
CA, 63, 15525g (1965)

118. HU, C. T.
Univ. Lanchow, Taiwan
SEPARATION OF METAL CATIONS BY REVERSED-PHASE PAPER CHROMATOGRAPHY.
J. Lanchow Univ., 1965, 59 (1965) (in Chinese)

119. HU, C. T. and LIU, C. L.
Soochow Univ., Wuhsien, Taiwan
REVERSED-PHASE PAPER CHROMATOGRAPHIC SEPARATION OF NIOBIUM AND
TANTALUM.
Ko Hsueh Tung Pao, N° 4, 349–50 (1965) (in Chinese)
CA, 65, 19289h (1966)
see also: Scientia Sinica, 14, 1536–7 (1965)

120. HU, C. T.
Univ. Lanchow, Taiwan
THE PAPER CHROMATOGRAPHIC BEHAVIOUR OF SELENIUM, TELLURIUM, COPPER,
CADMIUM, BISMUTH, AND NOBLE METALS.
Ko Hsueh Tung Pao, 17, 166–9 (1966) (in Chinese)

121. ISHIBASHI, M., KOMAKI, H., and DEMIZU, M.
A NEW SIMPLE METHOD OF ANALYSING METALS.
Mitsubitshi Denki Giho, 39, 907–11 (1965) (in Japanese)

122. KAWAMURA,S.,FUJIMOTO,T., and IZAWA, M.
Chemistry Division, National Inst. of Radiological Sciences, Anagawa-cho, Chiba,Japan
INORGANIC CHROMATOGRAPHY ON CELLULOSE ACETATE PRETREATED PAPER
IMPREGNATED WITH 2-THENOYLTRIFLUOROACETONE.
J. Chromatog., 34, 72–6 (1968)

123. KENNEDY, D. C. and FRITZ, J. S.
Inst. for Atomic Research and Dept. of Chemistry, Iowa State Univ., Ames, Iowa,USA
SULPHOXIDES AS SOLVATING REAGENT FOR THE SEPARATION OF METAL IONS.
Talanta, 17, 823–35 (1970)

124. KERTES, A. S. and BECK, A.
Hebrew Univ., Jerusalem, Israel
METALLIC NITRATES IN PAPER CHROMATOGRAPHY. IV. SYSTEMS CONTAINING
TRIBUTYL PHOSPHATE AND NITRIC ACID.
J. Chromatog., 3, 195–6 (1960)

125. KNOCH, W. and LAHR, H.
Inst. für Kern- und Radiochemie der Techn. Hochschule Braunschweig, BRD (FRG)
THE CHROMATOGRAPHIC SEPARATION OF AMERICIUM – CURIUM ON PAPER
Radiochimica Acta, 4, 114 (1965)

126. KNOCH, W., MUJU, B., and LAHR, H.
Institut für Kern-und Radiochemie der Technischen Hochschule Braunschweig, BRD (FRG)
THE SEPARATION OF SOME FISSION PRODUCTS AND ACTINIDES ON PAPER TREATED
WITH LIQUID ANION EXCHANGER
J. Chromatog., 20, 122–30 (1965)

127. KOROTKIN, Yu. S.
Joint Inst. for Nuclear Research, Dubna, USSR
SEPARATION OF ACTINIDE ELEMENTS AND THEIR VALENCE FORMS BY PAPER
AND THIN-LAYER PARTITION CHROMATOGRAPHY.
Radiokhimiya, 13, 137–40 (1971) (in Russian)
CA, 74, 150693 (1971)

128. KOROTKIN. Yu. S.
Joint Inst. for Nuclear Research, Dubna, USSR
STUDY OF TRANSURANIUM ELEMENTS HYDROLYSIS.
I. STUDY OF HYDROLYSIS BY PARTITION CHROMATOGRAPHY (UO_2^{++}, PuO_2^{++}).
JINR-P6-6402. 1972. 15 p. (in Russian, with abstract in English)

129. KOROTKIN, Yu. S.
Joint Inst. for Nuclear Research, Dubna, USSR
STUDY OF TRANSPLUTONIUM ELEMENT HYDROLYSIS.
II. HYDROLYSIS OF AMERICIUM(III) IN PRESENCE OF IONS WITH POSITIVE AND
NEGATIVE HYDRATION ENERGY.
JINR-P6-6403. 1972. 17 p. (in Russian with abstract in English)
KFK-tr-411. 1972. 17 p. (in German)

130. KOROTKIN, Yu. S.
Joint Inst. for Nuclear Research, Dubna, USSR

STUDY OF TRANSURANIUM ELEMENT HYDROLYSIS.III.
AMERICIUM AND CURIUM HYDROLYSIS IN SOLUTIONS OF PURE NITRIC ACID.
JINR-P6-6404. 1972. 10 p. (in Russian with abstract in English)
KFK-tr-412. 1972. 10 p. (in German)

131. KURODA, R., OGUMA, K., and NAKAMURA, K.
Fac. Eng., Chiba Univ., Chiba, Japan
REVERSED-PHASE THIN-LAYER CHROMATOGRAPHY OF A NUMBER OF
METAL IONS IN THE SYSTEM OF HYDROCHLORIC ACID-AMMONIUM THIOCYANATE.
Nippon Kagaku Zasshi, 91, 1155–8 (1970) (in Japanese)
CA, 74, 130785 (1971)
NSA, 25, 18337 (1971)

132. LEENE, H. R., DE VRIES, G., and BRINKMAN, U. A. Th.
Chemistry Laboratorium, Free Univ., Amsterdam, The Netherlands
THE USE OF TRI-N-ALKYLAMINE AND –ARSINE OXIDE IN REVERSED-PHASE
EXTRACTION CHROMATOGRAPHY.
J. Cromatog., 57, 173–80 (1971)
[correction published in J. Chromatog., 59, 516 (1971)]

133. LEENE, H. R., DE VRIES, G. and BRINKMAN, U. A. Th.
Chem. Lab., Free Univ., Amsterdam, The Netherlands
REVERSED-PHASE EXTRACTION CHROMATOGRAPHY USING SOLUTIONS OF NITRIC ACI
AS ELUANTS.
J. Chromatogr., 80, 221–32 (1973)

134. LEVAGGI, A. D.
Dept. of Chem. and Appl. Chem.,Univ. of Salford, Salford, Lancs., Great Britain
CHROMATOGRAPHIC STUDIES OF LIQUID ANION EXCHANGE BEHAVIOUR.
Thesis, 1969. 131 p.

135. LITEANU, C. and UNGUREANU, A.
Inst. Chem., Acad. RSR, Cluj. Roumania
CHROMATOGRAPHIC BEHAVIOUR OF METALS IONS ON PAPER TREATED WITH
PHOSPHORIC ESTERS. UTILISATION OF TRIALLYL PHOSPHATE FOR THE SEPARA-
TION OF THE MIXTURE: UO_2^{++}, Cu^{++} AND Fe^{3+}.
Rev. Roum. Chim., 13, 899–903 (1968)
CA, 69, 113143 (1968)
NSA, 23, 17546 (1969)

136. LITEANU, C. and UNGUREANU, A.
Dept. Chem., Babes-Bolyai Univ., Cluj, Roumania
CHROMATOGRAPHIC BEHAVIOUR OF METAL IONS ON PAPER TREATED WITH
PHOSPHORIC ACID ESTERS. II. REVERSED-PHASE CHROMATOGRAPHY OF THE
IONS $(AuCl_4)^-$, $(PtCl_6)^{2-}$, $(PtCl_4)^{2-}$, $(PdCl_6)^{2-}$ AND $(PdCl_4)^{2-}$ ON PAPER TREATED WITH
TRIALLYL PHOSPHATE.
Rev. Roum. Chim., 16, 1769–73 (1971) (in German)
CA, 76, 80569 (1972)
See also: Lucr. Conf. Nat. Chim. Anal., 3rd 1971, 4, 133–7 (in Roumanian)

137. LYLE, S. J. and NAIR, V. C.
Univ. of Kent, Canterbury, Great Britain
REVERSED-PHASE PARTITION CHROMATOGRAPHIC SEPARATIONS WITH 2-ETHYL-
HEXYL DIHYDROGEN PHOSPHATE AND DI-2-ETHYL–HEXYL HYDROGEN PHOSPHATE.
Talanta, 16, 813–21 (1969)

138. MARK P. and HECHT, F.
Anal. Inst., University of Vienna, Austria
THIN-LAYER CHROMATOGRAPHY OF INORGANIC IONS II.
Mikrochim. Acta, 5–6, 970–5 (1963) (in German)

139. McCORMICK, D., GRAHAM, R. J. T., and BARK, L. S.
Dept. of Pure and Appl. Chem., Univ. of Salford, Salford, Lancs., Great Britain
INORGANIC THIN-LAYER CHROMATOGRAPHY. III. THIN-LAYER CHROMATOGRAPHY
OF SOME FIRST ROW TRANSITION METAL IONS ON CELLULOSE IMPREGNATED
WITH TRI-n-OCTYLAMINE (TNOA) HYDROCHLORIDE.
Int. Symp. Chromatogr. Electrophor. IV, Brussels, Belgium (1966) pp. 199–206.
Presses Académiques Européennes, Bruxelles, 1968

140. MUCHOVA, A. and JOKL. V.
Pharmaceutical Faculty, Komenski University,   Bratislava, ČSSR
THIN-LAYER CHROMATOGRAPHY OF INORGANIC IONS.
II. CHROMATOGRAPHY IN SILICA GEL (IMPREGNATED WITH LIQUID ANION EXCHAN-
GERS) – AQUEOUS INORGANIC ACID-SALT SOLUTIONS SYSTEMS.
Chem. Zvesti, 25, 280–91 (1971) (in German)
CA, 76, 18188 (1972)

141. MUCHOVA, A. and JOKL, V.
Pharmaceutical Faculty, Komenski University, Bratislava, ČSSR
THIN-LAYER CHROMATOGRAPHY OF INORGANIC IONS. III. CHROMATOGRAPHY
IN THE SYSTEM: SILICA GEL (IMPREGNATED WITH LIQUID ANION EXCHANGERS)
AQUEOUS SOLUTIONS OF ORGANIC COMPLEXING AGENTS.
Chem. Zvesti, 26, 289–302 (1972) (in German)
NSA, 26, 58571 (1972)

142. MUCHOVA, A. and JOKL, V.
Pharmaceutical Faculty, Komenski University, Bratislava, ČSSR
THIN-LAYER CHROMATOGRAPHY OF INORGANIC IONS. IV. CHROMATOGRAPHY IN
THE SYSTEM: SILICA GEL (IMPREGNATED WITH LIQUID CATION EXCHANGERS)
–AQUEOUS HYDROCHLORIC ACID, NITRIC ACID, AND POTASSIUM THIOCYANATE
SOLUTIONS.
Chem. Zvesti, 26, 303–8 (1972) (in German)
NSA, 26, 58570 (1972)

143. MUCHOVA, A. and JOKL, V.
Pharmaceutical Faculty, Komenski Univ., Bratislava, ČSSR
THIN-LAYER CHROMATOGRAPHY OF INORGANIC IONS. CHROMATOGRAPHY IN
THE SYSTEM SILICA GEL (IMPREGNATED WITH TBP) – AQUEOUS SOLUTIONS
OF LiCl, NaBr AND NaI.
Chem. Zvesti, 27, 348–54 (1973) (in German)

144. NAGAI, H.* and KIYOSHIMA, H.**
* Dept. Chem.,Kumamoto Univ., Kurokami-machi, Kumamoto-shi, Japan
** Dept. Chem., Defense Acad. Hashirimizu, Yokasuka-shi, Kanagawa-ken, Japan
CHROMATOGRAPHIC SEPARATION OF LANTHANIDES ON A CIRCULAR FILTER
PAPER IMPREGNATED WITH OXINE AS THE IMMOBILE PHASE.
Bunseki Kagaku, 20, 1289–92 (1971) (in Japanese)

145. OKAMOTO, K. and TOMINAGA, T.
Kwansei-Gakuin Univ., Nishinomiya, Japan
SPECTROPHOTOMETRIC DETERMINATION OF RARE EARTHS IN STEEL AFTER

SEPARATION BY EXTRACTION WITH HIGH MOLECULAR WEIGHT QUATERNARY
AMMONIUM SALT.
Bunseki Kagaku, 20, 870–8 (1971) (in Japanese)
CA, 75, 157920 (1971)

146. O'LAUGHLIN, J. W. and BANKS, C. V.
Ames Lab., Ames, Iowa, USA
SEPARATION OF VARIOUS CATIONS BY REVERSED-PHASE PARTITION CHROMATO-
GRAPHY USING NEUTRAL ORGANOPHOSPHORUS COMPOUNDS.
IS–737. May 1963. 145 p.

147. O'LAUGHLIN, J. W. and BANKS, C. V.
Inst. for Atomic Research and Dept. of Chemistry, Iowa State University, Ames, Iowa, USA
SEPARATION OF VARIOUS CATIONS BY REVERSED-PHASE PARTITION CHROMATOG-
RAPHY USING NEUTRAL ORGANOPHOSPHORUS COMPOUNDS.
Anal. Chem., 36, 1222–9 (1964)

148. O'LAUGHLIN, J.W., KAMIN, G. J., and BANKS, C. V.
Inst. for Atomic Research and Dept. of Chemistry, Iowa State Univ., Ames, Iowa, USA.
REVERSED-PHASE PAPER CHROMATOGRAPHY ON THE RARE EARTHS, THORIUM,
AND URANIUM USING METHYLENEBIS (DI-n-HEXYLPHOSPHINE OXIDE) AS THE
STATIONARY PHASE.
(J. Chromatog., 21 460–70 (1966)

149. O'LAUGHLIN, J. W., FERGUSON, J. W., RICHARD, J. J., and BANKS, C. V.
Inst. for Atomic Research and Dept. of Chemistry, Iowa State Univ., Ames, Iowa, USA.
REVERSED-PHASE PARTITION CHROMATOGRAPHY ON THE RARE EARTHS USING
METHYLENEBIS(DI(2-ETHYLHEXYL) PHOSPHINE OXIDE) AS THE STATIONARY
PHASE.
J. Chromatog., 24 376–82(1966)

150. P'ANG, S.W. and LIANG, S. C.
Acad. Sinica, Peking, China
SEPARATION OF RARE EARTH ELEMENTS BY (REVERSED-PHASE) PAPER CHRO-
MATOGRAPHY.
Ko Hsueh Tung Pao, N° 11, 46 (1962)
CA, 58, 8394c (1963)

151. P'ANG, S. W. and LIANG, S. C.
Acad. Sinica, Peking, China
SEPARATION OF RARE EARTH ELEMENTS BY MEANS OF REVERSED-PHASE PAPER
CHROMATOGRAPHY I. THE BIS(2-ETHYLHEXYL) HYDROGEN PHOSPHATE-SODIUM
SULPHATE SYSTEM.
Hua Hsueh Hsueh Pao, 29, 319–23(1963)
Anal. Abstr., 12, 68 (1965)

152. P'ANG, S. W., LUI, C. C., and LIANG, S. C.
Acad. Sinica, Peking, China
CHROMATOGRAPHIC SEPARATION, DETECTION, AND DETERMINATION OF SCANDIUM
Hua Hsueh Pao, 30, (2) 160–5 (1964)
CA, 61, 7695b (1964)

153. P'ANG, S. W. and LIANG, S. C.
Acad.Sinica, Peking, China
SEPARATION OF RARE EARTH ELEMENTS BY MEANS OF REVERSED-PHASE PAPER
CHROMATOGRAPHY. II. THE HDEHP-OXALIC ACID SYSTEM.
Hua Hsueh Hsueh Pao, 30, N° 2, 237–40 (1964)
CA, 61, 10016a (1964)

154. P'ANG, S. W. and LIANG, S. C.
Acad. Sinica, Peking, China
SEPARATION OF RARE EARTH ELEMENTS BY MEANS OF REVERSED-PHASE PAPER
CHROMATOGRAPHY. III. TNOA-NH₄NO₃ SYSTEM.
Ko Hsueh Tung Pao, N° 2, 156 (1964)
CA, 61, 15324f (1964)

155. PIERCE, T. B. and PECK, P. F.
AERE, Harwell, Berks. Great Britain
SINTERED POLYMERIC CARRIERS IN PARTITION CHROMATOGRAPHY AND ELEC-
TROPHORESIS.
Analyst, 89,662–9 (1964)

156. PIERCE, T. B. and FLINT, R. F.
AERE, Harwell, Berks., Great Britain
THE RAPID SEPARATION OF RARE EARTH MIXTURES BY THIN-LAYER CHRO-
MATOGRAPHY.
Anal. Chim. Acta, 31,595–7 (1964)
CA, 62, 4592d (1965)

157. PIERCE, T. B. and FLINT, R. F.
Analytical Chemistry Branch, AERE, Harwell, Great Britain
SEPARATION OF SOME INORGANIC IONS BY REVERSED-PHASE PARTITION CHRO-
MATOGRAPHY ON THIN LAYERS.
J. Chromatog., 24, 141–7 (1966)

158. PRZESZLAKOWSKI, S. and SOCZEWINSKI, E.
Katedra Chemii Nieorganicznej, Wydzial Farmaceutyczny Akademii Medycznej, Lublin, Poland
CHROMATOGRAPHY OF ANIONIC BROMIDE AND IODIDE COMPLEXES OF METALS
ON PAPER IMPREGNATED WITH TRI-n-OCTYLAMINE.
Chem. Anal. (Warsaw), 11,895–903 (1966) (in Polish)
NSA, 21, 5880 (1967)

159. PRZESZLAKOWSKI, S.
Katedra Chem. Nieorg., Akad. Med., Lublin, Poland
CHROMATOGRAPHY OF ANIONIC THIOCYANATE COMPLEXES ON PAPER IMPREG-
NATED WITH HIGHER TERTIARY AMINES.
Chem. Anal. (Warsaw), 12, 57–66 (1967) (in Polish)
CA, 67, 50041 (1967)

160. PRZESZLAKOWSKI, S.
Katedra Chem. Nieorg., Akad. Med., Lublin, Poland
CHROMATOGRAPHY OF CERTAIN IONS ON PAPER IMPREGNATED WITH LIQUID
ANION-EXCHANGERS, WITH HYDROFLUORIC ACID SOLUTIONS AS MOBILE PHASE.
Chem. Anal. (Warsaw), 12, 321–8 (1967) (in Polish)
CA, 68, 26505 (1968)

161. PRZESZLAKOWSKI, S.
Katedra Chem. Nieorg., Akad. Med., Lublin, Poland
CHROMATOGRAPHY OF HALOGENIC AND THIOCYANATE METAL COMPLEXES ON A FILTER PAPER IMPREGNATED WITH LIQUID ANION EXCHANGERS: AMBERLITE LA 2 AND TRI-n-OCTYLAMINE.
Chem. Anal. (Warsaw), 12,1071−85 (1967) (in Polish)

162. PRZESZLAKOWSKI, S.
Katedra Chem. Nieorg., Akad. Med., Lublin, Poland
POSSIBLITY OF THE DETERMINATION OF A COMPOSITION OF EXTRACTABLE COMPLEXES OF METALS WITH AMINES OF HIGH MOLECULAR WEIGHT BY PAPER CHROMATOGRAPHY.
Rocz. Chem., 41, 1681−8 (1967) (in Polish)
CA, 68,111009 (1968)
NSA, 22, 14569 (1968)

163. PRZESZLAKOWSKI, S., SOCZEWINSKI, E., and FLIEGER, A.
Katedra Chem. Nieorg., Akad. Med., Lublin, Poland
CHROMATOGRAPHY OF SOME METAL IONS ON PAPER IMPREGNATED WITH TRI-n-OCTYLAMINE SALTS BY USING ACETIC, FORMIC, AND CHLOROACETIC ACIDS AS MOBILE PHASES.
Chem. Anal. (Warsaw), 13, 841−8 (1968) (in Polish)
CA, 70, 63767 (1969)

164. PRZESZLAKOWSKI, S. and FLIEGER, A.
Dept. of Inorg. and Anal. Chem., Medical School, Lublin, Poland
THE QUALITATIVE CHROMATOGRAPHIC ANALYSIS OF THE PLATINUM METALS AND GOLD ON PAPER IMPREGNATED WITH AMBERLITE LA-1 HYDROCHLORIDE.
J. Chromatog., 81, 129−35 (1973)

165. ROZEN. A. M., MURINOV, U. I., and NITIKIN, M. K.
EXTRACTION CAPACITY OF SULFOXIDES. EXTRACTIVE SEPARATION OF ELEMENTS BY REVERSED-PHASE PAPER CHROMATOGRAPHY.
Radiokhimiya, 14,754−5(1972) (in Russian)

166. SASTRI, M. N. and RAO, A. P.
Dept. of Chemistry, Andhra Univ.,Waltair, S. India
CHROMATOGRAPHIC SEPARATIONS OF SOME ELEMENTS ON PAPERS IMPREGNATED WITH A LIQUID CATION EXCHANGER.
Z. Anal. Chem., 196,166−9(1963)

167. SASTRI, M. N., RAO, A. P., and SARMA, A. R. K.
Dept. of Chemistry, Andhra Univ., Waltair, S.India
CHROMATOGRAPHIC BEHAVIOUR OF METAL IONS ON PAPERS TREATED WITH PRIMARY , SECONDARY, TERTIARY, AND QUATERNARY AMINES.
J. Chromatog., 19,630−3(1965)

168. SCARGILL, D.
United Kingdom Atomic Energy Authority, Research Group,. AERE, Harwell, Berks., Great Britain
CHROMATOGRAPHIC METHODS FOR THE SEPARATION OF NITRATO- AND NITRO-COMPLEXES OF NITROSYL-RUTHENIUM.
AERE-R-4583. April 1964. 19 p.

169. SCHMITT, M. A. S.
Iowa State Univ., Ames, Iowa, USA
SOME ANALYTICAL APPLICATIONS OF ESTERS OF PHOSPHOROTHIOIC AND
PHOSPHORODITHIOIC ACIDS.
(Thesis)
IS-T-144. Febr. 1967. 41 p.
NSA, 21, 30347 (1967)

170. SHIMIZU, T. and ISHIKURA, R.
Dept. of Chemistry, Faculty of Education, Gunma Univ., Maebashi, Gunma, Japan
THE REVERSED-PHASE THIN-LAYER CHROMATOGRAPHIC BEHAVIOUR OF THE RARE
EARTHS, THORIUM, URANIUM, AND ZIRCONIUM WITH A HIGH MOLECULAR WEIGHT
AMINE IN SULFURIC ACID AND AMMONIUM SULFATE SYSTEMS.
J. Chromatog., 56, 95−102 (1971)

171. SIJPERDA, W. S. and DE VRIES, G.
Free Univ., Amsterdam, The Netherlands
MINERAL ANALYSIS BY MEANS OF THIN-LAYER CHROMATOGRAPHY USING
LIQUID ION EXHANGERS. PART II. QUALITATIVE ANALYSIS OF SULPHIDE ORE
MINERALS.
Geol. Mijnbouw, 45, 315−8 (1966)

172. SIJPERDA, W. S. and DE VRIES, G.
Free Univ., Amsterdam, The Netherlands
MINERAL ANALYSIS BY MEANS OF THIN-LAYER CHROMATOGRAPHY USING LIQUID
ION EXCHANGERS. PART III. DATA ON NON-SULPHIDIC MINERALS.
Geol. Mijnbouw, 47,197−8(1968)

173. SING, S., CHENG, S. F., and PANG, S. W.
PAPER PARTITION CHROMATOGRAPHY USING THE SYSTEM DI-(2-ETHYLHEXYL)
PHOSPHORIC ACID-HCl AND TRIBUTYL PHOSPHATE-HCl.
Yuan Tsu Neng, 1964, 854−61 (in Chinese)

174. SOCZEWINSKI, E., PRZESZLAKOWSKI, S., and FLIEGER, A.
Akad. Med., Lublin, Poland
CHROMATOGRAPHY OF CATIONS OF SOME METALS ON PAPAR IMPREGNATED WITH
FATTY ACIDS.
Ann. Univ. Mariae Curie-Sklodowska, Sect. AA, 21 (2), 7−19, (1966) (in Polish)
CA, 69, 32645 (1968)

175. SOCZEWINSKI, E., ROJOWSKA, M., and PRZESZLAKOWSKI, S.
Akad. Med., Lublin, Poland
CHROMATOGRAPHY OF SOME METAL IONS ON PAPER IMPREGNATED WITH TRIISO-
OCTYL PHOSPHATE AND BIS (2-ETHYLHEXYL) PHOSPHATE.
Ann. Univ. Mariae Curie-Sklodowska, Sect. AA, 21 (3), 21−30 (1966)
CA, 69, 32643, (1968)

176. STARER MENDES, H.
Lab. Physiol. Chim., Fac. Sci., Paris, France
ROLE OF AMINES IN THE SEPARATION OF SODIUM AND POTASSIUM IONS. CHRO-
MATOGRAPHIC STUDY.
C. R. Acad. Sci., Ser. C., 273, 342−5 (1971) (in French)
CA, 75, 126028 (1971)

177. TESTA, C.
CISE, Milan, Italy
CHROMATOGRAPHY OF SOME CATIONS BY MEANS OF PAPER TREATED WITH A LIQUID
ANION EXCHANGER.
J. Chromatog., 5, 236—43 (1961)

178. TS'AI, C., CHOU, H. and YUAN, C.
CHROMATOGRAPHIC ANALYSIS AND LIQUID-LIQUID EXTRACTION.
Yan Tzu Neng, N° 4, 334—41 (1965)
NSA, 20, 29355 (1966)

179. WAKSMUNDZKI, A. and PRZESZLAKOWSKI, S.
Dept. of Inorg. Chemistry, Medical Academy, Lublin, Poland
CHROMATOGRAPHY OF SOME METAL IONS ON PAPER IMPREGNATED WITH HIGH-MO-
LECULAR WEIGHT AMINES AND DEVELOPED WITH THIOCYANATE SOLUTIONS.
K. MAČEK, I. M. HAIS, eds., Stationary Phase in Paper and Thin-Layer Chromatography.
Proceedings of the 2nd Symposium Held at Liblice, June 10—12, 1964. Elsevier Publishing Co.,
Amsterdam, London, New York. 1965. pp. 199—201

180. WAKSMUNDZKI, A. and PRZESZLAKOWSKI, S.
Akad. Med., Lublin, Poland
CHROMATOGRAPHY OF ANIONIC CHLORIDE COMPLEXES OF METALS ON PAPER
IMPREGNATED WITH TRI-n-OCTYLAMINE AS LIQUID ANION EXCHANGER.
Chem. Anal. (Warsaw), 11, N° 1, 159—68 (1966) (in Polish)
CA, 64, 18380a (1966)

181. WERNER, G.
Karl Marx Univ., Leipzig, DDR (GDR)
SEPARATION OF RARE EARTHS BY LIQUID ION-EXCHANGE. I. INVESTIGATION OF
THE ION-EXCHANGE PROPERTIES OF DINONYLNAPHTALENESULFONIC ACID BY
REVERSED-PHASE PAPER CHROMATOGRAPHY.
Z. Chem., 5, 147—8 (1965) (in German)
CA, 63, 2422b (1965)

182. WERNER, G.
Karl Marx Univ., Leipzig, DDR (GDR)
SEPARATION OF RARE EARTHS BY LIQUID ION EXCHANGE. II. SEPARATION OF
THORIUM, RARE EARTHS, AND SCANDIUM BY REVERSED-PHASE PAPER CHROMA-
TOGRAPHY.
Z. Chem., 5, 311—2 (1965) (in German)
CA, 63, 17115f (1965)

183. WERNER, G.
Karl Marx Univ., Leipzig, DDR (GDR) ;DR)
REVERSED-PHASE PAPER-CHROMATOGRAPHIC SEPARATION OF RARE EARTHS.
Angew. Chem. (Intern. Edition), 77, 920 (1965)

184. WERNER G.
Karl Marx-Univ. Leipzig, DDR (GDR)
SEPARATION OF THE RARE EARTHS BY LIQUID ION EXCHANGE. IV. SEPARATION OF
RARE EARTHS BY REVERSED-PHASE PAPER CHROMATOGRAPHY AND COMPLEX
ELUTION.
J. Chromatog., 22, 400—6 (1966) (in German)

185. WERNER, G.
Karl Marx Univ., Leipzig, DDR (GDR)
SEPARATION OF THE RARE EARTHS BY SOLVENT EXTRACTION AND REVERSED-
PHASE CHROMATOGRAPHY FROM COMPLEX AQUEOUS SOLUTIONS.
Solvent Extr. Chem., Proc. Int. Conf., Gothenburg, Sweden 1966, 54–9 (1967)
CA, 69, 39173 (1968)

186. WERNER, G., HANNIG, R., DEDEK, W., and HOLZAPFEL, H.
Karl-Marx-Univ. Leipzig, DDR (GDR)
THE SEPARATION OF CALCIUM AND STRONTIUM BY LIQUID ION EXCHANGE.
Proc. Anal. Chem. Conf., 3rd 1970, 1, 75–80 (1970) Edited by Buzas, I., Akad. Kiadó,
Budapest, Hungary
CA, 74, 49272 (1971)

187. WHITAKER, L. A.
Univ. of the Pacific. Stockton, California, USA
A REVERSED-PHASE THIN-LAYER CHROMATOGRAPHIC SEPARATION OF THE LIGHT
RARE EARTHS WITH A SPECTROPHOTOMETRIC DETERMINATION.
Thesis. 1967. 120 p.
NSA, 23, 6216 (1969)

188. YAGODIN, G. A. and CHEKMAREV, A. M.
Mosk. Khim. – Tekhnol. Inst. im. Mendeleeva, Moscow, USSR
EXTRACTION SEPARATION OF ELEMENTS BY REVERSED-PHASE CHROMATOGRAPHY.
Radiokhimiya, 11, 234–6 (1969) (in Russian)
CA, 71, 76939 (1969)

189. YAGODIN, G. A., CHEKMAREV, A. M. and DINEGRIBOVA, O. A.
Mosk. Khim. – Tekhnol. Inst. im. Mendeleeva, Moscow, USSR
REVERSED-PHASE PAPER CHROMATOGRAPHY AS A METHOD FOR STUDYING THE
EXTRACTION SEPARATION OF ELEMENTS WITH SIMILAR PROPERTIES.
Khim. Protsessov Ekstr., Mater. Konf. Khim. Ekstr., 3rd 1969 (Publ. 1972) 268–73 (in Rus-
sian)
CA, 78, 37477 (1973)

6. ORIGINAL PUBLICATIONS ON
COLUMN EXTRACTION CHROMATOGRAPHY

190. ABRÃO, A.
Inst. de Energia Atomica, Sao Paulo, Brasil
REVERSED-PHASE CHROMATOGRAPHY: SEPARATION CADMIUM-INDIUM IN THE
SYSTEM Al_2O_3– TRI-n-OCTYLAMINE-HCl.
IEA-241. 1971. 9 p. (in Portuguese)

191. AKAZA, I.
Dept. of Chemistry, Fac. of Science, Kanazawa Univ., Kanazawa, Japan
SEPARATION OF THE ALKALI METALS AS POLYIODIDES BY REVERSED-PHASE PAR-
TITION CHROMATOGRAPHY.
Bull. Chem. Soc. Japan, 39, 585–95 (1966)
CA, 65. 1354d (1966)

192. AKAZA, I.
Dept. of Chemistry, Fac. of Science, Kanazawa Univ., Kanazawa, Japan
SEPARATION OF ALKALINE EARTH METALS AS THEIR TTA-COMPLEXES BY REVERSED-PHASE PARTITION CHROMATOGRAPHY.
Bull. Chem. Soc. Japan, 39, 980–9 (1966)
CA, 65, 14406a (1966)

193. AKAZA, I., KIBA, TOSHIYASU, and TABA, M.
Dept. of Chemistry, Fac. of Science, Kanazawa Univ., Kanazawa, Japan
THE SEPARATION OF HEXACYANOFERRATE(II) AND HEXACYANOFERRATE(III) IONS BY REVERSED-PHASE PARTITION CHROMATOGRAPHY.
Bull. Chem. Soc. Japan, 42, 1291–6 (1969)

194. AKAZA, I., KIBA, TOSHIYASU, and KIBA, TOMOE
Dept. of Chemistry, Fac. of Science, Kanazawa Univ., Kanazawa, Japan
THE SEPARATION OF GOLD, PLATINUM, AND PALLADIUM BY REVERSED-PHASE PARTITION CHROMATOGRAPHY.
Bull. Chem. Soc. Japan, 42, 2063–7 (1970)

195. AKAZA, I., TAJIMA, T., and KIBA, TOSHIYASU
Fac. Sci., Kanazawa Univ., Kanazawa, Japan
SEPARATION SCHEME FOR THE ANALYSIS OF METAL IONS USING COLUMN EXTRAC-TION CHROMATOGRAPHY. GROUPING OF NINETEEN METAL IONS INTO SIX FRACTIONS.
Bull. Chem. Soc. Japan., 46, 1199–204 (1973)

196. AKERMAN, K., KOZAK, Z., and WIATER, D.
SORPTION OF URANIUM ON ACTIVATED CARBON AND SILICA GEL IMPREGNATED WITH AMINES.
Przemysl Chem., 42, 26–8 (1963)
CA, 58, 13171c (1963)

197. ALBINI, A.*, GIACOLETTO, C.*, MEZZADRI, M. G.**, and TRIULZI, C.
* CISE, Milan, Italy
** Zoological Inst., Univ. of Parma, Italy
RADIOCHEMICAL STUDIES FOR RADIOACTIVITY DETERMINATIONS ON MARINE SEDIMENT SAMPLES: ^{90}Sr, ^{144}Ce, ^{147}Pm, and ^{155}Eu.
CISE-108. 1968. 77 p.

198. ALBINI, A., BATTAGLIA, A., QUAINI, L., and TRIULZI, C.
CISE, Milan, Italy
MEASUREMENTS OF ^{90}Sr, ^{144}Ce, ^{147}Pm, ^{95}Zr, AND ^{106}Ru IN MONTHLY FALLOUT SAMPLES COLLECTED AT SEGRATE.
Energia Nucleare, 19, 257–62 (1972) (in Italian)

199. ALBINI, A., BONFANTI, G., CATTANEO, C.,
GIACOLETTO, C. and TRIULZI, C.
CISE, Milan, Italy
METHODS FOR THE DETERMINATION OF THE MAIN FISSION PRODUCTS CONTAINED IN FALL-OUT SAMPLES
CISE-N-158. 1972. 31 p., 5 figs.

200. ALI, S. A. and EBERLE, S. H.
Inst. für Radiochemie, Kernforschungszentrum Karlsruhe, BRD (FRG)
NEW VARIATION IN THE LANTHANIDES-ACTINIDES SEPARATION.
Inorg. Nucl. Chem. Letters, 7, 153–9 (1971) (in German)

201. ALIMARIN, I. P. and BOLSHOVA, T. A.
Moscow M.V. Lomonosov State Univ., Moscow, USSR
SEPARATION OF TRACES OF GALLIUM FROM ZINC BY PARTITION CHROMATOGRAPHY.
Zh. Anal. Khim., 21, 411–4 (1966) (in Russian)
CA, 65, 4629h (1966)

202. ALIMARIN, I. P., ERSHOVA, N. I., and BOLSHOVA, T. A.
Moscow M. V. Lomonosov State Univ., Moscow, USSR
CHROMATOGRAPHIC SEPARATION AND EXTRACTION OF GALLIUM FROM SOME
METAL CHLORIDE SOLUTIONS BY TRIBUTYL PHOSPHATE.
Vestn. Mosk. Univ., Ser. II, 22, 51–4(1967) (in Russian)
CA, 67, 121951 (1967)

203. ALIMARIN, I. P., BOLSHOVA, T. A., ERSHOVA, N. I., and KISEZEVA, A. D.
Moscow M. V. Lomonosov State Univ., Moscow, USSR
USE OF PARTITION CHROMATOGRAPHY FOR CONCENTRATION OF GALLIUM
TRACES.
Vestn. Mosk. Univ., Ser. II, 22, 55–8(1967) (in Russian)
CA, 68, 8958 (1967)

204. ALIMARIN, I. P. and BOLSHOVA, T. A.
Moscow M. V. Lomonosov State Univ., Moscow, USSR
APPLICATION OF PARTITION CHROMATOGRAPHY FOR CONCENTRATION OF TRACE
IMPURITIES
Z. Chem., 8, 411–4(1968) (in German)
CA, 70, 33967 (1969)

205. ALIMARIN, I. P., BOLSHOVA, T. A., ERSHOVA, N. I., and POLINSKAYA, M. B.
Moscow M. V. Lomonosov State Univ., Moscow, USSR
USE OF PARTITION CHROMATOGRAPHY FOR CONCENTRATING TRACES OF ELE-
MENTS.
Zh. Anal. Khim., 24, 26–30 (1969) (in Russian)
Ca, 70, 92853 (1969)

206. ALIMARIN, I. P., ERSHOVA, N. I., and BOLSHOVA, T. A.
Moscow M. V. Lomonosov State Univ., Moscow, USSR
COMPARATIVE STUDY OF THE CHROMATOGRAPHIC BEHAVIOUR AND EXTRACTION
OF GALLIUM, IRON(III), COPPER, COBALT, AND ALUMINIUM IN AN AMMONIUM
THIOCYANATE-HYDROGEN CHLORIDE-TRIBUTYL PHOSPHATE SYSTEM.
Vestn. Mosk. Univ., Khim., 24, 79-84 (1969) (in Russian)
CA, 73, 31201 (1970)

207. ALIMARIN, I. P., MIKLISHANSKII, A. Z., and YAKOLEV, Yu. V.
V.I. Vernadski Inst. of Geochemistry and Anal. Chem. of the Academy of Sciences of the
USSR, Moscow, USSR
NEUTRON ACTIVITION ANALYSIS OF RARE EARTH IMPURITIES IN EUROPIUM
OXIDE.
J. Radioanal. Chem., 4, 75-80 (1970)

208. ALIMARIN, I. P., BOLSHOVA, T. A., DOLZHENKO, T. G., and KANDYBINA, G. M.
Moscow M. V. Lomonosov State Univ., Moscow, USSR
USE OF TRIS (2-ETHYLHEXYL) PHOSPHATE AND BIS(2-ETHYLHEXYL) PHENYL
PHOSPHATE IN EXTRACTION AND PARTITION CHROMATOGRAPHY FOR THE SE-
PARATION OF INDIUM AND GALLIUM.
Zh. Anal. Khim., 26, 741–5 (1971) (in Russian)
Ca, 75. 53667 (1971)

209. ALIMARIN, I. P., BOLSHOVA, T. A., DOLZHENKO, T. G., and FERNANDES-GOMES, M. M.
Moscow M. V. Lomonosov Univ., Moscow, USSR
EXTRACTION AND CHROMATOGRAPHIC BEHAVIOR OF INDIUM, CADMIUM, AND ZINC IN THE HYDROBROMIC ACID-TRIBUTYL PHOSPHATE SYSTEM.
Zh. Anal. Khim., 26, 1213—5 (1971) (in Russian)
CA, 75, 80617 (1971)

210. ALIMARIN, I. P., GIBALO, I. M., and MYMRIK, N. A.
Moscow M. V. Lomonosov Univ., Moscow, USSR
EXTRACTION-CHROMATOGRAPHIC SEPARATION OF NIOBIUM AND TITANIUM IN THE TRIOCTYLAMINE-HYDROCHLORIC ACID SYSTEM.
Vestn. Mosk. Univ., Khim., 12,457—60(1971) (in Russian)
CA, 76, 28147 (1972)

211. ALIMARIN, I. P., YAKOVLEV, Yu. V., and STEPANETS, O. V.
V. I. Vernadsky Inst. of Geochem. and Anal. Chem.
USSR Academy of Sciences, Moscow, USSR
THE USE OF DISPLACEMENT EXTRACTION CHROMATOGRAPHY IN ACTIVATION ANALYSIS.
J. Radioanal. Chem., 11, 209—19 (1972)

212. ALIMARIN, I. P., GIBALO, I. M., LAPENKO, L. A.
Moscow M. V. Lomonosov Univ., Moscow, USSR
SEPARATION OF NIOBIUM (V), TITANIUM(IV) AND MOLYBDENUM(VI) BY CHROMA-TOGRAPHIC EXTRACTION IN THE TRIBUTYL PHOSPHATE-HYDROCHLORIC ACID SYSTEM.
Vestn. Mosk. Univ. Khim., 13,435—38 (1972) (in Russian)
CA, 78, 11124 (1973)

213. ALIMARIN, I. P., BOLSHOVA, T. A., and BAKHAREVA, G. A.
Moscow M. V. Lomonosov Univ., Moscow, USSR
SEPARATION OF INDIUM(III) AND ANTIMONY(V) BY THE EXTRACTION CHROMATO-GRAPHY METHOD.
Zh. Anal. Khim., 28,1300—3(1973) (in Russian)

214. ALY, H. F. and LATIMER, R. M.
Nucl. Chem. Dept., Atomic Energy Establishment, Cairo, UAR
EXTRACTION OF TRANSPLUTONIUM ELEMENTS BY DI-2-ETHYLHEXYL PHOSPHORIC ACID I. STABILITY CONSTANTS OF CURIUM, BERKELIUM, CALIFORNIUM AND EIN-STEINIUM WITH SOME HYDROXYCARBOXYLIC ACIDS.
Radiochim. Acta., 4.27—31(1970)

215. ALY, H. F. and EL-HAGGAN, M. A.
Nuclear Chem. Dept., Atomic Energy Establishment, Cairo, UAR
SYNERGISM IN EXTRACTION CHROMATOGRAPHY.
I. SEPARATION OF EUROPIUM FROM SAMARIUM.
Radiochem. Radioanal. Letters, 3,249—53(1970)

216. ALY, H. F. and RAIEH, M.
Nucl. Chem. Dept., Atomic Energy Establ. Cairo, UAR
SYNERGISM IN EXTRACTION CHROMATOGRAPHY,
PART II. FACTORS AFFECTING COLUMN PERFORMANCE FOR SEPARATION OF EUROPIUM FROM SAMARIUM AND CURIUM FROM CALIFORNIUM.
Anal. Chim. Acta, 54,171—6 (1971)

217. ALY, H. F. and EL-HAGGAN, M. A.
Nucl. Chem. Dept., Atomic Energy Establ., Cairo, UAR.
PRODUCTION OF CARRIER FREE SCANDIUM RADIOISOTOPES FROM NEUTRON
IRRADIATED POTASSIUM TITANIUM OXALATE TARGET.
Microchimica Acta (Wien), 1971, 4–8(1971)

218. ALY, H. F. and ABDEL-HAMID, M. M.
Nucl. Chem. Dept., Atomic Energy Establ., Cairo, UAR
NEW METHOD FOR ISOLATION OF CARRIER FREE SULFUR-35 AND PHOSPHOROUS-32
FROM NEUTRON ACTIVATED POTASSIUM CHLORIDE.
Microchemical Journal, 17, 215–9(1972)

219. ALY, H. F. and ABDEL-RASSOUL, A. A.
Nucl. Chem. Dept., Atomic Energy Establ. Cairo, UAR
SEPARATION OF CURIUM-CALIFORNIUM BY EXTRACTION CHROMATOGRAPHY.
Z. Anorg. Allg. Chem., 387, 252–8(1972)

220. ANDREEV, P. F., DANILOV, L. T., and KESHISHYAN, G. O.
APPLICATION OF COMPLEX-FORMING CHROMATOGRAPHY TO THE PROCESS OF
CONCENTRATING MICRO-QUANTITIES OF LEAD AND OTHER METALS FROM SOLU-
TIONS.
Zhur. Priklad. Khim., 34, 2419–26(1961)
CA, 56, 8362f (1962)

221. ANDREEV, P. F., ANDREEVA, I. V., DANILOV, L. T., and KESHISHYAN, G. O.
NEW PRINCIPLES AND METHODS FOR CONCENTRATION OF ELEMENTS FROM
ULTRADILUTE SOLUTIONS AND SEPARATION OF HEAVY METALS OF SIMILAR PRO-
PERTIES.
Tr. po Radiats. Gigiene, Leningr. Nauchn.– Issled. Inst. Radiats. Gigieny 1964(2), 218–21
(in Russian)
CA, 64, 4239h (1966)

222. ARTYUKHIN, P. I., STARTSEVA, E. A., MITYAKIN, Yu. L., IVANOV, I. M., GINDIN, L. M.,
and GLUKHOV, G. G.
EXTRACTION OF ELEMENTS FROM NITRATE SOLUTIONS BY TRIALKYLBENZYL-
AMMONIUM NITRATE, AND EXTRACTION-CHROMATOGRAPHIC SEPARATION OF
PLATINUM(IV), RHENIUM(VII); MOLYBDENUM(VI), AND TUNGSTEN(VI).
Radiokhimiya, 14, 150–1 (1972) (in Russian)
CA, 76, 159024 (1972)

223. BAETSLÉ, L. H., LIEVENS, F., DEMILDT, A. C, and MONSECOUR, M.
Dept. Chimie, CEN-SCK, Mol. Belgium
SOME ACTUAL ANALYTICAL PROBLEMS IN NUCLEAR ENERGY.
Paper presented at "Journées Internationales de Chimie Analytique", Brussels 14–17 Nov. 1967.
Industrie Chimique Belge, T. 33 N° special (1968) (in French)

224. BARBANO, P. G. and RIGALI, L.
Camen, San Piero a Grado, Pisa, Italy
SEPARATION OF AMERICIUM FROM RARE EARTHS BY REVERSED-PHASE PARTITION
CHROMATOGRAPHY.
J. Chromatog., 29, 309–11 (1967)
NSA, 21, 39014 (1967)

225. BECKER, R. and HECHT, F.
Reaktorzentrum Seibersdorf and Anal. Inst. of the Univ. of Vienna, Austria

RADIOCHEMICAL INVESTIGATIONS ON THE OPTIMIZATION OF THE EXTRACTION CHROMATOGRAPHY WITH POLYTRIFLUOROMONOCHLOROETHYLENE AS SUPPORT.
Proc. 6th Int. Symp. für Mikrochemie, Graz, 1970, Vol E. pp. 5–10 (in German)

226. BECKER, R. and HECHT, F.
Reatktorzentrum Seibersdorf and Anal. Inst. of the Univ. of Vienna, Austria
RADIOCHEMICAL ATTEMPTS AT THE OPTIMALIZATION OF THE EXTRACTION CHROMATOGRAPHY WITH POLYTRIFLUOROMONOCHLORETHYLENE (VOLTALEF) AS CARRIER.
Microchim. Acta, 1973,624–40(1973) (in German)

227. BELCHER, R., MAJER, J. R., and ROBERTS, G. A. H.
Dep. of Chem., Univ. of Birmingham, Great Britain
REDOX REACTIONS ON COLUMNS.
Talanta, 14,1245–51(1967)

228. BERANOVA, H. and NOVAK, M.
Czechosl. Acad. Sciences, Řež, ČSSR
PARTITION CHROMATOGRAPHY OF URANIUM AND FISSION PRODUCTS ON A POROUS STYRENE-DIVINYLBENZENE COPOLYMER SWOLLEN IN TRIBUTYL PHOSPHATE SOLUTION.
Collection Czech. Chem. Commun., 30, .073–81(1965)
CA, 63, 2590a (1965)

229. BERANOVA, H. and TEJNECKY, M.
Inst. of Nuclear Research, Prague, ČSSR
PARTITION CHROMATOGRAPHY OF URANIUM AND FISSION PRODUCTS ON A POROUS STYRENE-DIVINYLBENZENE COPOLYMER SWOLLEN IN TRIBUTYL PHOSPHATE SOLUTION. II. CONTRIBUTION TO THE STUDY OF THE EXTRACTION MECHANISM OF NITRIC ACID AND URANYL NITRATE.
Collect. Czech. Chem. Commun., 32, 437–43 (1967)
NSA, 21, 19746 (1967)

230. BERANOVA, H. and TEJNECKY, M.
Czechosl. Acad. Sciences, Řež, ČSSR
PARTITION CHROMATOGRAPHY OF URANIUM AND FISSION PRODUCTS ON POROUS STYRENE-DIVINYLBENZENE COPOLYMER SWOLLEN IN TRIBUTYL PHOSPHATE SOLUTION. IV. EQULIBRIUM DISTRIBUTION OF URANIUM OF URANYL NITRATE BETWEEN SWOLLEN GELS AND THE AQUEOUS PHASE. COLUMN DISTRIBUTION OF URANIUM AND FISSION PRODUCTS.
UJV-2567-Ch. 1971. 19 p.

231. BERANOVA, H. and TEJNECKY, M.
Inst. of Nuclear Research,
Czechosl. Acad. Sciences, Řež, ČSSR
PARTITION CHROMATOGRAPHY OF URANIUM AND FISSION PRODUCTS ON POROUS STYRENE-DIVINYLBENZENE COPOLYMER.
Coll. Czech. Chem. Commun., 37, 3579–89 (1972)

232. BITTNER, M., MIKULSKI, J., and SHEGLOUSKI, S.
THE SEPARATION OF SOME ELEMENTS OF THE II, III, IV, V, AND VI GROUPS BY MEANS OF EXTRACTION CHROMATOGRAPHY IN WORKS ON NUCLEAR SPECTROSCOPY.
Paper presented at the 19th Conference on Nuclear Spectroscopy and the Structure of the Atomic Nucleus, Yerevan, USSR, 27 Jan. – 4 Febr. 1969.

233. BLOURI, J. and REVEL, G.
Centre d'Etude de Chimie Métallurgique du C.N.R.S., Vitry-sur-Seine, France
APPLICATION OF COLUMN CHROMATOGRAPHY WITH NATURAL CELLULOSE
IMPREGNATED WITH TRI-n-OCTYLPHOSPHINE OXIDE FOR THE SEPARATION OF
SOME CATIONS.
J. Radioanal. Chem., 10, 121–4 (1972) (in French)
CA, 77, 42696 (1972)

234. BOGEN, D. C.
Health and Safety Laboratory, U. S. Atomic Energy Commission, New York, USA
RAPID DETERMINATION OF STRONTIUM-90 IN URINE.
Health Phys., 14, 131–3 (1968)

235. BOLSHOVA, T. A., ALIMARIN, I. P., and LITVINCHEVA, A. S.
Moscow M. V. Lomonosov State Univ., Moscow, USSR
SEPARATION OF SMALL AMOUNTS OF INDIUM AND GALLIUM BY PARTITION
CHROMATOGRAPHY ON A FLUOROPLAST-4 COLUMN.
Vestn. Mosk. Univ., Ser. II, 21, N° 6, 59–63 (1966) (in Russian)
CA, 67, 39757 (1967)

236. BOLSHOVA, T. A., DOLZHENKO, T. G., GRISHINA, K. V., and ALIMARIN, I. P.
Moscow M. V. Lomonosov State Univ., Moscow, USSR
EXTRACTION CHROMATOGRAPHIC SEPARATION OF INDIUM, ZINC, AND CADMIUM
FROM SOLUTIONS OF ALKALI METAL BROMIDES.
Zh. Anal. Khim., 28, 259–62 (1973) (in Russian)
CA, 73, 168121 (1973)

237. BONNEVIE-SVENDSEN, M. and MARTINI, V.
Institutt for Atomenergi, Kjeller, Norway
ANALYSIS OF THE CHEMICAL STATE OF PLUTONIUM AND FISSION PRODUCTS IN
PROCESS FEED SOLUTIONS
KR-81. Feb. 1966. 18 p.

238. BOSHOLM, J. and GROSSE-RUYKEN, H.
Techn. Univ., Dresden, DDR (GDR)
PARTITION CHROMATOGRAPHY OF RARE EARTHS BY BIS(2-ETHYLHEXYL)
PHOSPHORIC ACID II. SEPARATION OF MODEL MIXTURES.
J. prakt. Chem., 26, N° 1–2, 83–9 (1964)
CA, 62, 4588 g (1965)

239. BOSHOLM, J., GROSSE-RUYCKEN, H., LEBEDEV, N. A., HERRMANN, E., and KHALKIN,
V. A.
(GDR and USSR)
CONCENTRATION OF IMPURITIES IN HIGH-GRADE RARE-EARTH SAMPLES BY PAR-
TITION CHROMATOGRAPHY.]
Paper presented at the XXth Int. Congress on Pure and Appl. Chem., Moscow, USSR, 12–18 July 1965
CONF-650723, E93, p. 176 (1965)

240. BRANDŠTETR, J., WANG, T. S, GAVRILOV, K. A., GWOŹDŹ, E., MALY, J., and
TAUBE, M.
Joint Inst. for Nuclear Research, Dubna, Moscow, USSR
THE EXTRACTION PROPERTIES OF FERMIUM AND MENDELEVIUM (TBP-HNO$_3$
AND TBP-HCl).
JINR-P-1075 (1962)

Also published in Radiokhimiya, 6, 26–35 (1964) (in Russian)
NSA, 18, 23725 (1964)

241. BRAUN, T. and FARAG. A. B.
Inst. Inorg. and Anal. Chem.,L. Eötvös University, Budapest, Hungary
REVERSED-PHASE FOAM CHROMATOGRAPHY; SEPARATION OF PALLADIUM,
BISMUTH AND NICKEL IN THE TRIBUTYL PHOSPHATE-THIOUREA-PERCHLORIC ACID
SYSTEM.
Anal. Chim. Acta, 61,265–76 (1972)

242. BRAUN, T. and FARAG, A. B.
Inst. of Inorg. and Anal. Chem.,L. Eötvös Univ.,Budapest, Hungary
A NEW METHOD FOR COLUMN PACKING IN REVERSED-PHASE CHROMATOGRAPHY.
Anal. Chim. Acta, 62,476–80 (1972)

243. BRAUN, T. and FARAG, A. B.
Inst. of Inorg. and Anal. Chem.,L. Eötvös Univ.,Budapest, Hungary
FOAM CHROMATOGRAPHY. SOLID FOAMS AS SUPPORTS IN COLUMN CHROMATOG-
RAPHY.
Talanta, 19,828–30 (1972)

244. BRAUN, T., FARAG. A. B., and KLIMES-SZMIK, A.
Inst. of Inorg. and Anal. Chem.,L. Eötvös Univ.,Budapest Hungary
REVERSED-PHASE FOAM CHROMATOGRAPHY. REDOX REACTIONS ON OPEN-CELL
POLYURETHANE FOAM COLUMNS SUPPORTING TETRACHLOROHYDROQUINONE.
Anal. Chim. Acta, 64,71–6 (1973)

245. BRAUN, T.*, HUSZÁR, E.*, and BAKOS, L.**
* Inst. of Inorg. and Anal. Chem., L. Eötvös Univ.,Budapest, Hungary
** Central Research Inst. for Physics, Budapest, Hungary
REVERSED-PHASE FOAM CHROMATOGRAPHY. SEPARATION OF TRACE AMOUNTS
OF COBALT FROM NICKEL IN THE TRI-n-OCTYLAMINE-HYDROCHLORIC ACID SYSTEM.
Anal. Chim. Acta, 64,77–84 (1973)

246. BRAUN, T. and FARAG, A. B.
Inst. of Inorg. and Anal. Chem.,L. Eötvös Univ.,Budapest, Hungary
REVERSED-PHASE FOAM CHROMATOGRAPHY. CHEMICAL ENRICHMENT AND
SEPARATION OF GOLD IN THE TRIBUTYLPHOSPHATE-THIOUREA-PERCHLORIC
ACID SYSTEM
Anal. Chim. Acta, 65,115–26 (1973)

247. BRAUN, T. and FARAG, A. B.
Inst. of Inorg. and Anal. Chem.,L. Eötvös Univ.,Budapest, Hungary
PULSED COLUMN REDOX TECHNIQUES WITH FLEXIBLE FOAM FILLINGS.
Anal. Chim. Acta, 65,139–44 (1973)

248 BRAUN, T., BAKOS, L., and SZABÓ, Z. S.
Inst. of Inorg. and Anal. Chem.,L. Eötvös Univ.,Budapest, Hungary
REVERSED PHASE FOAM CHROMATOGRAPHY. SEPARATION OF IRON FROM COP-
PER, COBALT AND NICKEL IN THE TRI-n-BUTYL PHOSPHATE-HYDROCHLORIC ACID
SYSTEM.
Anal. Chim. Acta, 66,57–66 (1973)

249. BUIJS, K., MAINO, F., MULLER, W., REUL, J., and TOUSSAINT, J. Cl.
Euratom, Inst. for Transuranium Elements, Karlsruhe, BRD (FRG)
GRAM-SCALE PURIFICATION OF ^{244}Cm BY EXTRACTION CHROMATOGRAPHY.
Angew. Chem., 83, 766 (1971) (in German)
Angew. Chem., Int. Engl. Ed. 10,727–8 (1971)

250. BUIJS, K., MULLER, W., REUL, J., TOUSSAINT J. Cl.
Euratom, Inst. for Transuranium Elements, Karlsruhe, FGR
THE SEPARATION AND PURIFICATION OF AMERICIUM AND CURIUM ON THE
MULTIGRAM SCALE.
Paper presented at the Symp. on the Chemisty of Transuranium Elements, Moscow (USSR),
4-8 Sept. 1972

251. BUKOVSKAYA, A. V., KATYKHIN, G. S., and NIKITIN, M. K.
THE APPLICATION OF PARTITION CHROMATOGRAPHY TO THE SEPARATION AND
DETERMINATION OF PLATINUM IN GOLD-PLATINUM ALLOYS.
Working Experience of Testing Analytical Laboratories, Moscow 1969, pp. 129–136, (in Russian)

252. BUTECELEA. M. and CRAIN, M.
Cent. Phys. Chem., Bucarest, Roumania.
SPECTROGRAPHIC DETERMINATION OF MICROGRAM AMOUNTS OF YTTRIUM AND
SOME RARE EARTHS IN URANIUM OXIDES.
Rev. Roum. Chim., 16, 128–90 (1971)

253. CAMERA, V., BARASSI, G., and LEGROS, J. J.
Commission of the European Communities, Joint Nuclear Research Centre – ISPRA, Italy
RAPID AND SIMPLE METHOD FOR DETERMINATION OF TOTAL ALPHA RADIOAC-
TIVITY IN URINE.
EUR-4675i. 1971. 14 p. (in Italian)

254. CAMERA, V.
Commission of the European Communities, Joint Nuclear Research Centre – ISPRA, Italy
ISOLATION AND DETERMINATION OF PROTACTINIUM-231 IN URINE .
EUR 4870i.
1972. 20 p. (in Italian)

255. CARRIT, D. E.
Chesapeake Bay Institute, John Hopkins University, Annapolis, Md., USA
SEPARATION AND CONCENTRATION OF TRACE METALS FROM NATURAL WATERS,
PARTITION CHROMATOGRAPHIC TECHNIQUE.
Anal. Chem., 25,1927–8 (1953)

256. CATTIN, G. and HUGON, J.
CEA, Centre of Pierrelatte, France
PRECISE DETERMINATION OF URANIUM.
IAEA-SM-149/46
Paper presented at the IAEA Symposium on Analytical Methods in the Nuclear Fuel
Cycle, Vienna, 29 Nov.–3 Dec. 1971.

257. CERRAI, E. and TESTA, C.
CISE, Milan, Italy
THE APPLICATION OF CELLULOSE POWDER TREATED WITH TRI-n-OCTYLPHOSPHINE
OXIDE (TOPOC) TO COLUMN CHROMATOGRAPHY.
Energia Nucleare, 8, 510–8 (1961)

258. CERRAI, E. and TESTA, C.
CISE, Milan, Italy
THE USE OF TRI-n-OCTYLAMINE-CELLULOSE IN CHEMICAL SEPARATIONS.
J. Chromatog., 6,443- 51 (1961)

259. CERRAI, E., TESTA, C., and TRIULZI, C.
CISE, Milan, Italy
SEPARATION OF RARE EARTHS BY COLUMN CHROMATOGRAPHY WITH CELLULOSE
POWDER TREATED WITH DI(2-ETHYLHEXYL) ORTHOPHOSPHORIC ACID. PART I.
Energia Nucleare (Milan), 9, 193-9 (1962)

260. CERRAI, E. and TESTA, C.
CISE, Milan, Italy
SEPARATIONS BY REVERSED-PHASE COLUMN PARTITION CHROMATOGRAPHY WITH
KEL-F SUPPORTING TRI-n-OCTYLPHOSPHINE OXIDE.
J. Chromatog.,9, 216- 23 (1962)

261. CERRAI, E., TESTA, C.,and TRIULZI, C.
CISE, Milan, Italy
SEPARATION OF RARE EARTHS BY COLUMN CHROMATOGRAPHY WITH CELLULOSE
POWDER TREATED WITH DI(2-ETHYLHEXYL) ORTHOPHOSPHORIC ACID. PART II.
Energia Nucleare (Milan), 9,377--84 (July 1962)

262. CERRAI, E. and TESTA, C.
CISE, Milan, Italy
ELECTRON-EXCHANGE PROCESSES ON SIMPLE COLUMNS OF KEL-F SUPPORTING
TETRACHLOROHYDROQUINONE.
Anal. Chim. Acta, 28, 205--16 (1963)

263. CERRAI, E. and TESTA, C.
CISE, Milan, Italy
SEPARATION OF RARE EARTHS BY MEANS OF SMALL COLUMNS OF KEL-F SUPPORTING
DI(2-ETHYLHEXYL) ORTHOPHOSPHORIC ACID.
J. Inorg. Nucl. Chem., 25, 1045-50 (1963)

264. CERRAI, E.*,   PELATI, L. **, and TRIULZI, C.*
* CISE, Milan, Italy
** University of Parma, Italy
RADIOCHEMICAL STUDIES FOR RADIOACTIVITY DETERMINATIONS IN MARINE
PLANKTON.
CISE Report No. 95. 1963. 23 p., 6 tables, 8 figs.

265. CERRAI, E., HAINSKI, Z., ROSSI, G., and TRUCCO, R.
CISE, Milan, Italy
CHEMICAL SEPARATION AND SPECTROGRAPHIC DETERMINATION OF TRACE
AMOUNTS OF RARE EARTHS IN SAP MATERIAL. PART I.
Energia Nucleare (Milan), 11, 9-17 (1964)

266. CERRAI, E. and GHERSINI, G.
CISE, Milan, Italy
REVERSED-PHASE EXTRACTION CHROMATOGRAPHY WITH PAPER AND COLUMNS
SUPPORTING AN EXTRACTANT SELECTIVE FOR COPPER.
Analyst, 94, 599-604 (1969)

267. CESARANO, C., PUGNETTI, G., and TESTA, C.
Casaccia Nuclear Center of CNEN, Rome, Italy
SEPARATION OF CESIUM-137 FROM FISSION PRODUCTS BY MEANS OF A KEL-F COL-
UMN SUPPORTING TETRAPHENYLBORON.
J. Chromatog., 19, 589—93 (1965)

268. CHENG, H. S., KE, C. H., LIN, C. Yu., and WINCHESTER, J.
National Tsing Hua Univ., Hsinchu, Taiwan
CHARACTERISTICS OF BIS(2-ETHYLHEXYL) ORTHOPHOSPHORIC
ACID CHROMATOGRAPHIC COLUMN FOR RARE-EARTH SEPARATION.
J. Chinese Chem. Soc. (Taiwan), 10, 80—5 (1963)

269. CHUDINOV, E. G. and PIROZHKOV, S. V.
Kurchatov Atomic Energy Institute, Moscow, USSR
THE SEPARATION OF BERKELIUM(III) FROM CERIUM(III).
J. Radioanal. Chem., 10, 41—6 (1972)

270. CLINGMAN, A. L. and PARRISH, J. R.
SELECTIVE LIQUID ION-EXCHANGERS.
J. Appl. Chem. (London), 13, N° 5, 193—8 (1963)
CA, 59, 3341c (1963)

271. COLVIN, C. A.
Atlantic Richland Hanford Co., Richland, Wash., USA
RECOVERY OF PALLADIUM FROM NUCLEAR WASTE SOLUTIONS USING A PACKED
COLUMN OF TRICAPRYL MONOMETHYL AMMONIUM CHLORIDE ON AN INERT
SUPPORT.
ASH—SA—28. 1969. 12 p.
CONF-690606-2

272. CRAWLEY, R. H. A.
English Elec. Co., Whetstone, Great Britain
SEPARATION OF ZIRCONIUM AND HAFNIUM BY REVERSED-PHASE CHROMATOG-
RAPHY.
Nature, 197, 377—8 (1963)
CA, 58, 9609d (1963)

273. CROUCH, E. A. C. and FRYER, J. R.
AERE, Harwell, Berks., Great Britain
THE FAST SEPARATION OF THE RARE EARTHS.
Paper presented at the Discussion on Nuclear Chemistry, Oxford, Sept. 1962.
AERE-M-1078, p. 91 (1962)

274. DAHMER, L. H.
Iowa State Univ., Ames, Iowa, USA
CHROMATOGRAPHIC SEPARATIONS OF NIOBIUM, TANTALUM, MOLYBDENUM AND
TUNGSTEN.
Thesis
IS-T-113. Nov. 1966. 79 p.
NSA, 21, 8189 (1967)

275. DELLE SITTE, A. and TESTA C.
Radiotoxicological Lab., Medical Service CNEN, CSN Casaccia, Rome, Italy

THE APPLICATION OF COLUMN REDOX-EXTRACTION CHROMATOGRAPHY TO THE SEPARATION OF ACTINIDE ELEMENTS.
Paper presented at the 7th Radiochemical Conference at Mariánské Lázné, ČSSR, 24–28 April 1973

276. Delle Sitte, A.
Radiotoxicological Lab., Medical Service CNEN, CSN Casaccia, Rome, Italy
ANALYTICAL APPLICATIONS OF NEOTRIDECANO-HYDROXAMIN ACID AS STATIONARY PHASE IN EXTRACTION CHROMATOGRAPHY.
J. Radioanal. Chem., 14, 45–51 (1973)

277. DENIG, R., TRAUTMANN, N., and HERRMANN, G.
Inst. for Inorganic and Nuclear Chemistry, Univ. Mainz, BRD (FRG)
SEPARATION OF FISSION PRODUCTS BY EXTRACTION CHROMATOGRAPHY.
AED-CONF. 1966-306-6 (in German)

278. DENIG, R., TRAUTMANN, N., and HERRMANN, G.
Inst. for Inorganic and Nuclear Chemistry, Univ. Mainz, BRD (FRG)
RAPID EXTRACTION BY QUASI-SOLID EXTRACTING AGENTS.
Z. analyt. Chem., 216, 41–50 (1966) (in German)
CA, 64, 10470g (1966)

279. DENIG, R.
Inst. for Inorganic and Nuclear Chemistry, Univ. Mainz, BRD (FRG)
SEPARATION OF FISSION PRODUCTS BY EXTRACTION CHROMATOGRAPHY.
Thesis. 1967. 185 p. (in German)

280. DENIG, R., TRAUTMANN, N., and HERRMANN, G
Inst. for Inorganic and Nuclear Chemistry, Univ. Mainz, BRD (FRG)
SEPARATION OF FISSION PRODUCTS BY EXTRACTION CHROMATOGRAPHY.
Angew. Chem., 79, 247 (1967) (in German)

281. DENIG, R., and TRAUTMANN, N.
Inst. for Inorganic and Nuclear Chemistry, Univ. Mainz, BRD (FRG)
SEPARATION OF SOME PARENT-DAUGHTER PAIRS BY EXTRACTION CHROMATOGRAPHY
AED-CONF 68-196-014 (1968) (in German)

282. DENIG, R., TRAUTMANN, N., and HERRMANN, G.
Inst. for Inorganic and Nuclear Chemistry Univ. Mainz, BRD (FRG)
SEPARATION OF FISSION PRODUCTS BY EXTRACTION CHROMATOGRAPHY.
I. DETERMINATION OF THE PARTITION COEFFICIENTS.
J. Radioanal. Chem., 5, 223–31 (1970) (in German)

283. DENIG, R., TRAUTMANN, N., and HERRMANN, G.
Inst. for Inorganic and Nuclear Chemistry, Univ. Mainz, BRG (FRG)
SEPARATION OF FISSION PRODUCTS BY EXTRACTION CHROMATOGRAPHY.
II. SEPARATION OF MOTHER-DAUGHTER PAIRS.
J. Radioanal. Chem., 6, 57–65(1970) (in German)

284. DENIG, R., TRAUTMANN, N., and HERRMANN, G.
Inst. for Inorganic and Nuclear Chemistry, Univ. Mainz, BRD (FRG)

SEPARATION OF FISSION PRODUCTS BY EXTRACTION CHROMATOGRAPHY.
III. SEPARATION OF A FISSION PRODUCT MIXTURE.
J. Radioanal. Chem., 6, 331–43(1970) (in German)

285. DEPTULA, C., MAKOWSKI, H., and WIZA, J.
PREPARATION OF CERIUM-141 AND PRASEODYMIUM-143 FROM CERIUM DIOXIDE
IRRADIATED IN THE EWA REACTOR.
Nukleonika, 16,31–5 (1971) (in Polish)
CA, 75, 57406 (1971)

286. DE WET, W.J. and CROUCH, E. A. C.
AERE, Harwell, Berks., Great Britain
A SEQUENTIAL ANALYTICAL SCHEME FOR THE CARRIER-FREE SEPARATION OF
MICROGRAM QUANTITIES OF INDIVIDUAL FISSION PRODUCT AND ACTINIDE ELE-
MENTS FROM HIGHLY BURNT-UP FUELS.
J. Inorg. Nucl. Chem., 27,1735–44 (1965)

287. DIETRICH, W. C., CAYLOR, J. D., and JOHNSON, E. E.
Union Carbide, Oak Ridge, Tenn., USA
SEPARATION OF URANIUM(VI) FROM URINE BY A TRI-n-OCTYLPHOSPHINE OXIDE
AND AN AUTOMATION OF THE PROCEDURE.
Y-1322 (USAEC). 1960. p. 1–37.
CA, 55, 9539e (1961)

288. D'OLIESLAGER, W. and D'HONT, M.
Lab. of Radiochem., Univ. of Leuven, Belgium
SEPARATION OF FISSION PRODUCT MOLYBDENUM WHEN USING STABLE MO-
LYBDENUM-100 AS THE BURN-UP INDICATOR.
Meded. Kon. Vlaam. Acad. Wetensch., Lett. Schone Kunsten Belg., K1. Wetensch., 32 (5), 33 p.
(in Dutch)
CA, 75, 13830 (1971)

289. D'OLIESLAGER, W. and D'HONT, M.
Lab. of Radiochem., Univ. of Leuven, Belgium
MOLYBDENUM-COMPLEX EQUILIBRIA FOR LIQUID–LIQUID EXTRACTION.
J. Nucl. Mat., 44,185–93 (1972)

290. DOLZHENKO,T.G.,BOLSHOVA,T. A., and ALIMARIN, I. P.
Moscow M. V. Lomonosov State Univ., Moscow, USSR
EFFECT OF TEMPERATURE ON THE EXTRACTION-CHROMATOGRAPHIC DISTRIBUTION
OF INDIUM, ZINC, AND CADMIUM USING THE TRIBUTYL PHOSPHATE-HYDROBROMIC
ACID SYSTEM.
Zh. Anal. Khim., 28, 263–6 (1973) (in Russian)
CA, 78, 154481c (1973)

291. DZHELEPOV, B. S., KATYKHIN, G. S., MAIDANYUK, V. K., and FEOKTISTOV, A. I.
THE SPECTRA OF INTERNAL CONVERSION ELECTRONS, AND OF POSITRONS EMIT-
TED DURING THE DECAY OF ^{184}Re.
Izvest. Akad. Nauk S.S.S.R.; Ser. Fiz., 26,1030–4 (Aug. 1962) (in Russian)
NSA, 16, 33831 (1962)

292. EL-GARHY, M., ALY, H. F., and EL-REEFY
Nuclear Chemistry Dept., Atomic Energy Establishment, Cairo, UAR
SEPARATION OF CARRIER-FREE MANGANESE-54 AND IRON-59 USING REVERSED-
PHASE PARTITION CHROMATOGRAPHY.
Isotopenpraxis, 6, 478–80 (1970)
NSA, 25, 37442 (1971)

293. ERSHOVA, N.I., BOLSHOVA, T.A., and ALIMARIN, I. P.
Moscow M. V. Lomonosov State Univ., Moscow, USSR
EXTRACTION CHROMATOGRAPHIC BEHAVIOUR OF IRON(III) IN THE SYSTEM OF
TRIBUTYL PHOSPHATE(TBP)-HYDROCHLORIC ACID AND ITS SALTS.
Vestn. Mosk. Univ., Khim., 23, 76−9(1968) (in Russian)
CA, 69, 92453 (1968)

294. ERSHOVA, N. I., BOLSHOVA, T. A., ALIMARIN, I. P., and MOROZOVA, L. I.
Moscow M. V. Lomonosov State Univ., Moscow, USSR
EFFECT OF TEMPERATURE ON CHROMATOGRAPHIC EXTRACTION OF GALLIUM,
IRON(III), ZINC, CADMIUM, AND COPPER IN HYDROCHLORIC ACID-TRIBUTYL
PHOSPHATE AND AMMONIUM THIOCYANATE TRIBUTYL PHOSPHATE SYSTEMS.
Vestn. Mosk. Univ., Khim., 12, 348−50(1971) (in Russian)
CA, 75, 91533 (1971)

295. ERSHOVA, N. I., BOLSHOVA, T. A., and ALIMARIN, I. P.
Moscow M. V. Lomonosov State Univ., Moscow, USSR
THE EFFECT OF LARGE AMOUNTS OF VARIOUS CHLORIDES ON THE CONCENTRA-
TION, FROM THIOCYANATE MEDIUM, OF TRACES OF GALLIUM, IRON(III), COBALT
AND COPPER BY AN EXTRACTION-CHROMATOGRAPHIC METHOD.
Zh. Anal. Khim., 26, 243−5(1971) (in Russian)

296. ERSHOVA, Z. V., MARKOV, V. K., TSVETAEVA, N. E., IVANOVA, L. A., FEDOSEEV, D. A.
EROFEEVA, M. N., ZEMLYANUKHINA, N. A., KERMANOV, V. P., GOLUTVINA, M. M.,
NOVIKOVA, N. Ya., and KULICHENKO, M. N.
State Atomic Energy Committee, Moscow, USSR
METHODS OF DETERMINING RADIONUCLIDE CONTENT AT OR BELOW MPC LEVEL
IN WASTE WATER, ENVIRONMENTAL OBJECTS AND BIOLOGICAL SAMPLES.
Proc. Fourth Int. Conf. Geneva 1971, Vol. 11, pp. 641−62.
P/451 (in Russian)

297. ESCHRICH, H.
Institutt for Atomenergi, Kjeller, Norway
SEPARATION OF URANIUM, NEPTUNIUM, PLUTONIUM, AND AMERICIUM BY RE-
VERSED-PHASE PARTITION CHROMATOGRAPHY.
KR-11. 1961. 5 p.

298. ESCHRICH, H.
Institutt for Atomenergi, Kjeller, Norway
SEPARATION OF NEPTUNIUM(IV), -(V), AND -(VI) BY EXTRACTION CHROMATOGRAPH
Z. anal. Chem., 226, 100−14 (1967)

299. ESCHRICH, H. and HUNDERE, I.
Eurochemic, Mol, Belgium
SEPARATION OF IMPURITIES FROM PLUTONIUM BY EXTRACTION CHROMATOGRAPH
Eurochemic Technical Report ETR-222. 1968. 18 p.

300. ESCHRICH, H. and HUNDERE, I.
Eurochemic, Mol, Belgium
A NEW METHOD FOR THE ALPHA ACTIVITY SPECIFICATION ANALYSIS OF URANIUM
FINAL PRODUCTS.
Eurochemic Technical Report ETR-226. 1968. 31 p.

301. ESCHRICH, H. and HANSEN, P.
Eurochemic, Mol, Belgium

THE APPLICATION OF OPEN CAPILLARY COLUMNS TO EXTRACTION CHROMA-
TOGRAPHIC SEPARATIONS.
Eurochemic Technical Report ETR-239. 1969. 24 p.

302. ESCHRICH, H.
Eurochemic, Mol, Belgium
THE DISTRIBUTION OF U(IV) AND U(VI) BETWEEN TRIBUTYL PHOSPHATE AND
NITRIC ACID AND THEIR SEPARATION BY EXTRACTION CHROMATOGRAPHY.
Eurochemic Technical Report ETR-249. 1969. 22 p., 4 tables, 13 figs.

303. ESPAÑOL, C. E. and MARAFUSCHI, A. M.
Laboratorio de Radioisotopes, Facultad de Ingenieria. Univ. Buenos Aires, Argentina
APPLICATION OF REVERSED-PHASE COLUMN CHROMATOGRAPHY TO THE DETER-
MINATION OF TUNGSTEN IN STAINLESS STEELS BY ACTIVATION ANALYSIS.
J. Chromatog., 29, 311–5(1967)
NSA, 21, 38919 (1967)

304. ESPAÑOL, C. E. and MARAFUSCHI, A. M.
Laboratorio de Radioisotopes, Facultad de Ingenieria, Univ. Buenos Aires, Argentina
APPLICATION OF Ge(Li) DETECTOR TO THE DETERMINATION OF MANGANESE IN
ZIRCONIUM ALLOYS BY ACTIVATION ANALYSIS.
Radiochim. Acta, 9, 165–6(1968)
NSA, 23, 76 (1969)

305. FIDELIS, I. and SIEKIERSKI, S
Inst. of Nuclear Research, Polish Academy of Sciences, Warsaw, Poland
SEPARATION OF HEAVY RARE EARTHS BY REVERSED-PHASE PARTITION CHRO-
MATOGRAPHY.
Report N° 169/V. Sept. 1960. 9 p.
(NP-9643)
Later published in
J. Chromatog., 5, 161–5 (1961)

306. FIDELIS, I., GWOŹDŹ, R., and SIEKIERSKI, S.
Inst. of Nuclear Research, Polish Academy of Sciences, Warsaw, Poland
SEPARATION OF PROTACTINIUM FROM THORIUM BY REVERSED-PHASE PARTI-
TION CHROMATOGRAPHY.
Nukleonika, 8, 245–8(1963) (in Russian)

307. FIDELIS, I., GWOŹDŹ, R., and SIEKIERSKI, S.
Inst. of Nuclear Research, Polish Academy of Sciences, Warsaw, Poland
THE SEPARATION OF ARSENIC FROM GERMANIUM BY REVERSED-PHASE PARTITION
CHROMATOGRAPHY.
Nukleonika, 8, 319–26 (1963)

308. FIDELIS, I., GWOŹDŹ, R., and SIEKIERSKI, S.
Inst. of Nuclear Research, Polish Academy of Sciences, Warsaw, Poland
SEPARATION OF CARRIER-FREE ^{199}Au FROM PLATINUM BY REVERSED-PHASE
PARTITION CHROMATOGRAPHY.
Nukleonika, 8, 327–31(1963) (in Russian)

309. FIDELIS, I. and SIEKIERSKI, S.
Inst. of Nuclear Research, Polish Academy of Sciences, Warsaw, Poland

USE OF 2-ETHYLHEXYL PHENYLPHOSPHONIC ACID IN REVERSED-PHASE PARTITION
CHROMATOGRAPHY.
J. Chromatog., 17, 542–8 (1965)

310. FIELDS, P. R., AHMAD, I., BARNES, R. F., SJOBLOM, R. K., and HORWITZ, E. P.
Chem. Div., Argonne Nat. Lab., Argonne, Illinois, USA
NUCLEAR PROPERTIES OF Md-254, Md-255, Md-256, Md-257 AND Md-258.
Nuclear Physics, A 154, 407–16 (1970)

311. FLETCHER, W., FRANKLIN, R., and GOODALL, G. C.
UKAEA, Development and Engineering Group, Capenhurst, Ches., Great Britain
SPECTROGRAPHIC DETERMINATION OF TRACE IMPURITIES IN URANIUM COM-
POUNDS USING REVERSED-PHASE PARTITION CHROMATOGRAPHY.
TID-7629. 1961. pp. 276–84

312. FOLDZINSKA, A. and DYBCZYNSKI, R.
Inst. of Nuclear Research, Warsaw, Poland
RADIOCHEMICAL SCHEME FOR THE SEPARATION AND GAMMA-SPECTROMETRIC
DETERMINATION OF La, Ga, Sc, AND Hf IN FIRE RESISTANT MATERIALS BY
THE METHOD OF NEUTRON ACTIVATION ANALYSIS.
Paper presented at the 7th Radiochemical Conference at Mariánské Lázné, CSSR, 24-28
Apr. 1973

313. FRAZEE, R. T.
Iowa State Univ., Ames, Iowa, USA
REVERSED-PHASE CHROMATOGRAPHIC SEPARATION OF ZIRCONIUM AND HAF-
NIUM.
Thesis
IS-T-2. Feb. 1965. 41 p.

314. FRITZ, J. S. and HEDRICK, C. E.
Inst. for Atomic Research and Dept. of Chem., Iowa State Univ., Ames, Iowa, USA
SEPARATION OF IRON BY REVERSED-PHASE CHROMATOGRAPHY.
Anal. Chem., 34, 1411–4 (1962)

315. FRITZ, J. S. and HEDRICK, C. E.
Inst. for Atomic Research and Dept. of Chem., Iowa State Univ., Ames, Iowa, USA
SEPARATION OF MOLYBDENUM IN STEELS BY REVERSED-PHASE CHROMATOG-
RAPHY.
Anal. Chem., 36, 1324–6 (1964)

316. FRITZ, J. S. and FRAZEE, R. T.
Inst. for Atomic Research and Dept. of Chem., Iowa State Univ., Ames, Iowa, USA
REVERSED-PHASE CHROMATOGRAPHIC SEPARATION OF ZIRCONIUM AND HAF-
NIUM.
Anal. Chem., 37, 1358–61 (1965) (see also reference N° 313)

317. FRITZ, J. S., GILLETTE, R. K., and MISHMASH, H. E.
Inst. for Atomic Research and Dept. of Chem., Iowa State Univ., Ames, Iowa, USA
EXTRACTION OF METAL IONS WITH ISOOCTYL THIOGLYCOLATE.
Anal. Chem., 38, 1869–72 (1966)

318. FRITZ, J. S. and LATWESEN, G.
Inst. for Atomic Research and Dept. of Chem., Iowa State Univ., Ames, Iowa, USA

442

SEPARATION OF TIN FROM OTHER ELEMENTS BY PARTITION CHROMATOGRAPHY.
Talanta, 14,529–36 (1967)

319. FRITZ, J. S. and DAHMER, L. H.
Inst. for Atomic Research and Dept. of Chem., Iowa State Univ., Ames, Iowa, USA
COLUMN CHROMATOGRAPHIC SEPARATION OF NIOBIUM, TANTALUM, MOLYBDENUM,
AND TUNGSTEN.
Anal. Chem., 40, 20–3 (1968)

320. FRITZ, J. S. and LATWESEN, G. L.
Inst. for Atomic Research and Dept. of Chem., Iowa State Univ., Ames, Iowa, USA
A SEPARATION SCHEME FOR THE ANALYSIS OF MULTICOMPONENT SAMPLES.
Talanta, 17,81–91(1970)

321. FRITZ, J. S. and KENNEDY, D. C.
Inst. for Atomic Research and Dept. of Chem., Iowa State Univ., Ames, Iowa, USA
EXTRACTION CHROMATOGRAPHY OF URANIUM WITH DIOCTYL SULPHOXIDE.
Talanta, 17,837–43(1970)

322. FRITZ, J. S., FRAZEE, R. T., and LATWESEN, G. L.
Inst. for Atomic Research and Dept. of Chem., Iowa State Univ. Ames, Iowa, USA
COLUMN CHROMATOGRAPHIC SEPARATION OF GALLIUM, INDIUM AND TALLIUM.
Talanta, 17,857–64(1970)

323. FRITZ, J. S. and TOPPING, J. J.
Inst. for Atomic Research and Dept. of Chem., Iowa State Univ., Ames, Iowa, USA
CHROMATOGRAPHIC SEPARATION OF VANADIUM, TUNGSTEN AND MOLYBDENUM
WITH A LIQUID ANION-EXCHANGER.
Talanta, 18,865–72(1971)

324. FRITZ, J. S., BEUERMAN, D. R., and RICHARD, J. J.
Inst. for Atomic Research and Dept. of Chem., Iowa State Univ., Ames, Iowa, USA
CHROMATOGRAPHIC SEPARATION OF COPPER WITH AN ALPHA-HYDROXYOXIME.
Talanta, 18, 1095–1102 (1971)

325. FRITZ, J. S. and BEUERMAN, D. R.
Inst. for Atomic Research and Dept. of Chem., Iowa State Univ., Ames, Iowa, USA
CHROMATOGRAPHIC SEPARATION OF MOLYBDENUM USING AN ALIPHATIC ALPHA-
HYDROXYOXIME.
Anal. Chem., 44,692–4(1972)

326. GAVRILOV, K. A., GWOŹDŹ, E., STARÝ, J., and WANG, T. S.
Joint Inst. for Nuclear Research, Dubna, Moscow, USSR
INVESTIGATION OF THE SOLVENT EXTRACTION OF CALIFORNIUM, FERMIUM,
AND MENDELEVIUM.
Talanta, 13,471–6(1966)

327. GIBALO, I. M. and MYMRIK, N. A.
Moscow State Univ., USSR
SEPARATION OF NIOBIUM, TUNGSTEN, MOLYBDENUM, AND ZIRCONIUM BY PAR-
TITION CHROMATOGRAPHY.
Zh. Anal. Khim., 25,1744–7(1970) (in Russian)
NSA, 24, 50457 (1970)

328. GIBALO, I. M. and MYMRIK, N. A.
Moscow State Univ., USSR
CHROMATOGRAPHIC SEPARATION OF NIOBIUM FROM ACCOMPANYING ELEMENTS .
Zh. Anal. Khim., 26, 918—21 (1971) (in Russian)
NSA, 25, 37484 (1971)
CA, 75, 53642 (1971)

329. GIRARDI, F. and MERLINI, M.
Euratom Joint Nuclear Research Centre, Ispra, Italy
STUDIES ON THE DISTRIBUTION OF TRACE ELEMENTS IN A MOLLUSK FROM A
FRESHWATER ENVIRONMENT BY ACTIVATION ANALYSIS.
EUR-474e. 1963. 26 p.

330. GIRARDI, F., MERLINI, M., PAULY, J., and PIETRA, R.
Euratom Joint Nuclear Research Centre, Ispra, Italy
PROGRESS TOWARDS AUTOMATED RADIOCHEMICAL SEPARATIONS.
Proc. of IAEA Symp. on Radiochemical Methods of Analysis, Salzburg, 19-23 Oct. 1964,
Vol.2, pp. 3—14, IAEA, Vienna, 1965

331. GOLINSKI, M.
Institute of Nuclear Research, Warsaw-Zeran, Poland
EXTRACTION OF TIN AND INDIUM WITH TRIBUTYL PHOSPHATE FROM HYDRO-
CHLORIC ACID SOLUTIONS.
Solvent Extraction, Proc. of Int. Solvent Extr. Conf. ISEC 1971, The Hague, The Netherlands
19-23 Apr. 1971, Vol. 2, pp. 603—15.
Soc. for Chem. Industry, London, 1971

332. GORYANSKAYA, G. P., KAPLAN, B. Ya., KOVALIK, I. V., MERISOV, Yu. I., and
NAZAROVA, M. G.
State Scient. Res. Des. Inst. Rare Met. Ind., Moscow, USSR
USE OF EXTRACTION CHROMATOGRAPHY FOR QUANTITATIVE SEPARATION OF
RARE METALS FROM ADMIXTURES OF OTHER ELEMENTS.
1. SEPARATION OF TANTALUM.
Zh. Anal. Khim., 27, 1498—502 (1972) (in Russian)
CA, 77, 172309 (1972)

333. GOURISSE, D. and CHESNÉ, A.
Centre d'Etudes Nucléaires de Fontenay-aux-Roses, France
ANALYTICAL SEPARATION OF TRACES OF NEPTUNIUM FROM HIGHLY CONCENTRA-
TED URANIUM SOLUTIONS USING REVERSED-PHASE CHROMATOGRAPHY IN THE
TLA-HNO₃ SYSTEM.
CEA-R 3245. 1967. 31 p. (in French)

334. GOURISSE, D. and CHESNÉ, A.
Centre d'Etudes Nucléaires de Fontenay-aux-Roses, France
USE OF TRILAURYLAMINE IN PARTITION CHROMATOGRAPHY. PART I. THE SEPARA-
TION OF NEPTUNIUM FROM CONCENTRATED URANIUM SOLUTIONS.
Anal. Chim. Acta, 45, 311—9 (1969) (in French)

335. GOURISSE, D. and CHESNÉ, A.
Centre d'Etudes Nucléaires de Fontenay-aux-Roses, France
USE OF TRILAURYLAMINE IN PARTITION CHROMATOGRAPHY. PART II. SIMUL-
TANEOUS SEPARATION OF NEPTUNIUM AND PLUTONIUM FROM URANIUM.
Anal. Chim. Acta, 45, 321—5 (1969) (in French)

336. GROSS, D. J.
US Army Nuclear Defence Laboratory, Edgewood Arsenal, Maryland, USA
REVERSED-PHASE CHROMATOGRAPHIC SEPARATION AND LIQUID SCINTILLATION
DETERMINATION OF PLUTONIUM.
NDL-TR-103.1968. 25 p.

337. GROSSE-RUYKEN, H. and BOSHOLM, J.
Techn. Univ., Dresden, DDR (GDR)
PARTITION CHROMATOGRAPHY OF RARE EARTHS BY BIS(2-ETHYLHEXYL) PHOSPHO-
RIC ACID.I. EFFECT OF THE WORKING CONDITIONS ON THE SEPARATIVE EFFI-
CIENCY OF THE COLUMN.
J. Prakt. Chem., 25, n° 1–2, 79–87 (1964)
CA, 61, 12602f (1964)

338. GROSSE-RUYKEN, H. and BOSHOLM, J.
Techn. Univ., Dresden, DDR (GDR)
PARTITION CHROMATOGRAPHY OF RARE EARTHS BY BIS(2-ETHYLHEXYL)
PHOSPHORIC ACID. III. DETERMINATION OF YTTRIUM IN DYSPROSIUM AND TER-
BIUM CONCENTRATES.
J. Prakt. Chem., 30, n° 1–2,77–81 (1965) (in German)
CA, 64, 5752c (1966)

339. GROSSE-RUYKEN, H. and BOSHOLM, J.
Techn. Univ., Dresden, DDR (GDR)
PARTITION CHROMATOGRAPHY OF RARE EARTHS BY BIS(2-ETHYLHEXYL)
PHOSPHORIC ACID.IV. ACTIVATION ANALYTICAL DETERMINATION OF TRACES OF
HOLMIUM IN DYSPROSIUM AND OF TRACES OF LUTETIUM IN YTTERBIUM.
Kernenergie, 8, 224–6 (1965) (in German)

340. GROSSE-RUYKEN, H., BOSHOLM, J., and REINHARDT, G.
Techn. Univ., Dresden, DDR (GDR)
PARTITION CHROMATOGRAPHY OF RARE EARTHS BY BIS(2-ETHYLHEXYL)
PHOSPHORIC ACID.V. PREPARATION OF RADIOCHEMICAL PURE GADOLINIUM AND
DYSPROSIUM OXIDE.
Isotopenpraxis, 1, 124–5 (1965) (in German)

341. GWOŹDŹ, R. and SIEKIERSKI, S.
Inst. of Nuclear Research, Polish Academy of Sciences, Warsaw, Poland
SEPARATION OF VARIOUS OXIDATION STATES OF PLUTONIUM BY REVERSED-
PHASE PARTITION CHROMATOGRAPHY.
Report N° 168/V. 1960. 8 p.
Also published in Nukleonika, 5, 671–5 (1960)
CA, 60, 11355a (1964)

342. HAMLIN, A. G. and ROBERTS, B. J.
UKAEA, Capenhurst Works, Ches., Great Britain
SEPARATION OF URANIUM BY REVERSED-PHASE PARTITION CHROMATOGRAPHY.
Nature, 185,527–8 (1960)

343. HAMLIN, A. G., ROBERTS, B. J., LOUGHLIN, W., and WALKER, S. G.
UKAEA, Capenhurst Works, Ches., Great Britain
SEPARATION OF URANIUM BY REVERSED-PHASE PARTITION CHROMATOGRAPHY ON
A KEL-F COLUMN.
Anal. Chem., 33, 1547–52 (1961)

344. HAMLIN, A. G. and ROBERTS, B. J.
UKAEA, London, Great Britain
IMPROVEMENTS IN OR RELATING TO THE PRODUCTION OF URANIUM VALUES.
British Patent 900, 113
Filing Date: Nov. 25, 1960, Appl. Date: Dec. 17, 1960
Publ. Date: July 4, 1962
NSA, 16, 27149 (1962)

345. HANSEN, P. G., NIELSEN, H. L. and WILSKY, K.
Research Establishment Ris ø, Denmark
SEARCH FOR THE ISOMERIC STATE OF ^{144}Pr.
Paper presented at the Discussion on Nuclear Chemistry, Oxford, Sept. 1962
AERE-M-1078 p.63—4(1962)

346. HANSEN, R. K.
Iowa State Univ., Ames, Iowa, USA
SEPARATION AND DETERMINATION OF TRACE METAL IMPURITIES IN RARE-EARTH-
OXIDES.
Thesis
IS-T-88. 1966. 57 p.

347. HAYES, T. J. and HAMLIN, A. G.
UKAEA, Selwick, Great Britain
REVERSED-PHASE PARTITION CHROMATOGRAPHY OF URANIUM.
Analyst, 87, 770—7 (1962)
CA, 58, 921 f (1963)

348. HEDRICK, C. E.
Iowa State Univ., Ames, Iowa, USA
SEPARATIONS BY REVERSED-PHASE CHROMATOGRAPHY.
Univ. Microfilms (Ann Arbor, Mich.), Order N° 65-3795. 107 p.
Dissertation Abstr., 25, n° 10, 5523 (1965)

349. HEDRICK, C. E. and LERNER, M. W.
New Brunswick Lab., New Brunswick, New Jersey, USA
RAPID SEPARATION OF URANIUM BY REVERSED-PHASE CHROMATOGRAPHY AND
MEASUREMENT BY OXIDIMETRIC TITRATION AFTER FERROUS REDUCTION IN
PHOSPHORIC ACID.
NBL-250. April 1970. pp. 2—8.

350. HERMANN, A. and HUEBENER, S.
Zentralinstitut für Kernforschung, Rossendorf, DDR (GDR)
INVESTIGATION OF SOME CHROMATOGRAPHIC SUPPORTS FOR EXTRACTION CHRO-
MATOGRAPHY OF URANIUM AND FISSION PRODUCTS WITH TRIBUTYL PHOSPHATE.
Paper presented at the 7th Radiochemical Conference at Mariánské Lázné, CSSR, 24—28 Apr. 1973

351. HERMANN, A. and HUEBENER, S.
Zentralinstitut für Kernforschung, Rossendorf, DDR (GDR)
INVESTIGATIONS ON SUPPORT MATERIALS FOR THE EXTRACTION CHROMATO-
GRAPHY AND THE SEPARATION OF URANIUM, PLUTONIUM AND FISSION PROD-

446

UCTS ON TRIBUTYL PHOSPHATE/POLYTETRAFLUOROETHYLENE COLUMNS.
ZfK 252. 1973. 37 p. (in German)

352. HERRMANN, E., GROSSE-RUYKEN, H., LEBEDEV, N. A., and KHALKIN, V. A.
Joint Inst. for Nuclear Research, Dubna, USSR
EXTRACTION OF NEUTRON-DEFICIENT ISOTOPES OF THE RARE EARTH ELEMENTS
OF THE CERIUM SUBGROUP FROM ERBIUM EXPOSED TO 680 MEV PROTONS.
JINR-P-1455. 1963. 13 p.
NSA, 18, 10182 (1964)
Also published in Radiokhimiya, 6, 756–62 (1964) (in Russian)
CA, 62, 8612d (1965)

353. HERRMANN, E.
Techn. Univ., Dresden, DDR (GDR)
SOME PRINCIPLES OF EXTRACTION CHROMATOGRAPHIC MACRO-MICRO SEPARATION
OF RARE EARTH ELEMENTS.
Paper presented at the Conference on Radiochemistry, Bratislava, 1966
CA, 68, 56187 (1968)

354. HERRMANN, E., PFREPPER, G., and CHRISTOV, D.
Joint Inst. for Nuclear Research, Dubna, USSR
NEW ISOTOPES ^{141}Sm AND ^{140}Sm.
Radiochim. Acta, 7, 10–12 (1967)

355. HERRMANN, E.
Techn. Univ., Dresden, DDR (GDR)
A CONTRIBUTION TO THE SEPARATION OF RARE EARTHS BY EXTRACTION
CHROMATOGRAPHY WITH DI(2-ETHYLHEXYL) PHOSPHORIC ACID (HDEHP). I. SILICA
GEL AS A SUPPORT MATERIAL FOR THE STATIONARY PHASE.
J. Chromatog., 38, 498–507 (1968) (in German)
Paper presented at the Radioanalytical Conference at Stary Smokovec, 23-26 April 1968

356. HERRMANN, G.
Inst. for Inorg. and Nucl. Chem., Univ. of Mainz, BRD (FRG)

TECHNIQUES FOR RAPID CHEMICAL SEPARATIONS.
Paper presented at the Third Int. Transplutonium Element Symposium, Argonne, (Oct. 1971)
AED-CONF-71-386-002. 1971. 8 p., 10 figs, 1 table.

357. HEUNISCH, G. W.
Research and Development Dept., Continental Oil Company, Ponca City, Okla., USA
SEPARATION OF URANIUM FROM PHOSPHORIC ACID BY REVERSED-PHASE COLUMN
CHROMATOGRAPHY.
Anal. Chim. Acta, 45, 133–6 (1969)

358. HOFFMAN, D. C., MICHELSEN, O. B., and DANIELS, W. R.
Los Alamos Sci. Lab., Los Alamos, New Mexico, USA
BETA-DECAY OF SOME CERIUM ISOTOPES FAR TO THE NEUTRON-RICH SIDE OF
STABILITY.
USAEC Report LA-DC-7869. 1965. 21 p.
CONF. 660817-2

359. HOLCOMB, H. P.
Savannah River Lab., E. I. du Pont de Nemours and Co., Aiken, South Carolina, USA

DETERMINATION OF ^{243}Am AND ^{252}Cf IN CURIUM PROCESS STREAMS.
Paper presented at the 155th ACS National Meeting, San Francisco, California, USA, April 1-5, 1968

360. HONJO, T. and KIBA, Tashiyasu
Fac. Sci., Kanazawa Univ., Kanazawa, Japan
ANALYTICAL APPLICATION OF SULFUR ANALOGS OF β-DIKETONES. II. SEPARA-
TION OF MERCURY(II), COBALT(II), AND ZINC(II) AS THEIR STTA (1,1,1-TRIFLUORO-
4-(2-THIENYL)-4-MERCAPTOBUT-3-EN-2-ONE) COMPLEXES BY EXTRACTION CHROMATO-
GRAPHY.
Bull. Chem. Soc. Jap., 46, 1694–8 (1973)
CA, 79. 70677 (1973)

361. HORNBECK, R. F.
Lawrence Radiation Lab. Univ. of California, Livermore, Calif., USA
GRAFT COPOLYMERS AS SUBSTRATES IN COLUMN EXTRACTIONS. I. NON-CHROMATO-
GRAPHIC SEPARATIONS ON EXTRACTION COLUMNS.
J. Chromatog., 30, 438–46 (1967)

362. HORNBECK, R. F.
Lawrence Radiation Lab., Univ. of California, Livermore, Calif., USA
GRAFT COPOLYMERS AS SUBSTRATES IN COLUMN EXTRACTIONS. II. CHARACTER-
ISTICS OF HYDROPHILIC-ORGANOPHILIC COLUMNS IN CHROMATOGRAPHIC SEP-
ARATIONS.
J. Chromatog., 30, 447–58 (1967)

363. HORWITZ, E. P., ORLANDINI, K. A., and BLOOMQUIST, C. A. A.
Argonne Nat. Lab., Argonne, Illinois, USA
SEPARATION OF AMERICIUM AND CURIUM BY EXTRACTION CHROMATOGRAPHY
USING A HIGH-MOLECULAR-WEIGHT QUATERNARY AMMONIUM NITRATE.
Inorg. Nucl. Chem. Letters, 2, 87–91 (1966)
NSA, 20, 29337 (1966)

364. HORWITZ, E. P., BLOOMQUIST, C. A. A., ORLANDINI, K. A., and HENDERSON, D. J.
Argonne Nat. Lab., Argonne, Illinois, USA
THE SEPARATION OF MILLIGRAM QUANTITIES OF AMERICIUM AND CURIUM BY
EXTRACTION CHROMATOGRAPHY.
Radiochim. Acta, 8, 127–32 (1967)

365. HORWITZ, E. P., SAURO, L. J., and BLOOMQUIST, C. A. A.
Argonne Nat. Lab., Argonne, Illinois, USA
THE EXTRACTION CHROMATOGRAPHY OF CALIFORNIUM AND EINSTEINIUM WITH A
HIGH MOLECULAR WEIGHT QUATERNARY AMMONIUM NITRATE.
J. Inorg. Nucl. Chem., 29, 2033–40 (1967)

366. HORWITZ, E. P., BLOOMQUIST, C. A.A., BUZZELL, J. A., and HARVEY, H. W.
Argonne Nat. Lab., Argonne, Illinois, USA
THE SEPARATION OF MICROGRAM QUANTITIES OF ^{252}Cf AND ^{248}Cm BY EXTRAC-
TION CHROMATOGRAPHY IN A HIGH LEVEL CAVE.
ANL-7546. 1969. 16 p.

367. HORWITZ, E. P., BLOOMQUIST, C. A. A., and GRIFFIN, H. E.
Argonne Nat. Lab., Argonne, Illinois, USA
THE PREPARATION OF HIGH-PURITY ^{242}Cm IN MULTICURIE QUANTITIES.
ANL-7569. 1969. 12 p.

448

368. HORWITZ, E. P., BLOOMQUIST, C. A. A., and HENDERSON, D. J.
Argonne Nat. Lab., Argonne, Illinois, USA
THE EXTRACTION CHROMATOGRAPHY OF CALIFORNIUM, EINSTEINIUM, AND
FERMIUM WITH DI(2-ETHYLHEXYL) ORTHOPHOSPHORIC ACID.
J. Inorg. Nucl. Chem., 31, 1149–66 (1969)

369. HORWITZ, E. P., BLOOMQUIST, C. A. A., HENDERSON, D. J., and NELSON, D. E.
Argonne Nat. Lab., Argonne, Illinois, USA
THE EXTRACTION CHROMATOGRAPHY OF AMERICIUM, CURIUM, BERKELIUM AND
CALIFORNIUM WITH DI(2-ETHYLHEXYL) ORTHOPHOSPHORIC ACID.
J. Inorg. Nucl. Chem., 31, 3255–71 (1969)

370. HORWITZ, E. P. and BLOOMQUIST, C. A. A.
Argonne Nat. Lab., Argonne, Illinois, USA
THE TRACER CHEMISTRY OF TRIVALENT MENDELEVIUM WITH DI(2-ETHYLHEXYL)
ORTHOPHOSPHORIC ACID.
Inorg. Nucl. Chem. Letters, 5, 753–59 (1969)

371. HORWITZ, E. P.
Argonne Nat. Lab., Argonne, Illinois, USA
HIGH SPEED-HIGH EFFICIENCY SEPARATIONS OF THE TRANSPLUTONIUM ELEMENTS
USING LIQUID-LIQUID CHROMATOGRAPHY.
Paper presented at the Third Int. Transplutonium Element Symposium, Argonne, Ill., Oct.
20–22, 1971

372. HORWITZ, E. P. and BLOOMQUIST, C. A. A.
Argonne Nat. Lab., Argonne, Illinois, USA
PREPARATION, PERFORMANCE, AND FACTORS AFFECTING BAND SPREADING
OF HIGH EFFICIENCY EXTRACTION CHROMATOGRAPHIC COLUMNS FOR ACTINIDE
SEPARATIONS.
J. Inorg. Nucl. Chem., 34, 3851–71 (1972)

373. HORWITZ, E. P. and BLOOMQUIST, C. A. A.
Argonne Nat. Lab., Argonne, Illinois, USA
HIGH SPEED-HIGH EFFICIENCY SEPARATION OF THE TRANSPLUTONIUM ELEMENTS
BY EXTRACTION CHROMATOGRAPHY.
J. Inorg. Nucl. Chem., 35, 271–84 (1973)

374. HUEBENER, S. and HERMANN, A.
Zentralinstitut für Kernforschung, Rossendorf, DDR (GDR)
SEPARATION OF PLUTONIUM AND URANIUM FROM EACH OTHER AND FROM
FISSION PRODUCTS BY EXTRACTION CHROMATOGRAPHY.
Paper presented at the 7th Radiochemical Conference, Mariánské Lázné, CSSR,
24-28 Apr. 1973

375. HUFF, E. A.
Argonne Nat. Lab., Argonne, Illinois, USA
TRACE IMPURITY ANALYSIS OF THORIUM-URANIUM AND PLUTONIUM-THORIUM-
URANIUM ALLOYS BY ANION EXCHANGE PARTITION CHROMATOGRAPHY.
Anal. Chem., 37, 533–6 (1965)
CA, 63, 3599h (1965)

376. HUFF, E. A. and KULPA, S. J.
Argonne Nat. Lab., Argonne, Illinois, USA

TRACE IMPURITY ANALYSIS OF PLUTONIUM-URANIUM-ZIRCONIUM ALLOYS BY
ANION EXCHANGE-PARTITION CHROMATOGRAPHY.
Anal. Chem., 38, 939–40, (1966)
CA, 65, 7988a (1966)

377. HUFF, E. A.
Argonne Nat. Lab., Argonne, Illinois, USA
PARTITION CHROMATOGRAPHIC STUDIES OF AMERICIUM, YTTRIUM, AND THE
RARE EARTHS IN THE TRICAPRYLMETHYLAMMONIUM THIOCYANATE - AM-
MONIUM THIOCYANATE SYSTEM.
J. Chromatog., 27, 229–36 (1967)

378. HULET, E. K.
Institutt for Atomenergi, Kjeller, Norway
AN INVESTIGATION OF THE EXTRACTION CHROMATOGRAPHY OF AMERICIUM(VI)
AND BERKELIUM(IV).
KR-56. 1963. 13 p.
Later published in J. Inorg. Nucl. Chem., 26, 1721–1 (1964)

379. HULET, E. K., LOUGHEED, R.W., BRADY, J. D., STONE, R. E., and COOPS, M. S.
Lawrence Radiation Laboratory, University of California, Livermore, Calif., USA
MENDELEVIUM DIVALENCY AND OTHER CHEMICAL PROPERTIES.
Science, 158, 486–7 (1967)

380. HUNT, L. Ph.
Iowa State Univ., Ames, Iowa, USA
NEUTRON ACTIVATION ANALYSIS OF TRACE RARE EARTHS IN HOLMIUM OXIDE.
Thesis
IS-T-120. Febr. 1967. 90 p.
NSA, 21, 30346 (1967)

381. ITANI, S. M.
Iowa State Univ., Ames, Iowa, USA
CHROMATOGRAPHIC SEPARATION AND SPECTROPHOTOMETRIC DETERMINATION
OF URANIUM.
Thesis
IS-T-385. May 1970. 32 p.

382. JASKOLSKA, H.
Inst. Badan Jad., Warsaw, Poland
SEPARATION OF CALCIUM, STRONTIUM, AND BARIUM BY REVERSED-PHASE PAR-
TITION CHROMATOGRAPHY WITH BIS(2-ETHYLHEXYL) PHOSPHATE.
Chem. Anal. (Warsaw), 14, 285–92 (1969) (in Polish)
CA, 71, 27106 (1969)

383. JENTZSCH, D., OESTERHELT, G., RÖDEL, E., and ZIMMERMANN, H. G.
Bodenseewerke Perkin-Elmer, Ueberlingen/Bodensee, BRD (FRG)
NEW TECHNIQUE FOR ION EXCHANGE CHROMATOGRAPHY.
Z. Anal. Chem., 205, 237–43 (1964) (in German)
CA, 62, 6e (1965)

384. JOON, K.
Institutt for Atomenergi, Kjeller, Norway
MASKING OF NITROSYL-RUTHENIUM NITRATES AND NITRITES IN NITRIC ACID
SOLUTIONS.
KR-119. 1967. 6 p.

385. JOON, K.
Institutt for Atomenergi, Kjeller, Norway
RuNO—NITRO COMPLEXES. CONVERSION RATES AND EXTRACTION COEFFICIENTS.
A. S. KERTES, and Y. MARCUS (eds) Solvent Extraction Research, pp. 323—31
John Wiley & Sons, Inc., 1969

386. JOON, K., DEN BOEF, R., HAALAND, J., STIJFHOORN, D. E., and DE WIT, R.
Institutt for Atomenergi, Kjeller, Norway
CHEMICAL ANALYSIS OF HIGH BURN-UP PuO_2 FUEL. PART I. METHODS.
KR—140. 1970. 25 p.

387. JOON, K., DEN BOEF, R., and DE WIT, R.
Institutt for Atomenergi, Kjeller, Norway
DETERMINATION OF ^{147}Pm IN IRRADIATED NUCLEAR FUEL.
J. Radioanal. Chem., 8, 101—6 (1971)

388. KALININ, S. K., YAKOVLEVA, G. A., NIKITIN, M. K., and KATYKHIN, G. S.
ISOLATION OF PLATINUM USING PARTITION CHROMATOGRAPHY ON FTOROPLAST-4.
Proc. Project Sci. Res. Inst. "Gipronickel" (Leningrad), 33, 192—4 (1967)

389. KALININ, S. K., KATYKHIN, G. S., NIKITIN, M. K., and YAKOVLEVA, G. A.
Leningrad State Univ., Leningrad, USSR
PARTITION CHROMATOGRAPHY ON FLUOROPLAST-4. SEPARATION OF TRACE
AMOUNTS OF BISMUTH FROM NICKEL.
Zh. Anal. Khim., 23, 1481—4 (1968) (in Russian)
CA, 70, 25380 (1969)

390. KALYAMIN, A. V.
ISOLATION OF BISMUTH, MERCURY AND SILVER FROM COMPLEX MIXTURES OF
THE ELEMENTS BY THE METHOD OF PARTITION CHROMATOGRAPHY.
Radiokhimiya, 5, 749—51 (1963)

391. KAMIN, G. J., O'LAUGHLIN, J. W., and BANKS, C. V.
Inst. for Atomic Research and Dept. of Chemistry, Iowa State Univ., Ames, Iowa, USA
SEPARATION AND DETERMINATION OF ZIRCONIUM IN NIOBIUM USING METHYLENE-
BIS (DI-N-HEXYLPHOSPHINE OXIDE).
J. Chromatog., 31, 292—5, (1967)
CA, 68, 26645 (1968)

392. KAMINSKI, E. E.
Dept. of Chemistry, Iowa State Univ., Ames, Iowa, USA
ANION EXCHANGE SEPARATIONS OF METAL IONS IN THIOCYANATE MEDIA.
IS-T-377. 1970. 101 p.
NSA, 25, 4137 (1971)

393. KAMINSKI, E. E. and FRITZ, J. S.
Dept. of Chemistry, Iowa State Univ., Ames, Iowa, USA
METAL ION SEPARATIONS IN THIOCYANATE MEDIA USING A LIQUID ANION
EXCHANGER.
J. Chromatog., 53, 345—53 (1970)

394. KARYAKIN, A. V., ANIKINA, L. I., and LE VIET BINH
Inst. Geochem. Anal. Chem., Moscow, USSR
DETERMINATION OF SMALL AMOUNTS OF GADOLINIUM IN EUROPIUM OXIDE
USING THE LUMINESCENCE OF CRYSTAL PHOSPHORS.
Zh. Anal. Khim., 24,1156–9(1969) (in Russian)
CA, 71, 119307 (1969)

395. KATYKHIN, G. S. and MOSKVIN, L. N.
PARTITION CHROMATOGRAPHY ON FTOROPLAST-4.SEPARATION OF La FROM Ba
AND Y FROM Sr.
Programme and Abstracts of Reports, 13th Conference on Nuclear Spectroscopy,
Izd. AN SSSR, Moscow-Leningrad, 1963, p. 28

396. KATYKHIN, G. S.
PARTITION CHROMATOGRAPHY ON FTOROPLAST-4.SEPARATION OF METALS FROM
CYCLOTRONIC TARGETS (W, Au, Os).
Programme and Abstracts of Reports, 13th Conference on Nuclear Spectroscopy,
Izd. AN SSSR, Moscow-Leningrad 1963, p. 28

397. KATYKHIN, G. S. and NIKITIN, M. K.
PARTITION CHROMATOGRAPHY ON FTOROPLAST-4.SEPARATION OF Y, Zr, Nb, Mo
AND Cd FROM A SILVER-TARGET.
Programme and Abstracts of Reports, 15th Conference on Nuclear Spectroscopy,
Izd. "Nauka", Leningrad, 1965, pp. 72–3

398. KATYKHIN, G. S. and NIKITIN, M. K.
PARTITION CHROMATOGRAPHY ON FTOROPLAST-4. SEPARATION OF BISMUTH
FROM LEAD.
Programme and Abstracts of Reports.,15th Conference on Nuclear Spectroscopy,
Izd. "Nauka", Leningrad, 1965 p. 73.

399. KATYKHIN, G. S., NIKITIN, M.K., SERGEEV, V. P., and KALININ, S. K.
EXTRACTION AND SEPARATION OF METALS OF THE PLATINUM GROUP IN THE
FORM OF COMPLEXES WITH TIN CHLORIDE; PARTITION CHROMATOGRAPHY ON
FTOROPLAST-4.
The Application of Extraction and Ion Exchange to the Analysis of Noble Metals,
Moscow, 1967, pp. 39–46.

400. KATYKHIN, G. S. and NIKITIN, M. K.
APPLICATION OF POROUS FLUOROPLAST-4 IN DISTRIBUTION COLUMN CHROMA-
TOGRAPHY.
Radiokhimiya, 10,474–5(1968) (in Russian)

401. KATYKHIN, G. S. and NIKITIN, M. K.
RAPID SEPARATION OF THE RHODIUM FRACTION FROM AN IRRADIATED CAD-
MIUM-TARGET.
Paper presented at the 19th Conference on Nuclear Spectroscopy and the Structure of
the Atomic Nucleus,
Yerevan, USSR, 27 Jan. – 4 Febr. 1969

402. KATYKHIN, G. S. and NIKITIN, M. K.
SEPARATION OF THE TECHNETIUM-FRACTION FROM AN IRRADIATED CADMIUM-
TARGET.

Paper presented at the 19th Conference on Nuclear Spectroscopy and the Structure of the
Atomic Nucleus,
Yerevan, USSR, 27 Jan.–4 Febr. 1969

403. KENNEDY, D. C.
Ames Lab., Iowa, USA
SULFOXIDES AS SOLVENT EXTRACTION REAGENTS FOR THE ANALYTICAL SEPA-
RATION OF METAL IONS.
Thesis
IS-T-293. May 1969. 156 p.
NSA, 23, 43382 (1969)

404. KOMORI, T., YOSHIDA, H., GUNJI, K., TOIDA, K., and TAMURA, S.
JAERI, Tokai Mura, Naka-gun, Ibaraki, Japan
DETERMINATION OF RARE EARTH ELEMENTS IN NUCLEAR FUEL MATERIALS BY
ISOTOPE DILUTION METHOD. II. DETERMINATION OF CERIUM, GADOLINIUM, DYS-
PROSIUM, ERBIUM, AND YTTERBIUM.
Bunseki Kagaku, 15, 589–94 (1966) (in Japanese)
CA, 66, 71535 (1967)
English translation: NASA TT F–13, 177. 1970. 15 p.

405. KOOI, J., BODEN, R., and WIJKSTRA, J.
CEN-SCK, Mol, Belgium
SEPARATION OF AMERICIUM (CURIUM), BERKELIUM, AND CALIFORNIUM BY EXTRAC-
TION CHROMATOGRAPHY.
J. Inorg. Nucl. Chem., 26,2300–2(1964)
CA, 62, 4588b (1965)

406. KOOI, J. and BODEN, R.
CEN-SCK, Mol, Belgium
SIMPLE QUANTITATIVE SEPARATION OF BERKELIUM FROM CERIUM.
Radiochim. Acta, 3, 226 (1964)

407. KOOI, J., GANDOLFO, J. M., WÄCHTER, N., and WIJKSTRA, J.
Euratom, Chem. Dept., Ispra, Italy
BODEN, R., HECQ, R., VANHOOF, E., and LEYNEN, M.
CEN-SCK, Mol, Belgium
ISOLATION OF MICROGRAM AMOUNTS OF BERKELIUM AND CALIFORNIUM FROM
AMERICIUM.
EUR 2578.e. 1965. 46 p.

408. KOOI, J.
Euratom, Chem. Dept., Ispra, Italy
APPLICATION OF EXTRACTION CHROMATOGRAPHY IN THE PROCESSING OF IRRADI-
ATED AMERICIUM.
Radiochim. Acta, 5, 91–9 (1966)

409. KOPTA, S., NIEWODNICZANSKI, H., PETRYNA, T., and STRONSKI, I.
Inst. of Nucl. Physics, Kracow, Poland
PRODUCTION OF CARRIER-FREE RADIOISOTOPES FROM TARGETS IRRADIATED
WITH DEUTERONS OR ALPHA-PARTICLES IN THE U-120 CRACOW CYCLOTRON.
PART 2. PREPARATION OF RADIOISOTOPES OF GADOLINIUM, SAMARIUM,
AND THULIUM FROM TARGETS OF SAMARIUM, NEODYMIUM, AND HOLMIUM.
INP-No. 663/PL/C. 1968. 18 p.
Also published in Nukleonika, 14, 789–97 (1969)

410. KOROTKIN, Yu. S.
Joint Inst. for Nucl. Research, Dubna, USSR
RARE EARTH AND TRANSURANIUM ELEMENT SEPARATION BY PARTITION CHRO-
MATOGRAPHY WITH TETRABUTYLHYPOPHOSPHORIC ACID.
JINR-P6-6401.1972. 7 p. (in Russian, with abstract in English)
KFK-tr-409. 1972. 3 p., 4 figs. (in German)

411. KOSYAKOV, V. N. and ERIN, E. A.
Inst. Atomnoi Energii, Moscow, USSR
SEPARATION OF TRANSPLUTONIUM AND RARE EARTH ELEMENTS BY EXTRAC-
TION WITH D2EHPA FROM DTPA SOLUTIONS.
IAE-2090. 1971. 15 p. (in Russian)
NSA, 26, 35989 (1972)

412. KOTLINSKA-FILIPEK, B. and SIEKIERSKI, S.
Inst. of Nuclear Research, Polish Academy of Sciences, Warsaw, Poland
THE SEPARATION OF TRACE AMOUNTS OF PROMETHIUM, EUROPIUM, AND TERBIUM
FROM MACROAMOUNTS OF ERBIUM.
Nukleonika, 8, 607–16(1963) (in Polish)

413. KREFELD, R., ROSSI, G., and HAINSKI, Z.
Euratom, Chemistry Dept., Ispra, Italy
ANALYSIS OF PURIFIED URANIUM COMPOUNDS FOR NANOGRAM QUANTITIES OF
BERYLLIUM AND CERTAIN RARE EARTHS.
Mikrochim. Acta, n° 1, 133–40 (1965)

414. KROEBEL, R. and MEYER, A.
Farbenfabriken Bayer AG, Leverkusen, BRD (FRG)
PROCEDURE FOR THE PREPARATION OF ADSORBENTS.
German Patent Office, Offenlegungsschrift 2162951. Appl. date: 18 Dec. 1971. Publ. date: 20
June 1973 (in German)

415. KŘTIL, J., BEZDEK, M., and MENCL, J.
Nuclear Research Inst., Řež ČSSR
DETERMINATION OF RADIOCERIUM IN FISSION PRODUCTS BY EXTRACTION CHRO-
MATOGRAPHY USING DI-(2-ETHYLHEXYL) PHOSPHORIC ACID.
J. Radioanal. Chem., 1, 369–78(1968)
(also published as UJV 1978. April 1968. 9 p.)

416. KŘTIL, J. and BEZDEK, M.
Nuclear Research Inst., Řež.,ČSSR
DETERMINATION OF THE PERCENT BURN-UP OF FUEL BY A RADIOCHEMICAL METHOD
USING CESIUM-137 AND CERIUM-144.
Jad. Energ. 16, (6),181–8,(1970) (in Czech.)
CA, 73, 51289 (1970)

417. KŘTIL, J., BULOVIC, V.*, and MAXIMOVIC, Z.*
Nuclear Research Inst.,Řež.,ČSSR
*Boris Kidric Inst. of Nuclear Sciences, Beograd-Vinca, Yugoslavia
THE DETERMINATION OF PERCENT BURN-UP OF THE NUCLEAR FUEL ON THE BASIS
OF THE MEASURING OF THE STABLE ISOTOPES OF NEODYMIUM.
Jaderna Energie, 17, 10–5(1971)

418. KŘTIL, J., BULOVIC, V.*, MENCL, J., and MORAVEC, A.
Nuclear Research Inst.,Řež, ČSSR
*Boris Kidric Inst. of Nuclear Sciences, Beograd- Vinca, Yugoslavia

SEPARATION OF THE STABLE ISOTOPES OF NEODYMIUM FROM IRRADIATED URA-
NIUM FOR BURN-UP DETERMINATION.
Radiochem. Radioanal. Letters, 9,335–42(1972)

419. KUZMIN, N. M., VLASOV, V. S., and BOKOVA, T. A.
RADIOACTIVATION ANALYSIS OF ARSENIC.II. EXTRACTION-CHROMATOGRAPHIC
SEPARATION OF MACRO- AND MICROELEMENTS.
Zh. Anal. Khim., 27,1807–10(1972) (in Russian)

420. KÝRŠ, M. and KADLECOVA, L.
Nuclear Research Inst.,Řež, ČSSR
SEPARATION OF ALKALI METALS BY EXTRACTION CHROMATOGRAPHY WITH DI-
PICRYLAMINE AND NITROBENZENE.
J. Radioanal. Chem., 1,103–12(1968)

421. LATWESEN, G.
Inst. for Atomic Research and Dept. of Chem. Iowa State Univ., Ames, Iowa, USA
THE SEPARATION OF TIN FROM OTHER ELEMENTS BY REVERSED-PHASE CHRO-
MATOGRAPHY.
Thesis
IS-T-116. Aug. 1966. 54 p.

422. LEE, D. A.
ORNL, Oak Ridge, Tenn., USA
SEPARATION OF LITHIUM FROM OTHER ALKALI METAL IONS BY REVERSED-PHASE
CHROMATOGRAPHY.
J. Chromatog., 26,342–5(1967)

423. LESZKO, M. and KARCZ, W.
Univ. Jagiellonian, Krakow, Poland
SEPARATION OF COPPER, LEAD, ZINC, AND CADMIUM BY EXTRACTION CHROMATOG-
RAPHY USING AMBERLITE LA-2.
Zesz. Nauk. Uniw. Jagiellon., Pr. Chem., 1970, N° 15,183–8,(1970) (in Polish)
CA, 74, 49275 (1971)

424. LEVIN, V. I., KOZLOVA, M. D., and MALININ, A. B.
PREPARATION OF CARRIER-FREE [111]Ag. THE FORMATION OF [111]Ag AND [110m]Ag IN
THE NEUTRON IRRADIATION OF Pd.
Radiokhimiya, 7,673–7(1965) (in Russian)
CA, 64, 16970h (1966)

425. LEVIN, V. I., KURCHATOVA, L. N., and MALININ, A. B.
Inst. of Biophysics, Ministry of Public Health, USSR
SEPARATION OF YTTRIUM FROM STRONTIUM.
U.S.S.R. Patent 225, 343. Appl. Date: 16 May 1966, Publ. Date: 29 Aug. 1968.
CA, 70, 39355 (1969)

426. LEVIN, V. I., KURCHATOVA, L. N., and MALININ, A. B.
SEPARATION OF YTTRIUM-88 FROM A STRONTIUM TARGET WITHOUT A CARRIER.
Radiokhimiya, 11, 210–4 (1969) (in Russian)
CA, 71, 66396 (1969)

427. LIESER, K. H. and BERNHARD, H.
Lehrstuhl für Kernchemie, Technische Hochschule Darmstadt, BRD (FRG)
SEPARATION OF CALCIUM AND STRONTIUM BY COLUMN CHROMATOGRAPHY.
Z. Anal. Chem., 219,401–8(1966)

428. LIS, S., JOZEFOWICZ, E. T., and SIEKIERSKI, S.
Dept. of Radiochem. and Dept. of Reactor Physics, Inst. of Nucl. Research, Warsaw, Poland
SEPARATION OF NEPTUNIUM FROM NEUTRON IRRADIATED URANIUM BY REVERSED-PHASE PARTITION CHROMATOGRAPHY IN THE TBP-HNO$_3$ SYSTEM.
J. Inorg. Nucl. Chem., 28, 199–207 (1966)

429. LOVERIDGE, B. A. and OWENS, A. F.
Chem. Division, AERE Harwell, Berks., Great Britain
THE DETERMINATION OF ^{65}Zn IN HARWELL EFFLUENT.
AERE-R-3945. 1962. 17 p.

430. LOVERIDGE, B. A., GORDON, M. S., and WEAVER, J. R.
Chem. Division, AERE, Harwell, Berks., Great Britain
THE DETERMINATION OF ACTINIUM-227 IN EFFLUENT.
AERE-R-4747. 1964. 25 p.

431. LUGININ, V. A. and TSERKOVNITSKAYA, I. A.
Vestn. Leningrad. Univ., USSR
SEPARATION OF URANIUM(IV) AND (VI) BY THE METHOD OF PARTITION COLUMN CHROMATOGRAPHY.
Radiokhimiya, 12, 898–9(1970) (in Russian)
NSA, 26, 1955 (1972)

432. MA, H-C., NI, C-M., and LIANG, S-C.
Inst. of Chemistry, Academia Sinica, Peking, China
REVERSED-PHASE PARTITION CHROMATOGRAPHY. SEPARATION OF THORIUM, SCANDIUM, YTTRIUM, AND LANTHANIDE ELEMENTS.
Ko Hsueh Tung Pao, 1964, 64 (1964) (in Chinese)

433. MA, H-C., NI, C-M., and LIANG, S-C.
Inst. of Chemistry, Academia Sinica, Peking, China
THE SEPARATION OF THORIUM, SCANDIUM AND HEAVY LANTHANONS BY MEANS OF REVERSED-PHASE CHROMATOGRAPHY.
Hua Hsueh Hsueh Pao, 30, 388–93(1964) (in Chinese)
also Scientia Sinica, 14, 1176–83(1965)

434. MA, H-C. and NI, C-M.
Inst. of Chemistry, Academia Sinica, Peking, China
THE SEPARATION AND DETERMINATION OF TRACE AMOUNTS OF URANIUM IN WASTE WATER.
Yuan Zi Neng, 1965, 159–62(1965) (in Chinese)
CA, 64, 9422a (1966)

435. McCLENDON, L. T. and LAFLEUR, P. D.
Nat. Bureau of Standards, Anal. Chem. Div., Washington, D.C., USA
DETERMINATION OF RARE EARTHS IN STANDARD REFERENCE MATERIAL GLASS USING NEUTRON ACTIVATION ANALYSIS AND REVERSED-PHASE CHROMATOGRAPHY.
AED-CONF-72-347-051. 1972. 6 p., 1 fig., 9 ref.
Paper presented at the Int. Conf. on Modern Trends in Activation Analysis, Saclay, France, October 2-6, 1972

436. MALTSEVA, N. S. and GAVRILOV, K. A.
Joint Inst. for Nucl. Research, Dubna, USSR

CONCENTRATION OF CESIUM AND FRANCIUM BY DISTRIBUTION AND ADSORPTION CHROMATOGRAPHY.
Radiokhimiya, 14,773−8(1972) (in Russian)
Also published in J. Radioanal. Chem., 10, 173−80 (1972)

437. MALVANO, R., GROSSO, P., and ZANARDI, M.
Centro Ricerche Nucleari, Sorin, Saluggia, Italy
SINGLE-STEP RADIOCHEMICAL SEPARATIONS BY COLUMN PROCEDURES IN ACTIV-ATION ANALYSIS.
Anal. Chim. Acta, 41, 251−8 (1968)

438. MAPPER, D. and FRYER, J. R.
AERE, Harwell, Didcot, Great Britain
THE USE OF DITHIZONE-CELLULOSE ACETATE COLUMNS OF INCREASED CAPACITY IN THE DETERMINATION OF INDIUM BY PIERCE AND PECK'S NEUTRON ACTIV-ATION METHOD.
Analyst, 87, 297 (1962)
CA, 57, 7894a (1962)

439. MARKOV, V.
Vernadsky Inst. of Geochem. and Anal. Chem., Moscow, USSR
CHROMATOGRAPHIC SEPARATION OF URANIUM AND PLUTONIUM FROM REPRO-CESSING PLANT LIQUID WASTE FOLLOWED BY AUTOMATIC MEASUREMENT OF PLUTONIUM AMOUNTS BY PROBE TYPE ALPHA DETECTOR AND THE URANIUM BY PHOTOMETRIC METHODS.
Final report for the period 15 Dec. 1969−14 Dec. 1971
IAEA-R-880-F. 1972. 66 p.

440. MARSH, S. F.
Idaho Nuclear Corp., Idaho Falls, Idaho, USA
SEPARATION OF LANTHANIDE FISSION PRODUCTS FROM NUCLEAR FUELS BY EXTRACTION CHROMATOGRAPHY AND CATION EXCHANGE FOR ISOTOPE DILUTION MASS SPECTROMETRIC ANALYSIS.
Anal. Chem., 39,641−5(1967)

441. MARSH, S. F.
The Babcock and Wilcox Comp.,Lynchburg, Virginia, USA
RAPID SEQUENTIAL SEPARATION OF URANIUM, PLUTONIUM, AMERICIUM-CURIUM, CESIUM, AND NEODYMIUM FOR NUCLEAR FUEL BURN-UP ANALYSIS.
Paper presented at the 155th ACS National Meeting, San Francisco, California, USA, April 1−5, 1968

442. MAŠTALKA, A.
Czechoslovak Acad. of Sciences, Řež, ČSSR
DETERMINATION OF TANTALUM IN ROCKS.
Chem. Listy, 59, 1446−9 (1965) (in Czech)
CA, 64, 10395f (1966)

443. MAŠTALKA, A.
Inst. of Nuclear Research, Řež, ČSSR
ISOLATION OF CERIUM SPALLATION PRODUCTS.
Radiochem. Conf. Abstr. Papers, Bratislava, 1966, pp 11−2.
CA, 68, 25631 (1968)

444. MAŠTALKA, A., KHALKIN, V. A., and LEBEDEV, N. A.
Czechoslovak Acad. of Sciences, Řež, ČSSR
Joint Inst. of Nucl. Research, Dubna, USSR
RADIOCHEMICAL ISOLATIONS. IV. SEPARATION OF TARGET SUBSTANCE FROM THE
PRODUCTS OF CERIUM SPALLATION.
Collect. Czech. Chem. Commun., 32, 1604—8 (1967) (in German)
CA, 67, 28114 (1967)

445. MAŠTALKA, A., KHALKIN, V. A., and LEBEDEV, N. A.
Czechoslovak Acad. of Sciences, Řež, ČSSR
Joint Inst. of Nucl. Research, Dubna, USSR
RADIOCHEMICAL ISOLATIONS. V. THE ISOLATION OF LANTHANUM, BARIUM, CESIUM,
AND TELLURIUM FROM THE PRODUCTS OF CERIUM SPALLATION.
Collect. Czech. Chem. Commun., 33, 2332—6 (1968)
CA, 69, 48495 (1968)

446. MICHELSEN, O. B. and HOFFMAN, D. C.
Institutt for Atomenergi, Kjeller, Norway
RAPID SEPARATION OF CERIUM AND MOLYBDENUM FROM FISSION PRODUCTS.
Radiochim. Acta, 6,165—7(1966)
NSA, 21, 17729 (1967)

447. MICHELSEN, O. B. and SMUTZ, M.
Inst. for Atomic Research and Dept. of Chem. Engin., Iowa State Univ., Ames, Iowa, USA
EXTRACTION Of SOME RARE EARTH ELEMENTS WITH DI(2-ETHYLHEXYL) PHOS-
PHORIC ACID AT HIGH ACIDITIES.
IS-2586. 1970. 11 p.
CONF-710417-6
CA, 77, 106063 (1972)

448. MICHELSEN, O. B. and SMUTZ, M.
Inst. for Atomic Research and Dept. of Chem. Engin., Iowa State Univ., Ames, Iowa, USA
SEPARATION OF YTTRIUM, HOLMIUM, AND ERBIUM WITH DI(2-ETHYLHEXYL)
PHOSPHORIC ACID IN CHLORIDE AND NITRATE SYSTEMS.
J. Inorg. Nucl. Chem., 33,265—78 (1971)
Also published in Solvent Extraction. Proc. Int. Solvent Extraction Conf., ISEC 1971, The Hague,
19-23 Apr. 1971.
Vol. 2, pp. 939—49, Soc. Chem. Ind., London, 1971

449. MIKHAILICHENKO, A. I. and PIMENOVA, R. M.
SEPARATION OF RARE EARTH ELEMENTS BY PARTITION CHROMATOGRAPHY WITH
UTILIZATION OF BIS(2-ETHYLHEXYL) ORTHOPHOSPHATE SORBED ON SILICA GEL.
Zh. Prikl. Khim., 42,1010—16(1969) (in Russian)
NSA, 23, 40808 (1969)

450. MIKULSKI, J. and STRONSKI, I.
Inst. of Nuclear Physics, Krakow, Poland
THE SEPARATION OF ZINC, MANGANESE, AND COBALT FROM IRON BY REVERSED
PHASE CHROMATOGRAPHY.
Nukleonika, 6,295—8(1961)
CA, 60, 7432f (1964)

451. MIKULSKI, J. and STRONSKI, I.
Inst. of Nuclear Physics, Krakow, Poland
THE SEPARATION OF TIN, TELLURIUM, AND ANTIMONY BY REVERSED-PHASE
CHROMATOGRAPHY.
Nukleonika, 6,775–8(1961) (in Russian)
CA, 57, 11a (1962)

452. MIKULSKI, J., GAVRILOV, K. A., and KNOBLOCH, V.
Inst. of Nuclear Physics, Krakow, Poland
PARTITION CHROMATOGRAPHY OF RARE EARTH AND TRANSURANIUM ELEMENTS
IN THE SYSTEM TBHP-TBPP/NITRIC ACID.
Nukleonika, 9,785–94(1964) (in Russian)
NSA, 19, 540 (1965)

453. MIKULSKI, J., GAVRILOV, K. A., and KNOBLOCH, V.
Inst. of Nuclear Physics, Krakow, Poland
Inst. of Nucl. Research, Czechoslovak Academy of Sciences, CSSR
PARTITION CHROMATOGRAPHY OF TRANSURANIUM ELEMENTS IN THE SYSTEM
TBHP-TBPP/PERCHLORIC ACID. II.
Nukleonika, 10,81–7(1965) (in Russian)
NSA, 19, 28250 (1965)

454. MIKULSKI, J. and STRONSKI, I.
Inst. of Nuclear Physics, Krakow, Poland
THE RADIOCHEMICAL SEPARATION OF SOME METALS BY PARTITION CHROMATO-
GRAPHY WITH REVERSED PHASES ON TEFLON IN THE SYSTEM TRI-n-OCTYLAMINE-
ELECTROLYTE.
J. Chromatog., 17,197–200(1965)

455. MIKULSKI, J. and STRONSKI, I.
Inst. of Nuclear Physics, Krakow, Poland
SEPARATION OF SOME METALS BY EXTRACTION CHROMATOGRAPHY USING NEW
NEUTRAL PHOSPHORO-ORGANIC COMPOUNDS.
Nature, 207,749–50 (1965)

456. MIKULSKI, J.
Inst. of Nuclear Physics, Krakow, Poland
THE RADIOCHEMICAL SEPARATION OF SOME METAL IONS BY EXTRACTION CHRO-
MATOGRAPHY ON POWDERED POLYTETRAFLUORO-ETHYLENE IN THE SYSTEM
TRI-n-OCTYLAMINE-ELECTROLYTE.
Nukleonika, 11,57–9(1966)
NSA, 20, 27055 (1966)

457. MIKULSKI, J.
Inst. of Nuclear Physics, Krakow, Poland
STATIC AND DYNAMIC EXTRACTION OF SOME METALS IN THE SYSTEMS TETRA-
BUTYLHYPOPHOSPHATE-TETRABUTYLPYROPHOSPHATE (1:1) AND TETRABUTYL-
PYROPHOSPHATE/MINERAL ACIDS.
INP 591/C. 1968. 82 p. (in Polish)

458. MINCZEWSKI, J. and ROZYCKI, C.
Politech., Warsaw, Poland
THE THIOCYANATE METHOD OF NIOBIUM DETERMINATION. IV. DETERMINATION
OF NIOBIUM IN MOLYBDENITE BY REVERSED-PHASE PARTITION CHROMATOGRAPHY.

Chem. Anal. (Warsaw), 10,965–71(1965) (in Polish)
CA, 64, 14955e (1966)

459.   MIRZA,*M.Y. and KOCH,**G.
        * Pakistan Atomic Energy Commission
        ** Kernforschungszentrum, Inst. für Heisse Chemie, Karlsruhe, BRD (FRG)
        THE SEPARATION OF AMERICIUM AND CURIUM FROM FISSION PRODUCTS BY EXTRAC-
        TION CHROMATOGRAPHY USING DI(2-ETHYLHEXYL)PHOSPHORIC ACID.
        Proc. 6th Int. Symp. für Mikrochemie, Graz 1970, Vol. E, p. 169–74

460.   MONSECOUR, M. R. and DEMILDT, A. C.
        SCK-CEN, Mol, Belgium
        DETERMINATION BY NEUTRON ACTIVATION ANALYSIS OF THE BURN-UP INDICATOR
        NEODYMIUM-148 IN IRRADIATED URANIUM DIOXIDE-PLUTONIUM DIOXIDE.
        Anal. Chem., 41,27–31(1969)

461.   MOORE, F. L. and JURRIAANSE, A.
        Anal. Chem. Div. ORNL, Oak Ridge, Tenn., USA
        SEPARATION OF CALIFORNIUM FROM CURIUM AND BERKELIUM BY EXTRACTION
        CHROMATOGRAPHY.
        Anal. Chem., 39, 733–6 (1967)

462.   MOORE, F. L.
        Anal. Chem. Div. ORNL, Oak Ridge, Tenn., USA
        NEW EXTRACTION CHROMATOGRAPHIC METHOD FOR RAPID SEPARATION OF AMER-
        ICIUM FROM OTHER TRANSURANIUM ELEMENTS.
        Anal. Chem., 40, 2130–3 (1968)

463.   MOORE, F. L.
        Anal. Chem. Div. ORNL, Oak Ridge, Tenn., USA
        ISOLATION AND PURIFICATION OF AMERICIUM FROM OTHER 5f AND 4f ELEMENTS
        BY EXTRACTION CHROMATOGRAPHY.
        U. S. Patent 3,615,268.Appl. Date: 29 Nov. 1968.
        Publ. Date: 26 Oct. 1971

464.   MOSKVIN, L. N.
        PARTITION CHROMATOGRAPHY ON FLUOROPLAST-4. SEPARATION OF EUROPIUM
        FROM RARE EARTHS.
        Radiokhimiya, 5,747–9(1963) (in Russian)
        NSA, 18, 23723 (1964)

465.   MOSKVIN, L. N.
        PARTITION CHROMATOGRAPHY ON TEFLON(POLYTETRAFLUOROETHYLENE). THE
        ISOLATION OF TRACE AMOUNTS OF SELENIUM FROM A SrO TARGET IRRADIATED
        WITH FAST PROTONS.
        Radiokhimiya, 6,110–1(1964) (in Russian)
        NSA, 18, 23731 (1964)

466.   MOSKVIN, L. N. and PREOBRAZHENSKII, B. K.
        PARTITION CHROMATOGRAPHY ON FLUOROPLAST-4. I. GRADIENT ELUTION SEPA-
        RATION OF THE LIGHT RARE EARTHS.
        Radiokhim. Metody Opred. Mikroelementov, Akad. Nauk SSSR, Sb Statei,85–93(1965) (in
        Russian)
        NSA, 20, 20621 (1966)
        CA, 64, 6044f (1966)

467. MOSKVIN, L. N. and TOMILOV, S. B.
PARTITION CHROMATOGRAPHY ON FLUOROPLAST-4. II. SEPARATION OF TRACE
AMOUNTS OF Eu FROM Tb₂O₃ TARGETS IRRADIATED WITH FAST PROTONS.
Radiokhim. Metody Opred. Mikroelementov, Akad. Nauk SSSR, Sb. Statei, 93–5 (1965) (in Russian)
NSA, 20, 20622 (1966)
CA, 64, 6044h (1966)

468. MOSKVIN, L. N. and NOVIKOV, V. T.
PARTITION CHROMATOGRAPHY ON FLUOROPLAST-4. III. RAPID SEPARATION OF
LANTHANUM FROM CERIUM.
Radiokhim. Metody Opred. Mikroelementov, Akad. Nauk SSSR, Sb Statei, 95–6 (1965) (in Russian)
NSA, 20, 20623 (1966)
CA, 63, 14019e (1965)

469. MOSKVIN, L. N., SUR, L. I., TSARITSYNA, L. G., and MARTYS, G. T.
PARTITION CHROMATOGRAPHIC METHODS OF ISOLATION AND SEPARATION OF Mo,
W AND Re AND THEIR APPLICATION TO THE ANALYSIS OF THE RADIOISOTOPIC
PURITY OF RADIOACTIVE PREPARATIONS OF SUCH ELEMENTS.
Radiokhimiya, 9, 377–81 (1967) (in Russian)
CA, 68, 101448 (1968)

470. MOSKVIN, L. N. and TSARITSYNA, L. G.
SEPARATION OF ACTINIUM AND RADIUM FROM A THORIUM TARGET BOMBARDED
BY 600 MeV PROTONS.
At. Energ., 24, 383–4 (1968) (in Russian)
CA, 69, 23830 (1968)

471. MOSKVIN, L. N. and TSARITSYNA, L. G.
CONTINUOUS CHROMATOGRAPHIC SEPARATION OF MULTICOMPONENT MIXTURES
OF SUBSTANCES IN LIQUID-LIQUID PARTITION CHROMATOGRAPHY.
Radiokhimiya, 10, 740–3 (1968) (in Russian)
CA, 70, 71298 (1969)

472. MOSKVIN, L. N. and TSARITSYNA, L. G.
CONTINUOUS CHROMATOGRAPHIC SEPARATION OF A MULTICOMPONENT MIXTURE
OF SUBSTANCES IN LIQUID-LIQUID PARTITION CHROMATOGRAPHY. II. OPERATION
OF THE DEVICE AND POSITION OF THE MAXIMA OF THE EFFLUENT CURVES.
Radiokhimiya, 12, 730–6 (1970) (in Russian)
NSA, 26, 1966 (1972)

473. MOSKVIN, L. N. and TSARITSYNA, L. G.
CONTINUOUS CHROMATOGRAPHIC SEPARATION OF A MULTICOMPONENT MIXTURE
OF SUBSTANCES IN LIQUID-LIQUID PARTITION CHROMATOGRAPHY. III. BLOW-UP OF
THE AREAS OF SEPARATED SUBSTANCES.
Radiokhimiya, 12, 737–42 (1970) (in Russian)
CA, 74, 91445 (1971)

474. MOSKVIN, L. N. and KALININ, N. N.
GROUP SEPARATION OF FISSION PRODUCTS BY CHROMATOGRAPHIC METHOD.
Atomnaya Energiya, 29(6), 458–61 (1970) (in Russian)

475. MÜLLER, W., MAINO, F., and TOUSSAINT, J. Cl.
EURATOM, Inst. for Transuranium Elements, Karlsruhe, BRD (FRG)

PROCESSING OF IRRADIATED [241]Am TARGETS BY ION EXCHANGE AND EXTRACTION.
EUR 4409e. 1970. 26 p. 14 fig.

476. MÜLLER, W.
EURATOM, Inst. for Transuranium Elements, Karlsruhe, BRD (FRG)
EXTRACTION OF TRANSURANIUM ELEMENTS BY ALKYLAMMONIUM SALTS.
Angew. Chem., 83, 625 (1971) (in German)
Angew. Chem., Int. Ed. Engl., 10, 580–1 (1971)
NSA, 26, 1960 (1972)

477. MÜLLER, W.
EURATOM, Inst. for Transuranium Elements, Karlsruhe, BRD (FRG)
ON THE EXTRACTION OF TRIVALENT METALS BY ALKYLAMMONIUM SALTS.
Chemiker Ztg., 95, 499-507 (1971) (in German)

478. MÜNZE, R.[*], GROSSE-RUYKEN, H.[**], and WAGNER, G.[*]
[*] Zentralinstitut für Kernforschung, Rossendorf, DDR (GDR)
[**] Sektion Chemie der Techn. Universität, Dresden, DDR (GDR)
STUDIES ON TERNARY FISSION OF URANIUM. PART 6: DETECTION OF THE FISSION
CHAIN [166]Dy-[166]Ho-[166]Er AND FISSION YIELD DETERMINATION.
Kernenergie, 12, 380–2 (1969) (in German)

479. MURRAY, R. W. and PASSARELLI, R. J.
Dept. of Chemistry, Univ. of North Carolina, Chapel Hill, N. C., USA
REVERSED-PHASE THIN-LAYER AND COLUMN CHROMATOGRAPHY WITH ALKYL
AMINE AND QUATERNARY AMMONIUM SALT STATIONARY PHASES.
Anal. Chem., 39, 282–5(1967)

480. MYMRIK, N. A. and GIBALO, I. M.
Moscow M. V. Lomonosov State Univ., Moscow, USSR
STATE OF NIOBIUM IN THE TRI-n-OCTYLAMINE-HYDROCHLORIC ACID SYSTEM
STUDIED BY EXTRACTION CHROMATOGRAPHY.
Vestn. Mosk. Univ., Khim.,12,738–40(1971) (in Russian)
CA, 76, 104394 (1972)

481. NARBUTT, J. and SMUŁEK, W.
Dept.of Radiochem.,Inst.of Nucl.Research,Warsaw, Poland
PREPARATION OF [237]U AND [239]Np TRACES FROM REACTOR IRRADIATED URANIUM
BY EXTRACTION CHROMATOGRAPHY.
Paper presented at the Second Italian-Polish Meeting on "Some Chemical, Ceramic and Metal-
lurgical Aspects of Nuclear Fuels", held in Rome, 4–13 Oct. 1972
Nukleonika, 18,375–82(1973)

482. NAZAROV, A. S. and GROMOV, B. V.
PROCESS OF REEXTRACTION STUDIED BY A CHROMATOGRAPHIC METHOD.
Tr. Mosk. Khim.- Tekhnol.Inst.N°47,159–64(1964)(in Russian)
CA, 64, 10709a (1966)

483. NEEF, B. and GROSSE-RUYKEN, H.[*]
Zentralinstitut für Festkörperphysik und Werkstofforschung der Akademie der Wissen-
schaften der DDR, Dresden, DDR (GDR)
[*] Sektion Chemie der Techn. Univ., Dresden, DDR (GDR)

SEPARATION OF MANGANESE, IRON, COBALT, NICKEL, COPPER, ZINC AND CAD-
MIUM BY REVERSED-PHASE CHROMATOGRAPHY USING TRI-n-OCTYLAMINE AS THE
STATIONARY PHASE AND APPLYING A GRADIENT ELUTION TECHNIQUE.
J. Chromatog., 79,275—85(1973) (in German)

484. NELSON, F.
Chemistry Div., ORNL, Oak Ridge, Tenn., USA
ION-EXCHANGE PROCEDURES. VIII. SEPARATION OF SILVER FROM A NUMBER OF
ELEMENTS BY PARTITION CHROMATOGRAPHY.
J. Chromatog.,20, 378—83 (1965)

485. NGUYEN, K. N., RESHETNIKOVA, L. P., NOVOSELOVA, A.V., and BOLSHOVA, T. A.
EXTRACTION-CHROMATOGRAPHIC BEHAVIOUR OF BERYLLIUM, MANGANESE, AND
ZINC IN THE TRIBUTYL PHOSPHATE-AMMONIUM THIOCYANATE SYSTEM.
Vestn. Mosk. Univ., Khim.,13,342—4(1972) (in Russian)
CA, 77, 121718 (1972)

486. O'LAUGHLIN, J. W., KAMIN, G. J., BERNER, D. L., and BANKS, C. V.
Inst. for Atomic Research and Dept. of Chem., Iowa State Univ., Ames, Iowa, USA
REVERSED-PHASE PARTITION CHROMATOGRAPHIC SEPARATION OF SEVERAL AL-
KALINE EARTHS FROM EACH OTHER AND FROM YTTRIUM.
Anal. Chem., 36, 2110—2(1964)
CA, 62, 4d (1965)

487. O'LAUGHLIN, J. W. and BANKS, C. V.
Inst. for Atomic Research and Dept. of Chem., Iowa State Univ., Ames, Iowa, USA
SOLVENT EXTRACTION PROPERTIES OF METHYLENEBIS(DIALKYLPHOSPHINE OXIDES).
Solvent Extr. Chem., Proc. Int. Conf., Gothenburg, Sweden, 1966, pp.270—7(1967)
CA, 69, 70346 (1968)

488. O'LAUGHLIN, J. W. and JENSEN, D. F.
Inst. for Atomic Research and Dept. of Chem., Iowa State Univ., Ames, Iowa, USA
SEPARATION OF SEVERAL RARE EARTHS BY EXTRACTION CHROMATOGRAPHY
USING BIFUNCTIONAL PHOSPHINE OXIDES.
J. Chromatog., 32,567—76(1968)

489. OVERMAN, R. F.
Savannah River Lab., E. I. du Pont de Nemours and Comp., Aiken, S. C.,USA
RAPID SEPARATION OF BERKELIUM FROM RARE EARTH FISSION PRODUCTS AND
TRIVALENT ACTINIDES.
Anal. Chem., 43,600—1(1971)

490. PAGLIAI, V. and POZZI, F.
C.N.E.N. Programma EUREX, Rome, Italy
DETERMINATION OF METALLIC IMPURITIES IN SOLUTIONS OF HIGHLY ENRICHED
URANIUM BY ATOMIC ABSORPTION, AFTER SEPARATION OF URANIUM BY
EXTRACTION CHROMATOGRAPHY.
RT/CHI (72) 10. 1972. 12. p., 3 figs.

491. PANESKO, J. V.
Atlantic Richfield Hanford Co. Richland, Wash., USA
PROCESS FOR RECOVERY OF RHODIUM, PALLADIUM, AND TECHNETIUM FROM
AGED REPROCESSING WASTES AT HANFORD.
ARH-733. 1968. 65 p.

492. PEPPARD, D. F., BLOOMQUIST, C. A. A., HORWITZ, E. P., LEWEY, S., and MASON, G. W.
Chem. Div., Argonne Nat. Lab., Argonne, Ill., USA
ANALOGOUS ACTINIDE(III) AND LANTHANIDE(III) TETRAD EFFECTS.
J. Inorg. Nucl. Chem., 32, 339–43 (1970)

493. PERRICOS, D. C. and THOMASSEN, J. A.
Inst. for Atomenergi, Kjeller, Norway
RADIOCHEMICAL SEPARATION METHODS FOR NEPTUNIUM-239 AND MOLYBDENUM-99
USED FOR THE DETERMINATION OF MICROSCOPIC REACTOR LATTICE PARAMETERS.
KR-83. 1964. 34 p. (NORA-16)

494. PETIT-BROMET, M.
CEA, Fontenay-aux-Roses, France
PARTITION CHROMATOGRAPHY SEPARATION USING TRILAURYLAMINE ADSORBED
ON A SOLID SUPPORT. BEHAVIOUR OF THE URANYL ION.
CEA-R 3469. 1968. 80 p., 16 figs. (in French)

495. PFREPPER, G., HERRMANN, E., and CHRISTOW, D.
Joint Inst. for Nucl. Research, Dubna, USSR
EXTRACTION-CHROMATOGRAPHIC COLUMN METHOD FOR RAPID AND CONTINUOUS
SEPARATION OF SHORT-LIVED RADIONUCLIDES FROM PHTHALOCYANINE COMPLEXES.
JINR-P6-4623. 1969. 16 p. (in Russian)
Also published in Radiochim. Acta, 13, 196–200 (1970) (in German)

496. PHILLIPS, G. and MILNER, G. W. C.
AERE, Harwell, Berks., Great Britain
THE APPLICATION OF CONTROLLED-POTENTIAL COULOMETRY TO THE DETERMI-
NATION OF PLUTONIUM.
Paper presented at the SAC Conference, Nottingham, July 1965
Proceedings of the SAC Conference, Nottingham 1965 W. Heffer and Sons Ltd., Cambridge,
1965, pp. 240–53

497. PIERCE, T. B.
AERE, Harwell, Berks., Great Britain
THE APPLICATION OF CHELATE SYSTEMS TO PARTITION CHROMATOGRAPHY.
PRELIMINARY PAPER
Anal. Chim. Acta, 24, 146–52 (Feb. 1961)

498. PIERCE, T. B. and PECK, P. F.
AERE, Harwell, Berks., Great Britain
THE PREPARATION OF CARRIER-FREE [115]In BY A REVERSED-PHASE PARTITION
TECHNIQUE.
J. Chromatog., 6, 248–51(1961)

499. PIERCE, T. B. and PECK, P. F.
AERE, Harwell, Berks., Great Britain
A RAPID METHOD FOR DETERMINING INDIUM BY NEUTRON ACTIVATION.
Analyst, 86, 580–4 (1961)

500. PIERCE, T. B. and PECK, P. F.
AERE, Harwell, Berks., Great Britain

USE OF DI(2-ETHYLHEXYL) ORTHOPHOSPHORIC ACID FOR THE SEPARATION OF
THE ELEMENTS LANTHANUM-GADOLINIUM BY REVERSED-PHASE PARTITION
CHROMATOGRAPHY.
Nature, 194, 84 (1962)
CA, 57, 1516i (1962)

501. PIERCE, T. B. and PECK, P. F.
AERE, Harwell, Berks., Great Britain
USE OF DI(2-ETHYLHEXYL) ORTHOPHOSPHORIC ACID FOR THE SEPARATION OF
THE RARE EARTHS BY REVERSED-PHASE PARTITION CHROMATOGRAPHY.
Nature, 195, 597 (1962)
CA, 57, 1183f (1962)

502. PIERCE, T.B. and HOBBS, R. S.
AERE, Harwell, Berks., Great Britain
THE SEPARATION OF THE RARE EARTHS BY PARTITION CHROMATOGRAPHY WITH
REVERSED PHASES. I. BEHAVIOUR OF COLUMN MATERIAL.
J. Chromatog., 12, 74-80(1963)

503. PIERCE, T. B., PECK, P. F., and HOBBS, R. S.
AERE, Harwell, Berks.,Great Britain
THE SEPARATION OF THE RARE EARTHS BY PARTITION CHROMATOGRAPHY WITH
REVERSED PHASES. II. BEHAVIOUR OF INDIVIDUAL ELEMENTS ON HDEHP-CORVIC
COLUMNS.
J. Chromatog., 12,81-8(1963)

504. PIERCE, T. B. and HOBBS, R. S.
AERE, Harwell, Berks., Great Britain
THE SEPARATION OF THE RARE EARTHS BY PARTITION CHROMATOGRAPHY WITH
REVERSED PHASES. III. THE USE OF BATCH EXTRACTIONS TO OBTAIN DISTRIBU-
TION DATA.
AERE-R-4358. May 1963. 10 p.

505. PIERCE, T. B. and PECK, P. F.
AERE, Harwell, Berks., Great Britain
THE EXTRACTION OF ZINC BY DITHIZONE RETAINED ON CELLULOSE ACETATE.
AERE-R-4969. 1965. 7 p.

506. PIERCE, T. B. and HENRY, W. M.
AERE, Harwell, Berks., Great Britain
THE USE OF TRI-ISO-OCTYLAMINE AS AN ANION EXCHANGER FOR PARTITION
CHROMATOGRAPHY.
J. Chromatog., 23,457-64(1966)

507. PLUYM, A.
Hoger Rijksinst. voor Techn. Onderwijs en Kernenergiebedrijven-SCK, Mol, Belgium
THE RADIOCHEMICAL SEPARATION OF NEODYMIUM AND AMERICIUM BY REVERSED-
PHASE PARTITION CHROMATOGRAPHY.
Thesis. June 1970 (in Dutch)

508. POHLANDT, Chr., EWEN, M., and GERGHTY, A.
Nat. Inst. for Metallurgy, Johannesburg, South Africa
SEPARATION AND DETERMINATION OF NOBLE METALS BY REVERSED-PHASE
EXTRACTION CHROMATOGRAPHY.
Report NIM-994. 1971. 6 p.
CA, 75, 29605 (1971)

509. POHLANDT, Chr., EWEN, M., and GERGHTY, A.
Nat. Inst. for Metallurgy, Johannesburg, South Africa
SEPARATION OF NON-VOLATILE NOBLE METALS BY REVERSED-PHASE EXTRACTION
CHROMATOGRAPHY.
Report NIM-1245. 1971. 18 p.
CA, 75, 122430 (1971)

510. POHLANDT, Chr. and STEELE, T. W.
Nat. Inst. for Metallurgy, Johannesburg, South Africa
SEPARATION OF THE NON-VOLATILE NOBLE METALS BY REVERSED-PHASE EXTRACTION
CHROMATOGRAPHY.
Talanta, 19, 839–50(1972)

511. POHLANDT, Chr., ROBERT, R. V. D., and VALLÉE, A. M.
Nat. Inst. for Metallurgy, Johannesburg, South Africa
CHROMATOGRAPHIC SEPARATION AND DETERMINATION OF NOBLE METALS
IN MATTE-LEACH RESIDUES.
Rep. NIM-1500. 1973. 15 p.
CA, 78, 168127 (1973)

512. POLI, A.
CNEN, Laboratorio Operazione Calde, Rome, Italy
RELATION BETWEEN THE CHROMATOGRAPHIC BEHAVIOUR OF A METAL ION
DURING CHROMATOGRAPHIC EXTRACTION (HDEHP) AND THE IONIC PARAMETERS
OF THE SPECIE IN SOLUTION.
RT/CHI (70) 32, 1970. 23 p. (in Italian)
CA, 75, 25805 (1971)

513. PREOBRAZHENSKII, B. K. and KATYKHIN, G. S.
CHROMATOGRAPHY IN EXTRACTION PROCESSES.
Radiokhimiya, 4, 536–40(1962) (in Russian)
NSA, 17, 20113 (1963)

514. PREOBRAZHENSKII, B. K., KALYAMIN, A. V., and MIKHALCHA, I.
THE SEPARATION OF GOLD AND THALLIUM FROM A COMPLEX MIXTURE OF
ELEMENTS BY MEANS OF PARTITION CHROMATOGRAPHY.
Radiokhimiya, 6, 111–2 (1964) (in Russian)
NSA, 18, 23732 (1964)

515. PREOBRAZHENSKII, B. K., KALYAMIN, A. V., LILOVA, O. M., MOSKVIN, L. N., and
USIKOV, B. S.
GRANULAR POROUS POLY(PERFLUOROETHYLENE) RESIN FOR CHROMATOGRAPHY.
PREPARATION PROCESS AND PROPERTIES.
Radiokhimiya, 10, 375–7 (1968) (in Russian)
CA, 69, 52781 (1968)

516. PREOBRAZHENSKII, B. K., MOSKVIN, L. N., KALYAMIN, A. V., LILOVA, O. M., and
USIKOV, B. S.
USE OF COMPACT POROUS POLY(PERFLUOROETHYLENE) RESIN IN PARTITION
CHROMATOGRAPHY.
Radiokhimiya, 10, 377–9 (1968) (in Russian)
CA, 69, 52780 (1968)

466

517. PSZONICKA, M. and SIEKIERSKI, S.
      Inst. of Nucl. Research, Warsaw, Poland
      EFFECT OF THE SUPPORT ON THE SEPARATION EFFICIENCY OF COLUMNS IN
      REVERSED-PHASE CHROMATOGRAPHY.
      Chem. Anal. (Warsaw), 17,1321–6(1972) (in Polish)

518. RAICHEV, Kh. and KHALKIN, V. A.
      Joint Inst. for Nucl. Research, Dubna, USSR
      EFFECT OF COLUMN SHAPE ON ION-EXCHANGE CHROMATOGRAPHY PROCESSES.
      JINR-P6-3952. 1968. (in Russian)
      Also published in Radiokhimiya, 11, 259–60 (1969) (in Russian)

519. RAICHEV, Kh. and KHALKIN, V. A.
      Joint Inst. for Nucl. Research, Dubna, USSR
      EXTRACTION-CHROMATOGRAPHIC METHODS FOR SEPARATING POLONIUM FROM
      A PROTON-IRRADIATED BISMUTH OXIDE TARGET.
      JINR-P12-4181. 1968. 9 p. (in Russian)
      Also published in Radiokhimiya, 12, 778–9 (1970) (in Russian)

520. RAICHEV, Kh. and GESHEVA, M.
      Inst. of Nucl. Research and Nucl. Energy, Sofia, Bulgaria
      THE ISOLATION OF MÖSSBAUER SOURCE ^{161}Tb FROM AN IRRADIATED GADOLI-
      NIUM TARGET.
      Paper presented at the 7th Radiochemical Conference, Mariánské Lázné, CSSR, 24-28 Apr. 1973.

521. RAIS, J., KRTIL, J., and CHOTIVKA, V.
      Inst. of Nuclear Research, Řež, ČSSR
      EXTRACTION AND SEPARATION OF ^{137}Cs AND ^{86}Rb BY MEANS OF 4-t-BUTYL-2
      (α-METHYLBENZYL)PHENOL.
      Talanta, 18, 213–8 (1971)

522. RAMALEY, L. and HOLCOMBE, W. A.
      Univ. of Arizona, Tuscon, USA
      CHROMATOGRAPHIC SEPARATION OF THE HALIDES.
      Anal. Lett., 1,143–51(1967)

523. RAO, A. P. and SASTRI, M. N.
      Dept. of Chemistry, Andhra Univ., Waltair, India
      SEPARATION OF SOME ELEMENTS ON SILICA COLUMNS TREATED WITH A LIQUID
      CATION EXCHANGER.
      Z. Anal. Chem., 207,409–10(1965)

524. RICCATO, M. T. and HERRMANN, G.
      Inst. of Inorg. and Nucl. Chem.,Univ. Mainz, BRD (FRG)
      ELUTION CHROMATOGRAPHY OF RARE EARTHS AT HIGH FLOW RATES.
      Radiochim. Acta, 14,107–8(1970) (in German)
      NSA, 24, 48157 (1970)

525. RICCATO, M. T.
      Inst. of Inorg. and Nucl.Chem., Univ. Mainz, BRD (FRG)
      ON THE FAST SEPARATION OF RUTHENIUM FROM FISSION PRODUCTS OF URANIUM.
      Radiochim. Acta, 15, 3–7 (1971) (in German)

526. RIDER, B. F., RUIZ, C. P., PETERSON, J. P. Jr., and LUKE, P. S. Jr.
General Electric Co., Vallecitos Atomic Lab., San Jose, Calif., USA
ACCURATE NUCLEAR FUEL BURN-UP ANALYSES.
Quarterly Progress Report No. 6 March-May 1963
GEAP-4278. 1963. 29 p.

527. RIDER, B. F., PETERSON, J. P. Jr., and RUIZ, C. P.
General Electric Co., Vallecitos Atomic Lab., San Jose, Calif., USA
DETERMINATION OF NEODYMIUM-148 IN IRRADIATED UO_2 AS A MEASUREMENT
OF BURN-UP.
Trans. Amer. Nucl. Soc., 7, 350–1 (1964)

528. RIDER, B. F., PETERSON, J. P. Jr., and RUIZ, C. P.
General Electric Co., Atomic Power Equipment Dept., Pleasanton, Calif., USA
THE HALF-LIFE OF ^{149}Nd.
J. Inorg. Nucl. Chem., 28, 266–8 (1966)
NSA, 20, 15779 (1966)

529. ROBERTS, G. A. H.
Esso Research Centre, Abingdon, Berks., Great Britain
DETERMINATION OF IRON, NICKEL, CHROMIUM AND VANADIUM BY MEANS OF REDOX
AND ION-EXCHANGE COLUMNS.
Talanta, 15, 735–40 (1968)

530. SAMSAHL, K., WESTER, P. O., and LANDSTRÖM, O.
AB Atomenergi, Stockholm, Sweden
AN AUTOMATIC GROUP SEPARATION SYSTEM FOR THE SIMULTANEOUS DETER-
MINATION OF A GREAT NUMBER OF ELEMENTS IN BIOLOGICAL MATERIAL.
Anal. Chem., 40, 181–7 (1968)

531. SAMSAHL, K.
AB Atomenergi, Nyköping, Sweden
AUTOMATED NEUTRON-ACTIVATION ANALYSIS OF BIOLOGICAL MATERIAL WITH
HIGH RADIATION LEVELS.
Analyst, 93, 101–6 (1968)

532. SAMSAHL, K.
AB Atomenergi, Nyköping, Sweden
HIGH-SPEED, AUTOMATIC RADIOCHEMICAL SEPARATIONS FOR ACTIVATION ANAL-
YSIS IN THE BIOLOGICAL AND MEDICAL RESEARCH LABORATORY.
Sci. Total Environ., 1, 65–74 (1972)

533. SANTORI, G. and TESTA, C.
Radiotoxicological Lab.,
Medical Service CNEN, CSN Casaccia, Rome, Italy
DETERMINATION OF ^{237}Np IN URINE BY REVERSED-PHASE PARTITION CHROMA-
TOGRAPHY.
J. Radioanal. Chem., 14, 37–43 (1973)

534. SARAIYA, V. N., VENKATESWARLU, K. S., and SHANKAR, J.
Chem. Div., Bhabha At. Res. Cent., Bombay, India
ANTAGONISTIC EFFECTS IN REVERSE-PHASE PARTITION AND SOLVENT EXTRAC-
TION OF RUTHENIUM(III) FROM HYDROCHLORIC ACID SOLUTIONS.
Indian J. Chem., 11, 490–3 (1973)
CA, 79, 70652 (1973)

535. SASTRI, M. N., RAO, A. P., and SARMA, A. R. K.
Andhra Univ., Waltair, India
SEPARATION OF METAL IONS BY REVERSED-PHASE COLUMN PARTITION CHROMA-
TOGRAPHY WITH KEL-F SUPPORTING METHYL-TRICAPRYLAMMONIUM CHLORIDE
(ALIQUAT-336).
Indian J. Chem., 4, 287–9(1966)
CA, 65, 16043f (1966)

536. SCHMID, E. R.
Analyt. Inst., Univ. Vienna, Austria
EXTRACTION-CHROMATOGRAPHIC SEPARATION OF PLUTONIUM(III), URANIUM(IV)
AND URANIUM(VI).
Microchim. Acta, 1970, 301–12 (1970) (in German)

537. SCHMID, E. R.
Analyt. Inst., Univ. Vienna, Austria
DETERMINATION OF THE DISTRIBUTION RATIOS OF U(IV) BY MEANS OF EXTRAC-
TION CHROMATOGRAPHY WITH THE SYSTEM UNDILUTED TRI-n-BUTYL PHOSPHATE-
AQUEOUS NITRIC ACID.
Proc. 6th Int. Symp. für Mikrochemie, Graz, 1970, Vol. E, pp. 233–8(in German)
Also published in
Monatsh. Chem., 101, 1856–60 (1970) (in German)
CA, 74, 47171 (1971)

538. SCHMID, E. R. and PFANNHAUSER, W.
Anal. Inst., Univ. Vienna, Austria
SEPARATION OF URANIUM(IV) BY EXTRACTION CHROMATOGRAPHY FROM SOLU-
TIONS RESULTING FROM NUCLEAR TECHNOLOGY.
Microchim. Acta, 1971, 250–61(1971) (in German)
CA, 75, 44484 (1971)

539. SCHMID, E. R.
Anal. Inst., Univ. Vienna, Austria
EXTRACTION AND EXTRACTION CHROMATOGRAPHY OF URANIUM(IV) WITH
PHOSPHORIC ACID ESTERS AND ALKYLSUBSTITUTED AMMONIUM NITRATES FROM
NITRIC ACID SOLUTIONS.
Microchim. Acta, 1972, 544–70(1972) (in German)

540. ŠEBESTA, F.
Dept. of Nucl. Chem., Techn. Univ., Prague, ČSSR
EXTRACTION CHROMATOGRAPHY USING CHELATING AGENTS.
I. SYSTEM ZINC-DITHIZONE-CARBON TETRACHLORIDE.
J. Radioanal. Chem., 6, 41–6 (1970)

541. ŠEBESTA, F.
Dept. of Nucl. Chem., Techn. Univ., Prague, ČSSR
EXTRACTION CHROMATOGRAPHY USING CHELATING AGENTS.
II. SEPARATION OF SOME RADIOISOTOPES BY DITHIZONE.
J. Radioanal. Chem., 7, 41–7 (1971)

542. ŠEBESTA, F. and LAZNICKOVA, A.
Dept. of Nucl. Chem., Techn. Univ., Prague, ČSSR
EXTRACTION CHROMATOGRAPHY USING CHELATING AGENTS.
III. N-BENZOYL-N-PHENYLHYDROXYLAMINE AS A CHELATING AGENT.
J. Radioanal. Chem., 11, 221–9(1972)

543. ŠEBESTA, F.*, POSTA, S.*, and RANDA, Z.**
* Dept. of Nucl. Chem., Techn. Univ., Prague, ČSSR
** Inst. of Mineral Raw Materials, Kutna Hora, ČSSR
EXTRACTION CHROMATOGRAPHY USING CHELATING AGENTS.IV. SEPARATION OF
Re(VII) AND Tc(VII), AND ITS APPLICATION IN ACTIVATION ANALYSIS.
Radiochem. Radioanal. Letters, 11, 359–65 (1972)

544. ŠEBESTA, F. and POSTA, S.
Dept. of Nucl. Chem., Techn. Univ., Prague, ČSSR
EXTRACTION CHROMATOGRAPHY USING CHELATING AGENTS. V. SEPARATION OF
Hf(IV), Nb(V), Ta(V) AND Pa(V) WITH N-BENZOYL-N-PHENYLHYDROXYLAMINE.
Radiochem. Radioanal. Letters, 14, 183–91 (1973)

545. ŠEBESTA, F. and STARÝ, J.
Dept. of Nucl. Chem., Techn. Univ., Prague, ČSSR
A GENERATOR FOR PREPARATION OF CARRIER-FREE ^{224}Ra.
Paper presented at the 7th Radiochemical Conference at Mariánské Lázné, ČSSR, 24-28
Apr. 1973

546. SEKIZUKA, Y., KOJIMA, T., YANO, T., and UENO, KEIHEI
Dept. of Org. Synthesis, Fac. of Eng. Kyushu University, Kukuoka, Japan
ANALYTICAL APPLICATION OF ORGANIC REAGENTS IN HYDROPHOBIC GEL MEDIA. I.
GENERAL PRINCIPLE AND THE USE OF DITHIZONE GEL.
Talanta, 20, 979–85 (1973)

547. SHESTAKOV, B. I., SHESTAKOVA, I. A., and MOSKVIN, L. N.
PREPARATION OF THORIUM-234 (UX) BY PARTITION CHROMATOGRAPHY.
Radiokhimiya, 11, 471–3 (1969) (in Russian)
CA, 71, 130888 (1969)

548. SHIMIZU, T. and IKEDA, K.
Dept. of Chem., Fac. of Education, Gumma Univ., Maebashi, Gumma, Japan
THE REVERSED-PHASE EXTRACTION CHROMATOGRAPHY OF THE RARE EARTHS
THORIUM, URANIUM AND ZIRCONIUM WITH A HIGH-MOLECULAR-WEIGHT AMINE
IN SULPHURIC ACID AND AMMONIUM SULPHATE MEDIA.
J. Chromatog., 85, 123–8 (1973)

549. SHMANENKOVA, G.I., ZEMSKOVA, M. G., MELAMED, Sh. G., PLESHAKOVA, G. P., and
SHUKOV, G. V.
Moscow M. V. Lomonosov State Univ., Moscow, USSR
CHEMICAL-SPECTRAL AND CHEMICAL-LUMINESCENCE DETERMINATION OF MICRO-
QUANTITIES OF RARE EARTH ELEMENTS USING EXTRACTION CHROMATOGRAPHY.
Zavod. Lab., 35, 897–902 (1969) (in Russian)
NSA, 23, 49708 (1969)

550. SHMANENKOVA, G. I., MIKHEEVA, L. M., and ELFIMOVA, G. I.
Moscow M. V. Lomonosov State Univ., Moscow, USSR
SEPARATION OF GALLIUM FROM ALUMINIUM AND IRON BY EXTRACTION CHRO-
MATOGRAPHY.
Zh. Anal. Khim., 27, 2161–4 (1972) (in German)
CA, 78, 66503 (1973)

551. SHULEPNIKOV, M. N., SHMANENKOVA, G. I., YAKOVLEV, Yu. V., and DOGADKIN,
N. N.

Inst. of Geochem. and Anal. Chem., Academy of Sciences of the USSR, Moscow, USSR
RADIOACTIVATION DETERMINATION OF IMPURITIES IN EPITAXIAL GALLIUM
ARSENIDE FILMS.
Zh. Anal. Khim., 26, 1167–70 (1970) (in Russian)

552. SIEKIERSKI, S. and KOTLINSKA, B.
Polish Academy of Sciences, Inst. Nucl. Research, Warsaw, Poland
SEPARATING ZIRCONIUM AND NIOBIUM BY REVERSED-PHASE PARTITION CHROMA-
TOGRAPHY.
Atomnaya Energ., 7, 160–2(1959) (in Russian)
NSA, 13, n° 22, 19873 (1959),
CA, 54, 9584h (1960)
Also published in Kernenergie, 3, 482–4 (1960) (in German)

553. SIEKIERSKI, S. and FIDELIS, I.
Polish Academy of Sciences, Inst. of Nucl. Research, Warsaw, Poland
SEPARATION OF SOME RARE EARTHS BY REVERSED-PHASE PARTITION CHROMA-
TOGRAPHY.
INR-129/V. 1960. 10 p.
(NP-8663)
Also published in J. Chromatog., 4, 60–4 (1960)

554. SIEKIERSKI, S. and SOCHACKA, R. J.
Polish Academy of Sciences, Inst. of Nucl. Research, Krakow, Poland
SEPARATION OF CALCIUM FROM SCANDIUM BY REVERSED-PHASE CHROMATOGRAPHY.
INR-262/V. 1961. 7 p.
CA, 56, 14903h (1962)

555. SIEKIERSKI, S. and SOCHACKA, R. J.
Dept. of Radiochem.,Inst. of Nucl. Research, Warsaw, Poland
REVERSED-PHASE PARTITION CHROMATOGRAPHY WITH DI(2-ETHYLHEXYL) ORTHO-
PHOSPHORIC ACID AS THE STATIONARY PHASE. II. FACTORS AFFECTING THE
HEIGHT OF THE PLATE.
J. Chromatog., 16, 385–95(1964)

556. SILVA, R. J., SIKKELAND, T., NURMIA, M., and GHIORSO, A.
Lawrence Radiation Lab., Berkeley, Calif., USA
HULET, E. K.
Lawrence Radiation Lab., Livermore, Calif., USA
DETERMINATION OF THE No(II)-No(III) POTENTIAL FROM TRACER EXPERIMENTS.
J. Inorg. Nucl. Chem., 31, 3405–9(1969)

557. SILVA, R.
ORNL, Oak Ridge, Tennessee, USA
TRAUTMANN, N.
Univ. of Mainz, Mainz, BRD (FRG)
GHIORSO, A., HARRIS, J., and NURMIA, M.
Lawrence Radiation Lab., Berkeley, California, USA
AUTOMATED SYSTEM FOR RAPID CHEMICAL SEPARATIONS OF HEAVY ACTINIDE
AND TRANSACTINIDE ELEMENTS.
Paper presented at the Third Int. Transplutonium Element Symposium, Argonne, III., Oct.
20-22, 1971

558. SMALL, H.
Physical Research Lab., Dow Chemical Co., Midland, Mich., USA

GEL-LIQUID EXTRACTION. THE EXTRACTION AND SEPARATION OF METAL SALTS USING TRIBUTYL PHOSPHATE GELS.
J. Inorg. Nucl. Chem., 18 ,232–44 (1961)

559. SMALL, H.
Dow Chemical Co., Midland, Mich., USA
SOLVENT EXTRACTION PROCESS FOR THE RECOVERY OF URANIUM AND RARE EARTH METALS FROM AQUEOUS SOLUTIONS.
U. S. Patent 3, 102, 782
Filing Date: March 23, 1959.
Publ. Date: Sept. 3, 1963

560. SMUŁEK, W. and SIEKIERSKI, S.
Dept. of Radiochem.,Inst. of Nucl. Research, Warsaw, Poland
SEPARATION OF THE ALKALI METALS BY REVERSED-PHASE PARTITION CHROMATOGRAPHY.
J. Chromatog., 19,580–8(1965)

561. SMUŁEK, W. and ZELENAY, K.
Dept. of Radiochem.,Inst. of Nucl. Research, Warsaw, Poland
THE APPLICATION OF REVERSED-PHASE PARTITION CHROMATOGRAPHY TO THE SEPARATION OF MICRO-AMOUNTS OF COBALT FROM IRRADIATED NICKEL.
INR-667/V. 1965. 12 p.
AED-CONF. 66-094-34. 1966
CA, 65, 8288a (1966)

562. SMUŁEK, W.
Inst. of Nucl. Research, Warsaw, Poland
THE SEPARATION OF CARRIER-FREE ^{54}Mn FROM IRRADIATED IRON BY REVERSED-PHASE PARTITION CHROMATOGRAPHY.
Nukleonika, 11,635–41(1966)
NSA, 21, 10295 (1967)

563. SMUŁEK, W.
Inst. of Nucl. Research, Warsaw, Poland
SEPARATION OF EUROPIUM FROM IRRADIATED UO_2-Mg FUEL SAMPLES FOR BURN-UP DETERMINATION.
Nukleonika, 14,521–36(1969)

564. SMUŁEK,W.* and MOSZYNSKA, K.**
*   Institute of Nucl. Research, Warsaw, Poland
** Institute of Nucl. Research, Swierk, Poland
SEPARATION OF YTTRIUM FROM IRRADIATED URANIUM SAMPLES BY EXTRACTION CHROMATOGRAPHY.
Nukleonika, 14, 1123–30 (1969)

565. SOCHACKA, R. J. and SIEKIERSKI, S.
Dept. of Radiochem.,Inst. of Nucl. Research, Warsaw, Poland
REVERSED-PHASE PARTITION CHROMATOGRAPHY WITH DI(2-ETHYLHEXYL) ORTHO-PHOSPHORIC ACID AS THE STATIONARY PHASE.
I. SEPARATION OF RARE EARTHS.
J. Chromatog., 16,376–84(1964)

566. SPECHT, S., NOLTE, R. F., and BORN, H. – J.
Inst. für Radiochemie der Techn. Univ., München, BRD (FRG)

THE INFLUENCE ON THE EFFICIENCY OF THE EXTRACTION-CHROMATOGRAPHIC
SYSTEM BY THE KIND OF THE EXTRACTANT AND THE GEOMETRY OF THE SUPPORT.
Paper presented at the 7th Radiochemical Conference, Mariánské Lázné, CSSR, 24–28 Apr. 1973

567. SPÉVÁČKOVA, V. and KŘIVANEK, M.
Nucl. Research Inst., Czechoslovak Academy of Sciences, Řež, ČSSR
SEPARATION OF COBALT FROM MIXTURES OF IRON RADIOISOTOPES BY MEANS
OF EXTRACTION CHROMATOGRAPHY.
Radiochem. Radioanal. Letters, 3,63–9(1970)

568. SPÉVÁČKOVA, V. and KŘIVANEK, M.
Nucl. Research Inst., Czechoslovak Academy of Sciences, Řež, ČSSR
DITHIZONE AS A STATIONARY PHASE IN REVERSED-PHASE CHROMATOGRAPHY
USED IN ACTIVATION ANALYSES.
Proc. Anal. Chem. Conf. 3rd 1970, 1,121–6(1970)
Edited by Buzás, I., Akad. Kiadó, Budapest, Hungary
CA, 74, 49263 (1971)

569. SPÉVÁČKOVA, V. and KŘIVANEK, H.
Nucl. Research Inst., Czechoslovak Academy of Sciences, Řež, ČSSR
DETERMINATION OF GOLD IN PLATINUM BY NEUTRON ACTIVATION ANALYSIS.
Paper presented at the 7th Radiochemical Conference at Mariánské Lázné, ČSSR, 24-28
Apr. 1973

570. STAPLES, B. A., HUNTER, B. R., and SCOTT, J. C.
Allied Chemical Corp., Idaho Falls, Idaho, USA
DETERMINATION OF RADIOIRON IN SAMPLES CONTAINING HIGH LEVELS OF
RADIOACTIVITY.
AED-CONF. 72-268-001. 1972. 8 p., 3 fig., 5 ref.
Paper presented at the 27th Ann. Northwest Regional Meeting of the American Chemical Soc.,
Corvallis, Ore., June 15–16, 1972.

571. STEPANETS, O. V., YAKOVLEV, Yu. V., and ALIMARIN, I. P.
Inst. Geochem. Anal. Chem., Moscow, USSR
RADIOACTIVATION DETERMINATION OF IMPURITIES IN YTTRIUM AND MOLYB-
DENUM BY EXTRACTION DISPLACEMENT CHROMATOGRAPHY.
Zh. Anal. Khim., 25,1906–11 (1970) (in Russian)
CA, 74, 60484 (1971)

572. STRONSKI, I.
Inst. of Nucl. Physics, Krakow, Poland
RADIOCHEMICAL SEPARATION OF TERBIUM, HAFNIUM, AND THORIUM FROM
URANIUM, AND OF TERBIUM FROM HAFNIUM AND PROTACTINIUM, AND OF HAF-
NIUM FROM PROTACTINIUM AND TUNGSTEN USING REVERSED-PHASE PARTITION
CHROMATOGRAPHY WITH THE SYSTEM TBP/HCl.
Kernenergie, 8,175–8(1965) (in German)

573. STRONSKI, I.
Inst. of Nucl. Physics, Krakow, Poland
RADIOCHEMICAL STUDIES ON THE EXTRACTION AND STABILITY OF METAL CHE-
LATES AND ON THE SEPARATION OF METAL SALTS BY EXTRACTION CHROMA-
TOGRAPHY.
INP-427/C. 1965. 119 p. (in Polish)
NSA, 20, 10759 (1966)

574. STRONSKI, I.
Inst. of Nucl. Physics, Krakow, Poland
SEPARATION OF SOME METAL HALIDES BY EXTRACTION CHROMATOGRAPHY IN THE
SYSTEM TRI-n-BUTYL PHOSPHATE-HYDROCHLORIC ACID.
Radiochim. Acta, 6,163−4(1966)

575. STRONSKI, I., BITTNER, M., and KRUK, J.
Inst. of Nucl. Physics, Krakow, Poland
RADIOCHEMICAL SEPARATION OF SOME METAL SALTS BY EXTRACTION CHROMATO-
GRAPHY ON POWDERED POLYTETRAFLUOROETHYLENE IN THE SYSTEM: TRI-n-
OCTYLPHOSPHINE OXIDE − MINERAL ACIDS.
Nukleonika, 11,47−55(1966)
CA, 65, 6276h (1966)

576. STRONSKI, I., KEMMER, J., and KAUBISCH, N.
Technische Hochschule, München, BRD (FRG)
THE SEPARATION OF ^{228}Ra/^{288}Ac AND ^{212}Pb/^{212}Bi WITH THE HELP OF EXTRACTION
CHROMATOGRAPHY IN THE SYSTEM DI(-ETHYLHEXYL) PHOSPHATE-HYDROCHLORIC
ACID.
Z. Naturforsch., 23B, 135−7(1968) (in German)
NSA, 22, 14576 (1968)

577. STRONSKI, I.
Inst. of Nucl. Physics, Krakow, Poland
INVESTIGATION ON THE SEPARATION OF SOME LANTHANIDES AND ACTINIDES BY
EXTRACTION CHROMATOGRAPHY IN THE SYSTEM DI(2-ETHYLHEXYL) ORTHOPHOS-
PHORIC ACID - HYDROCHLORIC ACID.
INP-674/C. 1969. 18 p. (in German)
Also published in Radiochim. Acta, 13, 25−31 (1970) (in German)

578. STRONSKI, I.
Inst. of Nucl. Physics, Krakow, Poland
SOLVENT EXTRACTION AND SEPARATION OF SOME LANTHANIDES AND AMERICIUM
BY EXTRACTION CHROMATOGRAPHY IN THE SYSTEM ALIQUAT-336 − $LiNO_3$ AND
HNO_3.
INP-675/C. 1969. 15 p.
Also published in Chromatographia, 2, 285−8 (1969)

579. STRONSKI, I.
Inst. of Nucl. Physics, Krakow, Poland
RADIOCHEMICAL SEPARATION OF MICROAMOUNTS OF In-Cd, In-Sn-Te, AND Sb-Sn BY
EXTRACTION CHROMATOGRAPHY IN THE AMBERLITE LA-2−HYDROCHLORIC ACID
SYSTEM.
Radiochem. Radioanal. Letters, 5,113−7(1970)

580. SULCEK, Z. and SIXTA, V.
Ustred. Ustav Geol., Prague, CSSR
SEPARATION OF LEAD AND BISMUTH BY MEANS OF EXTRACTION CHROMATOGRAPHY.
Collect. Czech. Chem. Commun., 36,1561−8(1971) (in German)
CA, 75, 25822 (1971)

581. SWEET, Th. R., PARLETT, H. W., PORELL, A. L., KAHN, W., BLAIR, S. D., BALDECK, C.,
and GUENTHNER, Chr.

474

Ohio State Univ. Research Fond., Columbus, Ohio, USA
REVERSED-PHASE PARTITION CHROMATOGRAPHY OF EUROPIUM AND TERBIUM
WITH 2,2,6,6-TETRAMETHYL-3,5-HEPTANEDIONE   (H(thd)) AS THE STATIONARY
PHASE.
ARL-68-0210, pp. 89–98, 1968

582. TAUBE, M., GWOŹDŹ, E., GAVRILOV, K. A., MALY, J., BRANDSTETR, J., and WANG, T. S.
Joint Inst. for Nucl. Research, Dubna, Moscow, USSR
EXTRACTION OF MENDELEVIUM AND FERMIUM IN TRIBUTYL PHOSPHATE - NITRIC
ACID SYSTEM.
Nukleonika, 7, 479–82 (1962) (in Russian)
CA, 58, 9671c (1963)

583. TERADA, K. and KIBA, TOSHIYASU
Dept. of Chem., Fac. of Science, Kanazawa Univ., Kanazawa, Japan
A NEUTRON-ACTIVATION ANALYSIS OF RHENIUM IN ROCKS AND SEDIMENTS BY
EXTRACTION CHROMATOGRAPHY WITH PYRIDINE-KEL-F COLUMN.
Paper presented at the Int. Congress on Analytical Chemistry, Kyoto, Japan, 3 April 1973

584. TESTA, C.
CISE, Milan, Italy
TRI-n-OCTYLAMINE AS LIQUID ANION EXCHANGER FOR CHROMATOGRAPHIC SEPA-
RATION OF RARE EARTHS ON PAPER OR CELLULOSE POWDER.
Anal. Chem., 34, 1556–9 (1962)

585. TESTA, C.
CNEN, Casaccia, Italy
DETERMINATION OF THORIUM IN URINE BY MEANS OF KEL-F COLUMN SUPPORTING
TRI-n-OCTYLPHOSPHINE OXIDE.
Radiol. Health Safety Mining Milling Nucl. Mater., Symposium, Vienna, August 1963.
Proceedings, Vol. 2, pp. 489–501, IAEA, Vienna, 1964.
CA, 61, 13800e (1964)

586. TESTA, C. and CESARANO, C.
CNEN, Casaccia, Italy
SEPARATION OF CESIUM-137 FROM FISSION PRODUCTS BY MEANS OF A KEL-F COL-
UMN SUPPORTING DIPICRYLAMINE.
J. Chromatog., 19, 594–8 (1965)

587. TESTA, C. and MASI, G.
CNEN, Casaccia, Italy
DETERMINATION OF ENRICHED URANIUM IN THE URINE BY MEANS OF ITS CHRO-
MATOGRAPHICAL ISOLATION ON A KEL-F COLUMN IMPREGNATED WITH TRIBUTYL
PHOSPHATE AND SUBSEQUENT ALPHA COUNTING.
Minerva Nucl., 9, 22–5 (1965)

588. TESTA, C.
CNEN, Casaccia, Italy
ISOLATION AND DETERMINATION OF PLUTONIUM-239 IN THE URINE BY A COLUMN
OF POLY-TRIFLUOROCHLOROETHYLENE SUPPORTING TRI-n-OCTYLAMINE.
Minerva Fisiconucl. (Torino), 10, 202–9 (1966) (in Italian)

589. TESTA, C., DE ROSA, D., and SALVATORI, A.
CNEN, Rome, Italy
DETERMINATION OF RADIOTOXICOLOGY OF NATURAL THORIUM, ENRICHED URA-
NIUM, AND TOTAL BETA ACTIVITY BY MEANS OF REVERSED-PHASE PARTITION
CHROMATOGRAPHY.
RT/PROT (68) 6. 1968. 84 p. (in Italian)

590. TESTA, C.
CNEN, Casaccia, Italy
NEW RADIOTOXICOLOGICAL METHODS USED AT THE ITALIAN NUCLEAR CENTRE
OF CASACCIA.
RT/PROT (69) 44. 1969. 25 p.

591. TESTA, C. and SANTORI, G.
CNEN, Casaccia, Italy
A NEW METHOD FOR THE DETERMINATION OF STRONTIUM-90 IN URINE.
Energia Nucleare (Milan), 17,320−4(1970) (in Italian)

592. TESTA, C.
Radiotoxicology Lab.,Medical Service, CNEN,CSN-Casaccia, Rome, Italy
COLUMN REVERSED-PHASE PARTITION CHROMATOGRAPHY FOR THE ISOLATION
OF SOME RADIONUCLIDES FROM BIOLOGICAL MATERIALS.
Anal. Chim. Acta, 50,447−55(1970)

593. TESTA, C.
Medical Service, CNEN, CSN-Casaccia, Rome, Italy
INDIRECT METHODS USED AT CNEN FOR THE EVALUATION OF INTERNAL CONTAMI-
NATION.
Assessment of Radioactive Contamination in Man, Proc. IAEA-WHO-Symposium, Stockholm,
Sweden, pp. 405−21.
IAEA STI/PUB-290, Vienna 1972

594. TESTA, C.[*] and STACCIOLI,L.[**]
* Medical Service, CNEN, CNS-Casaccia, Rome, Italy
** Inst. of Inorg.Chem., Fac. of Pharmacy, Univ. of Urbino, Italy
NEW APPLICATIONS OF COLUMN REVERSED-PHASE PARTITION CHROMATOGRAPHY
WITH MICROTHENE-710 SUPPORTING SOME SELECTIVE EXTRACTANTS.
Analyst, 97,527−32(1972)

595. THOMASSEN, J. A.[*] and WINDSOR, H. H.[**]
* Institutt for Atomenergi, Kjeller, Norway
** Brookhaven Nat. Lab., New York, USA
A REVERSED-PHASE PARTITION CHROMATOGRAPHY METHOD FOR THE SEPARATION
OF ^{239}Np USED IN DETERMINATION OF RESONANCE ESCAPE PROBABILITIES IN ^{235}U-^{238}U
SYSTEMS.
Paper presented at the Discussion on Nuclear Chemistry, Oxford, Sept. 1962
AERE-M-1078, pp.92−3(1962)

596. THOMASSEN, J. and WINDSOR, H. H.
Institutt for Atomenergi, Kjeller, Norway
MEASUREMENT OF RESONANCE ABSORPTION IN ^{238}U USING A CHEMICAL SEPARA-
TION TECHNIQUE FOR ISOLATION OF ^{239}Np.
KR-44. 1963. 37 p.

476

597. TOMAŽIČ, B. and SIEKIERSKI, S.
Dept. of Radiochem., Inst. of Nucl. Research, Warsaw, Poland
SEPARATION OF SOME FISSION PRODUCTS FROM URANIUM(VI) BY REVERSED-
PHASE PARTITION CHROMATOGRAPHY.
J. Chromatog., 21,98–104(1966)

598. TOMAŽIČ, B.
"Rudjer Boskovic" Institute, Zagreb, Yugoslavia
EXTRACTION OF TRACES OF CERIUM, EUROPIUM, TERBIUM AND LUTETIUM FROM
URANIUM(VI) SOLUTIONS WITH DI(2-ETHYLHEXYL) PHOSPHORIC ACID.
Anal. Chim. Acta, 49,57–65(1970)

599. TRAUTMANN, N.
Univ. Mainz, BRD (FRG)
DECAY STUDIES ON SHORT LIVING NEUTRON-RICH ZIRCONIUM, NIOBIUM AND
PROTACTINIUM NUCLIDES WITH RAPID SOLVENT EXTRACTION PROCEDURES.
Thesis
AED-DISS 67-124. 1967. 134 p. 52 figs. 11 tables (in German)

600. TRAUTMANN, N., DENIG, R., and HERRMANN, G.
Univ. of Mainz, BRD (FRG)
CAPTURE CROSS-SECTIONS OF (n,p),(n,pn) AND (n,α) REACTIONS OF URANIUM-238
WITH 15-MeV NEUTRONS AND DETECTION OF THORIUM-235.
Radiochim. Acta, 11,168–71(1969) (in German)

601. TSERKOVNITSKAYA, I. A. and LUGININ, V. A.
Vestn. Leningrad. Univ., USSR
SEPARATION OF VANADIUM(IV) AND VANADIUM(V) BY COLUMN PARTITION CHRO-
MATOGRAPHY.
Fiz., Khim.,162–4(1969) (in Russian)
CA, 71, 119214 (1969)

602. UENO, KAORU and HOSHI, M.
Japan Atomic Energy Res. Inst., Tokai-mura, Japan
THE SEPARATION OF ZIRCONIUM AND HAFNIUM BY TRIBUTYL PHOSPHATE-CELITE
REVERSED-PHASE PARTITION CHROMATOGRAPHY.
Bull. Chem. Soc. Japan, 39,2183–7(1966)

603. UENO, KAORU, SAEKI, M., and ISHIMORI, T.
JAERI, Tokai-mura, Ibaraki-ken, Japan
ISOLATION OF RADIONIOBIUM FROM GAMMA-IRRADIATED MOLYBDENUM OXIDE.
J. Nucl. Sci. Techn., 6,203–6(1969)

604. UENO, KEIHEI, YANO, T., and KOJIMA, T.
Dept. of Org. Synthesis, Fac. of Eng., Kyushu Univ., Fukuoka, Japan
THE ANALYTICAL APPLICATIONS OF ORGANIC REAGENTS IN HYDROPHOBIC GEL
MEDIA.
Anal. Lett., 5, 439–44 (1972)

605. USHATSKII, V. N., PREOBRAZHENSKAYA, L. D., and OMELCHEKO, M. B.
USE OF COLUMN PARTITION CHROMATOGRAPHY FOR SEPARATING NEPTUNIUM
VALENCE STATES.
Radiokhimiya, 14,892–3 (1972) (in Russian)
CA, 78, 92061 (1973)

606. VAN OOYEN, J. and BAC, R.,
Reactor Centrum Nederland, Petten, The Netherlands
BONNEVIE-SVENDSEN, N. and ESCHRICH, H.
Institutt for Atomenergi, Kjeller, Norway
EXTRACTION STUDIES ON SELECTED PROBLEMS IN REPROCESSING.
Proc. Third Int. Conf. Geneva 1964, Vol. 10, pp. 402–11. P/758
CA, 64, 18884f (1966)

607. VAN OOYEN, J.
Reactor Centrum Nederland, Petten, The Netherlands
QUATERNARY AMMONIUM NITRATES AS EXTRACTANTS FOR TRIVALENT ACTINIDES.
In Solvent Extr. Chem., (Proc. Int. Conf., Gothenburg, Sweden, 1966) pp. 485–92
(D. DYRSSEN, J.O. LILJENZIN, and J. RYDBERG, eds.)
North Holland Publ. Co., Amsterdam (1967) pp. 485–92

608. VAN OOYEN, J.
Reactor Centrum Nederland, Petten, The Netherlands
THE PROPERTIES OF TRILAURYLMETHYLAMMONIUM NITRATE (TLMANO$_3$) AS AN
EXTRACTANT FOR TRIVALENT ACTINIDES.
Thesis
RCN-113. 1970. 101 p.

609. WADE, M. A. and YAMAMURA, S. S.
Atomic Energy Division, Phillips Petroleum Co., Idaho Falls, Idaho, USA
COLUMN EXTRACTION-SPECTROPHOTOMETRIC DETERMINATION OF IRON.
Anal. Chem., 36,1861–2 (1964)
CA, 61, 12621f (1964)

610. WATANABE, K.
Japan Atomic Energy Research Institute, Tokai-mura, Ibaraki-ken, Japan
SEPARATION OF AMERICIUM AND SOME RARE EARTHS FROM LANTHANUM BY
REVERSED-PHASE CHROMATOGRAPHY.
J. Nucl. Sci. Techn. (Tokyo), 2,45–50(1965)

611. WATANABE, K.
Japan Atomic Energy Research Institute, Tokai-mura, Ibaraki-ken, Japan
SEPARATION OF PROMETHIUM AND AMERICIUM BY REVERSED-PHASE CHROMATO-
GRAPHY.
J. Nucl. Sci. Technol. (Tokyo), 2,112–4(1965)
NSA, 19, 22418 (1965)

612. WEHNER, H., AL–MURAB, S.*, and STOEPPLER, M.
Zentralinstitut für Analytische Chemie, Kernforschungsanlage Jülich, BRD (FRG)
* Nuclear Research Institute, Tiwaita, Bagdad-Irak
EXTRACTION CHROMATOGRAPHIC SEPARATION OF ^{239}Np FROM FISSION AND
ACTIVATION PRODUCTS IN THE DETERMINATION OF MICROGRAM AND SUBMICROGF
QUANTITIES OF URANIUM.
Radiochem. Radioanal. Letters, 13, 1–6 (1973)

613. WENZEL, U. and RIEDEL, H. J.
Kernforschungsanlage Jülich, BRD (FRG)
ACTINIDE COMPOSITION STUDIES ON THORIUM/URANIUM FUEL ELEMENTS WITH
KNOWN IRRADIATION DATA.
Proc. Symp. Anal. Methods in the Nuclear Fuel Cycle, Vienna, 29 Nov.-3 Dec., 1971,
pp. 549–62

614. WENZEL, U. and HERZ, D.
Inst. für Chem. Techn., Kernforschungsanlage Jülich, BRD (FRG)
ALPHA-SPECTROMETRY IN NUCLEAR FUEL ANALYSIS.
Paper presented at the 7th Radiochemical Conference, Mariánské Lázné, CSSR, 24-28
Apr. 1973

615. WINCHESTER, J. W.
Oak Ridge National Laboratory, Tenn., USA
A PRELIMINARY INVESTIGATION OF A CHROMATOGRAPHIC COLUMN SEPARATION
OF RARE EARTHS USING DI(2-ETHYLHEXYL)PHOSPHORIC ACID.
CF-58-12-43. Dec. 30, 1958. 10 p.

616. WINCHESTER, J. W.
Oak Ridge National Laboratory, Tenn., USA
RARE EARTH CHROMATOGRAPHY USING BIS(2-ETHYLHEXYL) ORTHOPHOSPHORIC
ACID.
CF-60-3-158. March 14, 1960. 9 p.

617. WINCHESTER, J. W.
Massachusetts Inst. of Techn., Cambridge, USA
RARE EARTH CHROMATOGRAPHY USING BIS(2-ETHYLHEXYL) ORTHOPHOSPHORIC
ACID.
J. Chromatog., 10, 502–6 (1963)

618. WINSTEN, W. A.
Dept. of Chem., Hofstra College, Hempstead, L. I., N. Y., USA
REVERSED-PHASE CHROMATOGRAPHY ON MICROPOROUS POLYETHYLENE
SUPPORTS.
Anal. Chem., 34, 1334–5 (1962)

619. WOIDICH, H. and PFANNHAUSER, W.
Forschungsinstitut der Ernährungswirtschaft, Wien, Austria
RAPID SEPARATION OF MERCURY FROM BIOLOGICAL MATERIALS BY EXTRAC-
TION CHROMATOGRAPHY WITH DITHIZONE.
Z. Anal. Chem., 261, 31 (1972) (in German)

620. YAKOVLEV, Yu. V. and STEPANETS, O. V.
Inst. Geochem. Anal. Chem., Moscow, USSR
SUBSTOICHIOMETRIC RADIOACTIVATION DETERMINATION OF CADMIUM AND
COPPER IN YTTRIUM BY DISPLACEMENT EXTRACTION CHROMATOGRAPHY.
Zh. Anal. Khim., 25, 578–9 (1970) (in Russian)
CA, 73, 31253 (1970)

621. YAKOVLEV, Yu. V., STEPANETS, O. V., and SUKHANOVSKAYA, A. I.
Inst. Geochem. Anal. Chem., Moscow, USSR
USE OF DISPLACEMENT EXTRACTION CHROMATOGRAPHY FOR THE CONCENTRA-
TION OF SOME MICRO-IMPURITIES IN PURE SUBSTANCES.
Zh. Anal. Khim., 28, 173–4 (1973) (in Russian)

622. YAMAMOTO, D. and HIRAOKA, T.
Kumamoto Univ., Kumamoto, Japan
SPUTTERING CHROMATOGRAPHY SEPARATION OF INORGANIC CATIONS.
Bunseki Kagaku, 18, 1019–21 (1969) (in Japanese)
CA, 71, 119220 (1969)

623. YAMAMURA, S. S.
Atomic Energy Division, Phillips Petroleum Co., Idaho Falls, Idaho, USA
DETERMINATION OF IRON(III) BY (ETHYLENEDINITRILO) TETRAACETIC ACID
REPLACEMENT TITRIMETRY FOLLOWING SELECTIVE SEPARATION BY COLUMN
EXTRACTION.
Anal. Chem., 36, 1858–61(1964)
CA, 61, 10031a (1964)

624. YOSHIDA, H., YONEZAWA, C., and GUNJI, K.
Japan Atomic Energy Research Inst., Tokai-mura, Naka-gun, Ibaraki-ken, Japan
SEPARATION OF MOLYBDENUM IN FISSION PRODUCTS BY SOLVENT EXTRACTION
AND EXTRACTION CHROMATOGRAPHY.
Bunseki Kagaku (Japan Analyst), 19, 818–23(1970) (in Japanese)

625. YUSAF, M., SULCEK, Z., and DOLEZAL, J.
Dept. Anal. Chem., Charles Univ., Prague, CSSR
EXTRACTION CHROMATOGRAPHY OF Pd(II)–THIOUREA COMPLEX.
Anal. Lett., 4, 119–24(1971)
CA, 75, 29648 (1971)

626. ZELENAY, T.
Dept. of Radiochem., Inst. of Nucl. Research, Warsaw, Poland
DETERMINATION OF FISSION PRODUCT NEODYMIUM-148 FOR BURN-UP.
Radiochem. Radioanal. Letters, 2, 33–40 (1960)

627. ZEMSKOVA, M. G., LEBEDEV, N. A., MELAMED, Sh. G., SAUNKIN, O. F., SUKHOV, G.
V., KHALKIN, V. A., HERRMANN, E., SHMANENKOVA, G. I.
Joint Inst. for Nucl. Research, Dubna, USSR, State Scientific-Research and Planning Inst. of
the Rare Metal Industry, Moscow, USSR
USE OF EXTRACTION CHROMATOGRAPHY FOR IMPROVING YTTRIUM ANALYSIS
SENSITIVITY TO RARE EARTH IMPURITIES.
Zavod. Lab., 33, 667–71(1967) (in Russian)
NSA, 21, 38879 (1967)
CA, 67, 104893 (1967)

628. ZIV, D. M., SHESTAKOV, B. I., and SHESTAKOVA, I. A.
SEPARATION OF LANTHANUM AND ACTINIUM BY PARTITION CHROMATOGRAPHY.
Radiokhimiya, 10, 738–40(1968) (in Russian)

629. ZHUKOVSKII, Yu, G., KATYKHIN, G. S., MARTYNOV, A. L., and NIKITIN, M. K.
PARTITION CHROMATOGRAPHY ON POLY(TETRAFLUOROETHYLENE). SEPARATION
OF MANGANESE-56 FROM IRRADIATED POTASSIUM PERMANGANATE.
Radiokhimiya, 10, 252–3(1968) (in Russian)
CA, 69, 48489 (1968)

7. TABULAR SURVEY OF EXPERIMENTAL DATA ON COLUMN EXTRACTION CHROMATOGRAPHIC INVESTIGATIONS

Note: — Hyphens between the various species chromatographed indicate an experimentally verified separation of these species (e. g. Zr-Nb; Ce, La-Cs, Ba; U-Pu-fission products).

    — Commas between the various species indicate that the chromatographic behaviour of these species has been investigated and a useful separation of all or some of them has not necessarily been achieved.

    — The valency state of the ions has in most cases been omitted.

    — The abbreviations used for extractants and support materials are explained in Paragraphs 8 and 9. (Abbreviations used for organic components in mobile phases are also contained and explained in Paragraph 8 at the end).

    — Hyphens between eluting agents indicate a solution mixture.

    — Mobile phases, separated by commas, are given in the sequence of their feeding to the column.

    — The mobile phase, given after the double point, indicates the eluent for the species considered.

    — Reference is made to the first given author only.

SPECIES	SUPPORT	STAT. PHASE (EXTRACTANT)	MOB. PHASE (ELUENT)	REFERENCE Nº	AUTHOR
In-Cd	Al_2O_3	TNOA	HCl for concentr. on column In:0.15-0.20 M HCl; Cd: retained on column	190	ABRAO
Li, Na, K, Rb, Cs	Kel-F	$PhNO_2$	H_2O; 1 M and 6 M HCl	191	AKAZA
Alkaline earth metals	Kel-F	1.5 M HTTA in MIBK	$HOAc-NH_4OAc-NH_4OH$ buffer Ba-Sr:pH 6.5 Sr-Ca:pH 5.5	192	AKAZA
Hexacyano-ferrate(II) Hexacyano-ferrate(III)	Kel-F	TBP	HCl (various concentrations)	193	AKAZA
Au-Pt-Pd Au-metals	Daiflon	TBP	HCl (various concentrations)	194	AKAZA
Fe(III), Hg(II), Sn(IV)-Bi(III), Cd(II), Ag(I), Zn(II), Sb(III), Cr(III), Al(III), Cu(II), Co(II), Ni(II), Pb(II), Mn(II), Mg(II), Ca(II), Sr(II), Ba(II)	Daiflon	TBP	2 M HCl Fe, Hg, Sn: retained on column; rest of ions eluted	195	AKAZA
Ag, Bi, Zn Cd-rest of ions eluted from preceding column above	Daiflon	$TNOA-CCl_4$	2 M HCl Ag, Bi, Zn, Cd: retained on column		
Sb-rest of ions eluted from preceding column above	Daiflon	TBP	6 M HCl Sb: retained on column		
Cr, Al-rest of ions eluted from preceding column above	Daiflon	Acetylacetone-$CHCl_3$	0.5 M CH_3COONH_4 (pH 5.0) Cr, Al: retained on column		

482

SPECIES	SUPPORT	STAT. PHASE (EXTRACTANT)	MOB. PHASE (ELUENT)	REFERENCE N°	AUTHOR
Cu, Co, Ni, Pb, Mn, Mg-rest of ions eluted from preceding column	Daiflon	HTTA-MIBK	0.5 M CH_3COONH_4 (pH 5) Cu, Co, Ni, Pb, Mn, Mg: retained on column		AKAZA
Ca, Sr, Ba	Daiflon	HTTA-MIBK	0.5 M CH_3COONH_4 (pH 7) Ca, Sr, Ba: retained on column		
U	Activated C	Amines (e.g. laurylamine)	H_2SO_4	196	AKERMAN
Ce, Pr-Nd, ^{147}Pm-^{155}Eu	Cellulose	HDEHP (75°C)	Ce, Pr: 0.35 M HCl Nd, Pm: 0.5 M HCl Eu: 1.5 M HCl	197	ALBINI
^{144}Ce-^{147}Pm-^{155}Eu	Cellulose	0.5 M HDEHP	0.1 M HCl	198	ALBINI
Ce-Nd, Pm-Eu	Cellulose	HDEHP (75°C)	(see Ref. [197]) DTPA-Lactic acid-HNO_3 (pH ~ 2.5)	199	ALBINI
Actinides-rare earths	Porous glass beads (100-200 mesh)	HDEHP	Am: 0.5 M DPTA-1 M lactic acid (pH ~ 2.5) Ce, Eu, Tb: 2 M HNO_3	200	ALI
Ga-Zn	Fluoroplast-4	TBP	3 M HCl	201	ALIMARIN
Ga-Li, K, Na, Mg, Ca, Zn, Cd	Teflon	TBP	0-7 M Cl^-	202	ALIMARIN
Ga-Al, Fe	Fluoroplast-4	TBP	3 M HCl-5% ascorbic acid	203	ALIMARIN
Ga-Al-Fe	Fluoroplast-4	TBP	Al: 7 M HCl Fe: 3 M HCl-5% ascorbic acid		
(Cu(II), Mn(II))			Ga: 0.1 M HCl (stripping)		
Traces Ga-Zn, Cd, Al, Fe, In-Ga	Fluoroplast-4	TBP	0.1 M HCl or H_2O 0.8 M HBr	204	ALIMARIN

SPECIES	SUPPORT	STAT. PHASE (EXTRACTANT)	MOB. PHASE (ELUENT)	REFERENCE N°	AUTHOR
Ga-Zn, Al, Cd, Cu; Ga, Fe(III)-Zn, Al, Cu	Fluoro-plast-4	TBP	4 M HCl and 5 M NaCl 4 M HCl 0.1 M HCl (stripping)	205	ALIMARIN
Al-Ga, Co, Cu, Fe(III)	Teflon-powder	TBP	NH_4SCN 3% NH_4OH, H_2O 0.5 N HCl	206	ALIMARIN
Eu(II)-other rare earths	Polyfluoro carbon powder PF-4	HDEHP	Eu: 0.2 M HCl-3 M NH_4Cl Rare earths: conc. HCl	207	ALIMARIN
Ga-In	Ftoro-plast-4	TEHP (DEHPP)	2 M HBr	208	ALIMARIN
Zn-In-Cd	Ftoro-plast-4	TBP	0.1-1.0 M HBr	209	ALIMARIN
Ti-Nb	Ftoro-plast-4	TNOA in benzene	Ti: 7-11 M HCl Nb: 4 M HNO_3-0.05% H_2O_2	210	ALIMARIN
Zn-Cd; Cd-Cu; Cu-Ag; Ag-Hg;	Ftoro-plast-4	Zn-DDTC in chloroform	Displacement by substoichiometric exchange using aqu. solution of displacing element (e.g. Cd displ. Zn etc.)	211	ALIMARIN
Mn-Co-Fe-Hg (traces)			Aqueous salt sol. at pH 6-7, Hg: 1 M KI at pH 1 by elution		
Ti(IV)-Nb(V)-Mo(VI)	Ftoro-plast-4	TBP	Ti: 9 M HCl Nb: 9-5 M HCl Mo: 1 M HCl	212	ALIMARIN
In(III)-Sb(V)	Ftoro-plast-4	TBP-isoamyl-alcohol; TBP-acetyl-acetone	4 M HCl	213	ALIMARIN
Cm-Bk-Es-Cf	SiO_2^+	HDEHP	1 M 2-methyl-lactic acid	214	ALY

SPECIES	SUPPORT	STAT. PHASE (EXTRACTANT)	MOB. PHASE (ELUENT)	REFERENCE N°	AUTHOR
Eu-Sm	SiO_2^+	HTTA,DBDECP and mixture of both	Acetic acid buffer solution pH 3	215	ALY
Sm-Eu; Cm-Cf	SiO_2^+	DBDECP-HTTA	HNO_3, pH 1.5	216	ALY
Neutron irrad. potassium titanium oxalate target (sol.)- Sc isotopes	Kieselguhr	HDEHP	Target mat. (+Ca): 0.15 M-1.5 M HNO_3 46-48 Sc: 12 M HNO_3 - 1 M HCl	217	ALY
$^{32}P(PO_4^{3-})$ $^{35}S(SO_4^{2-})$ KCl	Kieselguhr (Celite-365)	Amberlite LA-2	PO_4^{3-} :0.05 M KCl (H_2O) SO_4^{2-} : 1 M HCl	218	ALY
Cm-Cf	Kieselguhr (Celite-360)	Alamine-336	11 M LiCl in 0.02 M HCl	219	ALY
Pb-(Cu, Fe, Mn, Zn, Cd, Co)	Activated C	0.6% dithizone in $CHCl_3$	0.1 M $KNaC_4H_4O_6$	220	ANDREEV
Cd, Co, Cu, Mn, Pb, U, Zn (concentration)	Activated C	α-nitroso-β-naphthol dithizone, Na-diethyl-dithiocarbamate	pH regulated for selective retention ($\geqslant 90\%$) elution with dil. HCl	221	ANDREEV
B, Cr, Cu		Hydroxyquinoline polyvinyl alcohol	(Doubtful if extr. chr.)		
Pt(IV)-Mo(VI)-Re(VII)-W(VI)	Ftoro-plast-4	0.35 M TABAN (alkyl C_7-C_9) in toluene	Pt(IV):1 M HNO_3 Mo(VI): 6 M HNO_3 Re(VII):7 M HNO_3 W(VI): toluene	222	ARTYUKHIN
U, Pu, Cs, Sr, Ba, Al, rare earths, Nd, Am, Cm	SiO_2^+	HDEHP	Cs, Sr, Ba, Al: 0.1 M HCl rare earths, Nd, Am, Cm: 1 M HCl	223	BAETSLÉ
Rare earths Am, Cm-Nd	SiO_2^+	HDEHP	Nd: 0.2 M HCl		

SPECIES	SUPPORT	STAT. PHASE (EXTRACTANT)	MOB. PHASE (ELUENT)		REFERENCE N°	AUTHOR
Am-rare earths	SiO_2^+	Aliquat-336	$M\ NH_4SCN$	$M\ H_2SO_4$	224	BARBANO
			R.E.: 1	0.01		
			Am: 0.1	0.2		
^{140}La-^{144}Ce-^{147}Pm-$^{152+154}Eu$-^{170}Tm-^{241}Am	SiO_2^+	Aliquat-336	La: 9 Ce: 7 Pm: 5 Eu: 3 Tm: 0.1	0.01 0.01 0.01 0.01 0.05		
Zn	Voltalef	TBP	HCl		225	BECKER
^{65}Zn Evaluation of theoretical plate height as function of various param.	Voltalef-300 LD, 300 CHR, 302 UF, 300 LD-PL micro, 300 UF	TBP	HCl(25-60°C)		226	BECKER
Fe(III), V(V) (reduction)	Kel-F	Tetrachloro-hydroquinone (chloranil)	Fe(III) → Fe(II): 0.03 M HCl V(V) → V(IV): 1 M H_2SO_4		227	BELCHER
U, fission prod.	Styrene-divinyl benzene copolymer	TBP org. solv.	HNO_3 (various conc.)		228	BERANOVA
(U, HNO_3) studies on extraction mechanism	Styrene-divinyl benzene copolymer	5-50% TBP in dichloroethane	(HNO_3)		229	BERANOVA
U(VI)-^{90}Sr, ^{90}Y, ^{95}Zr, ^{95}Nb, ^{106}Ru, ^{144}Ce	Styrene-divinyl benzene copolymer	TBP	HNO_3 (various conc.)		230	BERANOVA
U(VI)-fission products	Styrene-divinyl benzene copolymer	TNOA, TIOA, TLA	HNO_3, HCl (various conc.)		231	BERANOVA

SPECIES	SUPPORT	STAT. PHASE (EXTRACTANT)	MOB. PHASE (ELUENT)	REFERENCE N°	AUTHOR
Th-U; Eu-U-Pa; Cd-In; Ni, Co, Ca, Sr, Sn-Fe; Sr-Y; Sn-Te	Ftoro-plast-4	Dodecylphosphor. acid-ethylether	HNO_3; HCl	232	BITTNER
Au-Sb-Tl Ni-Cr-Mn-Co Sc, Pa, Zr, Hf, Zn, Cd, Al, La, Cu, Ni, Cr, Mn, As	Cellulose	TOPO	HCl 8 M HCl HCl, H_2SO_4, H_3PO_4	233	BLOURI
$^{90}Sr-^{90}Y$	Glass micro beads	HDEHP	Sr: HNO_3 of pH 0.5 Y:6 M HNO_3	234	BOGEN
Ga-In	Fluoro-plast-4	TBP	HBr, HCl or KBr	235	BOLSHOVA
In-Cd-Zn; In-Cd	Fluoro-plast-4	TBP	1 M LiBr-0.5 M HBr; 3 M NaBr	236	BOLSHOVA
Pu(III)-(IV)-(VI); U(VI), fission products	SiO_2^+	TBP	HNO_3	237	BONNEVIE-SVENDSEN
Rare earths	SiO_2	HDEHP	0.1-4 M HCl	238	BOSHOLM
Light lanthani-des-Er(Dy, Tb, Gd)	SiO_2-gel	HDEHP	2.5 M HCl	239	BOSHOLM
Fm, Md; Cf	SiO_2^+	TBP	HNO_3; HCl (various conc.)	240	BRANDS-TETR
Ni-Pd	Polyure-thane foam (polyether type)	TBP	Ni:0.1 M $HClO_4$-3% thiourea-1% $NaClO_4$ Pd: H_2O	241	BRAUN
Bi-Pd	id.	id.	Bi: 0.5 M $HClO_4$ Pd: H_2O		
Ni-Bi-Pd	id.	id.	Ni: 0.1 M $HClO_4$-3% thiourea-1% $NaClO_4$ Bi:0.5 M $HClO_4$ Pd: H_2O		

SPECIES	SUPPORT	STAT. PHASE (EXTRACTANT)	MOB. PHASE (ELUENT)	REFERENCE N°	AUTHOR
Bi-Pd	Voltalef	TBP	Bi: 0.5 M $HClO_4$	242	BRAUN
Ni-Pd	Polyure-thane foam (polyether type)	TBP	Ni: 0.1 M $HClO_4$- 3% thiourea- 1% $NaClO_4$ Pd: H_2O	243	BRAUN
Ce(IV) → Ce(III)	Polyure-thane foam (polyether type)	Tetrachloro-hydroquinone (chloranil)	H_2SO_4, H_2O, ascorbic acid, H_2O	244	BRAUN
V(V) → V(III)			H_2SO_4, H_2O, ascorbic acid, H_2O		
Fe(III) → Fe(II) (redox reactions)			HCl, H_2O, ascorbic acid, H_2O		
Ni-Co	Polyure-thane foam (polyether type)	TNOA. HCl	Ni: 4 M HCl Co: 1 M HCl	245	BRAUN
Au(III)- Zn, Co, Ni, Fe(III), Sb(III), Cu(II), Bi(III), Pb(II)	Polyure-thane foam	TBP	0.1 M HCl contg. 3% thiourea+1% $NaClO_4$; Au retained on column	246	BRAUN
Ce(IV) → (III) V(V) → (IV) Fe(III) → (II) (redox col.)	Polyure-thane foam	Tetrachloro-hydroquinone (chloranil)	1 M H_2SO_4 1 M H_2SO_4 0.03 M HCl	247	BRAUN
Fe-Co, Cu, Ni (^{59}Fe-^{58}Co)	Polyure-thane foam	TBP	Co, Cu, Ni: 5 M HCl Fe: 0.01 M HCl	248	BRAUN
Cm-Pu, Ce-Fe, Ni, Pb-Am	SiO_2^+	Aliquat-336	Fe, Ni, Pb: 8 M $LiNO_3$-0.01 M HNO_3-0.5 M $KBrO_3$	249	BUIJS

SPECIES	SUPPORT	STAT. PHASE (EXTRACTANT)	MOB. PHASE (ELUENT)	REFERENCE N°	AUTHOR
			Cm: 3.5 M $LiNO_3$-0.01 M HNO_3-0.5 M $KBrO_3$ Am: 1.0 M HNO_3-0.5 M $KBrO_3$ (Pu, Ce: retained on column)		
Cm-rare earths-Li, K	SiO_2^+	HDEHP	Li, K: 1 M lactic acid, pH 2.8 Cm: 1 M lactic acid-0.5 M DTPA (Rare earths: retained on column)		
Cm-Pu, Ce, other impurities	SiO_2^+	Aliquat-336	(see Ref. [249])	250	BUIJS
Cm-rare earths, Li, K	SiO_2^+	HDEHP			
Pt-Ir, Pd, Rh	Ftoro-plast-4	TBP	HCl	251	BUKOVS-KAYA
U-Y, Eu, Sm, Gd, Dy	SiO_2-gel	0.5 M TNOA in benzene	Y, rare earths: 8 M HCl U: retained on column	252	BUTECE-LEA
Pu, Th, Pa, Np	Mithene 350/80 (polyethylene)	TOPO	0.3 M H_2SO_4-1 M HF	253	CAMERA
^{231}Pa	Mithene 350/8 (polyethylene)	TOPO	^{231}Pa: 0.2 M HF-6 M HCl	254	CAMERA
Pb, Zn, Mn(II), Cd, Co, Cu	Cellulose acetate	Dithizone in CCl_4	H_2O	255	CARRIT
U-impurities	Kel-F	TBP	Impurities: 5 M HNO_3 U: 10% $(NH_4)_2SO_4$-sol.	256	CATTIN

SPECIES	SUPPORT	STAT. PHASE (EXTRACTANT)	MOB. PHASE (ELUENT)	REFERENCE Nº	AUTHOR
Fe(III)-Co-Ni;	Cellulose	TOPO	7 M HCl, 3 M HCl, 0.5 M H_2SO_4	257	CERRAI
U(VI)-Th-La;			8 M HCl, 1 M HCl, 4 M H_3PO_4		
La-Th-Zr;			8 M HCl, 1 M HCl, 4 M H_3PO_4		
Al-Cu-Fe(III)-U(VI);			6 M HCl, 2 M HCl, 0.5 M H_2SO_4, 4 M H_3PO_4		
Pb-Bi-Fe(III);			6 M HCl, 10 M HCl, 0.5 M H_2SO_4		
Cu-Bi-Sc-U(VI);			0.2 M HNO_3 6 M HNO_3 0.1 M HCl 4 M H_3PO_4		
Y-Th-Sc;			8 M HCl, 1 M HCl, 4 M H_3PO_4		
Ni-Pd-Pt-Au;			8 M HCl, 2 M HNO_3, 14 M HNO_3		
Mn-Bi-Hg;			2 M HCl, 10 M HCl, 6 M HNO_3		
As(III)-Sb(III);			1 M HCl, 5 M HNO_3		
V(IV)-Ti;			8 M HCl, 1 M HCl		
Fe(II)-Fe(III);			2 M HCl, 0.5 M H_2SO_4		
Cr(III)-Cr(VI);			2 M HCl, 10 M HCl		
Fe, Ti-Th-U(VI);			2 M HNO_3, 1 M HCl, 4 M H_3PO_4		

SPECIES	SUPPORT	STAT. PHASE (EXTRACTANT)	MOB. PHASE (ELUENT)	REFERENCE N°	AUTHOR
Ni, Co-Fe(III);	Cellulose	TNOA	8 M HCl, 3 M HCl, 0.2 M HNO_3	258	CERRAI
La-U-Th;			10 M NH_4NO_3, 6 M HNO_3, 8 M HCl		
Th-Zr-U(VI);			10 M HCl, 6 M HCl, 0.05 M HNO_3		
La-Zr;			10 M NH_4NO_3		
Zr-Hf			8 M HCl+5% conc. HNO_3		
Eu, Tb, Y;	Cellulose	HDEHP	1.2 M HCl, 2 M HCl, 10 M HCl	259	CERRAI
Sr, Ce, Eu, Y, Yb			0.1 M HCl, 0.5 M HCl, 2 M HCl, 4 M HCl, 8 M HCl		
Cs, Sr, Ce, Tb, Yb			0.5 M $NaNO_3$(Cs), 0.1 M HCl, 0.8 M HCl, 3 M HCl, 10 M HCl (Yb)		
Ni, Co, Fe, U(VI)	Kel-F	TOPO-cyclo-hexane	8 M HCl, 1 M HCl, 0.5 M H_2SO_4, 4 M H_3PO_4	260	CERRAI
Cu-Zr			Cu: 8 M HCl, Zr: 0.5 M $H_2C_2O_4$		
V(IV), Ti(IV), U(VI);			8 M HCl, 2 M H_2SO_4, 4 M H_3PO_4		
Mn, Cu, U(VI);			4 M HCl, 1 M HCl, 4 M H_3PO_4		
Al, Cu, Fe(III), U(VI);			8 M HCl, 1 M HCl, 0.5 M H_2SO_4, 4 M H_3PO_4		

SPECIES	SUPPORT	STAT. PHASE (EXTRACTANT)	MOB. PHASE (ELUENT)	REFERENCE N°	AUTHOR
Fe(II)-Fe(III)			1 M HCl, 0.5 M H_2SO_4		
Ni-Co-Fe			Ni: 8 M HCl Co: 1 M HCl Fe: 0.5 M H_2SO_4		
U, Th-Na, Ca, Fe, Co, Ni, Zn, Mn, Cd, Al, Y, La;			imp.: 1 M HNO_3 Th: 0.5 M HCl U: 1 M H_3PO_4		
Rare earths; Cs, Sr, La, Ce, Nd, Gd, Tb, Tm; Cs, Sr, Y; La, Ce, Nd, Eu	Cellulose	HDEHP	HCl	261	CERRAI
Oxidation reactions: Fe(II) → Fe(III) Cu(I) → Cu(II) Sn(II) → Sn(IV) I(I) → I; reduction reactions Fe(III) → Fe(II) I → I(I) Ce(IV) → Ce(III) V(V) → V(IV) Cr(VI) → Cr(III)	Kel-F	Tetrachloro-hydroquinone (chloranil)	Acid sol.	262	CERRAI
Rare earths	Kel-F	HDEHP	HCl at 85°C	263	CERRAI
Y-rare earths	Cellulose	HDEHP	Ce: 0.25 M HCl Pm: 1 M HCl Y: 9 M HCl	264	CERRAI
Rare earths	Kel-F	TOPO-cyclo-hexane	0.2 and 6 M HNO_3	265	CERRAI
Cu-Fe(II), Fe(III), Co, Ni, Mn(II), Cr(III), Mo(VI), V(IV), W(VI)	Cellulose	Lix-64	$(NH_4)_2SO_4$ or NH_4Cl (various conc.) metal ions: 0.5 M $(NH_4)_2SO_4$, Cu: H_2SO_4	266	CERRAI

SPECIES	SUPPORT	STAT. PHASE (EXTRACTANT)	MOB. PHASE (ELUENT)	N°	REFERENCE AUTHOR
Fe(III)-Cu(II)			Fe: $(NH_4)_2SO_4$- 0.18 M NH_4HF_2, Cu: 3 N H_2SO_4		
Cs-fission products	Kel-F	TBP-amylacetate	F.P.: 0.01 M EDTA (pH 6.8); Cs: 1 M HCl-acetone	267	CESARANO
La-Ce-Pr	SiO_2^+	HDEHP	HCl (various conc.)	268	CHENG
Bk(III)-Ce(III)	SiO_2-gel	TNOA MDOA	6 M $NaNO_3$ 5 M $NaNO_3$	269	CHUDINOV
Ni-Cu Fe-Cu	SDVB	N-oleylhydroxyl-amine in CCl_4	Ni:Na-acetate buffer pH 4 Cu: 2 M HNO_3 Fe: 5 M HCl	270	CLINGMAN
Co, Ni, Cu, Fe(II), Fe(III)	SDVB	Heptadecane 2,3-dione dioxime	HCl HNO_3		
Cu, Ag, Hg, Bi, Pb, Co, Ni, Fe(III)	SDVB	t-Hexyldecyl mercaptan	HCl		
Ag, Hg, Cu, Co, Ni, Pb	SDVB	N-Hexadecyl-N'-phenylthiourea	Buffer solution NH_4Cl-NH_4OH, pH 8-9		
Pd-nuclear waste components	Plaskon	Aliquat-336	8 M HNO_3 and 1.5 M NH_4OH	271	COLVIN
Zr, Hf	SiO_2^+	TBP	20-25% HNO_3- 50% NH_4NO_3	272	CRAWLEY
All rare earths	SiO_2^+ (Polystyrene beads)	HDEHP	HCl (diluted) at 80°C	273	CROUCH
Nb, Ta, Mo, W	Tee Six Chromosorb W Chromosorb P Cellulose	MIBK	HCl, HF	274	DAHMER

SPECIES	SUPPORT	STAT. PHASE (EXTRACTANT)	MOB. PHASE (ELUENT)	REFERENCE N°	AUTHOR
Pu(VI) → (III) Pu(IV) → (III) Np(VI) → (V.) Np(V) → (IV); Np-Pu; U-Pu, Am-Np	Micro-thene-710	2,3-dichloro- 1,4-naphtahydroquino-ne+TOPO	Mineral acid	275	DELLE SITTE
Am-U-Th	Micro-thene-710	HX-70	Am: 1 M NH_4NO_3 (pH 2.9) U(VI):0.03 M HNO_3 (pH 1.6) Th: 2 M HNO_3	276	DELLE SITTE
Pu-Np-(Fe)			Pu(III): 2 M HNO_3 Np(IV):0.1 M oxalic acid Fe(III):6 M HCl;		
Np-Pu-(Fe)			Np(V):2 M HNO_3 Pu(III): 2 M HNO_3-0.15% hydroqui-none Fe(III):6 M HCl;		
Pu-Np			Pu(III):2 M HNO_3-0.15% hydroqui-none Np(IV): 0.1 M oxalic acid		
U, Np, fission products	Hostaflon C2	TBP HDEHP	8 M HNO_3 $(NaClO_4)$ 8 M HNO_3 (SO_2) 0.1-12 M HCl (at 40-60°C)	277	DENIG
Pa-U	Hostaflon C2	DIBC	9 M HCl $(KClO_3)$	278	DENIG
Te-Sb			9 M HCl $(KClO_3)$ Sb: H_2O		
Sb-fission products		(HDEHP)	9 M HCl $(KClO_3)$		
Zr-Nb Th-Pa Mo-Tc Zn-Sb Sb-Te	Hostaflon C2	TBP TBP HDEHP HDEHP HDEHP	HCl (various conc.)	279	DENIG

SPECIES	SUPPORT	STAT. PHASE (EXTRACTANT)	MOB. PHASE (ELUENT)	REFERENCE N°	AUTHOR
Fission products, Th, Pa, U, Np	Hostaflon C2	HDEHP TBP TOPO TNOA	0.1–12 M HCl (a separation scheme is indicated including also TBP-HNO_3+ $NaClO_3$, TBP-HNO_3+SO_2 and H_2O)	280	DENIG
Ac, Th, Pa, U, Np, fission products	Hostaflon C2	HDEHP TBP TNOA	HCl (various conc.)	281	DENIG
Constituents of a complex fission product mixture	Hostaflon C2 Voltalef-300	TBP TNOA HDEHP	(HCl of various conc.) determination of partition coeff. in view of extraction chromat. separations	282	DENIG
Mo-Tc	Hostaflon C2	HDEHP	Tc: 1 N HCl (60°C) Mo: 12 N HCl (30°C)	283	DENIG
Te-Sb	"	HDEHP	Te: 9 N HCl+$KClO_3$ (45°C) Sb: 0.4 N HCl (70°C)		
Zr-Nb	"	TBP	Zr: 6 N HCl (20°C) Nb: 2.5 N HCl (20°C)		
Th-Pa	"	TBP	Th: 9 N HCl Pa: 3 N HCl		
Fission product mixture				284	DENIG
Np, U, Zr-other fission products	PTFCE	TBP	8 N HNO_3-$NaClO_3$ (50°C)		
Nb, Sb, I-residual fission products	"	HDEHP	9 N HCl-$NaClO_3$ (50°C)		
Rare earths, Mo-residual fiss. pr.	"	HDEHP	0.1 N HCl (70°C)		
Te, Tc-resid. fiss. pr.	"	TBP	6 N HCl-$NaClO_3$ (50°C)		

SPECIES	SUPPORT	STAT. PHASE (EXTRACTANT)	MOB. PHASE (ELUENT)	REFERENCE N° AUTHOR
^{143}Pr-^{141}Ce	Teflon	HDEHP	Pr: 7 M HNO_3 Ce: 0.3 M HNO_3 (SO_2-satd.)	285 DEPTULA
Rare earths, actinides	SiO_2^+	HDEHP	0.15-6 M HCl (gradient elution) (80°C)	286 DE WET
U (sep. from urine)	Glass beads	TOPO	C_2H_5OH (stripping)	287 DIETRICH
U-Pu-fission products (+^{99}Mo)	SiO_2^+	TBP	FP, Pu(coll.): 3 M HCl Mo, Tc: 3 M HNO_3	288 D'OLIE-SLAGER
^{99}Mo-U, Pu, Am	SiO_2^+	TBP	U, Pu, Am: 3 M HCl Mo: 3 M HNO_3	
^{99}Mo-fission products	SiO_2^+	TBP	Fission products: 3 M HCl ^{99}Mo(^{99}Tc): 3 M HNO_3	289 D'OLIE-SLAGER
^{99}Mo-U-Pu-Am	SiO_2^+	TBP	U, Pu, Am: 3 M HCl Mo: 3 M HNO_3	
In-Cd-Zn	Fluoroplast-4	TBP	0.8 M HBr (Influence of temp. up to 60°C investig.)	290 DOL-ZHENKO
^{184}Re, W	Fluorinated polymeric resin (fluororesin 4)	Cyclohexanol	3 N H_2SO_4; 1 N H_2SO_4 1 N HCl	291 DZHELE-POV
^{54}Mn-^{59}Fe; ^{59}Co-^{59}Fe	Celite	Aliquat-336 xylene	Mn: 1 M HCl Fe: 0.1 M HCl Co: 0.1 M HCl Fe: 0.1 M HCl	292 EL-GARHY

SPECIES	SUPPORT	STAT. PHASE (EXTRACTANT)	MOB. PHASE (ELUENT)	REFERENCE No	AUTHOR
Fe(III)(sorption)	Teflon	TBP	Aqueous solutions of HCl, NaCl, KCl, $MgCl_2$, $CaCl_2$, $AlCl_3$	293	ERSHOVA, N. I.
Ga, Fe(III), Zn, Cd, Cu, Co	Teflon	TBP	4 M HCl $1 M NH_4 SCN$ (20^0-90^0C)	294	ERSHOVA, N. I.
Ga, Fe(III), Co, Cu (concentration)	Fluoro-plast-4	TBP	$1 M NH_4 SCN$ + large amounts NaCl, $HgCl_2$, $CaCl_2$, $CdCl_2$, $ZnCl_2$ and $AlCl_3$ (in sample sol.) elution: 5% aqu. NH_3, 0.5 M HCl	295	ERSHOVA, N. I.
Pu-Al, U, Np	Teflon	TNOA	Al, U, Np: 1.5 M HNO_3 Pu: 1.5 M HNO_3- 3% $(NH_4)_2 C_2 O_4$	296	ERSHOVA, Z. V.
U-Np-Pu-Am (Cr-VI); U-Np-Pu	SiO_2^+	TBP	HNO_3 and HNO_3-$(NH_2)_2$; $HNO_3 \cdot Fe(SO_3 NH_2)_2$	297	ESCHRICH
Np(IV), -(V), -(VI) (Pu-III, -IV, -VI) Am, Cm, U(VI)	SiO_2^+ Hostaflon C2; Voltalef-300 LD; Fluon; Corvic-R65/81	TBP	0.2 - 4 M HNO_3 (4^0-50^0C)	298	ESCHRICH
Pu-impurities	Voltalef-300 micro PL	TBP	Impurities: 5 M, 3 M, 1 M HNO_3	299	ESCHRICH

SPECIES	SUPPORT	STAT. PHASE (EXTRACTANT)	MOB. PHASE (ELUENT)	REFERENCE N°	AUTHOR
Am-Pu, Np-U	Voltalef-300 micro PL	TBP	Am: 3 M HNO_3 Pu, Np: 3 M HNO_3-0.1 M phenylhydrazine	300	ESCHRICH
U-metal ions	Polyethylene tubes	TBP	HNO_3-$NaNO_3$	301	ESCHRICH
Pu(III)-U(IV)-U(VI)	Voltalef-300 LD	TBP	0.7 M - 5 M HNO_3-0.1 M hydrazine	302	ESCHRICH
Cr(III)-W(IV)-Co(II)-Fe(III)	Kel-F	TNOA	Cr: 10 M HCl W: 7 M HCl + 1 M HF Co: 3 M HCl Fe: 1 M HNO_3	303	ESPAÑOL
Mn(Cu)-zirconium (Zircaloy)	Kel-F	TBP	10 M HCl	304	ESPAÑOL
Gd-Tb-Dy-Ho-Er-Tm-Yb-Lu	SiO_2^+	TBP	11.5-12.3-13.0 M HNO_3 (HCl)	305	FIDELIS
Pa, Th	SiO_2^+	TBP	HCl(5% $H_2C_2O_4$)	306	FIDELIS
As, Ge As(III)-As(V)	SiO_2^+	TBP	7-12 M HCl (H_2O)	307	FIDELIS
^{199}Au-Pt	SiO_2^+	TBP	Pt: 7 M HCl Au: 14 M HNO_3	308	FIDELIS
Rare earths	SiO_2^+	HEHØP	HCl; HNO_3 (various conc.)	309	FIDELIS
Es, Fm-Md	Celite	HDEHP	Es, Fm: 0.974 M HNO_3 Md: 2 M HNO_3 (45-75°C)	310	FIELDS
^{254}Fm-^{254}Md			Fm: 0.76 M after 10-20 min.: Fm: 0.974 M HNO_3 Md: 2 M HNO_3		
U sep. from 25 trace elements	Kel-F	TBP	HNO_3	311	FLETCHER

SPECIES	SUPPORT	STAT. PHASE (EXTRACTANT)	MOB. PHASE (ELUENT)	REFERENCE N° AUTHOR
La, Ga, Sc, Hf	Kel-F	TOPO	HBr	312 FOLD-ZINSKA
Zr, Hf	Teflon-6	MIBK	Zr: NH_4SCN-$(NH_4)_2SO_4$ Hf: 1.2 M $(NH_4)_2SO_4$ or 1 M H_2SO_4 Zr/Hf: 4 M NH_4SCN - 0.2 M $(NH_4)_2SO_4$	313 FRAZEE
Fe(III) sep. from many elements e.g. Cu(II), Zn(II), Ti(IV)	Haloport-F	2-octanone	6-8 M HCl	314 FRITZ
Mo sep. from other steel components Mo(VI)-W(VI)	Kel-F	MIBK	HCl(1:1)-1 M H_2SO_4 Mo(VI):HCl(1:1) Fe(III):H_2SO_4 dil. Mixture: HCl(1:1) 1 M H_2SO_4- 0.1 M HF	315 FRITZ
Zr, Hf	Teflon-6	MIBK	(see Ref. [313])	316 FRITZ
Pb(II)-Zn(II)	Teflon-6 (70-80 mesh)	IOTG	Zn: acetate buffer pH 4	317 FRITZ
Bi(III)-Zn(II)	Chromosorb W (80-100 mesh)(also laminar techn. appl.)	IOTG	Zn: 1 M HNO_3 Bi: diluted HCl	
Sn(IV) sep. from Bi, Cd, Cu, Pb, Hg, Zn; Sb(V)	Teflon-6	MIBK	8 M HCl	318 FRITZ
Sn(IV)-Mo(VI)	Chromosorb W (silaned)	IPE	1 M HCl- 3 M H_2SO_4	

SPECIES	SUPPORT	STAT. PHASE (EXTRACTANT)	MOB. PHASE (ELUENT)	REFERENCE N°	AUTHOR
Nb-Ta	Teflon-6 70 mesh	MIBK	Nb: 3 M HCl- 1.1 M HF Ta: 15% H_2O_2	319	FRITZ
W-Mo	Teflon-6 170 mesh	MIBK	W: 6 M HCl- 3 M HF - 1 M H_2SO_4 Mo: 3 M HCl- 10 M HF- 1 M H_2SO_4		
W-Mo-Nb	"	MIBK	W: 6 M HCl- 3 M HF - 1 M H_2SO_4 Mo: 3 M HCl- 10 M HF - 1 M H_2SO_4 Nb: 3 M HCl- 1 M HF		
W-Mo-Nb-Ta	"	MIBK	W: 6 M HCl- 1 M H_2SO_4 - 3 M HF Mo: 3 M HCl- 1 M H_2SO_4 - 10 M HF Ta: MIBK		
W-Mo-Nb-Ta	"	MIBK	W: 7 M HCl-2 M HF Mo: 6 M HCl-6 M HF Nb: 3 M HCl-1 M HF Ta: MIBK		
27 metal ions	Amberlyst XAD-2	IPE MIBK TOPO	HCl, H_2SO_4, HBr, HNO_3 (various conc.)	320	FRITZ
U-metal ions (e.g. Th, Zr, rare earths)	Amberlyst XAD-2	0.5 M DOSO in dichloroethane	$HClO_4$, H_2SO_4	321	FRITZ
Ga(III)-In(III)- Tl(III)	Amberlyst XAD-2	IPE	Ga(III): 5 M HBr In(III): 1 M HBr Tl(III): 3 M HNO_3	322	FRITZ
	Amberlyst XAD-2	MIBK	Ga(III): 3 M HBr In(III): 0.25 M HBr Tl(III): 3 M HNO_3		

SPECIES	SUPPORT	STAT. PHASE (EXTRACTANT)	MOB. PHASE (ELUENT)	REFERENCE N°	AUTHOR
V, W, Mo	Amberlyst XAD-2 60-80 mesh	Aliquat-336 (0.05 or 0.20 M) in toluene	$0.01-4 M H_2SO_4$ 0.5 or 1.0% H_2O_2	323	FRITZ
Cu, other metals	Amberlyst XAD-2 80-100 mesh	0.64 M DHDO or 0.54 M HEO in toluene	Sodium acetate-acetic acid sol. of pH 5	324	FRITZ
Mo, other metals	Amberlyst XAD-2 80-100 mesh	0.64 M DHDO or 0.54 M HEO in toluene	Metal ions: $0.01 N H_2SO_4$ Mo: NH_4OH, pH 12	325	FRITZ
Cf, Fm (Am, Cm)	SiO_2^+	HDEHP-toluene	HCl (various conc.)	326	GAVRILOV
Nb(V)-W(VI), Zr(IV) Nb(V)-Mo(VI), Zr(IV)	Teflon	0.5 M TNOA in benzene	HCl	327	GIBALO
Nb-other metals	Ftoro-plast	TNOA	HCl	328	GIBALO
Sc-Th-Co, Cr, Sr	Cellulose	HDEHP	Sc-Th: 10 M HCl	329	GIRARDI
Th, U Rare earths, Sc, Th, U (automation of procedure)	Teflon Teflon	TOPO HDEHP	HCl HCl	330	GIRARDI
$^{113}Sn-^{113m}In$	SiO_2^+	TBP	In: 3-6 M HCl	331	GOLINSKI
Ta-trace impurities	Teflon	TBP	1 M HF- 1 M HNO_3	332	GORYAN-SKAYA
U, Pu, fission products-Np	Kel-F	TLA	U, Pu, fission products: 1 M HNO_3-0.01 M $Fe(SO_3NH_2)_2$ Np: 0.5 M H_2SO_4-0.18 M HNO_3	333	GOURISSE

SPECIES	SUPPORT	STAT. PHASE (EXTRACTANT)	MOB. PHASE (ELUENT)	Nº	AUTHOR
U-Np	Kel-F	TLA	U: 1 M HNO_3- $Fe(SO_3NH_2)_2$ Np: 0.5 M H_2SO_4- 0.18 M HNO_3	334	GOURISSE
U-Pu-Np	Kel-F	TLA	U: 1 M HNO_3 Pu+Np: 0.5 M H_2SO_4 -0.18 M HNO_3	335	GOURISSE
Pu-fission products	Paper	TNOA	7.5 M HNO_3	336	GROSS
Eu (rare earths)	SiO_2^+	HDEHP	HCl (various conc.)	337	GROSSE-RUYKEN
Y sep. from Dy or Tb	SiO_2^+	HDEHP-toluene (1:1)	Dy+Tb: 1 M HCl Y: 1.2 M HCl	338	GROSSE-RUYKEN
Ho, Dy; Lu, Yb	SiO_2-gel	HDEHP-toluene (1:1)	0.1-4 M HCl	339	GROSSE-RUYKEN
Gd, Dy, Eu, Tb, Y	SiO_2-gel	HDEHP	0.1-2.5 M HCl	340	GROSSE-RUYKEN
Pu(III)-(IV)-(VI), Pu(colloidal); Pu-U	SiO_2^+	TOPO / TNOA	0.5-1.0 M HNO_3 / 3 M HNO_3-0.05 M H_2NOH. $RClO_4$	341	GWOŹDŹ, R.
U-impurities	Kel-F	TBP	HNO_3	342	HAMLIN
U-impurities	Kel-F	TBP	5.5 M HNO_3	343	HAMLIN
U-impurities	Kel-F-300	TBP	Impurities: 5.5 M HNO_3 U: H_2O	344	HAMLIN
Ce-Pr	SiO_2^+	DOPA	Pr: 6 M HNO_3+ 0.2 M $KBrO_3$ Ce(IV): retained on column	345	HANSEN, P. G.
Rare-earths-trace metal impurities	Kel-F 80-100 mesh	MHDPO	Impurities: 1 M HNO_3	346	HANSEN, R. K.

SPECIES	SUPPORT	STAT. PHASE (EXTRACTANT)	MOB. PHASE (ELUENT)	REFERENCE N°	AUTHOR
U sep. from Al, Co, Cu, Cr, Fe, Mg, Mn, Mo, Ni, Pb, Bi, Cd, Se, W and Nb	Kel-F	TBP	HCl or HNO_3	347	HAYES
U-impurities	Fluoro-carbon 80-90 mesh	TBP	U: H_2SO_4-H_2O_2 mixture at 90°C	349	HEDRICK
U, fission products	PTFE (5 types) Hyflo Super Cel	TBP	5 M HNO_3	350	HERMANN, A.
U-Pu-Am, fission products	PTFE	TBP	Am+f.p. 5 M HNO_3 Pu: 1.8 M HNO_3- 0.1 M H_2SO_3- 0.05 M HSO_3NH_2 U: H_2O	351	HERMANN, A.
Ce-group of lanthanides sep. from Er	SiO_2^+	HDEHP	2.9 M HCl (40°C)	352	HERR-MANN, E.
Rare earths	SiO_2^+	HDEHP	HCl	353	HERR-MANN, E.
^{140}Sm, ^{141}Er	SiO_2^+	HDEHP	0.73 M HCl, Sm: 6 M HCl	354	HERR-MANN, E.
Rare earths	SiO_2^+	HDEHP	HCl (various conc. and temperat.)	355	HERR-MANN, E.
^{235}U, fission products-Zr	Hostaflon C2	TBP	HNO_3	356	HERR-MANN, G.
^{239}Pu, fission products-Tc	Hostaflon C2	TBP	HNO_3		
U-PO_4^{3-}	SiO_2^+	TBP	PO_4^{3-}: $Al(NO_3)_3$ U: H_2O	357	HEUNISCH
Na, Ca, Fe, U	SiO_2^+	TBP	$Al(NO_3)_3$-H_2O		

SPECIES	SUPPORT	STAT. PHASE (EXTRACTANT)	MOB. PHASE (ELUENT)	REFERENCE N°	AUTHOR
Fission products, Ce-Pr	Hostaflon C2	HDEHP	Pr: 6 M HNO_3	358	HOFF-MAN
^{244}Cm-^{252}Cf-^{243}Am	SiO_2^+	HDEHP		359	HOLCOMB
Zn-Hg(II)-Co(II)	Kel-F	STTA in cyclo-hexane	Zn: aqu. sol. of pH 5.0 Hg: 1 M HCl Co: not eluted with 12 M HCl, removed by acetone with stat. phase	360	HONJO
Ce(IV), Cs, Ag	Graft co-polymers of styrene and various hydrophobic polymers as substra-tes with complexing agents such as HDEHP, thio-THPA and BAMBP modified Kel-F	See support	NaOH, HNO_3 (various conc.)	361	HORNBECK
Ce-Eu-Y-Yb (Nb, Tm, Lu)	Graft co-polymers Kel-F	HDEHP	HCl (various conc.)	362	HORNBECK
Am(III)-Cm(III)	SiO_2^+	Aliquat-336	3.5 M $LiNO_3$-0.01 M HNO_3	363	HORWITZ
Am(III)-Cm(III)	SiO_2^+	Aliquat-336	Cm: 3.6 M $LiNO_3$ 0.01 M HNO_3 Am: 1 M HNO_3	364	HORWITZ
Cf(III)-Es(III)	SiO_2^+	Aliquat-336	4.8 M $LiNO_3$-0.05 M HNO_3	365	HORWITZ
^{252}Cf-^{248}Cm	SiO_2^+	HDEHP	HNO_3 or HCl (diluted)	366	HORWITZ

504

SPECIES	SUPPORT	STAT. PHASE (EXTRACTANT)	MOB. PHASE (ELUENT)	REFERENCE Nº	AUTHOR
^{241}Am-^{248}Cm	SiO_2^+	Aliquat-336	Cm: 4 M $LiNO_3$- 0.01 M HNO_3 Am: 1 M HNO_3	367	HORWITZ
Cf, Es, Fm	SiO_2^+	HDEHP	HNO_3 or HCl (diluted)	368	HORWITZ
Am, Cm, Bk, Cf, Es	SiO_2^+	HDEHP	HNO_3 or HCl (diluted)	369	HORWITZ
Es(III)-Fm(III)-Md(III)	SiO_2^+	HDEHP	0.98 M HNO_3 (75ºC and 60ºC)	370	HORWITZ
Transplutonium elements	SiO_2^+	HDEHP Aliquat-336	HNO_3 sol.	371	HORWITZ
Cf, Cm	Celite (various particle sizes)	HDEHP (various mol/g Celite)	HNO_3 (various concentrations and temperatures)	372	HORWITZ
Cm-Cf	Celite	HDEHP	Cm:0.37 M HNO_3 Cf:1 M HNO_3	373	HORWITZ
Bk-Cf			Bk-Cf:0.37 M HNO_3		
Bk-Es-Fm			Bk-Es:0.4 M HNO_3 Fm: 1 M HNO_3		
Es-Fm-Md			Es-Fm:1 M HNO_3 Md:2 M HNO_3		
Am-Cm			0.17 M HNO_3		
Es-Cf			0.41 M HNO_3		
U-Pu-Am, fission products	PTFE	TBP	(see Ref. [351])	374	HÜBENER
Impurities-Th, U, alloys	Kel-F	TBP	8 M HNO_3	375	HUFF
Impurities-Pu-U, Zr	Plaskon, type CTFE-2300	95% TEHP + 5% HDEHP	Impurities: 8 M HNO_3 Pu:0.4 M HNO_3-0.02 M HF U+Zr:0.5 M HNO_3-3 M HF	376	HUFF

SPECIES	SUPPORT	STAT. PHASE (EXTRACTANT)	MOB. PHASE (ELUENT)	REFERENCE N°	AUTHOR
Rare earths, Am, Y	Plaskon-CTFE 2300	Aliquat-336	NH_4SCN(+acid)	377	HUFF
Am(VI), Cm(III), Bk(IV)	SiO_2^+	0.2-1.0 M HDEHP in n-heptane	0.1 M HNO_3 (10 M HNO_3)	378	HULET
Some mono- and divalent ions, Es, Fm, Md	Kel-F	HDEHP	Mono- and divalent ions: 0.1 M HCl; trivalent ions, Es, Fm, Md: 3 M HCl	379	HULET
Ho, rare earths	SiO_2^+	HDEHP	HNO_3 (various concentrations)	380	HUNT
U-25 cations and 6 anions	Chromosorb W	Alamine-336	0.05 M H_2SO_4	381	ITANI
Ba-Sr-Ca	SiO_2^+	HDEHP	0.5 M $NaNO_3$ Ba: pH 3-3.2 Sr: pH 2 Ca: pH 0.8	382	JASKOLSKA
Ni-Fe(III) Al-Fe(III)	Polyethylene capillary tubing	Amberlite LA-2	0.1 M HCl 0.1 M HCl	383	JENTZSCH
RuNO-nitrates, RuNO-nitrites	SiO_2^+	TBP	HNO_3 various conc. (+org. masking agents)	384	JOON
RuNO-nitro complexes (separation)	SiO_2^+	TBP	0.3 M-2 M-3.5 M-5 M-10 M HNO_3 in sequence	385	JOON
Ru-Ce-Nd-Pm-Sm-Eu	SiO_2^+	HDEHP	HNO_3 (various concentrations)	386	JOON
Ru-Ce-Nd-Pm-Sm-Eu	SiO_2^+	HDEHP	HNO_3 (various concentrations)	387	JOON
Pt-metals	Ftoroplast-4	TBP	0.5-11 M HCl	388	KALININ
Bi-Ni(+other metal ions)	Fluoroplast-4	DIAP	Ni+other ions: 0.5 M HNO_3 Bi: 13.5 M HNO_3	389	KALININ

506

SPECIES	SUPPORT	STAT. PHASE (EXTRACTANT)	MOB. PHASE (ELUENT)	REFERENCE N°	AUTHOR
Bi, Hg, Ag (Au, Tl, Re, Tc, W)	Teflon	TBP	0.5 M HCl, 5 M HCl, 11 M HCl	390	KALYA-MIN
Zr-Nb	Plaskon 3200	MHDPO	Zr: 10 M HF Nb: 1.25 M HF-3 M HNO_3	391	KAMIN
Ni-Co (Zn, Sn, Fe); Ni-Cu	Chromo-sorb W	Alamine-336	Ni: 0.1 M HNO_3 Co or Cu: methanol	392	KAMINSKI
		(see Ref. 392)		393	KAMINSKI
Gd-Eu	Fluoro-plast-4	HDEHP	Gd: HCl	394	KARYAKIN
^{140}Ba-^{140}La; ^{90}Sr-^{90}Y	Ftoro-plast-4	HDEHP	Ba: 0.01 M HNO_3 La: 2 M HCl or HNO_3 Y: 6 M HNO_3	395	KATYKHIN
W-Au, Os; Ta, W	Ftoro-plast-4	TNOA	W: 1 M HF Au, Os: 1 M HCl+ thiourea; $HF-NH_4Cl-NH_4F$	396	KATYKHIN
Y-Mo-Nb-Zr-Cd	Ftoro-plast-4	DOPA	Y: HNO_3 conc. Mo: HCl conc. Nb: $HNO_3-H_2O_2$ (1:1) Zr: > 4 M HF	397	KATYKHIN
Pb-Bi	Ftoro-plast-4	DOPA	Pb: 0.5 M HNO_3 Bi: HNO_3 conc.	398	KATYKHIN
Pt-group met.	Ftoro-plast-4	TBP	0.5-11 M HCl	399	KATYKHIN
Application of support	Fluoro-plast-4	HDEHP	--	400	KATYKHIN
Rh-Cd	Ftoro-plast-4	Dibutyl ether	Rh: 6 M HCl	401	KATYKHIN
Tc-Cd	Ftoro-plast-4	TBP	4.5 M H_2SO_4-3% H_2O_2 (sample solution) elution: 1 M NH_4OH	402	KATYKHIN

SPECIES	SUPPORT	STAT. PHASE (EXTRACTANT)	MOB. PHASE (ELUENT)	REFERENCE No	AUTHOR
U(VI)-50 metal ions	Paper	Sulfoxides	HCl and HNO_3 (various conc.)	403	KENNEDY
Rare earths	SiO_2^+	HDEHP	HCl (various concentrations)	404	KOMORI
Am, (Cm), Bk, Cf	SiO_2^+	HDEHP	0.5 M HCl	405	KOOI
Bk, Ce	SiO_2^+	HDEHP	0.3 M HCl (87°C)	406	KOOI
Am, Cm, Bk, Cf	SiO_2^+	HDEHP	0.1-1 M HCl	407	KOOI
Am, Bk, Cf	SiO_2^+	HDEHP	HCl	408	KOOI
Tm-Ho	Kel-F	HDEHP	HCl	409	KOPTA
Rare earths, transuranium element	Cellulose	TBHP	8 M HNO_3, 8 M $HClO_4$	410	KOROTKIN
Rare earths, transplutonium elements	SiO_2-gel	HDEHP	Cm: Na_5-DTPA in 1 M lactic acid (pH 3.2) Eu: 4 M HNO_3	411	KOSYAKOV
Pm, Eu, Tb, Er	SiO_2^+	DBP-CCl_4	0.7-13 M HNO_3	412	KOTLINSKA-FILIPEK
Be, La, Eu, Gd, Dy, Sm, Er-U	SiO_2	TNOA	8 M HCl	413	KREFELD
Preparation of col. materials	(SDVB)	(e.g. TBP, HDEHP, TNOA)	--	414	KROEBEL
Ce-fission products	SiO_2^+	HDEHP	Ce: 5 M HNO_3- 0.1 M ascorbic acid	415	KŘTIL
[144]Ce-U, other fission products	Chromosorb W	HDEHP	5 M HNO_3 (0.5 M $NaBrO_3$)	416	KŘTIL
Rare earths-fission products+ fuel components Nd-other rare earths Ce-other rare earths	SiO_2^+	HDEHP	HCl HNO_3-ascorbic acid	417	KŘTIL

SPECIES	SUPPORT	STAT. PHASE (EXTRACTANT)	MOB. PHASE (ELUENT)	REFERENCE N°	AUTHOR
^{147}Nd-other fission products	Chromosorb W	HDEHP	0.15 M HNO_3	418	KŘTIL
As-Au(III), Fe(III) Ga, In	Polychrome-1	o-xylene; decyl-alcohol	HCl or HCl-HBr	419	KUZMIN
Rb-Cs; Cs-alkali and alkali-earth metals	Teflon Teflex Ftoroplast	NH_4DPA in nitro-benzene	$(NH_4)_2CO_3$ (various conc.)	420	KÝRŠ
Sn-Mo(VI)	Teflon-6	MIBK	1 M HCl- 3 M H_2SO_4	421	LATWESEN
Sn-Sb(V) Sn-Bi, Cd, Cu, Hg, Pb, Zn	Teflon-6 Teflon-6	IPE MIBK	8 M HCl 8 M HCl		
Li-Na, K, Rb, Cs	Teflon	0.1 M DBM - 0.1 M TOPO in dodecane; 0.18 M DBM-50% TBP in dodecane	3.2 M NH_4OH Li: 0.6 M HCl	422	LEE
Cu-Pb-Zn-Cd	Cellulose powder	Amberlite LA-2	Cu, Pb: 2 M HCl Zn: 6 M HCl Cd: 10 M HCl	423	LESZKO
Pd-^{111}Ag	Teflon	TBP	Pd: 0.05 M HCl Ag: 4 M HCl	424	LEVIN
Y-Sr	Fluoroplast	HDEHP	0.1-0.15 M HCl	425	LEVIN
Sr(impurities)-Y(+Fe); Y-Fe	Fluoroplast	HDEHP TBP	(Sr: 0.1 M HCl) Y: 6 M HCl Y: 6 M HCl	426	LEVIN
Ca-Sr	Kel-F	HDEHP	Sr: 0.03 M HNO_3 Ca: 1 M HNO_3	427	LIESER
^{239}Np-U, fission prod.	SiO_2^+	TBP	HNO_3 (various conc.)	428	LIS
Zn (concentr.)	Cellulose acetate	Dithizone	Zn: HCl	429	LOVE-RIDGE
Am-impurities	Corvic 100-150 mesh	HDEHP	Am: 0.01 M HNO_3 (70-80°C)	430	LOVE-RIDGE

SPECIES	SUPPORT	STAT. PHASE (EXTRACTANT)	MOB. PHASE (ELUENT)	REFERENCE Nº	AUTHOR
U(IV)-U(VI)-Th	Fluoro-plast	HOEH HDEHP	U(VI): HCl U(IV): HCl-15% H_2O_2 Th: 6 M H_3PO_4	431	LUGININ
Sc, heavy lanth. - Th;	PTFE	TBP	Sc, R.E.: 4 M HNO_3 Th: H_2O	432	MA
Heavy lanth. - Sc	PTFE	TBP	R.E.: 8 M HCl Sc: 2 M HCl		
Sc, Sm, Lu-Th, Y	PTFE	TBP	Sc, R.E.: 4 M HNO_3 Th, Y: 8 M HCl	433	MA
U (concentr.)	PTFE	TBP	3 M HNO_3	434	MA
Rare earths	Corvic	HDEHP in cyclo-hexane	0.05, 0.1, 0.5, 1, 3, 5 and 7 M $HClO_4$	435	McCLEN-DON
Cs, Fr (con-centr.)	Not indi-cated	Not indicated	Impurities: H_2O, sat. with nitro-benzene Cs or Fr: 2 M HCl or HNO_3	436	MALTSEVA
Mo-alkali metals, alkaline earths, rare earths, As, Au, Co, Cr, Cu, Fe, Mn, PO_4^{3-}; Sb, Tc, Zn	Algoflon-F	α-benzoinoxime	2 M HCl	437	MALVANO
In	Cellulose	Dithizone-$CHCl_3$-CCl_4	NaOAc-HCl	438	MAPPER
Pu-U, fiss. products;	Fluoro-plast-4	TNOA	Pu: 1.5 M HNO_3-3% $(NH_4)_2C_2O_4$;	439	MARKOV
Pu-U, fiss. products;	Fluoro-plast-4	TNOA	Pu: 0.5 M H_2SO_4-0.2 M HNO_3		
U-fiss. prod.	Fluoro-plast-4	TBP	U: 0.001 M HNO_3		

SPECIES	SUPPORT	STAT. PHASE (EXTRACTANT)	MOB. PHASE (ELUENT)	REFERENCE N°	AUTHOR
Lanthanide fission products; trival. actinides	Chromosorb	HDEHP-(DIPB)	Lactic acid-Na DTPA, HNO_3-$KBrO_3$	440	MARSH
Zr, Th, U, Pu	Chromosorb	HDEHP	HNO_3 (sorption) HCl and HCl+I⁻ (sequential elution)	441	MARSH
Am, Cm-Nd			Am, Cm: glycolic acid, DTPA (sodium salt) Nd: diluted HNO_3		
Ta-impurities	Teflon	MIBK	10 M HF- 6 M H_2SO_4 - 2.2 M NH_4F	442	MAŠTALKA
Ce(IV)-target subst.	SiO_2^+	TBP	HNO_3-0.1% $NaBrO_3$	443	MAŠTALKA
Ce(IV)-Cs, Ba, La, Pr	SiO_2^+	TBP	Cs, Ba, La, Pr: 14 M HNO_3- 0.1% $NaBrO_3$	444	MAŠTALKA
Ce-spallation prod.	SiO_2^+	TBP	Spall. pr.: 14 M HNO_3- 0.1% $NaBrO_3$	445	MAŠTALKA
Ce-fission products Mo-fission prod.	Hostaflon C2 (60-100 mesh)	HDEHP	HNO_3	446	MICHELSEN
Ho-Y Y-Er	Plaskon-CTFE 2300	HDEHP	2.1 M HCl, conc. HCl	447	MICHELSEN
Ho-Y Y-Er	Plaskon-CTFE 2300	HDEHP	2.1 M HCl, conc. HCl	448	MICHELSEN
Rare earths	Silica gel	HDEHP	HNO_3	449	MIKHAILI-CHENKO
Mn(II), Zn, Co(II)-Fe(III)	SiO_2^+	TBP	HCl	450	MIKULSKI
Sn(II)-Sn(IV); Sn(II)-Te(IV)-Sb(V)	SiO_2^+	TBP	1 M and 11.7 M HCl	451	MIKULSKI

SPECIES	SUPPORT	STAT. PHASE (EXTRACTANT)	MOB. PHASE (ELUENT)	REFERENCE N⁰	AUTHOR
Pm, Eu, Tb, Tm; Am, Cm, Cf, Fm, Md	SiO_2^+	TBHP-TBPP (1:1)	HNO_3	452	MIKULSKI
Am, Cm, Fm, Md	SiO_2^+	TBHP-TBPP (1:1)	$HClO_4$	453	MIKULSKI
Ca, Mn(II), Fe(III), Co(II), Ni, Cu(II), Zn, Sn(II), Y, Cd, Th, Pa, U(VI) (as chlorides)	Teflon	TNOA	HCl	454	MIKULSKI
Mn(II), Fe(III), Co(II), Ni, Cu(II), Zn, Cd, Sb(V), Te(IV), Tb, U(VI)	SiO_2^+	TBPP; TBPP-TBHP	HCl (various conc.) 0.01-5 M HNO_3	455	MIKULSKI
Tb, Hf, Cd, Zn-U; Am-Pu(VI), Te(IV)-Sb(V); Ni-Co, Cu	Teflon	TNOA	HCl (various conc.)	456	MIKULSKI
Cs-Eu-Tb-Tm Cs-Am-Eu-Tb Cs-(Cm, Cf, Fm, Md)	Hyflo Super Cel	TBHP-TBPP(1:1)	14.5 M HNO_3 15.3 M HNO_3 14-15.3 M HNO_3	457	MIKULSKI
Nb, Mo, (Ti)	SiO_2	TBP	Mo, Ti: conc. HCl Nb: 6 M HCl - 0.03% H_2O_2	458	MINCZEW-SKI
Am, Cm-fission products	SiO_2^+	0.3 M HDEHP in Solvesso-100	F.P.(including partly Ru-Rh and Zr-Nb): 0.1 M HNO_3 F.P. (incl. rest of Zr-Nb): 0.1 N $H_2C_2O_4$-0.01 N HNO_3 Am+Cm: 0.05 M DTPA- 1 M lactic acid (pH 3) residual F.P.: remaining on column	459	MIRZA

SPECIES	SUPPORT	STAT. PHASE (EXTRACTANT)	MOB. PHASE (ELUENT)	REFERENCE N°	AUTHOR
U, Pu, Am, Cm, fission products, canning impurities	SiO_2^+	HDEHP	Cs, Sr, Ba, Al, Fe: 0.1 M HCl Rare earths, Am, Cm, Ru: 1 M HCl Ru: 0.1 M HCl	460	MONSE-COUR
Rare earths, Am, Cm, Ru	SiO_2^+	HDEHP	La, Pr-Am, Nd, Pm, Sm: 0.2 M HCl		
Cf(III), Cm(III), Bk(III), Ce(III)	Teflon	HDEHP	Dilute HNO_3	461	MOORE
Am(V)-Cm(III) [+Am(III)]- Cf(III)	Teflon	HDEHP	Am: 0.01 M HCl Cm(+Am(III)): 0.3 M HNO_3 Cf: 4 M HNO_3	462	MOORE
Am-other transuranium elements	Inert materials	HDEHP	HNO_3	463	MOORE
Eu from other rare earths	Fluoroplast-4	DOPA HDEHP DIAPA	0.2 M HCl - 3 M NH_4Cl; 11 M HCl	464	MOSKVIN
Se, Te, (As(V)	Teflon	TBP	13 M HCl; 5 M HCl; 1 M HCl	465	MOSKVIN
La-Ce-Pm-Eu	Fluoroplast-4	HDEHP	0.2 M HCl	466	MOSKVIN
Eu, Cr- rare earths Eu-La	Fluoroplast-4	DOPA HDEHP	HCl 2 M HCl	467	MOSKVIN
La-Ce	Fluoroplast-4	HDEHP	La: 0.5 M HNO_3 - 0.4 M $(NH_4)_2S_2O_8$ - 0.1 M $AgNO_3$ Ce(III): 6 M HCl	468	MOSKVIN
W-Mo	Teflon	MIBK	6 M HCl - 0.2 M tartaric acid	469	MOSKVIN
Re-Mo	Teflon	MIBK	idem		
Mo-Tc	Teflon	TBP	Tc: NH_4OH		
W-Mo-Re	Teflon	TBP	W: 3 M HCl - 0.2 M tartaric acid Mo: 0.1 M HCl - 0.2 M tartaric acid Re: H_2O		

SPECIES	SUPPORT	STAT. PHASE (EXTRACTANT)	MOB. PHASE (ELUENT)	N°	REFERENCE AUTHOR
Th, Pa, Zr, Hf, Nb-Ac, Ra, alkali, alk. earths, rare earths, Ac-La	Fluoro-plast-4	TBP HDEHP	6 M HNO_3 0.15-0.2 M HNO_3	470	MOSKVIN
(Theoretical considerat.)	Fluoro-plast-4 (porous block)	--	--	471	MOSKVIN
Mo-Tc(cont.sep.)	Fluoro-plast-4	TBP	Mo:0.1 M HCl - 0.2 M tartaric acid Tc:dil. (1:10) NH_4OH	472	MOSKVIN
W-Mo-Re	Fluoro-plast-4	TBP	W: 3 M HCl- 0.2 M tartaric acid Mo: 0.1 M HCl - 0.2 M tartaric acid Re: NH_4OH	473	MOSKVIN
Ce, La-Cs, Ba	Fluoro-plast-4	HDEHP	Cs, Ba:0.01 M HCOOH Ce, La:retained on column	474	MOSKVIN
Cm-Am	SiO_2^+	Aliquat-336	Cm:3.5 M $LiNO_3$ Am: 1 M HNO_3	475	MÜLLER
^{241}Am, ^{243}Am, ^{244}Cm	SiO_2^+	Tridecylmethyl-ammonium-nitrate	$LiNO_3$, HNO_3	476	MÜLLER
Pu(IV), Ce(IV)-Am, impurit.	SiO_2^+	Aliquat-336	Am, imp.: 8 M $LiNO_3$-0.01 M HNO_3-0.5 M $KBrO_3$ Pu, Ce:0.1 M HNO_3	477	MÜLLER (see Ref. [39] in respective publ.)
Rare earths	Silica gel	HDEHP	HCl(various conc.)	478	MÜNZE

514

SPECIES	SUPPORT	STAT. PHASE (EXTRACTANT)	MOB. PHASE (ELUENT)	REFERENCE No	AUTHOR
Ni(II)-Co(II)-Fe(III)	Cellulose or Kel-F	Aliquat-336	Ni: 8 M HCl Co: 3 M HCl Fe: 0.2 M HNO_3	479	MURRAY
Nb	Fluoroplast-4	TNOA	HCl	480	MYMRIK
Fiss.prod.-Np-U	Hyflo Super Cel	TBP	F.P.: 8 M HNO_3 (+0.05 M oxal. acid for Zr-95); Np: 4 M HNO_3 - 0.2 M $NaNO_2$ U: H_2O	481	NARBUTT
U(IV)(stripping)	Teflon	TNOA in diluent	0.2 M HF (stripping solution)	482	NAZAROV
Ni-Mn-Co-Cu-Fe-Zn-Cd	SiO_2-gel	TNOA	HCl, HCl-HNO_3, HNO_3 (various concentrations) temp. 25 or 60°C	483	NEEF
Ag sep. from Li, Na, K, Rb, Cs, Be, Mg, Ba, Ra, V, Cr, Mn, Ni, Cu, Al, In, Rh, Pb	PE Fluoropak	TIOTP-CCl_4	1-8 M HNO_3 (60°C)	484	NELSON
Be-Mn-Zn	Fluoroplast-4	TBP	NH_4SCN	485	NGUYEN
Ca, Mg-Y;	Kel-F	HDPM	3 M HNO_3, 1 M HCl(Y)	486	O'LAUGHLIN
Ca, Mg, Sr, (Ba)	Paper	HDPM	0.1-0.2 M $HClO_4$		
Tb-Dy-Ho	Plaskon	(MHDPO) MEHDPO	9 M HNO_3	487	O'LAUGHLIN
Rare earths	Kel-F, SiO_2^+, Plaskon	MHDPO, MEHDPO	HNO_3 (various concentrations)	488	O'LAUGHLIN

SPECIES	SUPPORT	STAT. PHASE (EXTRACTANT)	MOB. PHASE (ELUENT)	REFERENCE N°	AUTHOR
(Eu, Cm, Cf)-Ce(III)-Bk	SiO_2^+	HDEHP	Eu, Cm, Cf: 8 M HNO_3-0.2 M $NaBrO_3$ Ce(III):0.15 M HNO_3-10% H_2O_2 Bk(III):4 M HNO_3	489	OVERMAN
U(VI)-impurities (Al, Ca, Cd, Cu, Fe, Li, Mg, Na, Ni)	Kel-F	TBP	HNO_3 (0.1-2 M)	490	PAGLIAI
Pd-Purex waste solutions	Plaskon	MTC	Pd: H_2O	491	PANESKO
Actinides, Lanthanides (Theoretical considerations using known literature data)	SiO_2^+	HDEHP HEHØP	HCl, HNO_3 HCl, HNO_3	492	PEPPARD
^{99}Mo, ^{239}Np-U, fission products	Silaned glass powder; (Teflon)	0.5 M HTTA in xylene	0.5-1 M HCl; Np: strong acid	493	PERRICOS
U(VI), Fe(III), Cu(II), Sr, Cs	Kel-F	TLA	HCl (various concentrations)	494	PETIT-BROMET
Rare earths (phthalocyanine compl.) (mother-daughter sep.)	SiO_2-gel Teflon	Quinoline	1% EDTA-sol.	495	PFREPPER
Fe-Pu(U)	Kel-F	TBP	Fe: 5.5 M HNO_3-0.5 M HCl Pu(U): 1 M H_2SO_4	496	PHILLIPS
Cu	SiO_2	Dithizone-$CHCl_3$ Dithizone-CCl_4	(No elution carried out)	497	PIERCE
^{115}In, Cd	SiO_2	Dithizone	Dil. $HClO_4$	498	PIERCE
^{116}In-other active components	Cellulose acetate	Dithizone-solvent	1 M HCl	499	PIERCE

SPECIES	SUPPORT	STAT. PHASE (EXTRACTANT)	MOB. PHASE (ELUENT)	REFERENCE N°	AUTHOR
La, Pr, Nd, Pm, Sm, Eu, Gd	Corvic	HDEHP	Dil. $HClO_4$ (0.32 to 0.81 M)	500	PIERCE
All rare earths	Corvic	HDEHP	Dil. $HClO_4$ (0.195 to 6.87 M) at 60°C	501	PIERCE
Rare earths	Corvic	HDEHP	$HClO_4$ (various conc.) 60°C	502	PIERCE
Rare earths	Corvic	HDEHP	$HClO_4$ HCl (various conc.)	503	PIERCE
Nd, Pm, Sm, Eu, Gd, Tb, Dy, Ho, Er	Corvic	HDEHP	$HClO_4$ (various conc.)	504	PIERCE
Zn	Cellulose acetate	Dithizone-solvent	(No elution carried out)	505	PIERCE
As, Ge, In, Mn, Co, Cu, Zn;	Corvic R 51/83	TIOA	8.6 M HCl, 1 M HCl, H_2O; 5.9 M HCl; 3.1 M HCl, 0.55 M HCl, H_2O;	506	PIERCE
Zr-Nb; Mo-U			7.3 M HCl, 6.7 M H_2SO_4 (Nb); 0.6 M H_2O(U)		
Am-Nd	SiO_2^+	HDEHP	0.2 M HNO_3 (1 M NH_4SCN)	507	PLUYM
Noble metals	Porasil C	TBP	H_2SO_4-HCl (various conc.)	508	POHLANDT
Noble metals	Porasil C	TBP	H_2SO_4-HCl (various conc.)	509	POHLANDT
Noble metals	Kel-F Porasil C	TBP	H_2SO_4-HCl (various conc.)	510	POHLANDT
Au-other noble metals	Porasil C	TBP	Noble met.: 0.1 M HCl Au: 14 M HNO_3	511	POHLANDT
Actinides, rare earths	Al_2O_3	HDEHP	HCl, HNO_3 (various conc.)	512	POLI

SPECIES	SUPPORT	STAT. PHASE (EXTRACTANT)	MOB. PHASE (ELUENT)	REFERENCE N°	AUTHOR
Nb-Zr, In-Tl, Co-Fe(III)	Teflon (Fluoroplast-4)	$0.5\,M$ HTTA-C_6H_6 Isoamylacetate; TBP	Mineral acids (various conc.)	513	PREOBRA-ZHENSKII
Au, Tl, (Bi)	Teflon	Diethylether	Au: $0.5\,M$ HCl Tl: $0.2\,M$ HNO_3	514	PREOBRA-ZHENSKII
(Preparation of columns)	Ftoroplast	--	--	515	PREOBRA-ZHENSKII
(Preparation of columns)	Ftoroplast	--	--	516	PREOBRA-ZHENSKII
Eu-Cd	SiO_2^+ SiO_2-gel	HDEHP	$0.4\,M$ HCl	517	PSZONICKA
Er-Dy	SiO_2-gel	HDEHP	$0.4\,M$ HCl(50^oC)	518	RAICHEV
Bi-Po	Teflon	TBP	Bi: $6\,M$ HCl Po: $2\,M$ HF	519	RAICHEV
Gd-Tb	SiO_2-gel	HDEHP	Tb: $1.7\,M$ HCl	520	RAICHEV
^{86}Rb-^{137}Cs	Ftoroplast-4	t-BAMP in CCl_4	$> 0.5\,M$ NaOH	521	RAIS
F^-, Cl^-, Br^-, I^-	Kel-F	Aliquat-336	F: $0.1\,M$ Na acetate Cl: $0.1\,M$ $NaNO_3$ Br: $0.1\,M$ $NaNO_3$ I: $1\,M$ $NaNO_3$	522	RAMALEY
Fe(III), La, Th, Zr, Ti(IV), Zn, In, Ga, Cu(II), U(VI), Mn(II), Co, Be, Al, Sc, Cr(III)	SiO_2	DNNS	HCl; HNO_3 (various conc.) $0.5\,N$ H_2SO_4 Cr(III): HCl	523	RAO
Eu-Tb	Hostaflon C2	HDEHP	HCl (various conc.)	524	RICCATO
Ru-other fission products	Voltalef 300 Micro PL	HDEHP	$9\,M$ HCl($+Cl_2$)	525	RICCATO
^{144}La-^{144}Ce-^{147}Nd-^{147}Pm	Chromosorb-W	HDEHP	$0.3\,M$ HCl (various temperatures)	526	RIDER

SPECIES	SUPPORT	STAT. PHASE (EXTRACTANT)	MOB. PHASE (ELUENT)	REFERENCE N°	AUTHOR
Nd-other rare earths	Chromo-sorb-W	HDEHP	0.3 M HCl (various temperatures)	527	RIDER
Nd-impurities	Chromo-sorb-W	HDEHP	Impurities: 0.1-0.2 M HCl Nd: 0.3 M HCl (temp. 87°C)	528	RIDER
Fe(III) → (II) V(V) → (IV) Cr(VI) → (III) (redox col.)	Celite A	Tetrachlorohy-droquinone (chloranil)	0.03 M HCl 2 M HCl 2 M HCl	529	ROBERTS
Hf, Sc-metal ions, La, rare earths-metal ions	SiO_2^+	HDEHP	8 M HCl HCl, pH: 3.5 (acetate buffer)	530	SAMSAHL
Hf+Sc; Rare earths (concentrat.)	SiO_2^+	HDEHP	8 M HCl Dil. HCl(acetate buffer)	531	SAMSAHL
		-see Ref. [531]-		532	SAMSAHL
Np(IV)-urine	Micro-thene-710	TOPO in cyclo-hexane	Np: 2 M HCl+Cl_2 (as Np-V)	533	SANTORI
Ru(III)(study of antagonistic effects)		TOP; TOPO; TBPO; THA; Aliquat-336 to each:n-octanol, cyclo-hexanol and chloroform	HCl (various conc.)	534	SARAIYA
U, Fe, Co, Ni, Cu, Cd, La, Th, Zr	Kel-F	Aliquat-336	HCl	535	SASTRI
Pu(III)-U(IV)-U(VI)	Voltalef 300 Micro PL	TBP	4.8 M HNO_3 (0.1 M N_2H_4)	536	SCHMID
U(VI)-U(IV)	Voltalef 300 Micro PL	TBP	HNO_3 (various conc.)	537	SCHMID
U(IV)-U(VI); U(IV)-U(VI)-Pu(III), fiss.prod., corrosion prod.	Voltalef	HDEHP in xylene	HNO_3-0.1 M N_2H_4; Pu(III), FP, corr. pr.: HNO_3-0.1 M N_2H_4 U(IV): 11.7 M HNO_3 or 15% H_2SO_4-5% H_3PO_4	538	SCHMID

SPECIES	SUPPORT	STAT. PHASE (EXTRACTANT)	MOB. PHASE (ELUENT)	REFERENCE Nº	AUTHOR
U(IV), U(VI)	Voltalef	TBP DBP HDEHP Primene JM-T Amberlite LA-1 TLA Hyamine-1622	HNO_3 various conc.,(determination of partition ratios)	539	SCHMID
Zn	SiO_2^+	Dithizone in CCl_4	$H_2C_2O_4(-Na_2C_2O_4)$ various conc. $(HClO_4-NaClO_4)$	540	ŠEBESTA
Zn-Cd	SiO_2^+	Dithizone in CCl_4	Zn:$0.1 M H_2C_2O_4-$ $Na_2C_2O_4$,pH 4.1 Cd:$0.1 M H_2C_2O_4$	541	ŠEBESTA
Ag-Hg			Ag:2% KCNS-0.1 N H_2SO_4 Hg:6% KI in acetate buffer,pH 4		
Cd-Ag			Cd:0.1 M HNO_3 Ag:6% KI in acetate buffer,pH 4		
Pb-Bi-Po			Pb: 0.1 M HCl Bi: 0.3 M HCl Po: 1.5 M HCl		
$^{65}Zn-^{64}Cu$	Celite 545	HBPHA	Zn:$0.1 M HClO_4-$ $NaClO_4$ pH 2.5 Cu: $0.1 M HClO_4$	542	ŠEBESTA
$^{60}Co-^{59}Fe$			Co:0.05 M HCl-3% H_2O_2 Fe: 0.1 M oxalic acid		
$^{60}Co-^{64}Cu-^{59}Fe$			Co:0.01 M HCl-3% H_2O_2 Cu: 0.2 M HCl Fe:0.1 M oxalic acid		
U-^{234}Th			U:0.025 M HCl Th:0.5 M HCl		

SPECIES	SUPPORT	STAT. PHASE (EXTRACTANT)	MOB. PHASE (ELUENT)	REFERENCE No	AUTHOR
^{234}Th-^{233}Pa			Th: 4 M HCl Pa: 1 M oxalic acid		
Re-Tc	Chromo-sorb	HBPHA	Re: 5 M $HClO_4$ Tc: 0.05 M $HClO_4$	543	ŠEBESTA
$^{175+181}Hf$-^{182}Ta;	Teflon	HBPHA	Hf: 2 M HCl-0.1 M HF Ta: 2 M HF;	544	ŠEBESTA
^{233}Pa-^{182}Ta;			Pa: 2 M HCl - 0.1 M HF Ta: 2M HF		
^{233}Pa-^{95}Nb;			Pa: 2 M HCl - 0.1 M HF Nb: 2 M HF;		
^{95}Nb-^{182}Ta;			Nb: 2 M HCl - 0.4 M HF; Ta: 2 M HF;		
^{233}Pa-^{95}Nb-^{182}Ta			Pa: 6 M HCl - 0.05 M HF Nb: 2 M HCl - 0.4 M HF Ta: 2 M HF		
^{224}Ra-^{228}Th	Chromo-sorb-W	HDEHP	Ra: 0.01-0.1 M HCl Th: retained on col.	545	ŠEBESTA
Hg(II), Cd, Pb, Zn (concentr.)	Styrene-divinyl-benzene copoly-mer; Dibenzal sorbitol in dimeth-ylform-amide	Dithizone	pH 3.5-5.2 (for concentr.) Cd, Pb, Zn: 1 M HCl Hg: 0.4 M H_2SO_4 - 7% KBr	546	SEKIZUKA
(Fe-)^{234}, Th-U(VI)	Fluoro-plast-4	TBP	(Fe), Th: 6 M HCl U: 0.1 M HNO_3	547	SHESTA-KOV

SPECIES	SUPPORT	STAT. PHASE (EXTRACTANT)	MOB. PHASE (ELUENT)	REFERENCE No	AUTHOR
La(or Yb)-Th-Zr-U(VI)	SiO_2-gel	Amberlite LA-2 (SO_4^{2-})	La(Yb):0.025 M H_2SO_4-1 M $(NH_4)_2SO_4$ Th:0.1 M H_2SO_4 Zr: 4 M HCl U: 1 M $HClO_4$	548	SHIMIZU
Rare earths-base oxides	SiO_2^+	HDEHP	2.5 M HCl	549	SHMANEN-KOVA
Al-Ga	SiO_2-gel	TNOA	Al:2 M NH_4SCN, pH 2.5 Ga: 0.1 M HCl	550	SHMANEN-KOVA
Ga-Fe	Teflon	TNOA	Ga: 0.1 HCl Fe:1.5 M HI-2 M HCl		
Ga, As-Cr, Mn, Co, Ni, Cu, Zn;	SiO_2-gel	TBP	HCl	551	SHULEP-NIKOV
Cr-Mn-Co-Ni-Cu-Zn		TNOA	HCl		
Zr-Nb	SiO_2^+	TBP	4-5 M HNO_3-0.3% H_2O_2	552	SIEKIER-SKI
Ce-Nd-Pm; Ce-Pm-Sm-Eu-Gd-Tb; Eu-Tb-Y	SiO_2^+	TBP	15.8 M HNO_3 15.1 M HNO_3 11.5 M HNO_3	553	SIEKIER-SKI
Ca, Sc	SiO_2^+	TBP	6-10 M HCl	554	SIEKIER-SKI
Rare earths	SiO_2^+	HDEHP	HNO_3; HCl; H_2SO_4; (various conc. and temperatures)	555	SIEKIER-SKI
No(II)-No(III)	Kel-F	HDEHP	0.1 M and 3 M HNO_3	556	SILVA
		(Not available)		557	SILVA

SPECIES	SUPPORT	STAT. PHASE (EXTRACTANT)	MOB. PHASE (ELUENT)	REFERENCE No	AUTHOR
U(VI), Fe(III), Th, Y	SDVB	TBP-PCE	HNO_3-(various conc.) $NaNO_3$; NH_4NO_3; H_2O	558	SMALL
U; rare earths (recovery) Th-U; Y-Th	Cross-linked copolymers	Alkyl phosphates (e.g. TBP)	(Various patent examples)	559	SMALL
Li, Na, K, Rb, Cs	SiO_2^+	2,6,8-Trimethyl nonyl-4 phosphoric acid	$LiNO_3$-HNO_3; $LiCl$-HCl; H_2O	560	SMULEK
Co-Ni	SiO_2^+	TNOA	HCl(various conc.)(H_2O)	561	SMULEK
Mn-Fe	SiO_2^+	TNOA	HCl, HNO_3	562	SMULEK
Eu-U, fission products	SiO_2^+	HDEHP	HNO_3(various conc.)	563	SMULEK
Y-U, fission products	SiO_2^+	HDEHP	HNO_3(various conc.)	564	SMULEK
Rare earths	SiO_2^+	HDEHP	HCl, HNO_3	565	SOCHACKA
Eu; Sc	Micro-porous glass beads	HDEHP TBP	HCl HNO_3	566	SPECHT
55,59Fe-60Co	Macro-porous styrene-vinyl-benzene	Dithizone in CCl_4	Fe: acetate buffer pH 5.6 Co: 4 N HCl	567	SPÉVÀČ-KOVA
59Fe-60Co	Styrene-divinyl-benzene (10 %) macro-porous	Dithizone in CCl_4 (saturated sol.)	Fe: acetate buffer of pH 5.6 Co: 6 M HCl;	568	SPÉVÀČ-KOVA
Cu, Zn, Fe			Acetate buffer of pH 4.9 3 M HCl to elute the retained cations;		
Sc, Ca, Mn;					
Na, Mn(Pb),	"	Pb-dithizonate	Na, Mn: not retained		

SPECIES	SUPPORT	STAT. PHASE (EXTRACTANT)	MOB. PHASE (ELUENT)	REFERENCE No	AUTHOR
Zn, In, (Bi), Cu, Hg			Zn, In(Pb):Bi-dithizonate (as displacement agent)		
Pt-Au	Styrene-divinyl-benzene copolymer	Dithizone	Au: retained on col.	569	SPÉVÀČ-KOVA
Fe-separation from other elements	Fluoropak	2-octanone	Fe: H_2O	570	STAPLES
Cd, Ag, Hg, Mn, Co, Zn (radio-activation determin.)	Ftoro-plast-4	Zn-diethyldithio-carbamate in $CHCl_3$	H_2O adjusted to pH 5-6 (sample sol.) Mn+Co pass through; Cd, Ag, Hg(Cu) displace Zn from the complex; Zn: $Cd(NO_3)_2$-sol. Cd:$Cu(NO_3)_2$-sol. (Cu:$AgNO_3$-sol.) Ag:$Hg(NO_3)_2$-sol. Hg: 1 N KI at pH 1	571	STEPANETS
Tb, Hf, Th, - U; Tb - Hf- Pa; Hf - Pa - W	SiO_2^+	TBP	HCl	572	STRONSKI
Transition and actinide metal salts	SiO_2^+	TBP TOPO TNOA	HCl Mineral acid Electrolyte	573	STRONSKI
Ga-Zn; Os-W; Pb-Bi-Tl	Teflon	TBP	HCl(various conc.) 1 M NaOH	574	STRONSKI
Ni, Cu, Zn, Sr, Y, Pb, Bi, Th, U, Cd, Tb	Teflon	TOPO	HCl(various conc.) HNO_3(various conc.) 2 N H_2SO_4 9.1 M $HClO_4$	575	STRONSKI

SPECIES	SUPPORT	STAT. PHASE (EXTRACTANT)	MOB. PHASE (ELUENT)	REFERENCE No	AUTHOR
^{228}Ra-^{228}Ac; ^{228}Ra, ^{228}Ac, ^{212}Pb, ^{212}Bi	PTFCE	HDEHP	HCl (various conc.)	576	STRONSKI
Eu-Gd; Eu-Tb; Ho-Er; Tm-Lu; Yb-Lu; (Am+Cm)-Eu; (Am+Cm)-U(VI); (Am+Cm)-Pu(VI)	Kel-F	HDEHP	HCl (various conc.)	577	STRONSKI
Eu-Am; Nd-Pr; Sm-Pm; Gd-Eu; Er-Ho; Tm-Er; Yb-Tm; Lu-Tm	Kel-F	Aliquat-336	$LiNO_3$, HNO_3	578	STRONSKI
In(III)-Cd(II); In(III)-Sn(IV)-Te(IV); Sb(V)-Sn(IV)	SiO_2^+	Amberlite LA-2	HCl (various conc.)	579	STRONSKI
Pb-Bi	Voltalef	TBP	$HClO_4$-thiourea (various conc.)	580	SULCEK
Eu-Tb	Kel-F	H(thd)	0.5 M sodium acetate-acetic acid buffers (pH 6.5-7.05)	581	SWEET
Actinide and lanthanide elements	SiO_2^+	TBP	11.9-14.1 M HNO_3	582	TAUBE
Mo(VI), Ru(VI), Os(VI)-Tc(VII)-Re(VII)	Daiflon M-300	Pyridine-benzene (7:3)	Mo, Ru, Os: 4.5 M NaOH Tc(Os): 4.5 M NaOH-2.5% (N_2H_4) H_2SO_4 Re: H_2O	583	TERADA
Rare earths	Cellulose powder Paper (also laminar techn. applied)	TNOA	$LiNO_3$ sol. (+HNO_3)	584	TESTA

SPECIES	SUPPORT	STAT. PHASE (EXTRACTANT)	MOB. PHASE (ELUENT)	REFERENCE N°	AUTHOR
Th sep. from urine	Kel-F	TOPO	H_2O; (6 M HCl, 0.5 M HCl)	585	TESTA
[137]Cs-fission prod.	Kel-F	Dipicrylamine	0.01 M EDTA (pH 7.0) Cs: 0.5 M HNO_3	586	TESTA
U sep. from urine	Kel-F	TBP	7.5 M HNO_3; U: H_2O	587	TESTA
Pu(IV)-metal ions (in urine)	Kel-F	TNOA	Th, U: 1 M HNO_3 Pu: H_2SO_3 (as Pu(III)	588	TESTA
Th-U	Kel-F	TOPO 0.5 M in cyclohexane	Th: diluted H_2SO_4 U: 1 M HF	589	TESTA
Th-urine	Micro-thene	TOPO	0.3 M H_2SO_4	590	TESTA
U-urine	Micro-thene	TOPO	1 M HF		
U-Th-urine	Micro-thene	TOPO	Th: 3 M H_2SO_4 U: 1 M HF		
U, Th-Pu-urine	Kel-F	TNOA	Pu: H_2SO_3		
[90]Y-urine	Micro-thene	HDEHP	6 M HNO_3		
[90]Y-(urine+[90]Sr)	Micro-thene-710	HDEHP	0.3 M HNO_3(washing) Y: 6 M HNO_3	591	TESTA
Th, U, ^{90}Y-urine	Micro-thene-710	TOPO in cyclohexane; HDEHP in toluene or benzene	Various mineral acids	592	TESTA
Summary of Refs [584-592]				593	TESTA
Pu-urine	Micro-thene-710	TOPO in cyclohexane	Pu: 6 M HCl-0.1 M HI	594	TESTA
[55]Fe in fallout material	Micro-thene-710	HDEHP	Fe: 6 M HCl		
Fe(III)-H_3BO_3	Micro-thene-710	TMPD	Fe: 0.1 M HCl H_3BO_3: 0.1 M NaOH		

SPECIES	SUPPORT	STAT. PHASE (EXTRACTANT)	MOB. PHASE (ELUENT)	REFERENCE N^O	AUTHOR
U, fission products -Np	glass-powder (hydrophobic)	0.5 M HTTA in xylene or benzene	0.8 - 1 M HCl Np: 6 M HCl and alcohol	595 and 596	THOMAS-SEN
Fission prod. - U	SiO_2^+	HDEHP	HNO_3 (various conc.)	597	TOMAZIČ
U(IV)-Ce(III), Eu, Tb, Lu	SiO_2^+	HDEHP	HNO_3 (various conc.)	598	TOMAZIČ
Sb, Zr, Pa	Hostaflon C2	HDEHP DIBC TBP	HCl, HNO_3	599	TRAUT-MANN
Pa-U, Th, Np, fission products	Hostaflon C2	DIBC	9 M HCl-0.1 M oxalic acid 12 M HCl-0.5 M HF	600	TRAUT-MANN
V(IV)-V(V)	Fluoroplast-4	Benzohydroxamic acid in TBP	V(IV):0.5-5 M HCl $HClO_4$, H_2SO_4, 1-10 M H_3PO_4; V(V): conc. H_3PO_4	601	TSERKOV-NITSKAYA
Zr, Hf	SiO_2^+	TBP	5-8 M HCl	602	UENO, KAORU
Nb-Mo-Tc	Celite	TBP	Nb: 3 M HCl Mo: 1 M HCl Tc: 6 M HNO_3	603	UENO, KAORU
Hg(II)-Cd, Pb, Zn (concentr.)	(see Ref. [546])	Dithizone	Cd, Pb, Zn:0.1 M HCl Hg: stripped with methyl-ethyl ketone	604	UENO, KEIHEI
Np(IV)-Np(V)-Np(VI)	Fluoroplast-4	HDEHP	0.5-3 M HNO_3 or 0.5-2 M HCl	605	USHATSKII
Pu(III)-(IV)-(VI);	SiO_2^+ Teflon Corvic Kel-F	TBP	HNO_3 (various conc. and temp.); 0.20-4.98 M HNO_3; F.P.:6.1 M HNO_3 and 1.03 M HNO_3	606	van OOYEN
Pu(VI)-Cr(VI); U-Pu-fission prod;			Pu: 1.03 M HNO_3- 0.1 M $Fe(SO_3NH_2)_2$ U: H_2O;		
RuNO-nitrato compl.			0.4-10 M HNO_3 (4°, 21° and 22^OC)		

SPECIES	SUPPORT	STAT. PHASE (EXTRACTANT)	MOB. PHASE (ELUENT)	REFERENCE N^o	AUTHOR
Am, Cm	SiO_2^+	TLMANO$_3$	LiNO$_3$-HNO$_3$	607	van OOYEN
Am-Cm(Cf-Es)	SiO_2^+ Kel-F	Aliquat-336 TLMANO$_3$	4 M LiNO$_3$+HNO$_3$ at pH 2 6-7 N Al(NO$_3$)$_3$ at 10ºC and at room temperature	608	van OOYEN
Fe(III)sep. from Bi(III), Cd, Ce(III) Cr(III), Cu(II) Co(II), Hg(II), Mo(VI), Ni, Th, U(VI), V(IV); Zr, Zn	Fluoropak (Halo-port-F)	2-Octanone	1-7 M HCl	609	WADE
Am, Pr, Pm, Eu-La	SiO_2^+	HDEHP-xylene TOPO-xylene	0. 1 -1 M HNO$_3$	610	WATA-NABE
^{149}Pm, ^{151}Pm-Nd; ^{241}Am-Nd	SiO_2^+	HDEHP	0. 3-2. 1 M HNO$_3$	611	WATA-NABE
Np(IV)-fission and activation products	Poropak Q	HTTA in o-xylene	FP+AP: 1 M HCl Np: 10 M HNO$_3$	612	WEHNER
Th, U, Np, Pu-fission products	Voltalef UF 300	TBP	FP: 5. 5 M HNO$_3$ Actinides:0.01 M HNO$_3$	613	WENZEL
Fiss. prod. , Th, Np, Pu-U;	Voltalef	TBP	F. P. , Th, Np, Pu: 5. 5 M HNO$_3$ (0. 5 M H$_2$SO$_4$) 0. 025 M Fe(SO$_3$NH$_2$)$_2$ U:0. 01 M HNO$_3$;	614	WENZEL
Fiss. prod. -Zr, Nb, Th, U, Np, Pu			F.P.: 6 M HNO$_3$ Zr, Nb+actin. : 0. 01 M HNO$_3$		
Nd, Pm, Eu	Al$_2$O$_3$	HDEHP	Dil. HCl	615	WINCHES-TER

SPECIES	SUPPORT	STAT. PHASE (EXTRACTANT)	MOB. PHASE (ELUENT)	REFERENCE N°	AUTHOR
Rare earths	Al_2O_3	HDEHP	HCl	616	WINCHESTER
Rare earths	SiO_2	HDEHP	0.14-3.8 M HCl (at 25° and 70°C)	617	WINCHESTER
U(VI), Cu(II)	Polyethylene (microporous)	TBP	HNO_3	618	WINSTEN
Hg-impurities	Voltalef 300 Micro PL	0.001% Dithizone in $CHCl_3$	Impurit.: 3 M H_2SO_4 Hg-dithizonate: $CHCl_3$	619	WOIDICH
Cu-Cd-Y	Ftoroplast-4	Diethyldithiocarbamate in $CHCl_3$	(Successive displacement from the diethyldithiocarbamate complexes)	620	YAKOVLEV
		(not yet available)		621	YAKOVLEV
Hg, Co, Cu, Fe(III) Fe, Cu(I), Co	Cellulose	Phenyl 2-pyridyl-ketone-oxime; 8-hydroxyquinoline	Aqueous sol.	622	YAMAMOTO
Fe(III)-various ions	Fluoropak	2-Octanone	7 M HCl (acetone)	623	YAMAMURA
^{95}Zr-^{95}Nb-^{99}Mo	Chromosorb-W	α-benzoinoxime	Zr+Nb: 1 M HCl Mo: conc. HNO_3	624	YOSHIDA
Pd(Pt, Au, Ir)	Voltalef	TBP	0.1-0.5 M $HClO_4$- 1% $NaClO_4$-0.1-5% thiourea for Pd(II) retention, Pd(II): H_2O	625	YUSAF
Nd-U, fission products	SiO_2^+	HDEHP	HCl (various conc.)	626	ZELENAY
Y-lanthanides (La to Ho)	SiO_2^+	HDEHP	Lanthanides: 2.5 M HCl (at 40°C)	627	ZEMSKOVA

SPECIES	SUPPORT	STAT. PHASE (EXTRACTANT)	MOB. PHASE (ELUENT)	REFERENCE No	AUTHOR
^{140}Ba-^{140}La-^{228}Ac	Fluoro-plast-4	TBP	Ba and Ac: 10 M NH_4OH, 0.1 M HNO_3 La: 0.1 M HNO_3	628	ZIV
^{56}Mn-irradiated $KMnO_4$	Fluoro-plast-4	HDEHP	Conc. HCl or oxalic acid	629	ZHUKOV-SKY

8. STATIONARY PHASES USED IN COLUMN
EXTRACTION CHROMATOGRAPHY

(Abbreviations or trade names refer to those used in Paragraph 7)

N°	ABBREVIATION OR TRADE NAME	ORGANIC COMPOUND (STATIONARY PHASE)	REFERENCE
1	Alamine-336	Tertiary straight-chain alkylamine, average mol. wt. of 392	219, 381, 392, 393
2	Aliquat-336	Methyl-tri-n-alkylammonium chloride (mainly octyl and decyl), mol. wt. of approx. 475	224, 249, 250, 271, 292, 323, 363, 364, 365, 367, 371, 377, 475, 477, 479, 522, 534, 535, 578, 608
3	Amberlite LA-1	N-Dodecenyltrialkylmethylamine	539
4	Amberlite LA-2	N-Lauryltrialkylmethylamine	218, 383, 423, 548, 579
5	BAMBP	4-sec-Butyl-2-(α-methylbenzyl) phenol	361
6	DBDECP	N,N, dibutyl diethyl carbamyl phosphonate	215, 216
7	DBM	Dibenzoylmethane	422
8	DBP	Di-n-butyl phosphate	412, 539
9	DEHPP	Bis(2-ethylhexyl)-phenyl phosphate	208
10	DHDO	5,8-Diethyl-7-hydroxydodecan-6-oxime	324, 325
11	DIAP	Diisoamyl phosphoric acid	389
12	DIAPA	Diisoamyl phosphate	464
13	DIBC	Diisobutyl carbinol	278, 599, 600
14	DIPB	Diisopropyl benzene	440
15	DNNS	Dinonyl naphthalene sulphonic acid	523
16	DOPA	Di-n-octyl phosphoric acid	345, 397, 398, 464, 467
17	DOSO	Di-n-octyl sulphoxide	321
18	HBPHA	N-Benzoylphenylhydroxylamine	542, 543, 544

N⁰	ABBREVIATION OR TRADE NAME	ORGANIC COMPOUND (STATIONARY PHASE)	REFERENCE
19	HDEHP	Di(2-ethylhexyl) orthophosphoric acid	197, 198, 199, 200, 207, 214, 217, 223, 234, 238, 239, 249, 250, 259, 261, 263, 264, 268, 273, 277, 278, 279, 280, 281, 282, 283, 284, 285, 286, 310, 326, 329, 330, 337, 338, 339, 340, 352, 353, 354, 355, 358, 359, 361, 362, 366, 368, 369, 370, 371, 372, 373, 376, 378, 379, 380, 382, 386, 387, 394, 395, 400, 404, 405, 406, 407, 408, 409, 411, 414, 415, 416, 417, 418, 425, 426, 427, 430, 431, 435, 440, 441, 446, 447, 448, 449, 459, 460, 461, 462, 463, 464, 466, 467, 468, 470, 474, 478, 489, 492, 500, 501, 502, 503, 504, 507, 512, 517, 518, 520, 524, 525, 526, 527, 528, 530, 531, 532, 538, 539, 545, 549, 555, 556, 563, 564, 565, 566, 576, 577, 590, 591, 592, 593, 594, 597, 598, 599, 605, 610, 611, 615, 616, 617, 626, 627, 629
20	HDPM	Bis(di-n-hexylphosphinyl)methane	486
21	HEHØP	2-Ethylhexylphenyl phosphoric acid	309, 492
22	HEO	10-Hydroxyeicosan-9-oxime	324, 325
23	H(thd)	2,2,6,6-Tetramethyl-3,5-heptanedione	581
24	HX 70	Neotridecanohydroxamic acid	276

Nº	ABBREVIATION OR TRADE NAME	ORGANIC COMPOUND (STATIONARY PHASE)	REFERENCE
25	Hyamine-1622	Diisobutylphenoxyethyldimethylbenzyl ammonium chloride	539
26	IAA	Isoamyl acetate	513
27	IOTG	Isooctyl thioglycolate	317
28	IPE	Isopropyl ether	318, 320, 322, 421
29	Lix-64	Water-soluble mixture of substituted oximes, probably 2-hydroxy-5-dodecylbenzophenone oxime and 5,8-diethyl-7-hydroxy-6-dodecanone oxime	266
30	MDOA	Methyldioctylamine	269
31	MEHDPO	Methylenebis[di(2-ethylhexyl)-phosphine oxide]	487, 488
32	MHDPO	Methylenebis(di-n-hexylphosphine oxide)	346, 391, 487, 488
33	MIBK	Methylisobutyl ketone	192, 195, 274, 313, 315, 316, 318, 319, 320, 322, 421, 442, 469
34	MTC	Methyltricaprylylammonium chloride	491
35	NH$_4$DPA	Ammonium dipicrilamine	420
36	PCE	Perchlorethylene	558
37	PhNO$_2$	Nitrobenzene	191
38	Primene JM-T	Primary trialkylmethylamine with 18-24 C atoms and an average mol. wt. of 311	539
39	STTA	1,1,1-Trifluoro-4-(2-thienyl)-4-mercaptobut-3-en-2-one	360
40	TABAN	Trialkylbenzyl ammonium nitrate	222
41	t-BAMBP	4-tert-Butyl-2-(α-methylbenzyl) phenol	521
42	TBHP	Tetrabutyl hypophosphate	410, 452, 453, 455, 457

N°	ABBREVIATION OR TRADE NAME	ORGANIC COMPOUND (STATIONARY PHASE)	REFERENCE
43	TBP	Tri-n-butyl phosphate	193, 194, 195, 201, 202, 203, 204, 205, 206, 209, 212, 213, 225, 226, 228, 229, 230, 235, 236, 237, 240, 241, 242, 243, 246, 248, 251, 256, 267, 272, 277, 279, 280, 281, 282, 283, 284, 288, 289, 290, 293, 294, 295, 297, 298, 299, 300, 301, 302, 304, 305, 306, 307, 308, 311, 331, 332, 342, 343, 344, 347, 349, 350, 351, 356, 357, 374, 375, 384, 385, 388, 390, 399, 402, 414, 422, 424, 426, 428, 432, 433, 434, 439, 443, 444, 445, 450, 451, 458, 465, 469, 470, 472, 473, 481, 485, 490, 496, 508, 509, 510, 511, 513, 519, 536, 537, 539, 547, 551, 552, 553, 554, 558, 559, 566, 572, 573, 574, 580, 582, 587, 599, 601, 602, 603, 606, 613, 614, 618, 625, 628
44	TBPO	Tri-n-butylphosphine oxide	534
45	TBPP	Tetrabutyl pyrophosphate	452, 453, 455, 457
46	TEHP	Tri(2-ethylhexyl) phosphate	208, 376
47	THA	Tri-n-hexylamine	534
48	thio THPA	N,N',N''-Tri-n-hexylphosphoro-thioic triamide	361
49	TIOA	Triisooctylamine	231, 506
50	TIOTP	Triisooctyl thiophosphate	484

No	ABBREVIATION OR TRADE NAME	ORGANIC COMPOUND (STATIONARY PHASE)	REFERENCE
51	TLA	Trilaurylamine	231, 333, 334, 335, 494, 539
52	TLMANO$_3$	Trilaurylmethylammonium nitrate	607, 608
53	TMPD	2, 2, 4-Trimethylpentane-1, 3-diol	594
54	TNOA	Tri-n-octylamine	190, 195, 210, 231, 245, 252, 258, 269, 280, 281, 282, 296, 303, 327, 328, 336, 341, 396, 413, 414, 439, 454, 456, 480, 482, 483, 550, 551, 561, 562, 573, 584, 588, 590
55	TOP	Trioctyl phosphate	534
56	TOPO	Tri-n-octylphosphine oxide	233, 253, 254, 257, 260, 265, 275, 280, 287, 312, 320, 330, 341, 422, 533, 534, 573, 575, 585, 589, 590, 592, 593, 594, 610
57	HTTA	2-Thenoyltrifluoroacetone	192, 195, 215, 216, 493, 513, 595, 596, 612
58	Zn-DDTC	Zinc diethyldithiocarbamate	211, 571
59		Acetylacetone	195
60		Amines	196
61		Benzohydroxamic acid	601
62		α-Benzoinoxime	437, 624
63		Cyclohexanol	291
64		Decylalcohol	419
65		Dibutyl ether	401
66		2, 3-Dichloro-1, 4-naphtahydro-quinone	275
67		Diethyldithiocarbamate	620
68		Diethyl ether	514

N°	ABBREVIATION OR TRADE NAME	ORGANIC COMPOUND (STATIONARY PHASE)	REFERENCE
69		Dipicrylamine	586
70		Dithizone (Diphenylthiocarbazone)	220, 429, 438, 497, 498, 499, 505, 540, 541, 546, 567, 568, 569, 604, 619
71		Dodecyl phosphoric acid	232
72		Heptadecane-2,3-dione-dioxime	270
73		N-Hexadecyl-N'-phenylthiourea	270
74		t-Hexyldecylmercaptan	270
75		8-Hydroxyquinoline	221, 622
76		Na-diethyldithiocarbamate	221
77		α-Nitroso-β-naphthol dithizone	221
78		N-Oleylhydroxylamine	270
79		2-Octanone	314, 570, 609, 623
80		Phenyl 2-pyridyl-ketone-oxime	622
81		Pyridine	583
82		Quinoline	495
83		Sulfoxides	403
84		Tetrachlorohydroquinone (Chloranil)	227, 244, 247, 262, 529
85		Tridecylmethylammonium nitrate	476
86		2,6,8-Trimethylnonyl-4 phosphoric acid	560
87		o-Xylene	419
88	DTPA[x]	Diethylene triaminepentaacetic acid	200, 249, 250, 411, 440, 441, 459
89	EDTA[x]	Ethylene diaminetetraacetic acid	267, 495

[x]Organic compound in eluent (mobile phase)

9. SUPPORT MATERIALS USED IN COLUMN EXTRACTION CHROMATOGRAPHY

Nº	CHEMICAL COMPOUND	MATERIAL USED OR TRADE NAME	REFERENCE
1	Aluminium oxide	Alumina powder	190, 512, 615, 616
2	Carbon	Activated charcoal	196, 220, 221
3	Cellulose	Cellulose powder	197, 198, 199, 233, 257, 258, 259, 261, 264, 266, 274, 329, 336, 403, 410, 423, 438, 479, 486, 584, 622
4	Cellulose acetate	Cellulose acetate powder	255, 429, 499, 505
5	Glass	Beads or powder	200, 234, 287, 493, 566, 595, 596
6	Polyethylene (PE)	(in granular form)	484, 618
		Microthene	275, 276, 533, 590, 591, 592, 593, 594
		Mithene	253, 254
		Polyethylene capillary tubing	301, 383
7	Polystyrene	Amberlyst XAD-2	320, 321, 322, 323, 324, 325
8	Polytetrafluoro-ethylene (PTFE)	(in granular form)	350, 351, 374, 432, 433, 434
		Algoflon-F	437
		Fluon	298
		Fluoropak	484, 570, 609, 623
		Fluoroplast-4	201, 203, 204, 205, 235, 236, 290, 295, 389, 394, 425, 426, 431, 439, 464, 466, 467, 468, 470, 471, 472, 473, 474, 480, 485, 513, 547, 601, 605, 628, 629
		Fluororesin-4	291

No	CHEMICAL COMPOUND	MATERIAL USED OR TRADE NAME	REFERENCE
		Voltalef	225, 226, 242, 282, 298, 299, 300, 302, 525, 536, 537, 538, 539, 580, 613, 614, 619, 625
10	Polyurethane	Polyurethane foam	241, 243, 244, 245, 246, 247, 248
11	Poly(vinylchloride/ vinylacetate) copolymer	Corvic	298, 430, 435, 500, 501, 502, 503, 504, 506, 606
12	Silicium dioxide	SiO_2^+: silaned kieselguhr (diatomaceous earth)	214, 215, 216, 217, 223, 224, 237, 240, 249, 250, 268, 272, 273, 286, 288, 289, 297, 298, 305, 306, 307, 308, 309, 326, 331, 337, 338, 341, 345, 352, 353, 354, 355, 357, 359, 363, 364, 365, 366, 367, 368, 369, 370, 371, 378, 380, 382, 384, 385, 386, 387, 405, 406, 407, 408, 412, 415, 417, 428, 443, 444, 445, 450, 451, 452, 453, 455, 459, 460, 475, 476, 477, 488, 489, 492, 507, 530, 531, 532, 540, 541, 549, 552, 553, 554, 555, 560, 561, 562, 563, 564, 565, 572, 573, 579, 582, 598, 599, 602, 606, 607, 608, 610, 611, 626, 627
		Celite, Hyflo Super Cel	218, 219, 292, 310, 350, 372, 373, 457, 481, 529, 542, 603
		Chromosorb	440, 441, 543
		Chromosorb P	274
		Chromosorb W	274, 317, 318, 381, 392, 393, 416, 418, 526, 527, 528, 545, 624

No	CHEMICAL COMPOUND	MATERIAL USED OR TRADE NAME	REFERENCE
		Porasil C	508, 509, 510, 511
		SiO$_2$-gel: silica gel	239, 252, 269, 339, 340, 411, 449, 478, 483, 495, 518, 520, 548, 550, 551
		SiO$_2$: not specified if kieselguhr or silica gel	238, 413, 458, 497, 498, 523, 617
13	Styrene-divinylbenzene copolymer (SDVB)	(In granular form)	228, 229, 230, 231, 270, 546, 558, 559, 567, 568, 569, 604
		Levextrel (includes extractant)	414
		Porapak Q	612

10. AUTHOR INDEX

POHLANDT, Chr. 508, 509, 510, 511
POITRENAUD, C. 31
POLI, A. 512
POLINSKAYA, M. B. 205
PORELL, A. L. 581
POSTA, S. 543, 544
POZZI, F. 490
PREOBRAZHENSKAYA, L. D. 605
PREOBRAZHENSKII, B. K. 32, 466, 513, 514, 515, 516
PRZESZLAKOWSKI, S. 98, 99, 158, 159, 160, 161, 162, 163, 164, 174, 175, 179, 180
PSZONICKA, M. 517
PUGNETTI, G. 267

Q

Quaini, L. 198

R

RAICHEV, Kh. 518, 519, 520
RAIEH, M. 216
RAIS, J. 521
RAMALEY, L. 522
RANDA, Z. 543
RAO, A. P. 166, 167, 523, 535
REINHARDT, G. 340
RESHETNIKOVAN, L. P. 485
REUL, J. 249, 250
REVEL, G. 233
RICCATO, M. T. 524, 525
RICHARD, J. J. 149, 324
RIDER, B. P. 526, 527, 528
RIEDEL, H. J. 613
RIGALI, L. 224
ROBERT, R. V. D. 511
ROBERTS, B. J. 342, 343, 344
ROBERTS, G. A. H. 227, 529
RÖDEL, E. 383
ROJOWSKA, M. 175
ROSSI, G. 265, 413
ROZEN, A. M. 165
ROZYCKI, C. 458
RUIZ, C. P. 526, 527, 528

S

SAEKI, M. 603
SALVATORI, A. 589
SAMSAHL, K. 530, 531, 532

SANTORI, G. 533, 591
SARAIYA, V. N. 534
SARMA, A. R. K. 167, 535
SASTRI, M. N. 166, 167, 523, 535
SAUNKIN, O. F. 627
SAURO, L. J. 365
SCARGILL, D. 168
SCHMID, E. R. 536, 537, 538, 539
SCHMITT, M. A. S. 169
SCOTT, J. C. 570
ŠEBESTA, F. 540, 541, 542, 543, 544, 545
SEKIZUKA, Y. 546
SERGEEV, V. P. 399
SHANKAR, S. 534
SHEGLOWSKI, S. 232
SHERMA, J. 100
SHESTAKOV, B. I. 547, 629
SHESTAKOVA, I. A. 547, 629
SHIMIZU, T. 170, 548
SHMANENKOVA, G. I. 549, 550, 551, 627
SHUKOV, G. V. 549, 627
SHULEPNIKOV, M. N. 551
SIEKIERSKI, S. 305, 306, 307, 308, 309, 341, 412, 428, 517, 552, 553, 554, 555, 560, 565, 597
SIJPERDA, W. S. 90, 171, 172
SIKKELAND, T. 556
SILVA, R. J. 556, 557
SINEGRIBOVA, O. A. 189
SING, S. 173
SIXTA, V. 580
SJOBLOM, R. K. 310
SMALES, A. A. 33
SMALL, H. 558, 559
SMUŁEK, W. 15, 34, 481, 560, 561, 562 563, 564
SMUTZ, M. 447, 448
SOCHACKA, R. J. 554, 555, 565
SOCZEWINSKI, E. 158, 163, 174, 175
SPECHT, S. 566
SPEVÀČKOVA, V. 35, 567, 568, 569
STACCIOLI, L. 594
STAPLES, B. A. 570
STARER MENDES, H. 176
STARTSEVA, E. A. 222
STARÝ, J. 36, 326, 545
STEELE, T. W. 510
STEPANETS, O. V. 211, 571, 620, 621
STERRENBURG, P. J. J. 61
STOEPPLER, M. 612
STONE, R. E. 379
STRONSKI, I. 37, 38, 39, 409, 450, 451, 454, 455, 572, 573, 574, 575, 576, 577, 578, 579

Ref. 1 —   3 : Bibliographies
Ref. 4 — 42 : Reviews and Books,
Ref. 43 — 189 : Laminar Extraction Chromatography
Ref. 190 — 629 : Column Extraction Chromatography

ARGENTINA

University of Buenos Aires, 303, 304

AUSTRIA

Analytical Institute, University of Vienna, 24, 27, 138, 536, 537, 538, 539
Forschungsinstitut der Ernährungswirtschaft, Vienna, 619
Reaktorzentrum Seibersdorf, 225, 226

BELGIUM

CEN — SCK, Mol, 223, 405, 406, 407, 460
EURATOM, Mol, 407, 408
EUROCHEMIC, Mol, 1, 2, 299, 300, 301, 302
Hoger Rijksinstituut voor Technisch Onderwijs en Kernenergiebedrijven,
   Mol, 507
Laboratory for Radiochemistry, University of Louvain, 288, 289

BRASIL

Instituto de Energia Atomica, Sao Paulo, 190

BULGARIA

Institute of Nuclear Research and Nuclear Energy, Sofia, 520

CHINA

Academia Sinica, Peking, 150, 151, 152, 153, 154, 432, 433, 434
Northwestern University, Siun, Shensi, 77, 78, 79
Lanchow University, 111, 112, 113, 114, 115, 116, 117, 118, 120

CZECHOSLOVAKIA (ČSSR)

Institute of Nuclear Research, Czechoslovak Academy of Sciences,
Rez near Prague, 5, 35, 65, 228, 229, 230, 231, 415, 416, 417, 418, 420, 442, 443, 444, 445, 521, 567, 568, 569.
Department of Analytical Chemistry, CHARLES University, Prague, 625
Department of Nuclear Chemistry, University of Prague, 36, 540, 541, 542, 543, 544, 545
Institute of Mineral Raw Materials, Kutna Hora, 543
Pharmaceutical Faculty, Komenski University, Bratislava, 140, 141, 142, 143
Ustredni Ustav Geologicky, Prague, 580

DENMARK

Research Establishment Risø, 345

FEDERAL REPUBLIC OF GERMANY (FRG)
(Bundesrepublik Deutschland (BRD))

Bodenseewerke Perkin-Elmer & C°, GmbH, Ueberlingen/Bodensee, 383
EURATOM, Institute for Transuranium Elements, Karlsruhe, 249, 250, 475, 476, 477

Farbenfabriken Bayer AG, Leverkusen, 414
Institut für Anorganische Chemie und Kernchemie Johannes Gutenberg-Universität, Mainz 277, 278, 279, 280, 281, 282, 283, 284, 356, 524, 525, 557, 599, 600
Institut für Chemische Technologie, Kernforschungsanlage, Jülich, 614
Institut für Heisse Chemie, Kernforschungszentrum, Karlsruhe, 459
Institut für Kern- und Radiochemie der Techn. Hochschule Braunschweig, 125, 126
Institut für Radiochemie, Kernforschungszentrum, Karlsruhe, 200
Lehrstuhl für Kernchemie, Technische Hochschule, Darmstadt, 427
Technische Hochschule, München, 566, 576
Zentralinstitut für Analytische Chemie der Kernforschungsanlage, Jülich, 612, 613

FRANCE

CEA, Centre de Pierrelatte, 256
Centre d'Etude de Chimie Métallurgique du CNRS, Vitry-sur-Seine, 233
Centre d'Etudes Nucléaires de Fontenay-aux-Roses, 333, 334, 335, 494
Institut National des Sciences & Techniques Nucléaires, Saclay, 31
Laboratoire de Physiologie et de Chimie, Faculté des Sciences, Paris, 176
Université de Paris, 16

GERMAN DEMOCRATIC REPUBLIC (GDR)
(Deutsche Demokratische Republic (DDR))
Institut für Anorganische Chemie der Karl Marx-Universität, Leipzig, 109, 110, 181, 182,
    183, 184, 185, 186
Technische Universität, Dresden, 238, 337, 338, 339, 340, 355, 478, 483
Zentralinstitut für Festkörperphysik & Werkstofforschung der
    Akademie der Wissenschaften der DDR, Dresden, 483
Zentralinstitut für Kernforschung, Rossendorf bei Dresden, 350, 351, 353, 374, 478

GREAT BRITAIN

AERE, Harwell, Berks., 33, 155, 156, 157, 168, 273, 286, 429, 430, 438, 496, 497, 498,
    499, 500, 501, 502, 503, 504, 505, 506
British Cast Iron Research Association, Bordesley Hall,
    Alvechurch, Birmingham, 17
English Electricity Co., Whetstone, 272
Esso Research Centre, Abingdon, Berks., 529
Manchester Polytechnic, Department of Chemistry & Biology, Manchester, 47
Royal College of Advanced.Technology,Salford, Lancs., 92
United Kingdom Atomic Energy Authority, Development & Engineering Group, Capen-
    hurst, Ches., 311, 342, 343
United Kingdom Atomic Energy Authority, London, 344
United Kingdom Atomic Energy Authority, Salwick, 347
University of Birmingham, Department of Chemistry, 227
University of Kent, Canterbury, 137
University of Salford, Salford, Lancs., 44, 45, 46, 47, 48, 49, 50, 51, 93, 101, 102, 103,
    104, 105, 106, 134, 139

HUNGARY

L. Eötvös University, Institute of Inorganic and Analytical Chemistry, Budapest, 241, 242,
    243, 244, 245, 246, 247, 248

INDIA

Andhra University, Department of Chemistry, Waltair, 166, 167, 523, 535
Bhabha Atomic Research Centre, Bombay, 534
Visva-Bharati, Department of Chemistry, Santiniketan, 89

ISRAEL

Hebrew University, Jerusalem, 124

550

NEW ZEALAND

Auckland Industrial Development Division, D.S.I.R., Auckland, 18

NORWAY

Institutt for Atomenergi, Kjeller, 14, 237, 297, 298, 378, 384, 385, 386, 387, 446, 493, 595, 596

POLAND

Akademia Medyczna, Katedra Chemii Nieorganicznej i Analitycznej, Lublin, 98, 99, 158, 159, 160, 161, 162, 163, 164, 174, 175, 179, 180
Institute of Nuclear Research, Swierk, 564
Jagiellonian, University, Krakow, 423
Polish Academy of Sciences, Krakow, 37, 38, 39, 409, 450, 451, 452, 453, 454, 455, 456, 457, 552, 553, 554, 555, 572, 573, 574, 575, 577, 578, 579
Polish Academy of Sciences, Warsaw, 15, 34, 305, 306, 307, 308, 309, 312, 331, 341, 382, 412, 428, 481, 517, 560, 561, 562, 563, 564, 565, 597, 626
Politech. Warsaw, 458

ROUMANIA

Babes-Bolyai University, Cluj, 136
Centre of Physical Chemistry, Bucharest, 212
Institute of Chemistry, Academy of the SRR, Cluj, 135
University "Al. I. Cuza", Iasi, 96, 97

SOUTH AFRICA

National Institute for Metallurgy, Johannesburg, 508, 509, 510, 511

SWEDEN

AB Atomenergi, Nyköping, 531, 532
AB Atomenergi, Stockholm, 530

TAIWAN

National Taiwan University, Taipei, 80
National Tsing Hua University, Hsinchu, 268
Soochow University, Wuhsien, 119

U.A.R.

Atomic Energy Establishment, Cairo, 214, 215, 216, 217, 218, 219, 292

USA

Allied Chemical Corp., Idaho Falls, Idaho, 440, 570
Argonne National Laboratory, Argonne, Ill, 30, 310, 363, 364, 365, 366, 367, 368, 369, 370, 371, 372, 373, 375, 376, 377, 492
Atlantic Richland Hanford Co., Richland, Wash., 271, 491
The Babcock & Wilcox Co., Lynchburg, Virginia, 441
Brookhaven National Laboratory, New York, 595
Continental Oil Co., Ponca City, Okla., 357
The Dow Chemical Co., Midland, Michigan, 558, 559
E. I. du Pont de Nemours and Co., Aiken, South Carolina, 359, 489
General Electric Co., Pleasanton, Cal., 528
General Electric Co., San Jose, Cal., 526, 527
Hofstra College, Hempstead, L. I., New York, 618
Idaho Nuclear Corp., Idaho Falls, Idaho, 609
Iowa State University, Ames, Iowa, 3, 29, 52, 94, 95, 100, 123, 146, 147, 148, 149, 169, 274, 313, 314, 315, 316, 317, 318, 319, 320, 321, 322, 323, 324, 325, 348, 380, 381, 391, 392, 393, 403, 421, 448, 486, 487, 488
John Hopkins University, Annapolis, Md., 255
Lawrence Radiation Laboratory, Berkeley, Cal., 19, 556
Los Alamos Scientific Laboratory, Los Alamos, New Mexico, 358
Massachusets Institute of Technology, Cambridge, 617
National Bureau of Standards, Washington, D. C., 435
New Brunswick Laboratory, New Brunswick, New Jersey, 349
Ohio State University, Columbus, Ohio, 581
ORNL, Oak Ridge, Tenn., 422, 461, 462, 484, 615, 616
Phillips Petroleum Co., Idaho Falls, Idaho, 609, 623
Union Carbide, Oak Ridge, Tenn., 287
University of Arizona, Tucson, 522
University of California, Livermore, Cal., 361, 362, 379, 556
University of Cincinnati, Cincinnati, Ohio, 81
University of North Carolina, Chapel Hill, N. C., 479
University of the Pacific, Stockton, Cal, 187

US Army Nuclear Defence Laboratory, Edgewood Arsenal, Maryland, 336
US Atomic Energy Commission, New York, 234, 463

USSR

Academy of Sciences, Moscow, 207, 211, 439
Inst. Atomnoi Energii, Moscow, 411
Inst. of Geochem., Moscow, 43, 394, 551, 571, 620, 621
Kurchatov Atomic Energy Institute, Moscow, 269
Joint Inst. for Nuclear Research, Dubna, Moscow, 128, 129, 130, 240, 326, 332, 352, 354, 410, 436, 444, 495, 518, 519, 582, 627
Leningrad State University, Leningrad, 20, 21, 42, 389, 431, 601
Ministry of Public Health, 425
Moscow M. V. Lomonosov State University, Moscow, 4, 201, 202, 203, 204, 205, 206, 208, 209, 210, 212, 213, 235, 236, 290, 293, 294, 295, 327, 328, 480, 549, 550
Moscow D. I. Mendeleyev Institute of Chemical Technology, Moscow, 188, 189
State Atomic Energy Committee, Moscow, 296
Tr. po Radiats, Gigiene, Leningrad, 221

YUGOSLAVIA

Boris Kidric Institute of Nuclear Sciences, Beograd- Vinca, 84, 85, 86, 87, 88, 417, 418
Rudjer Boskovic Institute, Zagreb, Croatia, 598

WITHOUT ADDRESS:

BELGIUM

28

CHINA

26, 173, 178

GERMAN DEMOCRATIC REPUBLIC (GDR)
239

GREAT BRITAIN

270

HUNGARY

568

JAPAN

121,

POLAND

196, 285

USSR

32, 127, 165, 220, 222, 232, 239, 251, 291, 294, 388, 390, 395, 396, 397, 398, 399, 400, 401, 402, 419, 424, 426, 449, 464, 465, 466, 467, 468, 469, 470, 471, 472, 473, 474, 482, 485, 513, 514, 515, 516, 547, 605, 628, 629

YUGOSLAVIA

82, 83

SUBJECT INDEX*

*The underlined page numbers refer to detailed discussion of the entry

557

558

Double–double effect, 2, 6, <u>23</u>, 34, 197, 227, 228, <u>233</u>, 239, 242
 – in actinides, <u>236</u>
 – in lanthanides, <u>235</u>
 – theory of, 236
DPA (dipicrylamine), 181, 256, 264
DTPA (diethylenetriaminopentaacetate), 197, 217, 221, 222, 264
Dysprosium, Dy
 – separation from Gd, Tb, Ho, 245
 – separation from Ho, 241, 242,
 – separation from Y, 247

Electron exchange columns, <u>129</u>, <u>360</u>, <u>362</u>
Elution analysis, <u>59</u>, 61
 – gradient elution, 63
 – stepwise elution, 63
Elution curve, 20, 35, 199, 314
 – evaluation of, <u>64</u>
Elution volume, 19, 20, 34, 35, 52, 63, 65
Enflon, 145
Erbium, Er
 – separation from Ho, 242, 246
 – separation from La, Nd, Sm, Eu, Gd, Ho, 244
 – separation from Y, 247
 – synergistic effect, 40, 41
Einsteinium, Es
 – separation from Bk and Fm, 200
 – separation from Cm, Bk and Cf, 219
 – separation of Es, Bk, Cf from irradiated Pu, Am, Cm, 217
 – separation on Aliquat–336, 34
Ethylparathion, 395
Europium, Eu, 11, 12, 98, 128, 250, <u>271</u>
 – separation from Am, 211, 212
 – separation from Am, Cm, 246
 – separation from Ce, 244
 – separation from Cm(III), 247
 – separation from Er, 244
 – separation from Fm, 247
 – separation from Gd, 241, 242
 – separation from La, 246
 – separation from La, Ce, Nd, 242
 – separation from La, Ce, Pm, 242
 – separation from La, Nd, Sm, Gd, Ho, Er, 244
 – separation from Md(III), 247
 – separation from Pm, 246
 – separation from Tb, 242
 – separation from Tb, Tm, 245
 – separation from Tm, 246
Exchange constant, 29
Extraction chromatography
 – cellular plastics in, <u>344</u>
 – correlation with liquid–liquid extraction, 17
 – dynamic factors in, <u>10</u>
 – for actinides, <u>191</u>
 – inert supports in, <u>134</u>

 – in radiotoxicology, <u>279</u>
 – ion separations by, <u>175</u>
 – laminar techniques in, <u>366</u>
 – of fission products, <u>254</u>
 – of lanthanides, <u>226</u>
 – preconcentration and separation of trace metals by, <u>325</u>
 – stationary phases in, <u>68</u>, <u>304</u>
 – techniques in, <u>45</u>
 – theoretical aspects of, <u>1</u>
Extraction coefficient, 3–5, 9, 10, 194, 196, 245
Extraction constant, 22, 26, 27, 30
 – for metal chelates, <u>305</u>, <u>306</u>, <u>311</u>
Extraction curve, 27, 34, 39
Extraction kinetics, 13, 14
Extraction mechanism, 2, 88, 99, 111
 – for Am, 194
 – on HDEHP, 88
Extraction process
 – models for, 7
Extraction systems
 – cation-exchange, 30
 – chelate, <u>25</u>, <u>314</u>
 – ion-association, <u>32</u>, 230
 – solvation, <u>36</u>, 229
 – synergistic, <u>39</u>, <u>127</u>

$Fe(CN)_6^{4-}$, 38
 – separation of $Fe(CN)_6^{4-}$ and $Fe(CN)_6^{3-}$, 39, 185
Fermium, Fm, 205
 – separation from Bk and Es, 200
 – separation from Cf and Md, 218
 – separation from Eu, 247
Fission products
 – elution behaviour of, <u>262</u>
 – extractability of, <u>265</u>
 – recovery of, <u>275</u>, <u>276</u>
 – separation of, <u>255</u>, <u>257</u>, <u>261</u>, <u>315</u>
Flow rate determination, 58
Fluon, 144
Fluoroflex, 144
Fluoron, 144
Fluoropak, 125, 182
Fluoropak–80, 145
Fluoroplast, 344
Fluorothane, 144
Frontal analysis, <u>59</u>
Ftoroplast–3, 144
Ftoroplast–4, 118, 145, 155–157, 160, 161, 167, 168, 176–179, 242, 314, 318, 322, 326, 328, 330, 341
Ftoroplast–40, 96, 105, 109, 126, 181

Gadolinium, Gd
 – separation from Er, 244
 – separation from Eu, 241, 242
 – separation from La, Ce, Nd, Tb, Tm, 241

560

561

565